Three-Dimensional Contact Problems

SOLID MECHANICS AND ITS APPLICATIONS
Volume 93

Series Editor: G.M.L. GLADWELL
Department of Civil Engineering
University of Waterloo
Waterloo, Ontario, Canada N2L 3GI

Aims and Scope of the Series

The fundamental questions arising in mechanics are: *Why?*, *How?*, and *How much?* The aim of this series is to provide lucid accounts written by authoritative researchers giving vision and insight in answering these questions on the subject of mechanics as it relates to solids.

The scope of the series covers the entire spectrum of solid mechanics. Thus it includes the foundation of mechanics; variational formulations; computational mechanics; statics, kinematics and dynamics of rigid and elastic bodies: vibrations of solids and structures; dynamical systems and chaos; the theories of elasticity, plasticity and viscoelasticity; composite materials; rods, beams, shells and membranes; structural control and stability; soils, rocks and geomechanics; fracture; tribology; experimental mechanics; biomechanics and machine design.

The median level of presentation is the first year graduate student. Some texts are monographs defining the current state of the field; others are accessible to final year undergraduates; but essentially the emphasis is on readability and clarity.

For a list of related mechanics titles, see final pages.

Three-Dimensional Contact Problems

by

V.M. ALEXANDROV

Moscow State University,
Department of Mechanics and Mathematics,
Moscow, Russia

and

D.A. POZHARSKII

Rostov-on-Don State University,
Mechanics and Applied Mathematics Institute,
Rostov-on-Don, Russia

KLUWER ACADEMIC PUBLISHERS
DORDRECHT / BOSTON / LONDON

A C.I.P. Catalogue record for this book is available from the Library of Congress.

ISBN-13: 978-1-4020-0387-5 e-ISBN-13: 978-94-010-9893-9
DOI: 10.1007/978-94-010-9893-9

Published by Kluwer Academic Publishers,
P.O. Box 17, 3300 AA Dordrecht, The Netherlands.

Sold and distributed in North, Central and South America
by Kluwer Academic Publishers,
101 Philip Drive, Norwell, MA 02061, U.S.A.

In all other countries, sold and distributed
by Kluwer Academic Publishers,
P.O. Box 322, 3300 AH Dordrecht, The Netherlands.

Printed on acid-free paper

Contents

Preface

The mechanics of contact between deformable bodies, so-called contact mechanics, is an important and actively developing part of continuum mechanics. Realistic boundary conditions on a surface of deformable body may be formulated only as a result of a solution of a contact problem. Therefore, the determination of the stress–strain state of a body is also connected with a contact problem.

The characteristic feature of contact problems is that they have mixed boundary conditions and may thus be reduced to integral equations. This work focuses on such integral equations, properties of their kernels, suggesting a large number of beautiful asymptotic and numerical solutions.

In this monograph a systematic treatment is given for analytical and numerical methods and results for many nonclassical three-dimensional contact problems in the linear theory of elasticity. We consider semiinfinite elastic bodies, such as half-spaces, layers, holes, cylinders, spaces with cylindrical or spherical cavities, wedges, cones, half-spaces with spherical holes; and finite elastic bodies, such as circular plates, finite cylinders, spheres and spherical segments.

The monograph includes a vast amount of material accumulated during scores of years in numerous publications by the authors and their colleagues.

Methods introduced in the book can also be applied to crack and inclusion problems in fracture mechanics as well as to other problems in the disciplines of mechanics, applied mathematics, mathematical physics and engineering with mixed boundary conditions.

For formulae, figures and tables, we use double enumeration where the first number is that for the chapter. For special mathematical functions, functional spaces and operators, we usually use gothic script.

We wish to thank Prof. G.M.L. Gladwell (University of Waterloo, Canada) for his careful editing the book, and Prof. W.L. Wendland (University of Stuttgart, Germany) for his useful remarks.

Our work was supported by the Alexander von Humboldt Foundation (Germany) and Russian Foundation for Basic Research (99-15-96012).

V.M. Alexandrov and D.A. Pozharskii
Moscow, Stuttgart, 2001

Chapter 1

CONTACT PROBLEMS FOR A HALF-SPACE

1.1 Introduction

In this book we are concerned with contact problems within the classical (linear, infinitesimal) theory of elasticity. The theory of elasticity has a long history, stretching back to Galileo and Hooke. The history of Western European research up to the 1890's is detailed in Todhunter and Pearson (1893). This historical account is extended to the 1920's by Love's (1927) *Treatise on The Mathematical Theory of Elasticity.* He gives not only the history, but also the foundation of the theory, and its detailed application to a vast variety of problems. Love's analysis is built in large part on the work of Boussinesq (1885) and Cerruti (1882), it also draws on the work of mathematical physicists such as Lamb, Lord Rayleigh and Lord Kelvin; for the contact problems, it is based on the work of Hertz (1882). It is universally recognised that Hertz was far ahead of his time. His beautiful analysis of the contact between two elastic bodies remains to-day as the most important and far reaching example of a solution of a classical problem in elasticity theory. It is still the basis for many tests, and is used in situations which are far removed from the one originally considered by Hertz, for instance in the analysis of the contact between a wheel and a rail.

The first treatise devoted to *Contact Problems* is by Shtaerman (1949). He developed the theory of Hertz examining contact problems for an elastic half-plane and for a half-space. In 1953, Galin published his treatise; this was translated in 1961. In particular, he considered friction forces in the axisymmetric contact problem and the contact problem for a wedge-shaped punch. At the same time (1955) Lur'e wrote his treatise on *Three-Dimensional Problems of the Theory of Elasticity*, translated in 1964. He devoted one chapter to three-dimensional contact problems for a half-space and apllied potential theory to the cases of blunt, conical, circular and elliptic punches.

In 1951, Sneddon published his *Fourier Transforms.* This book was the first to bring the mathematical methods developed by Titchmarsh, Busbridge and Erdelyi for the treatment of integral equations, to a wider audience. Contact problems were treated here as examples of problems which may be reduced to the solution of a pair of integral equations. Sneddon (1966) extended this analysis in his *Mixed Boundary Value Problems in Potential Theory.*

Contact problems are treated in passing, in Timoshenko and Goodier (1951) and in Sokolnikoff (1956).

One of the most significant developments in the theory of elasticity in the 20th Century was the introduction of complex potentials by Kolosov, and their developement by Muskhelishvili (1953b), and his school. Muskhelishvili's analysis was applied solely to two-dimensional problems. In his monograph, Gladwell (1980) showed how the methods developed by Muskhelishvili could be combined with Sneddon's transform methods to solve a variety of contact problems in three dimensions.

All these books deal almost exclusively with idealized contact problems. The emphasis in the work is on the solution of the mixed boundary value problem which arise in various idealised contact problems. Johnson (1985) is concerned as much with experiment as with theory. He considers inelastic contact, and contact between rough bodies which cannot be idealized as Hertzian ellipsoids. Mossakovskii, Kachalovskaia and Golikova (1985) examined, in particular, the contact problem for an approximately circular punch by using the exact solution for the circular punch. Kalker (1990) was concerned with rolling contact, the contact encountered in the wheel–rail problem, and developed computational algorithms for the analysis of such problems.

In the first chapter we consider the contact problems for a half-space with strip, circular or elliptic contact region.

1.2 Elasticity Theory

In this book we will consider exclusively static problems for a homogeneous isotropic linearly-elastic body. Such a body is characterised by two elastic constants; we will usually choose the two constants as the shear modulus G (sometimes denoted by μ) and Poisson's ratio ν. These are related to the Young's modulus E and Lamé constant λ by

$$G = \frac{E}{2(1+\nu)}, \qquad \lambda = \frac{E\nu}{(1+\nu)(1-2\nu)} = \frac{2G}{1-2\nu}.$$

The governing equations for such a body fall into three groups: the stress equilibrium equations

$$\sigma_{ij,j} = 0,$$

where a repeated suffix denotes summation over $j = 1, 2, 3$, the comma denotes differentiation with respect to x_j, and where $(x_1, x_2, x_3) \equiv (x, y, z)$; the stress–strain equations (Hooke's law)

$$\sigma_{ij} = 2G\left(\varepsilon_{ij} + \frac{\nu}{1-2\nu}\delta_{ij}e\right), \quad e = \varepsilon_{ii}; \tag{1.1}$$

and the strain–displacement equations (Cauchy formulae)

$$\varepsilon_{ij} = \frac{1}{2}\left(u_{i,j} + u_{j,i}\right). \tag{1.2}$$

It is convenient to eliminate the stresses and strains, and to write the equilibrium equations in terms of the displacements u_i; the resulting equations are called Navier's equations:

$$u_{i,jj} + \frac{1}{1-2\nu}u_{j,ij} = 0.$$

These may be written in vector form by introducing $\mathbf{u} = (u_1, u_2, u_3) = (u, v, w)$:

$$\Delta\,\mathbf{u} + \frac{1}{1-2\nu}\,\mathbf{grad}\,\mathrm{div}\,\mathbf{u} = 0. \tag{1.3}$$

We will make frequent use of the Papkovich–Neuber solution of (1.3) in terms of a scalar harmonic function ϕ_0 and a vector harmonic function $\psi = (\psi_1, \psi_2, \psi_3)$:

$$\mathbf{u} = 4(1-\nu)\psi - \mathbf{grad}\,(\phi_0 + \mathbf{r}\cdot\psi), \tag{1.4}$$

where $\mathbf{r} = (x, y, z)$. Written in the component form, equation (1.4) is

$$u = 4(1-\nu)\psi_1 - \frac{\partial}{\partial x}(\phi_0 + x\psi_1 + y\psi_2 + z\psi_3),$$

$$v = 4(1-\nu)\psi_2 - \frac{\partial}{\partial y}(\phi_0 + x\psi_1 + y\psi_2 + z\psi_3),$$

$$w = 4(1-\nu)\psi_3 - \frac{\partial}{\partial z}(\phi_0 + x\psi_1 + y\psi_2 + z\psi_3). \tag{1.5}$$

We will need explicit expressions for the stresses, particularly $\tau_{xz} = \sigma_{13}$, $\tau_{yz} = \sigma_{23}$, $\sigma_z = \sigma_{33}$. After routine differentiation we find

$$\tau_{xz} = 4G(1-\nu)\left(\frac{\partial\psi_1}{\partial z} + \frac{\partial\psi_3}{\partial x}\right) - 2G\frac{\partial^2}{\partial x\,\partial z}(\phi_0 + x\psi_1 + y\psi_2 + z\psi_3), \tag{1.6}$$

$$\tau_{yz} = 4G(1-\nu)\left(\frac{\partial\psi_2}{\partial z} + \frac{\partial\psi_3}{\partial y}\right) - 2G\frac{\partial^2}{\partial y\,\partial z}(\phi_0 + x\psi_1 + y\psi_2 + z\psi_3), \tag{1.7}$$

$$\sigma_z = 8G(1-\nu)\frac{\partial\psi_3}{\partial z} - 2G\frac{\partial^2}{\partial z^2}(\phi_0 + x\psi_1 + y\psi_2 + z\psi_3) \tag{1.8}$$
$$+4G\nu\left(\frac{\partial\psi_1}{\partial x} + \frac{\partial\psi_2}{\partial y} + \frac{\partial\psi_3}{\partial z}\right).$$

1.3 Frictionless Contact Problems for a Half-Space

1.3.1. THE FORMULA FOR THE NORMAL DISPLACEMENT

We will be particularly interested in frictionless contact problems for the half-space $z \le 0$. For such problems the boundary conditions are

$$\tau_{xz}(x, y, 0) = 0 = \tau_{yz}(x, y, 0).$$

For these problems we may simplify the Papkovich–Neuber representation. We will make repeated use of the fundamental result
(A) *If* $\Delta\phi = 0$ *in the half-space* $z \le 0$, $\phi = 0$ *on* $z = 0$, *and* $\phi \to 0$ *as* $x^2 + y^2 + z^2 \to \infty$, *then* $\phi \equiv 0$ *for* $z \le 0$.
First we note that

$$\frac{\partial}{\partial z}(z\psi_3) = z\frac{\partial\psi_3}{\partial z} + \psi_3,$$

and then we rearrange the terms in τ_{xz}, τ_{yz} in (1.6), (1.7) to give the following conditions on $z = 0$:

$$2(1-\nu)\frac{\partial\psi_1}{\partial z} + (1-2\nu)\frac{\partial\psi_3}{\partial x} - \frac{\partial^2}{\partial x\,\partial z}(\phi_0 + x\psi_1 + y\psi_2) = 0, \tag{1.9}$$

$$2(1-\nu)\frac{\partial\psi_2}{\partial z}+(1-2\nu)\frac{\partial\psi_3}{\partial y}-\frac{\partial^2}{\partial y\,\partial z}(\phi_0+x\psi_1+y\psi_2)=0. \tag{1.10}$$

Differentiating (1.9) with respect to y, and (1.10) with respect to x and subtracting, we find

$$\frac{\partial^2\psi_1}{\partial y\,\partial z}-\frac{\partial^2\psi_2}{\partial x\,\partial z}=0 \quad \text{on } z=0.$$

Now (A) shows that this result holds for $z\le 0$, and this implies that

$$\frac{\partial\psi_1}{\partial z}=\frac{\partial\chi}{\partial x},\qquad \frac{\partial\psi_2}{\partial z}=\frac{\partial\chi}{\partial y}, \tag{1.11}$$

where χ is harmonic. Now we rewrite equations (1.9) and (1.10) as

$$\frac{\partial\phi_1}{\partial x}=0=\frac{\partial\phi_1}{\partial y}\quad \text{on } z=0,$$

where

$$\phi_1=2(1-\nu)\chi+(1-2\nu)\psi_3-\frac{\partial}{\partial z}(\phi_0+x\psi_1+y\psi_2).$$

The function ϕ_1 is constant on $z=0$, but it is not harmonic, so we cannot apply (A). But, on $z=0$, ϕ_1 is equal to the harmonic function

$$\phi_2=2(1-\nu)\chi+(1-2\nu)\psi_3-\frac{\partial\phi_0}{\partial z}-\mathbf{r}\cdot\mathbf{grad}\,\chi,$$

where we have used (1.11). Now the result (A) shows that $\phi_2\equiv 0$ in $z\le 0$. By adding and subtracting a term, we may rewrite ϕ_2 as

$$\phi_2=2(1-\nu)\chi+(1-2\nu)\psi_3-\frac{\partial\phi_0}{\partial z}+\frac{1-2\nu}{2(1-\nu)}\mathbf{r}\cdot\mathbf{grad}\,\psi_3 \tag{1.12}$$

$$-\frac{1}{2(1-\nu)}\mathbf{r}\cdot\mathbf{grad}\,[2(1-\nu)\chi+(1-2\nu)\psi_3].$$

The Papkovich–Neuber representation involves four harmonic functions ϕ_0, ψ_1, ψ_2, ψ_3, but there is flexibility in the representation: we may express one function in terms of the others. We use this flexibility to take

$$2(1-\nu)\chi+(1-2\nu)\psi_3=0.$$

With (1.11), we then have

$$\frac{\partial\psi_1}{\partial z}=-\gamma\frac{\partial\psi_3}{\partial x},\quad \frac{\partial\psi_2}{\partial z}=-\gamma\frac{\partial\psi_3}{\partial y},\quad \gamma=\frac{1-2\nu}{2(1-\nu)},$$

and (1.12) gives

$$\frac{\partial \phi_0}{\partial z} = \gamma \, \mathbf{r} \cdot \mathbf{grad}\, \psi_3 = \gamma \left(x \frac{\partial \psi_3}{\partial x} + y \frac{\partial \psi_3}{\partial y} + z \frac{\partial \psi_3}{\partial z} \right).$$

Now routine calculation shows that if we introduce

$$\phi = \frac{3 - 4\nu}{2(1-\nu)} \psi_3,$$

then w (1.5) and σ_z (1.8) may be written

$$w = 2(1-\nu)\phi - z\frac{\partial \phi}{\partial z},$$

$$\sigma_z = 2G\left(\frac{\partial \phi}{\partial z} - z\frac{\partial^2 \phi}{\partial z^2}\right).$$

In particular therefore

$$w(x, y, 0) = 2(1-\nu)\phi, \quad \sigma_z(x, y, 0) = 2G\frac{\partial \phi}{\partial z},$$

provided that

$$\lim_{z\to 0} z\frac{\partial \phi}{\partial z} = 0 = \lim_{z \to 0} z \frac{\partial^2 \phi}{\partial z^2}.$$

Now suppose that a normal stress is applied to the half-space $z \le 0$ over the region Ω of $z = 0$. Then

$$\sigma_z = -q(x, y), \quad (x, y) \in \Omega; \quad \sigma = 0, \quad (x, y) \in \omega, \tag{1.13}$$

where w is the complement of Ω in $z = 0$. We seek the potential ϕ as a double Fourier integral

$$\phi = \frac{1}{2\pi} \int_{-\infty}^{\infty}\int_{-\infty}^{\infty} \Phi(\alpha, \beta, z) \exp[-i(\alpha x + \beta y)]\, d\alpha\, d\beta.$$

The properties of the Fourier integral (Sneddon (1951)) ensure that ϕ decays to zero as $(x^2 + y^2) \to \infty$.

Since ϕ is harmonic, we have

$$\Phi'' - \rho^2 \Phi = 0,$$

where $' = d/dz$ and $\rho = (\alpha^2 + \beta^2)^{1/2}$. This has the solution

$$\Phi(\alpha, \beta, z) = C(\alpha, \beta) \exp(\rho z),$$

which vanishes as $z \to -\infty$.

Now we write the function $q(x,y)$, which has support Ω, as a double Fourier integral:

$$q(x,y) = \frac{1}{2\pi} \int_{-\infty}^{\infty} \int_{-\infty}^{\infty} Q(\alpha,\beta) \exp[-i(\alpha x + \beta y)]\, d\alpha\, d\beta,$$

which may be inverted to give

$$Q(\alpha,\beta) = \frac{1}{2\pi} \int_{\Omega} q(\xi,\eta) \exp[-i(\alpha\xi + \beta\eta)]\, d\Omega,$$

where $d\Omega = d\xi\, d\eta$. We now write the boundary condition (1.13) as

$$2G\rho C(\alpha,\beta) = -Q(\alpha,\beta),$$

so that

$$w(x,y,0) = -\frac{1}{2\pi\theta} \int_{\Omega} J(x,y,\xi,\eta)\, q(\xi,\eta)\, d\Omega, \quad \theta = \frac{G}{1-\nu}, \tag{1.14}$$

where

$$J(x,y,\xi,\eta) = \frac{1}{2\pi} \int_{-\infty}^{\infty} \int_{-\infty}^{\infty} \exp\{-i[\alpha(\xi - x) + \beta(\eta - y)]\} \frac{d\alpha\, d\beta}{\rho}.$$

We write this double integral in polar coordinates (ρ, ϑ) where $\alpha = \rho\cos\vartheta$, $\beta = \rho\sin\vartheta$, and put

$$\xi - x = R\cos\psi, \quad \eta - y = R\sin\psi \quad (R = \sqrt{(\xi - x)^2 + (\eta - y)^2}).$$

We find

$$J = \frac{1}{2\pi} \int_0^{\infty} \int_0^{2\pi} \exp[i\rho R\cos(\vartheta - \psi)]\, d\rho\, d\vartheta.$$

The integral is periodic, with period 2π, in ϑ, so that

$$J = \frac{1}{2\pi} \int_0^{\infty} \int_0^{2\pi} \exp[i\rho R\cos\vartheta]\, d\rho\, d\vartheta.$$

The imaginary part of the integrand is odd, so integrates to zero over $(0, 2\pi)$; thus

$$J = \frac{1}{2\pi}\int_0^\infty\int_0^{2\pi} \cos(\rho R\cos\vartheta]\,d\rho\,d\vartheta$$

$$= \int_0^\infty \mathcal{J}_0(\rho R)\,d\rho = \frac{1}{R}, \tag{1.15}$$

where we have used formulae 8.411.5 and 6.511.1 of Gradshteyn and Ryzhik (1994). Substituting J into (1.14), we find

$$w(x, y, 0) = -\frac{1}{2\pi\theta}\int_\Omega \frac{q(\xi,\eta)}{R}\,d\Omega. \tag{1.16}$$

This representation, due to Boussinesq, may also be found without using Fourier integrals, see for example section 2.5 in Gladwell (1980).

1.3.2. A STRIP PUNCH ON A HALF-SPACE

If the pressure distribution $q(x, y)$ is known, then equation (1.16) gives the normal displacement $w(x, y, 0)$ as an integral over Ω. In contact problems, we are given $w(x, y, 0)$ over Ω, and have to find $q(x, y)$ over Ω: equation (1.16) is an integral equation for $q(x, y)$. In the simplest case, in which Ω is a circle of radius a, the integral equation is best solved by bypassing it, and using a polar coordinate representation as discussed in section 1.4. Here we will first apply equation (1.16) to the case in which Ω is an infinite strip, i.e. $\Omega = (-a, a) \times (-\infty, \infty)$. However, if we simply assume that the pressure distribution $q(x, y)$ is constant in the y-direction, then the total force applied to the half-space will be infinite and the integral on the right-hand side of (1.16) will be infinite: the problem is not well posed. To circumvent this difficulty, we differentiate both sides of (1.16) with respect to x, and assume that $q(\xi, \eta) = q(\xi)$. Now

$$\frac{\partial w(x, y, 0)}{\partial x} = -\frac{1}{2\pi\theta}\int_{-a}^{a}(x-\xi)q(\xi)\,d\xi \int_{-\infty}^{\infty}\frac{d\eta}{[(x-\xi)^2+(y-\eta)^2]^{3/2}},$$

and since

$$\int_{-\infty}^{\infty}\frac{d\eta}{[(x-\xi)^2+(y-\eta)^2]^{3/2}} = \frac{2}{|x-\xi|^2},$$

we have

$$\frac{\partial w(x,y,0)}{\partial x} = \frac{1}{\pi\theta}\int_{-a}^{a}\frac{q(\xi)\,d\xi}{\xi - x}, \quad x \in (-a,a). \tag{1.17}$$

Equation (1.17) is an example of the singular equation

$$h(x) = \frac{1}{\pi}\int_{\Gamma}\frac{g(\xi)\,d\xi}{\xi - x}, \quad x \in \Gamma, \tag{1.18}$$

studied in detail by Muskhelishvili (1953a). Here Γ can be a segment of the real axis, more that one such segment, an arc of the (x,y) plane, etc. When $\Gamma = (-a,a)$, the equation has the general solution

$$g(x) = -\frac{(a^2-x^2)^{-1/2}}{\pi}\int_{-a}^{a}\frac{(a^2-\xi^2)^{1/2}\,h(\xi)\,d\xi}{\xi - x} + C(a^2-x^2)^{-1/2}, \tag{1.19}$$

where C is arbitrary. Thus solution has a square root singularity at each end $\pm a$ of Γ. We may choose C so that there is no singularity at one end, say $x = a$:

$$C = -\frac{1}{\pi}\int_{-a}^{a}\left(\frac{a+\xi}{a-\xi}\right)^{1/2} h(\xi)\,d\xi.$$

For that value of C, $g(x)$ becomes

$$g(x) = -\frac{1}{\pi}\left(\frac{a-x}{a+x}\right)^{1/2}\int_{-a}^{a}\left(\frac{a+\xi}{a-\xi}\right)^{1/2}\frac{h(\xi)\,d\xi}{\xi - x}.$$

In order for there to be no singularity at the other end, $x = -a$, we must have

$$\int_{-a}^{a}\frac{h(\xi)\,d\xi}{(a^2-\xi^2)^{1/2}} = 0. \tag{1.20}$$

Under this condition, $g(x)$ takes the form

$$g(x) = -\frac{(a^2-x^2)^{1/2}}{\pi}\int_{-a}^{a}\frac{h(\xi)\,d\xi}{(a^2-\xi^2)^{1/2}\,(\xi - x)}.$$

We now apply these results to equation (1.17). Suppose the punch is a cylindrical body with equation $z = f(x)$ with respect to the axes shown in

Fig. 1.1*a*. Suppose the punch is pressed into the surface by an amount δ, and is rotated through a small angle α, as shown in Fig. 1.1*b*. In order for the contact region to be $-a \le x \le a$, we must displace the origin by an amount c, to be determined. In the contact region $-a \le x \le a$ we have

$$w(x, y) = \delta + \alpha x - f(x - c) = \delta_0(x).$$

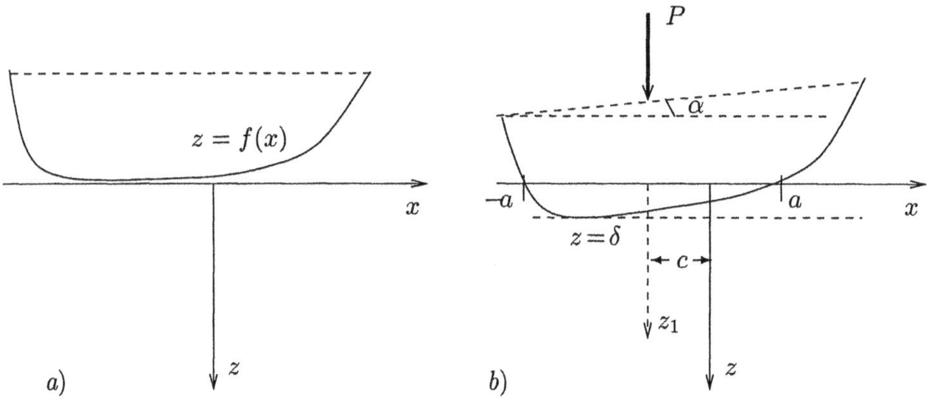

Figure 1.1: A punch and a half-space in 2-D: *a*) before contact; *b*) in contact

We note that one of the consequences of the unrealistic assumption that the punch extends to $\pm\infty$ in the y-direction, is that we cannot find δ. The integral equation for the contact pressure $q(x)$ is

$$\delta_0'(x) = \frac{1}{\pi\theta} \int_{-a}^{a} \frac{q(\xi)\, d\xi}{\xi - x}, \quad x \in (-a, a),$$

and according to (1.19)

$$q(x) = -\frac{\theta(a^2 - x^2)^{-1/2}}{\pi} \int_{-a}^{a} \frac{(a^2 - \xi^2)^{1/2}\, \delta_0'(\xi)\, d\xi}{\xi - x} + C(a^2 - x^2)^{-1/2}.$$

We may link C to the total load (per unit length) applied to the punch:

$$P = \int_{-a}^{a} q(x)\, dx = \pi\theta C.$$

Consider the case in which the punch is parabolic: $z = x^2/(2R_0)$. The condition that there be no pressure singularity at the ends is, from (1.20)

$$\int_{-a}^{a} \frac{\delta_0'(\xi)\, d\xi}{(a^2 - \xi^2)^{1/2}} = 0. \tag{1.21}$$

But

$$\delta_0'(\xi) = \alpha - (\xi - c)/R_0,$$

so that (1.21) gives $\alpha = -c/R_0$, $\delta_0'(\xi) = -\xi/R_0$. The total load P is

$$P = \pi\theta C = \frac{\theta}{R_0} \int_{-a}^{a} \xi \left(\frac{a+\xi}{a-\xi}\right)^{1/2} d\xi = \frac{\pi\theta a^2}{2R_0}.$$

The pressure distribution is

$$q(x) = \frac{\theta}{R_0}(a^2 - x^2)^{1/2}.$$

The moment of the pressure distribution about the origin is zero, showing that the total load P appears to act at the (new) origin $x = c = -\alpha R_0$.

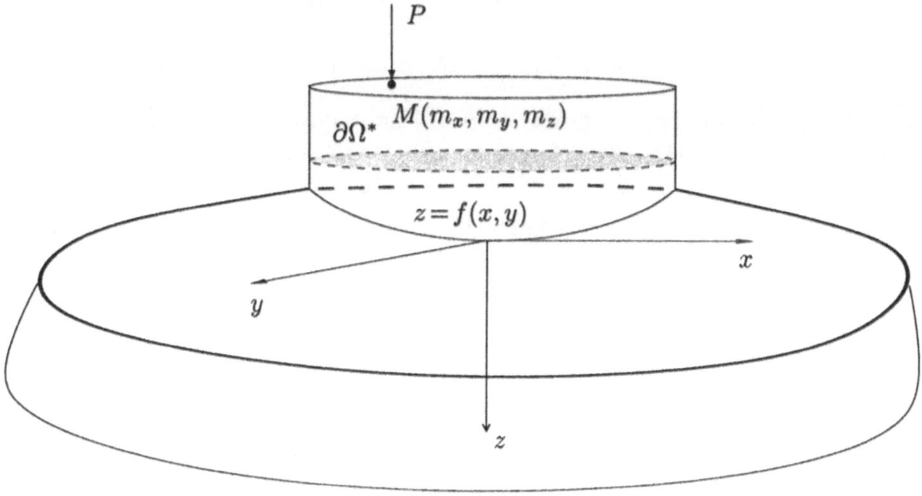

Figure 1.2: A punch and a half-space in 3-D

1.3.3. GENERAL FORMULATION OF THE CONTACT PROBLEM

Now consider the general problem of action of a rigid punch on an elastic half-space (Fig. 1.2). The punch is indented by the force P parallel to the z-axis and applied with the eccentricity $\mathbf{m} = (m_x, m_y)$. The punch is a cylindrical body with cross-section Ω^* and base surface $z = f(x,y)$. Suppose the equation of the contour of $\partial\Omega^*$ defined in the form

$$N^*(x,y) = 0 \quad ((x,y) \in \partial\Omega^*), \tag{1.22}$$

and N^* is normalized so that

$$N^*(x,y) > 0 \quad ((x,y) \in \Omega^*), \qquad \partial N^*/\partial n = 1 \quad ((x,y) \in \partial\Omega^*), \tag{1.23}$$

where n is the internal normal to $\partial\Omega^*$.

Example. *The normalized equation for a unit circle has the form*

$$\frac{1}{2}\left(1 - x^2 - y^2\right) = 0.$$

Let the region of contact of the punch with the half-space be $\Omega \subseteq \Omega^*$, and the normalized equation of its contour be

$$N(x,y) = 0 \quad ((x,y) \in \partial\Omega), \tag{1.24}$$

$$N(x,y) > 0 \quad ((x,y) \in \Omega), \qquad \partial N/\partial n = 1 \quad ((x,y) \in \partial\Omega), \tag{1.25}$$

where n is the internal normal to $\partial\Omega$. The origin of coordinates O in the plane $z = 0$ and the directions of the x- and y-axes may be chosen as convenient.

The boundary conditions have the form

no friction:

$$\tau_{xz}(x,y,0) = \tau_{yz}(x,y,0) = 0; \tag{1.26}$$

no load outside Ω:

$$\sigma_z(x,y,0) = 0 \quad ((x,y) \in \omega); \tag{1.27}$$

given displacement in Ω:

$$w(x,y,0) = -\delta(x,y) = -[\delta + \alpha x + \beta y - f(x,y)] \quad ((x,y) \in \Omega); \tag{1.28}$$

the displacements vanish at infinity:

$$\mathbf{u}\,(x,y,z) \to \mathbf{0} \text{ as } r = \sqrt{x^2 + y^2} \to \infty.$$

Here, $\delta + \alpha x + \beta y$ is the translation of the punch under action of the force P. Thus, the punch settles by the amount δ and rotates by the angles α and β around the x- and y-axes, respectively.

It is necessary to determine the distribution $\sigma_z(x, y, 0) = -q(x, y)$ of the contact normal stresses under the punch in the region of contact Ω and also the relations between P, m_x, m_y and δ, α, β.

To solve the boundary problem (1.26)–(1.28), we use formula (1.16), noting that the boundary conditions (1.26) and (1.27) will be satisfied. The satisfaction of (1.28) leads to the 2-D convolution integral equation of the first kind for the contact pressure $q(x, y)$:

$$\int_{\Omega} \frac{q(\xi, \eta)}{R}\, d\Omega = 2\pi\theta\delta(x, y) \quad ((x, y) \in \Omega). \tag{1.29}$$

After solving the integral equation (1.29), we can find the relations between P, m_x, m_y and δ, α, β using the equilibrium conditions of the punch:

$$P = \int_{\Omega} q(\xi, \eta)\, d\Omega, \tag{1.30}$$

$$M_x = Pm_y = \int_{\Omega} \eta q(\xi, \eta)\, d\Omega, \qquad M_y = Pm_x = \int_{\Omega} \xi q(\xi, \eta)\, d\Omega. \tag{1.31}$$

We close this section by noting that contact problems are essentially nonlinear, even when the contacting bodies are linearly elastic and we consider only infinitesimal deformations. We explain. In a static or dynamic problem involving the deformation of a single elastic body, the problem is linear in the sense that if a distribution load W produces a deformation field D, then the load αW will produce a deformation αD. If two load distributions W_1, W_2 produce deformation fields D_1, D_2 when applied separately, then the load distribution $\alpha W_1 + \beta W_2$ will produce a deformation $\alpha D_1 + \beta D_2$. The situation for contact problems is quite different: if there are two contacting bodies, then a change in the load distribution from W to αW may change the contact region, so that D will not change to αD; similar remark hold for the combination load $\alpha W_1 + \beta W_2$: again the combination load will change the contact region.

1.4 A Circular Punch on a Half-Space

1.4.1. THE METHOD OF PAIRED INTEGRAL EQUATIONS

We return to equations (1.26)–(1.29) and change to cylindrical polar coordinates (r, ϑ, z). Laplace's equation has a separable solution

$$\phi(r,\vartheta,z) = \mathcal{H}_n\{\xi^{-1}\Phi(\xi)\exp(\xi z); \xi \to r\}\cos(n\vartheta),$$

where

$$\mathcal{H}_n\{\xi^{-1}f(\xi); \xi \to r\} \equiv \int_0^\infty f(\xi)\mathcal{J}_n(\xi r)\, d\xi. \tag{1.32}$$

If therefore the displacement, over the circle $r \leq a$, has the form

$$w(r,\vartheta,0) = \delta_n(r)\cos(n\vartheta), \quad 0 \leq r \leq a,$$

and there is no loading outside the circle, then

$$\sigma_z(r,\vartheta,0) = \tilde{\sigma}_z \cos(n\vartheta) = 0, \quad r > a.$$

This leads to a pair of dual integral equations for Φ:

$$2(1-\nu)\mathcal{H}_n\{\xi^{-1}\Phi(\xi); r\} = \delta_n(r), \quad 0 \leq r \leq a; \tag{1.33}$$

$$\tilde{\sigma}_z(r) \equiv 2G\mathcal{H}_n\{\Phi(\xi); r\} = 0, \quad r > a. \tag{1.34}$$

This pair of equations is a particular case of a pair of equations of Titchmarsh type considered in great generality by Sneddon (1966). The general equations were first studied by Titchmarsh and Busbridge. Full bibliographic details may be found in Sneddon (1966).

Equations (1.33) and (1.34) involve two different Hankel transforms, $\mathcal{H}_n\{\xi^{-1}\Phi(\xi); r\}$ and $\mathcal{H}_n\{\Phi(\xi); r\}$, over two complementary intervals $[0, a]$, $[a, \infty)$, of the axis $[0, \infty)$. The secret to their solution is to apply operators to the two equations which reduce them to the *same*, Hankel type, operator, on the two complementary intervals. The required operators are so-called *fractional integration operators*, the simplest of which are the Abel operators

$$\mathcal{A}_1\{f(t); x\} = \left(\frac{2}{\pi}\right)^{1/2} \int_0^x \frac{f(t)\, dt}{(x^2-t^2)^{1/2}}, \tag{1.35}$$

$$\mathcal{A}_2\{f(t); x\} = \left(\frac{2}{\pi}\right)^{1/2} \int_x^\infty \frac{f(t)\, dt}{(t^2-x^2)^{1/2}}. \tag{1.36}$$

It is straightforward to show that the inverses of these operators are given by

$$\mathcal{A}_1^{-1}\{f(x);t\} = D_t\,\mathcal{A}_1\{xf(x);t\}, \quad D_t = d/dt, \tag{1.37}$$

$$\mathcal{A}_2^{-1}\{f(x);t\} = -D_t\,\mathcal{A}_2\{xf(x);t\}. \tag{1.38}$$

The Abel operators change $\mathcal{H}_0$ into cosine transform, according to

$$\mathcal{A}_1^{-1}\{\mathcal{H}_0\{\xi^{-1}\Phi(\xi);r\};x\} = \mathcal{F}_c\{\Phi(\xi);x\}, \tag{1.39}$$

$$\mathcal{A}_2\{r\mathcal{H}_0\{\Phi(\xi);r\};x\} = \mathcal{F}_c\{\Phi(\xi);x\}, \tag{1.40}$$

where

$$\mathcal{F}_c\{\Phi(\xi);x\} = \left(\frac{2}{\pi}\right)^{1/2}\int_0^\infty \Phi(\xi)\cos(x\xi)\,d\xi. \tag{1.41}$$

Now

$$(2/\pi)^{1/2}\cos(x\xi) = (x\xi)^{1/2}\mathcal{J}_{-1/2}(x\xi),$$

so that we may write equations (1.33), (1.34) as

$$\mathcal{A}_1^{-1}\{\mathcal{H}_0\{\xi^{-1}\Phi(\xi);r\};x\} = \hat{\mathcal{H}}_{-1/2}\{\Phi(\xi);x\}, \tag{1.42}$$

$$\mathcal{A}_2\{r\mathcal{H}_0\{\Phi(\xi);r\};x\} = \hat{\mathcal{H}}_{-1/2}\{\Phi(\xi);x\}, \tag{1.43}$$

where

$$\hat{\mathcal{H}}_{-1/2}\{\Phi(\xi);x\} = \int_0^\infty (x\xi)^{1/2}\,\mathcal{J}_{-1/2}(x\xi)\Phi(\xi)\,d\xi.$$

The relations (1.42), (1.43) are special cases of

$$x^{-n}\mathcal{A}_1^{-1}\{r^n\mathcal{H}_n\{\xi^{-1}\Phi(\xi);r\};x\} = \hat{\mathcal{H}}_{n-1/2}\{\Phi(\xi);x\}, \tag{1.44}$$

$$x^n\mathcal{A}_2\{r^{1-n}\mathcal{H}_n\{\Phi(\xi);r\};x\} = \hat{\mathcal{H}}_{n-1/2}\{\Phi(\xi);x\}, \tag{1.45}$$

which may be verified by referring to the integrals 6.683.6 and 6.592.10 in Gradshteyn and Ryzhik (1994). They are special cases of the more general results (2.5.1) and (2.5.2) of Sneddon (1966). In (1.44), (1.45)

$$\hat{\mathcal{H}}_{n-1/2}\{\Phi(\xi);x\} = \int_0^\infty (x\xi)^{1/2}\,\mathcal{J}_{n-1/2}(x\xi)\Phi(\xi)\,d\xi.$$

We now apply these results to equations (1.33), (1.34). We apply $x^{-n}\mathcal{A}_1^{-1}r^n$ to (1.33), and $x^n\mathcal{A}_2r^{1-n}$ to (1.34), noting that the first operator operates on the interval $(0, r)$, while the second operates on (r, ∞). Thus we obtain

$$2(1-\nu)\hat{\mathcal{H}}_{n-1/2}\{\Phi(\xi); x\} = x^{-n}\mathcal{A}_1^{-1}\{r^n\delta_n(r); x\} = p_n(x), \quad 0 \le x \le a,$$

$$x^n\mathcal{A}_2\{r^{1-n}\tilde{\sigma}_z(r); x\} = 2G\hat{\mathcal{H}}_{n-1/2}\{\Phi(\xi); x\} = 0, \quad x > a. \tag{1.46}$$

These equations may now be inverted to give

$$2(1-\nu)\Phi(\xi) = \hat{\mathcal{H}}_{n-1/2}\{p_n(x); \xi\}; \quad p_n(x) = 0, \quad x > a,$$

where we have used the fact that $\hat{\mathcal{H}}_{n-1/2}^{-1} = \hat{\mathcal{H}}_{n-1/2}$. To obtain $\tilde{\sigma}_z(r)$, we must invert (1.46) to give

$$\mathcal{H}_n\{\Phi(\xi); r\} = r^{n-1}\mathcal{A}_2^{-1}\{x^{-n}\mathcal{H}_{n-1/2}\{\Phi(\xi); x\}; r\}.$$

Thus

$$\tilde{\sigma}_z(r) = 2G\mathcal{H}_n\{\Phi(\xi); r\} = r^{n-1}\mathcal{A}_2^{-1}\{x^{-n}p_n(x); r\}.$$

Since $p_n(x) = 0$ when $x > a$, we may write

$$\tilde{\sigma}_z(r) = -\theta r^{n-1}\left(\frac{2}{\pi}\right)^{1/2}\frac{d}{dr}\int_r^a \frac{x^{1-n}p_n(x)\,dx}{(x^2-r^2)^{1/2}}, \tag{1.47}$$

where

$$p_n(x) = x^{-n}\left(\frac{2}{\pi}\right)^{1/2}\frac{d}{dx}\int_0^x \frac{r^{n+1}\delta_n(r)\,dr}{(x^2-r^2)^{1/2}}. \tag{1.48}$$

Here we have used equations (1.37), (1.38) for $\mathcal{A}_1^{-1}$ and $\mathcal{A}_2^{-1}$.

1.4.2. SPECIAL CASES

There are a number of important special cases.

A) $n = 0$, and the punch has a plane face:

$$\delta_0(r) = \delta, \quad p_0(x) = \left(\frac{2}{\pi}\right)^{1/2}\delta, \quad \tilde{\sigma}_z(r) = \frac{2\theta\delta}{\pi(a^2-r^2)^{1/2}}.$$

The total load on the circle is

$$P = 2\pi\int_0^a r\tilde{\sigma}_z(r)\,dr = 4\theta\delta a.$$

B) $n = 0$, and the punch is a paraboloidal with radius of curvarure R_0 at the tip:

$$\delta_0(r) = \delta - \frac{r^2}{2R_0}, \quad p_0(x) = \left(\frac{2}{\pi}\right)^{1/2} \left(\delta - \frac{x^2}{R_0}\right),$$

$$\tilde{\sigma}_z(r) = \frac{2\theta}{\pi}\left[\left(\delta - \frac{r^2}{R_0}\right)(a^2 - r^2)^{-1/2} + \frac{(a^2 - r^2)^{1/2}}{R_0}\right].$$

The normal stress $\tilde{\sigma}_z(r)$ must vanish at the edge, $r = a$, of the contact region; this requires $\delta = a^2/R_0$. Thus

$$\tilde{\sigma}_z(r) = \frac{4\theta(a^2 - r^2)^{1/2}}{\pi R_0},$$

and the total load is

$$P = \frac{8\theta a^3}{3R_0} = \frac{8\theta(\delta^3 R_0)^{1/2}}{3}.$$

C) $n = 1$, and the punch has a plane inclined face:

$$\delta_1(R) = \alpha r, \quad p_1(x) = \left(\frac{2}{\pi}\right)^{1/2} 2\alpha x,$$

so that

$$\tilde{\sigma}_z(r) = \frac{4\theta\alpha r}{\pi(a^2 - r^2)^{1/2}}.$$

The moment applied to the punch is

$$M = \int_0^{2\pi}\int_0^{a} \tilde{\sigma}_z(r) r^2 \cos^2\vartheta \, dr \, d\vartheta = \frac{8\theta\alpha a^3}{3}.$$

Other boundary value problems related to the circle $r = a$ on the surface of the half-space may be treated in a similar fashion. Thus, for example, in the so-called exterior problem, in which the displacement $w(r, \vartheta, 0) = \delta_n(r)\cos(n\theta)$ is specified for $r \geq a$, and there is no load on the interior $r < a$, the boundary conditions yield the pair of integral equations

$$\tilde{\sigma}_z(r) \equiv 2G\mathcal{H}_n\{\Phi(\xi); r\} = 0, \quad 0 \leq r < a;$$

$$2(1-\nu) = \delta_n(r)\mathcal{H}_n\{\xi^{-1}\Phi(\xi); r\}, \quad r \geq a.$$

Now we use analogues of (1.44), (1.45) , namely

$$x^{-n}\mathcal{A}_1\{r^{n+1}\mathcal{H}_n\{\Phi(\xi); r\}; x\} = \hat{\mathcal{H}}_{n+1/2}\{\Phi(\xi); x\},$$

$$x^n \mathcal{A}_2^{-1}\{r^{-n}\mathcal{H}_n\{\xi^{-1}\Phi(\xi);r\};x\} = \hat{\mathcal{H}}_{n+1/2}\{\Phi(\xi);x\},$$

and obtain

$$x^{-n}\mathcal{A}_1\{r^{n+1}\tilde{\sigma}_z(r);x\} = 2G\hat{\mathcal{H}}_{n+1/2}\{\Phi(\xi);x\} = 0, \quad 0 \le x \le a,$$

$$2(1-\nu)\hat{\mathcal{H}}_{n+1/2}\{\Phi(\xi);x\} = x^n\mathcal{A}_2^{-1}\{r^{-n}\delta_n(r);x\} = p_n(x), \quad x > a. \quad (1.49)$$

Now

$$\tilde{\sigma}_z(r) = 2G\,\mathcal{H}_n\{\Phi(\xi);r\} = \theta r^{-n-1}\mathcal{A}_1^{-1}\{x^n p_n(x);r\} \quad (1.50)$$

$$= \theta r^{-n-1}\left(\frac{2}{\pi}\right)^{1/2}\frac{d}{dr}\int_a^r \frac{x^{n+1}p_n(x)\,dx}{(r^2-x^2)^{1/2}}.$$

1.5 An Elliptic Punch on a Half-Space

The theory of contact between two linearly elastic bodies has its origin in the famous work of Hertz (1882). He showed that if the dimensions of the contact region were small compared to the radii of curvature of the contacting bodies, then the bodies could be regarded globally as half-spaces, locally as ellipsoids, in contact over an elliptic region. In fact the contact problem for two elastic bodies may be reduced to that of a rigid punch of ellipsoidal form indenting an elastic half-space. The full analysis may be found in Love (1927, sections 137–139). There are two important special cases of the general analysis: when the punch has a flat face, and when it has a parabolic face. In the first case the condition (1.28) shows that $w(x,y,0)$ is linear in x, y over Ω, in the second $w(x,y,0)$ is quadratic in x, y over Ω.

When the face is flat, the contact pressure distribution has the form

$$q(x,y) = p_1(x,y)\left(1 - \frac{x^2}{a^2} - \frac{y^2}{b^2}\right)^{-1/2},$$

where the contact ellipse has equation

$$\frac{x^2}{a^2} + \frac{y^2}{b^2} = 1,$$

and $p_1(x,y)$ is linear in x, y.

When the punch is parabolic, then

$$q(x,y) = C\left(1-\frac{x^2}{a^2}-\frac{y^2}{b^2}\right)^{1/2} = p_2(x,y)\left(1-\frac{x^2}{a^2}-\frac{y^2}{b^2}\right)^{-1/2}.$$

This suggests the result, proved by Galin (1961) and Lur'e (1964) that if the contact region is an ellipse, and if the displacement is a polynomial $w_N(x,y)$ of order N, in the contact region, then the contact pressure $q(x,y)$ has the form

$$q(x,y) = p_N(x,y)\left(1-\frac{x^2}{a^2}-\frac{y^2}{b^2}\right)^{-1/2},$$

where $p_N(x,y)$ is also a polynomial of order N in x, y. The relationships between the coefficients of the polynomials $w_N(x,y)$ and $p_N(x,y)$ are complicated. The relationships are perhaps found most easily by formulating the problem in terms of oblate spheroidal coordinates, as in Shail (1978), or as described in Gladwell (1980).

1.5.1. THE SOLUTION FOR A POLYNOMIAL BASE OF THE PUNCH

Here we follow the monographs by Galin (1961) and Lur'e (1964). They showed that if the contact region Ω between a punch and a half-space is an ellipse with semiaxes a and $b \le a$, and if the function $\delta(x,y)$ in the right-hand side of (1.29) is a polynomial, e.g.

$$\delta(x,y) = \sum_{i=0}^{k}\sum_{j=0}^{l} b_{ij}x^i y^j \qquad (k+l=N), \tag{1.51}$$

then the solution of the equation (1.18) has the form

$$q(x,y) = \left(1-\frac{x^2}{a^2}-\frac{y^2}{b^2}\right)^{-1/2}\sum_{i=0}^{m}\sum_{j=0}^{n} a_{ij}x^i y^j \qquad (m+n=N). \tag{1.52}$$

To express the unknown constants a_{ij} via the known constants b_{ij}, we substitute (1.52) into the left-hand side of (1.29) and calculate the integral

$$J(x,y) = \int_{\Omega} q(\xi,\eta)\frac{d\Omega}{R} = \sum_{i=0}^{m}\sum_{j=0}^{n} a_{ij}J_{ij}(x,y), \tag{1.53}$$

$$J_{ij}(x,y) = \int\limits_{\Omega} \frac{\xi^i \eta^j}{R} \left(1 - \frac{\xi^2}{a^2} - \frac{\eta^2}{b^2}\right)^{-1/2} d\Omega.$$

The way of calculation of the integrals $J_{ij}(x,y)$ was indicated by Shtaerman (1949). As a result, we have

$$J_{ij}(x,y) = 2\sum_{r=0}^{i}\sum_{s=0}^{j} \binom{i}{r}\binom{j}{s} \mathcal{B}\left(\frac{r+s+1}{2}, \frac{1}{2}\right)(1-e^2)^{(r-s+2j+1)/2} \tag{1.54}$$

$$\times \sum_{q=0}^{(r+s)/2} \binom{(r+s)/2}{q} (-1)^{(r-s)/2-q} a^{2q+1} \sum_{p=0}^{i+j-2q} \binom{i+j-2q}{p}$$

$$\times\, x^{i+j-2q-p} y^p S_{i-q+(j-p+s-r)/2,\,(j+p-s+r)/2}\,,$$

where $r+s$ is even, $j+p$ is even too, e is the eccentricity of the ellipse Ω, $\mathcal{B}(x,y)$ is the beta-function,

$$\binom{n}{k} = \frac{n!}{k!(n-k)!}$$

are the binomial coefficients, and

$$S_{mn} = \int\limits_0^{\pi/2} \frac{\cos^{2m}\varphi \sin^{2n}\varphi \, d\varphi}{(1-e^2\sin^2\varphi)^{m+n+1/2}}.$$

All the quantities S_{mn} may be expressed via the complete elliptic integrals $\mathcal{K}(e)$ and $\mathcal{E}(e)$.

One can see from (1.53) and (1.54) that $J(x,y)$ is a polynomial of degree N. As follows from the equation (1.29), equating the integral $J(x,y)$ to the expression $2\pi\theta\delta(x,y)$, where $\delta(x,y)$ has the form (1.51), we obtain the relation whose left- and right-hand sides are polynomials in x and y of degree N. Then equating the coefficients in this relation at the same powers of x and y, we come to a system of $(N+1)(N+2)/2$ linear algebraic equations for the $(N+1)(N+2)/2$ constants a_{ij}. By solving this system, we obtain by formula (1.52) the solution of the integral equation (1.29) for the case (1.51) and for the elliptic contact region Ω. The quantities of the force P and the moments M_x, M_y, acting onto the punch, must be determined by formulae (1.30) and (1.31) with the help of the integral

$$\int\limits_{\Omega} x^{2m} y^{2n} \left(1 - \frac{x^2}{a^2} - \frac{y^2}{b^2}\right)^{-1/2} d\Omega \tag{1.55}$$

$$= 2\pi a^{2m+1} b^{2n+1} \frac{(2m-1)!!\,(2n-1)!!}{(2m+2n+1)!!}.$$

For the particular case

$$\delta(x,y) = \delta + \alpha x + \beta y + Ax^2 + By^2 + Cxy,$$

the system of the linear algebraic equations has the form

$$a_{00}S_{00} + \frac{1}{2}a_{20}S_{01}b^2 + \frac{1}{2}a_{02}S_{10}b^2 = \frac{\theta\delta}{b},$$

$$a_{10}S_{10} = \frac{\theta\alpha}{b}, \qquad a_{01}S_{01}\frac{b^2}{a^2} = \frac{\theta\beta}{b}, \qquad a_{11}S_{11}\frac{3b^2}{a^2} = \frac{\theta C}{b},$$

$$a_{20}\left(S_{20} - \frac{1}{2}S_{11}\frac{b^2}{a^2}\right) + a_{02}\frac{b^2}{a^2}\left(S_{11}\frac{b^2}{a^2} - \frac{1}{2}S_{20}\right) = \frac{\theta A}{b},$$

$$a_{20}\left(S_{11} - \frac{1}{2}S_{02}\frac{b^2}{a^2}\right) + a_{02}\frac{b^2}{a^2}\left(S_{02}\frac{b^2}{a^2} - \frac{1}{2}S_{11}\right) = \frac{\theta B}{b},$$

where we write

$$S_{00} = \mathcal{K}(e), \qquad S_{10} = \frac{\mathcal{K}(e) - \mathcal{E}(e)}{e^2},$$

$$S_{01} = \frac{\mathcal{E}(e) - (1-e^2)\mathcal{K}(e)}{e^2(1-e^2)}, \qquad e^2 = 1 - \frac{b^2}{a^2},$$

$$S_{20} = \frac{-2(1+e^2)\mathcal{E}(e) + (2+e^2)\mathcal{K}(e)}{3e^4},$$

$$S_{11} = \frac{(2-e^2)\mathcal{E}(e) - 2(1-e^2)\mathcal{K}(e)}{3e^4(1-e^2)},$$

$$S_{02} = \frac{-2(1-2e^2)\mathcal{E}(e) + (1-e^2)(2-3e^2)\mathcal{K}(e)}{3e^4(1-e^2)^2}.$$

By substituting the constants a_{ij}, found from this system, into (1.52), we obtain

$$q(x,y) = \frac{\theta}{b}\left(1 - \frac{x^2}{a^2} - \frac{y^2}{b^2}\right)^{-1/2} \tag{1.56}$$

$$\times \left\langle \frac{1}{\mathcal{K}(e)}\left\{\delta - \frac{2a^2}{\psi}[A\psi_1 + B(1-e^2)\psi_2]\right\}\right.$$

$$+ \frac{\alpha e^2 x}{\psi_\alpha} + \frac{\beta e^2 y}{\psi_\beta} + \frac{4x^2}{\psi}\,[A\psi_{11} - B(1-e^2)\psi_{10}]$$

$$+\frac{4y^2}{\psi}(B\psi_{00} - A\psi_{01}) + \frac{Ce^4xy}{\psi_C}\Big\rangle,$$

where

$$\psi = -(1-e^2)\mathcal{K}^2 + 2(2-e^2)\mathcal{K}\mathcal{E} - 3\mathcal{E}^2,$$

$$\psi_1 = (1-e^2)\left(\frac{3}{2} - e^2\right)\mathcal{K}^2 - 2(1-e^2)\,\mathcal{K}\mathcal{E} + \frac{1}{2}\,\mathcal{E},$$

$$\psi_2 = \left(\frac{3}{2} - \frac{1}{2}\,e^2\right)\mathcal{K}^2 - 2\mathcal{K}\mathcal{E} + \frac{1}{2}\,\mathcal{E}^2,$$

$$\psi_\alpha = \mathcal{K} - \mathcal{E}, \quad \psi_\beta = \mathcal{E} - (1-e^2)\mathcal{K},$$

$$\psi_C = (2-e^2)\mathcal{E} - 2(1-e^2)\mathcal{K}, \quad \psi_{00} = \mathcal{K} - \left(1 + \frac{1}{2}\,e^2\right)\mathcal{E},$$

$$\psi_{10} = \psi_{01} = \mathcal{E} - \left(1 - \frac{1}{2}\,e^2\right)\mathcal{K},$$

$$\psi_{11} = (1-e^2)^2\mathcal{K} - \left(1 - \frac{3}{2}\,e^2\right)\mathcal{E}.$$

Here and later on, we omit the arguments in the functions $\mathcal{K}(e)$ and $\mathcal{E}(e)$.

By setting in (1.56) $A = B = C = 0$, we obtain the well-known solution of Lur'e (1964), for the planar sloping elliptic punch, in the form

$$q(x,y) = \frac{\theta}{b}\left(1 - \frac{x^2}{a^2} - \frac{y^2}{b^2}\right)^{-1/2}\left(\frac{\delta}{\mathcal{K}} + \frac{\alpha e^2 x}{\psi_\alpha} + \frac{\beta e^2 y}{\psi_\beta}\right). \quad (1.57)$$

By formulae (1.30), (1.31) and using the integral (1.55), we have the following relations, which connect the force P and the moments M_x, M_y with the settlement of the punch δ and the angles β, α of rotation of the punch around the x- and y-axes:

$$P = \frac{2\pi a\delta\theta}{\mathcal{K}}, \quad M_x = Pm_y = \frac{2\pi ab^2e^2\beta\theta}{3\psi_\beta}, \quad M_y = Pm_x = \frac{2\pi a^3e^2\alpha\theta}{3\psi_\alpha}. \quad (1.58)$$

Here, m_x and m_y are projections on the x- and y-axes of the force P arm counted from the punch symmetry center.

On the basis of (1.57), setting $y = -ba^{-1}\sqrt{a^2 - x^2}$, we obtain the condition for the absence of separation of the punch from the half-space surface in the form

$$\frac{\delta}{\mathcal{K}} + \frac{\alpha e^2 x}{\psi_\alpha} - \frac{\beta e^2 b\sqrt{a^2 - x^2}}{a\psi_\beta} \geq 0.$$

Taking account of (1.58), we rewrite the latter condition as follows:

$$1 + \frac{3m_x x}{a^2} - \frac{3m_y\sqrt{a^2 - x^2}}{ab} \geq 0. \tag{1.59}$$

By investigating the minimum the left-hand side of (1.59), we finally obtain the following condition for the absence of separation:

$$1 - 3\sqrt{\frac{m_x^2}{a^2} + \frac{m_y^2}{b^2}} \geq 0.$$

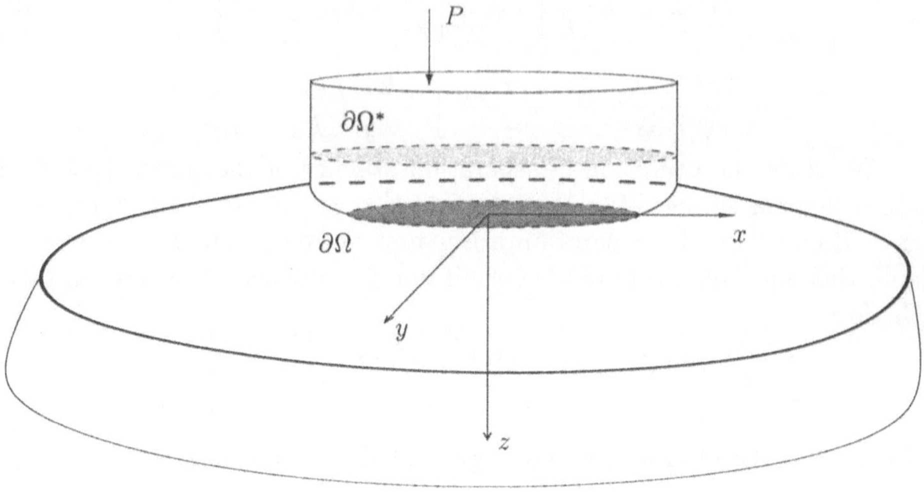

Figure 1.3: An elliptic paraboloid and a half-space in contact

1.5.2. THE SOLUTION FOR AN ELLIPTIC PARABOLOID

Consider the case when the punch is an elliptic paraboloid which is in contact with the surface $z = 0$ over an ellipse, as shown in Fig. 1.3. Let the symmetry planes of the cylinder and paraboloid coincide, the impressed force P act along the cylinder axis, and the contact area be fixed[1]. If in

[1]It is supposed that the edge of the punch, obtained by the intersection of the elliptic cylinder with the elliptic paraboloid, is cutting into the half-space surface and, therefore, the contact area is an ellipse with the semiaxes a and b.

(1.56) we set $\alpha = \beta = C = 0$, $A = -1/(2R_1)$, and $B = -1/(2R_2)$ (R_1 and R_2 are the radii of curvature at the elliptic paraboloid apex in the planes $y = 0$ and $x = 0$, respectively), then we have

$$q(x,y) = \frac{\theta}{b}\left(1 - \frac{x^2}{a^2} - \frac{y^2}{b^2}\right)^{-1/2}\left\langle \frac{1}{\mathcal{K}}\left\{\delta + \frac{a^2}{\psi}\left[\frac{\psi_1}{R_1} + \frac{(1-e^2)\psi_2}{R_2}\right]\right\}\right. \tag{1.60}$$
$$\left. - \frac{2x^2}{\psi}\left[\frac{\psi_{11}}{R_1} - \frac{(1-e^2)\psi_{10}}{R_2}\right] - \frac{2y^2}{\psi}\left(\frac{\psi_{00}}{R_2} - \frac{\psi_{01}}{R_1}\right)\right\rangle.$$

In accordance with formulae (1.30) and (1.55), we obtain the following equilibrium condition for the punch:

$$P = 2\pi a\theta\left\langle \frac{1}{\mathcal{K}}\left\{\delta + \frac{a^2}{\psi}\left[\frac{\psi_1}{R_1} + \frac{(1-e^2)\psi_2}{R_2}\right]\right\}\right. \tag{1.61}$$
$$\left. - \frac{2a^2}{3\psi}\left[\frac{\psi_{11}}{R_1} - \frac{(1-e^2)\psi_{10}}{R_2}\right] - \frac{2b^2}{3\psi}\left(\frac{\psi_{00}}{R_2} - \frac{\psi_{01}}{R_1}\right)\right\rangle.$$

We state the condition of separation absence of the punch base from the half-space surface. Let $R_1 > R_2$ and $l^2 = 1 - R_2/R_1$. If $l^2 > e^2$, then, as P diminishes, the separation arises first at the points $x = 0$, $y = \pm b$ and, therefore, due to (1.60) the condition for absence of separation takes the form

$$\frac{1}{\mathcal{K}}\left\{\delta + \frac{a^2}{\psi}\left[\frac{\psi_1}{R_1} + \frac{(1-e^2)\psi_2}{R_2}\right]\right\} - \frac{2b^2}{\psi}\left(\frac{\psi_{00}}{R_2} - \frac{\psi_{01}}{R_1}\right) \geq 0.$$

Taking account of (1.61), we can represent this condition as follows:

$$P \geq P_b = \frac{4\pi a^3\theta}{3\psi}\left\{\frac{1-e^2}{R_2}(\psi_{10} + 2\psi_{00})\right. \tag{1.62}$$
$$\left. - \frac{1}{R_1}[\psi_{11} + 2(1-e^2)\psi_{01}]\right\}.$$

If $l^2 < e^2$ or $R_1 < R_2$, then the separation arises first at the points $x = \pm a$, $y = 0$. Then on the basis of (1.60) the condition for absence of separation takes the form

$$\frac{1}{\mathcal{K}}\left\{\delta + \frac{a^2}{\psi}\left[\frac{\psi_1}{R_1} + \frac{(1-e^2)\psi_2}{R_2}\right]\right\} - \frac{2a^2}{\psi}\left[\frac{\psi_{11}}{R_1} - \frac{(1-e^2)\psi_{10}}{R_2}\right] \geq 0.$$

Taking account of (1.61), we can represent the latter condition as follows:

$$P \geq P_a = \frac{4\pi a^3\theta}{3\psi}\left\{\frac{1}{R_1}[(1-e^2)\psi_{01} + 2\psi_{11}]\right. \tag{1.63}$$

$$-\frac{1-e^2}{R_2}(\psi_{00}+2\psi_{10})\Big\}.$$

If $l^2 = e^2$, then, as P diminishes, the separation arises at once all along the edge of the punch, and, we can use (1.62) or, equivalently, (1.63) as the condition for absence of separation.

If $l^2 > e^2$ and $P = P_b$, $l^2 < e^2$ or $R_1 < R_2$ and $P = P_a$, then the contact pressure vanishes at the points $x = 0$, $y = \pm b$ or $x = \pm a$, $y = 0$, respectively. If $l^2 = e^2$ and $P = P_b = P_a$, then the contact pressure vanishes at once all along the edge of the punch.

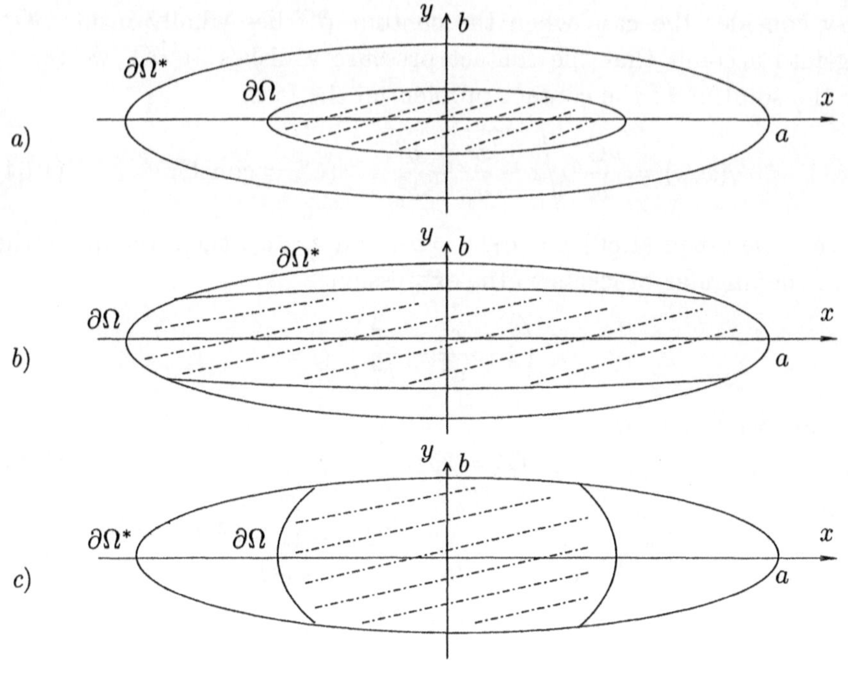

Figure 1.4: Contact domain Ω and punch shape Ω^*: *a*) $\Omega \subset \Omega^*$; *b*), *c*) $\Omega \subseteq \Omega^*$

When $l^2 = e^2$ and $P < P_a = P_b$, the contour $\partial\Omega$ of the contact area will lie wholly inside the contour $\partial\Omega^*$ of the punch edge (Fig. 1.4*a*). These contours do not intersect each other and the contact pressure equals zero on all the contour $\partial\Omega$.

A more complicated situation arises in the cases when $l^2 > e^2$ and $P_b^* < P < P_b$, and also when $l^2 < e^2$ or $R_1 < R_2$ and $P_a^* < P < P_a$. Here,

(Fig. 1.4b,c) the contact region is bounded by parts of $\partial\Omega^*$ and $\partial\Omega$. The quantities P_b^* and P_a^* will be determined below. The contact pressure is infinite at the parts of $\partial\Omega^*$ and vanishes at the parts of $\partial\Omega$. Such cases of the problem have not yet been investigated.

When $l^2 > e^2$ and $P = P_b^*$, $l^2 < e^2$ or $R_1 < R_2$ and $P = P_a^*$, then the contour $\partial\Omega$ of the contact area will lie wholly inside the contour $\partial\Omega^*$, having with it only two common points $x = \pm a$, $y = 0$ or $x = 0$, $y = \pm b$, respectively. The contact pressure vanishes over all the contour $\partial\Omega$. Finally, when $l^2 > e^2$ and $P < P_b^*$, $l^2 < e^2$ or $R_1 < R_2$ and $P < P_a^*$, the contour $\partial\Omega$ will also lie wholly inside $\partial\Omega^*$ having no common points with it. As before, the contact pressure vanishes over all the contour $\partial\Omega$.

Now consider the case when the contour $\partial\Omega$ lies wholly inside $\partial\Omega^*$. Taking into account that the contact pressure vanishes at $\partial\Omega$, we try to obtain the solution of the contact problem in the form

$$q(x,y) = \frac{\theta C^*}{b_1}\sqrt{1 - \frac{x^2}{a_1^2} - \frac{y^2}{b_1^2}} \qquad (C^* = \text{const}). \tag{1.64}$$

To do this, we set in (1.60) $a = a_1$, $b = b_1$, and $e = e_1$, then we equate the terms in the angular brackets to the expression

$$C^*\left(1 - \frac{x^2}{a_1^2} - \frac{y^2}{b_1^2}\right).$$

As a result, we find that

$$C^* = 2\delta/\mathcal{K}, \tag{1.65}$$

and also

$$\frac{1}{R_1} = \frac{2\delta(\mathcal{K} - \mathcal{E})}{a_1^2 e_1^2 \mathcal{K}}, \qquad \frac{1}{R_2} = \frac{2\delta[\mathcal{E} - (1 - e_1^2)\mathcal{K}]}{b_1^2 e_1^2 \mathcal{K}}. \tag{1.66}$$

Besides, by substituting the expression (1.64) for $q(x,y)$ into formula (1.30), calculating the integral with the help of (1.55) and taking account of (1.65), we obtain[2]

$$P = \frac{4\pi a_1 \theta \delta}{3\mathcal{K}}. \tag{1.67}$$

Formulae (1.66) and (1.67) allow us to determine, for specified values of P, R_1, and R_2, the quantities a_1, b_1, and δ. Excluding the quantity δ from formulae (1.66), we come to the following equation in e_1:

$$\frac{(1 - e_1^2)(\mathcal{K} - \mathcal{E})}{\mathcal{E} - (1 - e_1^2)\mathcal{K}} = \frac{R_2}{R_1}. \tag{1.68}$$

[2] In the well-known Hertz contact problem on local contact between two elastic solids (see, e.g. Lur'e (1964)) there are formulae which are similar to (1.65)–(1.67).

Note that e_1 depends only on the ratio R_2/R_1. The plot of the dependence of e_1^2 in R_2/R_1 was given by Lur'e (1964). By substituting δ from (1.67) into the first formula of (1.66), we find that

$$a_1 = \left[\frac{3PR_1(\mathcal{K} - \mathcal{E})}{2\pi\theta e_1^2}\right]^{1/3}, \tag{1.69}$$

and then $b_1 = a_1\sqrt{1 - e_1^2}$. Finally, we determine the quantity δ from (1.67).

All these results hold for $R_1 > R_2$. For $R_1 < R_2$, to save formulae (1.64), (1.65), and (1.67), one must interchange R_1 and R_2 in (1.66), (1.68), and also must assume that the semi-major axis of the punch edge is elongated along the y-axis and the semi-minor one along the x-axis.

The quantities P_b^* and P_a^* can be obtained by setting in (1.67) $a_1 = a$ and $a_1 = b$, respectively. At the same time, δ and e_1 are determined from (1.66) where we must set $a_1 = a$ or $a_1 = b$.

1.5.3. TWO IDENTICAL ELLIPTIC PUNCHES

We pass to the problem on embedding of two identical planar elliptic punches with semiaxes a and $b < a$ into a half-space. The scheme of the contact is shown in Fig. 1.5. The punches are impressed by identical forces P acting along the symmetry axes of the punches. It is clear that the distributions of the contact pressures under the punches will be identical, and if for the punch with the contact area Ω $(1 - x^2/a^2 - y^2/b^2 \leq 0)$ the contact pressure is $q(x, y)$, then for the punch with the contact area Ω_1 $(1 - x_1^2/a^2 - y_1^2/b^2 \leq 0)$ the indicated pressure will be $q(x_1, y_1) = q(-x + 2h, -y + 2g)$.

Taking account of the latter circumstance, the integral equation (1.29) for the case under consideration takes the form

$$\int\limits_{\Omega} q(\xi, \eta)\frac{d\Omega}{R} + \int\limits_{\Omega_1} q(-\xi + 2h, -\eta + 2g)\frac{d\Omega}{R} = 2\pi\theta\delta(x, y) \tag{1.70}$$

$$((x, y) \in \Omega \cup \Omega_1),$$

$$\delta(x, y) = \delta + \alpha x + \beta y \qquad ((x, y) \in \Omega),$$

$$\delta(x, y) = \delta + \alpha x_1 + \beta y_1 = \delta - \alpha(x - 2h) - \beta(y - 2g) \qquad ((x, y) \in \Omega_1).$$

Here, α and β are the angles of rotation of the punches around the y- and x-axes or y_1- and x_1-axes, respectively, arising due of an interaction between the punches.

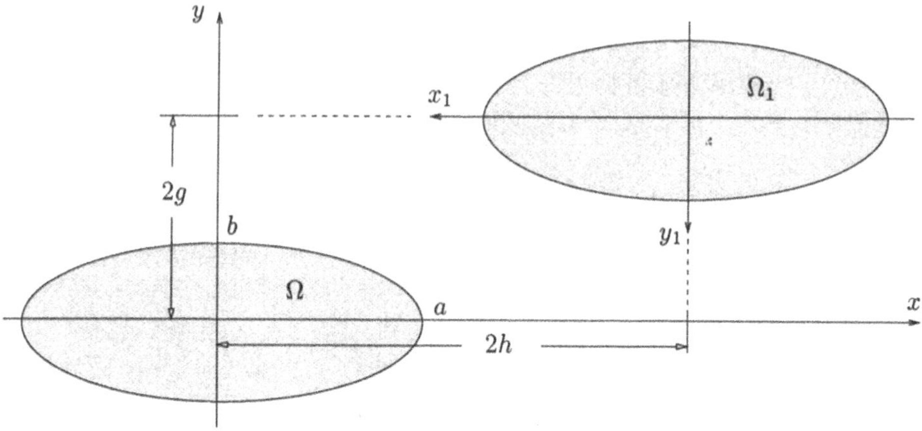

Figure 1.5: Two identical elliptic contact domains

By changing the variables $\xi_1 = -\xi + 2h$ and $\eta_1 = -\eta + 2g$ in the second integral of (1.70), we finally come to the following integral equation:

$$\int_{\Omega} q(\xi,\eta) M(x,y,\xi,\eta)\, d\Omega = 2\pi\theta(\delta + \alpha x + \beta y) \tag{1.71}$$

$$((x,y) \in \Omega, \qquad k = g/h),$$

$$M(x,y,\xi,\eta) = \frac{1}{\sqrt{(\xi - x)^2 + (\eta - y)^2}} + \frac{1}{\sqrt{(\xi + x - 2h)^2 + (\eta + y - 2kh)^2}}.$$

We suppose that the parameter k is of the order 1 and the parameter $\lambda = h/a$ is so large that we can neglect the terms of the order $O(\lambda^{-4})$ in the kernel $M(x,y,\xi,\eta)$ of equation (1.71), i.e., we can approximately present it in the form

$$M(x,y,\xi,\eta) \approx \frac{1}{\sqrt{(\xi - x)^2 + (\eta - y)^2}} + \frac{1}{2h\sqrt{1+k^2}} \tag{1.72}$$

$$\times \left[1 + \frac{\xi + x}{2h(1+k^2)} - \frac{(\xi + x)^2 m}{8h^2(1+k^2)} + \frac{k(\eta + y)}{2h(1+k^2)} - \frac{(\eta + y)^2 n}{8h^2(1+k^2)} \right.$$

$$\left. + \frac{3k(\xi + x)(\eta + y)}{4h^2(1+k^2)^2} \right],$$

$$m = 1 - \frac{3}{1+k^2}, \qquad n = 1 - \frac{3k^2}{1+k^2}.$$

Note that the function $q(\xi, \eta)$ must give zero moments M_x and M_y. Then, using (1.72), we rewrite the integral equation (1.71) as follows:

$$\int_\Omega q(\xi,\eta)\frac{d\Omega}{R} = 2\pi\theta(\tilde{\delta} + Ax^2 + By^2 + Cxy) \qquad ((x,y) \in \Omega), \tag{1.73}$$

where we have

$$\tilde{\delta} = \delta - \frac{1}{4\pi h\theta\sqrt{1+k^2}} \int_\Omega q(\xi,\eta)\Big[1 - \frac{\xi^2 m}{8h^2(1+k^2)} \tag{1.74}$$

$$- \frac{\eta^2 n}{8h^2(1+k^2)} + \frac{3k\xi\eta}{4h^2(1+k^2)^2}\Big]\, d\Omega,$$

$$A = \frac{mP}{32\pi h^3\theta(1+k^2)^{3/2}}, \qquad P = \int_\Omega q(\xi,\eta)\, d\Omega, \tag{1.75}$$

$$B = \frac{nP}{32\pi h^3\theta(1+k^2)^{3/2}}, \tag{1.76}$$

$$C = \frac{3kP}{16\pi h^3\theta(1+k^2)^{5/2}}, \tag{1.77}$$

and

$$\alpha = \frac{P}{8\pi h^2\theta(1+k^2)^{3/2}}, \tag{1.78}$$

$$\beta = \frac{kP}{8\pi h^2\theta(1+k^2)^{3/2}}. \tag{1.79}$$

The solution of the integral equation (1.73) is given by formula (1.56) in which one must set $\alpha = \beta = 0$ and $\delta = \tilde{\delta}$. By substituting this expression into formulae (1.74)–(1.77), we obtain a system of four algebraic equations in $\tilde{\delta}$, A, B and C. As a result, we find the asymptotic solution for large λ of the problem on embedding of two punches, i.e., the contact pressure $q(x, y)$ with an accuracy up to the terms of the order $O(\lambda^{-4})$. We then find with the same accuracy the expressions for δ, α, and β by using formulae (1.30), (1.78) and (1.79).

Note that in the similar way one can obtain the asymptotic solution for large λ of the problem on impression of two elliptic punches with parabolic bases. If we require for this case that the contact pressure $q(x, y)$ vanishes at the contour of the region Ω, then we obtain that the contact regions Ω and Ω_1 of the punches with the half-space will be elliptic only with the accuracy up to the terms $O(\lambda^{-4})$, moreover, we come to two relations which are similar to (1.66) and serve for determination of the semiaxes of Ω.

For $0 \leq h/g < 1$ and large g/a the investigation can be carried out similarly.

1.5.4. AN ELLIPTIC PUNCH HAVING A CIRCULAR HOLE

We consider the problem on impression of a planar punch whose contact area Ω is an ellipse with semiaxes a and $b < a$ having a circular hole of radius $R < b$ which does not touch the ellipse boundary (Fig. 1.6). We will assume that the vector arm $\mathbf{m}$ of the force P applied is such that the punch perfoms only a translation along the z-axis without rotations.

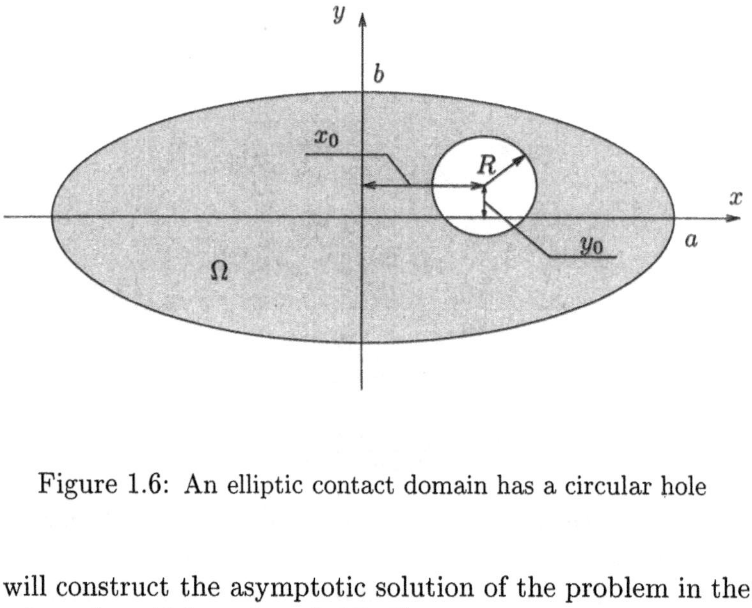

Figure 1.6: An elliptic contact domain has a circular hole

We will construct the asymptotic solution of the problem in the multiplicative form (see Aleksandrov (1967a))

$$q(x,y) = \frac{q_1(x,y)\, q_2(x,y)}{v(x,y)}, \tag{1.80}$$

where $q_1(x,y)$ is said to be the main field of the contact pressures and is determined as the solution (1.57) ($\alpha = \beta = 0$) for the problem for planar elliptic punch without a hole, $q_2(x,y)$ is said to be the perturbed field and is determined as the solution for the problem for planar punch which is a plane with a circular hole of radius R, and $v(x, y)$ is so-called the degenerate or correcting field and is determined as the first term in the asymptotics of the function $q_2(x, y)$ as $r/R \to \infty$ $(r = \sqrt{x^2 + y^2})$.

By using formulae (1.50), (1.49) for $n = 0$ for the ratio $\chi(x, y) = q_2(x, y)/v(x, y)$, we get the formula (Aleksandrov (1967a))

$$\chi(x, y) = \frac{2}{\pi}\left[\frac{R}{\sqrt{r^2 - R^2}} + \arccos\frac{R}{r}\right]. \tag{1.81}$$

For the case when x_0 and y_0 are the coordinates of the center of the circular hole, by using formulae (1.57), (1.80) and (1.81) for the problem under consideration, we finally obtain

$$q(x, y) = \frac{2\delta\theta}{\pi b K}\left(1 - \frac{x^2}{a^2} - \frac{y^2}{b^2}\right)^{-1/2}\left[\frac{R}{\sqrt{(x - x_0)^2 + (y - y_0)^2 - R^2}} + \arccos\frac{R}{\sqrt{(x - x_0)^2 + (y - y_0)^2}}\right]. \tag{1.82}$$

Formula (1.82) will have a fair accuracy if the minimal distance between the boundaries of the ellipse and circular hole satisfies the condition $h \geq 2R$. The settlement δ of the punch and the projections m_x and m_y of the impressed force arm on the x- and y-axes can be found by substituting (1.82) into formulae (1.30), (1.31) and calculating the double integrals.

Chapter 2

CONTACT PROBLEMS FOR LAYERS AND HALF-SPACES

Among the earlier publications, the contact problems for a layer were considered most completely in Gladwell (1980) and in Vorovich, Aleksandrov and Babeshko (1974). In this chapter, we concentrate on the case in which the layer is fixed along its base. Such problems have many applications. For example in biomechanics, the contact problem for a thin layer arises in the study of mechanoreceptors in tissue, including Pacinian corpuscles, SA II receptors, and chondrocytes (Haider and Holmes (1997)). These receptors are responsible for transducing a component of the deformation field into a neural or chemical signal.

In this chapter, we examine various contact problems for layers, the problem for a two-layer foundation, and the contact between a narrow rectangular punch and a half-space.

2.1 Equilibrium under Given Normal Loads

2.1.1. GENERAL SOLUTION

We consider the equilibrium of an infinite linearly elastic, isotropic and homogeneous layer ($|x| < \infty$, $|y| < \infty$, $0 \leq z \leq h$), subject to 3-D (non-axisymmetric) deformation.

As is well-known (Sneddon (1951), Lur'e (1964)), we can use double integral Fourier transformations. We ignore inertia effects and seek the solution of Navier's equations (1.3) in the form

$$\mathbf{u} = \frac{1}{2\pi} \int_{-\infty}^{\infty} \int_{-\infty}^{\infty} \mathbf{U}(\alpha, \beta, z) \exp[-i(\alpha x + \beta y)] \, d\alpha \, d\beta, \tag{2.1}$$

where $\mathbf{u} = (u, v, w)$ and $\mathbf{U} = (U, V, W)$.

By substituting (2.1) into (1.3), we obtain the following system of ordinary differential equations for U, V, and W:

$$(1-2\nu)U_z'' - i\alpha W_z' - [(1-2\nu)\gamma^2 + \alpha^2]U - \alpha\beta V = 0,$$

$$(1-2\nu)V_z'' - i\beta W_z' - [(1-2\nu)\gamma^2 + \beta^2]V - \alpha\beta U = 0,$$

$$2(1-\nu)W_z'' - i\alpha U_z' - i\beta V_z' - (1-2\nu)\gamma^2 W = 0.$$

The solution of this system can be presented in the form

$$U = \exp(\gamma z) \left[\kappa a_1 + \alpha z \left(\frac{\alpha}{\gamma} a_1 + \frac{\beta}{\gamma} a_2 + i a_3\right)\right] \tag{2.2}$$

$$+ \exp(-\gamma z) \left[\kappa b_1 - \alpha z \left(\frac{\alpha}{\gamma} b_1 + \frac{\beta}{\gamma} b_2 - i b_3\right)\right],$$

$$V = \exp(\gamma z) \left[\kappa a_2 + \beta z \left(\frac{\alpha}{\gamma} a_1 + \frac{\beta}{\gamma} a_2 + i a_3\right)\right] \tag{2.3}$$

$$+ \exp(-\gamma z) \left[\kappa b_2 - \beta z \left(\frac{\alpha}{\gamma} b_1 + \frac{\beta}{\gamma} b_2 - i b_3\right)\right],$$

$$W = \exp(\gamma z) \left[\kappa a_3 + i\gamma z \left(\frac{\alpha}{\gamma} a_1 + \frac{\beta}{\gamma} a_2 + i a_3\right)\right] \tag{2.4}$$

$$+ \exp(-\gamma z) \left[\kappa b_3 + i\gamma z \left(\frac{\alpha}{\gamma} b_1 + \frac{\beta}{\gamma} b_2 - i b_3\right)\right].$$

Here, $\kappa = 3 - 4\nu$, $\gamma = \sqrt{\alpha^2 + \beta^2}$; a_n and b_n ($n = 1, 2, 3$) are arbitrary constants which must be determined from the boundary conditions at $z = 0$ and $z = h$.

The formulae (2.1)–(2.4) provide the general solution (Sneddon (1951)) of Navier's equations (1.3). Then using Cauchy formulae (1.2), we can find the strains and, using Hooke's law (1.1), the stresses

$$\sigma_z = -\frac{iG}{\pi} \int_{-\infty}^{\infty} \int_{-\infty}^{\infty} \left\{ \exp(\gamma z) \left[2i(1-\nu)\gamma a_3 - (1-2\nu)(\alpha a_1 + \beta a_2) \right. \right. \tag{2.5}$$

$$-\gamma^2 z\left(\frac{\alpha}{\gamma}a_1+\frac{\beta}{\gamma}a_2+ia_3\right)\Bigg]+\exp(-\gamma z)\Bigg[-2i(1-\nu)\gamma b_3$$

$$-(1-2\nu)(\alpha b_1+\beta b_2)+\gamma^2 z\left(\frac{\alpha}{\gamma}b_1+\frac{\beta}{\gamma}b_2-ib_3\right)\Bigg]\Bigg\}\exp[-i(\alpha x+\beta y)]\,d\alpha\,d\beta,$$

$$\tau_{xz}=\frac{G}{2\pi}\int_{-\infty}^{\infty}\int_{-\infty}^{\infty}\Bigg\{\exp(\gamma z)\Bigg[\alpha(1+2\gamma z)\left(\frac{\alpha}{\gamma}a_1+\frac{\beta}{\gamma}a_2+ia_3\right)+\gamma\kappa a_1-i\alpha\kappa a_3\Bigg]$$

$$+\exp(-\gamma z)\Bigg[-\alpha(1-2\gamma z)\left(\frac{\alpha}{\gamma}b_1+\frac{\beta}{\gamma}b_2-ib_3\right)-\gamma\kappa b_1-i\alpha\kappa b_3\Bigg]\Bigg\}$$

$$\times\exp[-i(\alpha x+\beta y)]\,d\alpha\,d\beta, \tag{2.6}$$

$$\tau_{yz}=\frac{G}{2\pi}\int_{-\infty}^{\infty}\int_{-\infty}^{\infty}\Bigg\{\exp(\gamma z)\Bigg[\beta(1+2\gamma z)\left(\frac{\alpha}{\gamma}a_1+\frac{\beta}{\gamma}a_2+ia_3\right)+\gamma\kappa a_2-i\beta\kappa a_3\Bigg]$$

$$+\exp(-\gamma z)\Bigg[-\beta(1-2\gamma z)\left(\frac{\alpha}{\gamma}b_1+\frac{\beta}{\gamma}b_2-ib_3\right)-\gamma\kappa b_2-i\beta\kappa b_3\Bigg]\Bigg\}$$

$$\times\exp[-i(\alpha x+\beta y)]\,d\alpha\,d\beta. \tag{2.7}$$

2.1.2. BOUNDARY CONDITIONS

Consider the conditions on the faces of the layer.
The face $z=0$ is fixed:

$$u(x,y,0)=v(x,y,0)=w(x,y,0); \tag{2.8}$$

the face $z=h$ is subject to a normal load, and there are no friction forces:

$$\tau_{xz}(x,y,h)=\tau_{yz}(x,y,h)=0,\qquad \sigma_z(x,y,h)=-q^*(x,y), \tag{2.9}$$

$$q^*(x,y)=q(x,y)\quad (x,y)\in\Omega,\qquad q^*(x,y)=0\quad (x,y)\in\omega,$$

where Ω is a simply connected and bounded domain, with complement ω in the plane $z=h$ (Fig. 2.1). Note that $q^*(x,y)$ is obtained by extending $q(x,y)$ to the whole plane.

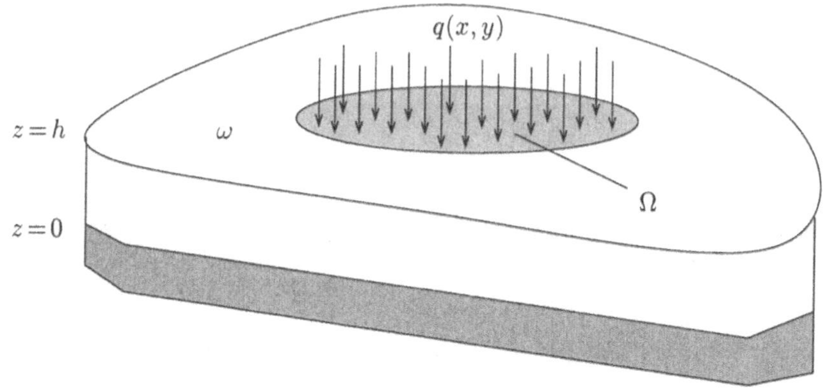

Figure 2.1: The elastic layer $0 \le z \le h$ under the normal load $q(x, y)$

Concerning the function $q(x, y)$, we suppose that $q(x, y) \in \mathcal{L}(\Omega)$ (absolutely integrable in Ω) and $q(x, y) \in \mathcal{C}(\Omega_0)$ (continuous in $\subset \Omega$). In this case we have the relations

$$Q(\alpha, \beta) = \frac{1}{2\pi} \int\limits_{\Omega} q(\xi, \eta) \exp[i(\alpha\xi + \beta\eta)] \, d\Omega \qquad (d\Omega = d\xi \, d\eta), \tag{2.10}$$

$$q^*(x, y) = \frac{1}{2\pi} \int\limits_{-\infty}^{\infty} \int\limits_{-\infty}^{\infty} Q(\alpha, \beta) \exp[-i(\alpha x + \beta y)] \, d\alpha \, d\beta.$$

To satisfy the boundary conditions (2.8), (2.9), we use formulae (2.1)–(2.7).

The boundary conditions (2.8) give

$$b_1 = -a_1, \qquad b_2 = -a_2, \qquad b_3 = -a_3. \tag{2.11}$$

From the boundary conditions (2.9), using (2.11), we find that

$$a_1 \frac{\alpha}{\gamma} \Big(\cosh(\gamma h) + 2\gamma h \sinh(\gamma h) + \kappa \frac{\gamma^2}{\alpha^2} \cosh(\gamma h) \Big) + a_2 \frac{\beta}{\gamma} [\cosh(\gamma h) \tag{2.12}$$

$$+ 2\gamma h \sinh(\gamma h)] + 2ia_3 [\gamma h \cosh(\gamma h) - (1 - 2\nu) \sinh(\gamma h)] = 0,$$

$$a_1 \frac{\alpha}{\gamma} [\cosh(\gamma h) + 2\gamma h \sinh(\gamma h)] + a_2 \frac{\beta}{\gamma} \Big(\cosh(\gamma h) + 2\gamma h \sinh(\gamma h) \tag{2.13}$$

$$+ \kappa \frac{\gamma^2}{\beta^2} \cosh(\gamma h) \Big) + 2ia_3 [\gamma h \cosh(\gamma h) - (1 - 2\nu) \sinh(\gamma h)] = 0,$$

$$-\alpha a_1[(1-2\nu)\sinh(\gamma h)+\gamma h\cosh(\gamma h)]-\beta a_2[(1-2\nu)\sinh(\gamma h) \tag{2.14}$$
$$+\gamma h\cosh(\gamma h)]+i\gamma a_3[2(1-\nu)\cosh(\gamma h)-\gamma h\sinh(\gamma h)]=\frac{Q(\alpha,\beta)}{4iG}.$$

From equations (2.12)–(2.14) we determine a_n $(n = 1, 2, 3)$ in the form

$$\frac{a_1}{\alpha}=\frac{a_2}{\beta}=-\frac{iQ(\alpha,\beta)}{G\gamma^2P_2(\gamma h)}[(1-2\nu)\sinh(\gamma h)-\gamma h\cosh(\gamma h)],$$

$$a_3=-\frac{Q(\alpha,\beta)}{G\gamma P_2(\gamma h)}[2(1-\nu)\cosh(\gamma h)+\gamma h\sinh(\gamma h)],$$

$$P_2(\gamma h)=2\kappa\cosh(2\gamma h)+1+\kappa^2+4\gamma^2h^2. \tag{2.15}$$

Thus, problem (2.8), (2.9) is solved. It can be shown that the displacements and stresses in the layer, which are determined by formulae (2.1)–(2.7), behave like $\exp(-Drh^{-1})$ when $r=\sqrt{x^2+y^2}\to\infty$, where $D>0$ and depends only on ν (see Table 2.1).

Later on, we will need the expression for the displacement w when $z = h$. Using formulae (2.1), (2.4) and the formulae for a_n, b_n, we have

$$w(x,y,h)=-\frac{1}{2\pi\theta}\int\limits_{-\infty}^{\infty}\int\limits_{-\infty}^{\infty}\frac{P_1(\gamma h)}{\gamma P_2(\gamma h)}Q(\alpha,\beta)\exp[-i(\alpha x+\beta y)]\,d\alpha\,d\beta, \tag{2.16}$$

$$\theta=G(1-\nu)^{-1},\qquad P_1(\gamma h)=2\kappa\sinh(2\gamma h)-4\gamma h. \tag{2.17}$$

By substituting formula (2.10) for $Q(\alpha,\beta)$ into (2.16) and transforming, we find that

$$w(x,y,h)=-\frac{1}{\pi^2\theta h}\int\limits_{\Omega}q(\xi,\eta)K_*\left(\frac{x-\xi}{h},\frac{y-\eta}{h}\right)d\Omega, \tag{2.18}$$

$$K_*(\sigma,\tau)=\int\limits_0^{\infty}\int\limits_0^{\infty}\frac{L(u)}{u}\cos(s\sigma)\cos(v\tau)\,ds\,dv\quad(u=\sqrt{s^2+v^2}). \tag{2.19}$$

Here, the function $L(u)$ has the form

$$L(u)=P_1(u)/P_2(u). \tag{2.20}$$

By changing the variables $s = u\cos\chi$, $v = u\sin\chi$ in formula of (2.19) and using the value of integral 8.411.4 in Gradshteyn and Ryzhik (1994)

$$\int\limits_0^{\pi/2}\cos(a\cos\chi)\cos(b\sin\chi)\,d\chi=\frac{\pi}{2}\mathcal{J}_0(\sqrt{a^2+b^2}),$$

where $\mathcal{J}_0(x)$ is the Bessel function, we may represent the expression for $w(x,y,h)$ in the alternative form

$$w(x,y,h) = -\frac{1}{2\pi\theta h}\int\limits_{\Omega} q(\xi,\eta)K\left(\frac{R}{h}\right)\,d\Omega, \qquad R = \sqrt{(\xi-x)^2+(\eta-y)^2}, \tag{2.21}$$

$$K(t) = \int\limits_0^\infty L(u)\mathcal{J}_0(ut)\,du, \qquad t = \frac{R}{h}. \tag{2.22}$$

2.1.3. FORMULATION OF THE CONTACT PROBLEM

Consider the problem of the interaction of a rigid punch with an elastic layer fixed along its foundation (Fig. 2.2). The punch is indented by the force P parallel to the z-axis and applied with the eccentricity $\mathbf{m} = (m_x, m_y)$. The punch is a cylindrical body with the cross-section Ω^* and base surface $z = f(x,y)$. The normalized equation of the contour $\partial\Omega^*$ of the lateral cylindrical surface of the punch has the form (1.22), (1.23). The region of contact of the punch with the layer is $\Omega \subseteq \Omega^*$, and the normalized equation of the contour $\partial\Omega$ has the form (1.24), (1.25). The origin of coordinates O on the lower side of the layer and the directions of the x- and y-axes may be chosen as convenient.

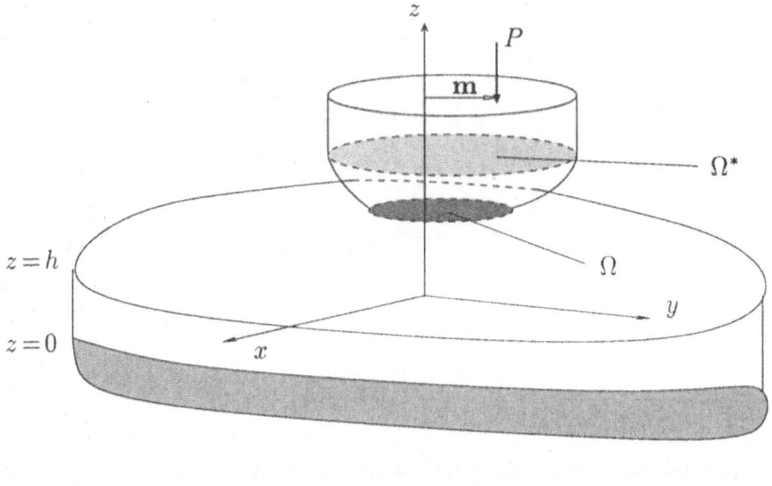

Figure 2.2: The elastic layer $0 \le z \le h$ in contact with a punch

The boundary conditions have the form
$z = 0$ is fixed:

$$u(x, y, 0) = v(x, y, 0) = w(x, y, 0) = 0, \tag{2.23}$$

no friction:

$$\tau_{xz}(x, y, h) = \tau_{yz}(x, y, h) = 0, \tag{2.24}$$

no load outside Ω:

$$\sigma_z(x, y, h) = 0 \quad ((x, y) \in \omega), \tag{2.25}$$

given displacement in Ω:

$$w(x, y, h) = -\delta(x, y) = -[\delta + \alpha x + \beta y - f(x, y)] \quad ((x, y) \in \Omega), \tag{2.26}$$

and the displacements vanish at infinity:

$$\mathbf{u}(x, y, z) \to 0 \text{ as } r = \sqrt{x^2 + y^2} \to \infty.$$

Here, $\delta + \alpha x + \beta y$ is the translation of the punch under action of the force P.

It is necessary to determine the distribution $\sigma_z(x, y, h) = -q(x, y)$ of the contact normal stress under the punch in the region of contact Ω and also the relations between P, m_x, m_y and δ, α, β.

To solve the boundary problem (2.23)–(2.26), we use formula (2.21), noting that the boundary conditions (2.23)–(2.25) will be satisfied. The satisfaction of (2.26) leads to the 2-D convolution integral equation of the first kind for the contact pressure $q(x, y)$:

$$\int_{\Omega} q(\xi, \eta) K\left(\frac{R}{h}\right) d\Omega = 2\pi\theta h \delta(x, y) \quad ((x, y) \in \Omega). \tag{2.27}$$

After solving the integral equation (2.27), we can find the relations between P, m_x, m_y and δ, α, β using the equilibrium conditions of the punch (1.30), (1.31).

2.1.4. PROPERTIES OF THE FUNCTION $L(u)$

In what follows, we will use certain properties of the function $L(u)$ given in (2.20) which occurs in the kernel $K(t)$ of the integral equation (2.27).

A) Since $L(u) = P_1(u)/P_2(u)$, and $P_1(u)$ is odd while $P_2(u)$ is even and bounded below by $(1 + \kappa)^2$ for real u, the function $L(u)$ is odd, and continuous for $|u| < \infty$. $L(u)$ has the asymptotic forms

$$L(u) = 1 + O(u^2 \exp(-2|u|)) \quad (|u| \to \infty), \tag{2.28}$$

$$L(u) = Au[1 + Bu^2 + O(u^4)] \quad (u \to 0), \tag{2.29}$$

$$A = \frac{4(\kappa - 1)}{(\kappa + 1)^2}, \qquad B = \frac{2(\kappa^2 - 5\kappa + 6)}{3(\kappa^2 - 1)}. \tag{2.30}$$

B) In the complex plane of the variable $z = u + iv$, the function $L(z)$ is meromorphic, i.e. it can be presented by the ratio (2.15a) of two entire fuctions of the exponential type and equal order. $L(z)$ has an enumerable set of zeros z_k, and poles ζ_k. Apart from the zero $z = 0$, those nearest to the real axes are

$$z_1 = \pm iC, \qquad \zeta_1 = \pm iD, \tag{2.31}$$

where C and D are the real positive numbers. The other zeros and poles are symmetric with respect to the real and imaginary axes. It may be verified that all the zeros and poles are simple and have asymptotic forms

$$\operatorname{Re} z_k = O(\ln k), \quad \operatorname{Re} \zeta_k = O(\ln k),$$

$$\operatorname{Im} z_k = O(k), \quad \operatorname{Im} \zeta_k = O(k), \quad (k \to \infty).$$

The values of the constants A, B, C, and D, which occur in (2.30) and (2.31), are given in Table 2.1.

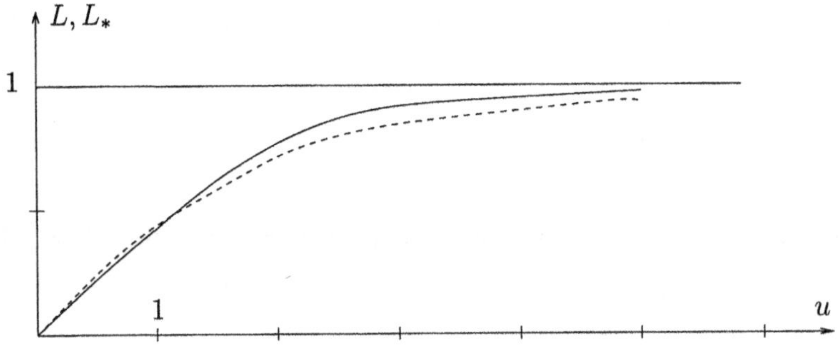

Figure 2.3: The functions $L(u)$ and $L_*(u)$ (the dashed line) for $\nu = 0.3$

In some cases later on, we will use the following approximation of the function $L(u)$ (see Fig. 2.3)

$$L_*(u) = \frac{u\sqrt{u^2 + E^2}}{u^2 + F^2}. \tag{2.32}$$

It is constructed so that the relations

$$\frac{E}{F^2} = A, \qquad \frac{1}{2E^2} - \frac{1}{F^2} = 2B \tag{2.33}$$

hold. Note that $L_*(u)$ has asymptotics

$$L_*(u) = 1 + O(u^{-2}) \quad (|u| \to \infty),$$

$$L_*(u) = Au[1 + Bu^2 + O(u^4)] \quad (u \to 0).$$

The values of the constants E and F and also the maximal for $|u| \in [0, \infty)$ relative error ε (in percent) of the approximation (2.32) are shown in Table 2.1.

Table 2.1: Characteristics of $L(u)$ and $L_*(u)$

ν	0.25	0.30	0.35	0.40
A	0.(4)	0.408163	0.355030	0.2(7)
B	0	0.0714286	0.239316	0.(6)
C	0.947747	0.882931	0.799674	0.686295
D	0.956380	0.912785	0.869512	0.826310
E	1.125000	0.925322	0.716408	0.517002
F	1.590990	1.505669	1.420522	1.364261
ε	11.2	10.3	9.2	12.4

2.2 Integral Equation for Layer Contact Problems

2.2.1. BASIC PARAMETERS AND GENERAL RESULTS

Introduce the dimensionless parameters λ and μ determined by the geometry of the contact region Ω and the thickness h of the layer. We will assume the region Ω is simply connected and bounded and that its contour $\partial\Omega$ has a continuous curvature.

The parameter λ is defined by the relation

$$\lambda = \frac{h}{a}, \qquad a = \frac{1}{2}\max_{\Omega} R,$$

where R is the distance between two points.

The parameter μ is defined as follows. Draw a normal to the contour $\partial\Omega$ at the point A. Let the normal intersect $\partial\Omega$ at the points B, C, and D (Fig. 2.4). We then measure the lengths of the segments between the point

A and the points of intersection. We denote the minimal length among the all such segments for the all points $A \in \partial\Omega$ as $2a_0$. Then we have

$$\mu = h[\inf(a_0, \rho_0)]^{-1},$$

where ρ_0 is the minimal curvature radius of $\partial\Omega$.

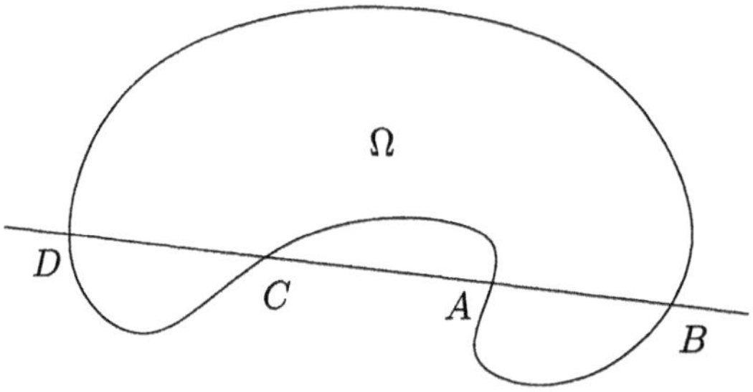

Figure 2.4: A normal to the contour $\partial\Omega$ at the point A

Note that for a convex region Ω, we always have $a_0 \geq \rho_0$, therefore, in this case $\mu = h/\rho_0$.

Generally, by comparing the parameters λ and μ, we find that always $0 < \lambda \leq \mu < \infty$ because $a \geq \inf(a_0, \rho_0)$, with equality only when the region Ω is a circle. We will use the parameter λ for a characteristic of a thick layer, and the parameter μ for a thin one.

Later on, we will need certain properties of the kernel $K(t)$ of the integral equation (2.27). Using the value of the integral (1.15), we rewrite formula (2.22) for $K(t)$ in the form

$$K(t) = t^{-1} - F(t),$$

$$F(t) = \int_0^\infty [1 - L(u)]\mathcal{J}_0(ut)\,du. \tag{2.34}$$

Now, on the basis of properties (2.28), (2.29) of the function $L(u)$ one can prove that in the complex plane $w = t + i\tau$ the function $F(t)$ is regular in the strip $|t| < \infty$, $|\tau| < 2$. Therefore, as $t \to 0$, the kernel $K(t)$ tends to an infinity as t^{-1}, and the function $F(t)$ is its regular part because the latter function is continuous with all its derivatives for $0 \leq |t| < \infty$.

If we expand the Bessel function $\mathcal{J}_0(ut)$ in (2.34) in powers of ut, we see that for $|t| < 2$ the function $F(t)$ can be represented by the absolutely convergent series

$$F(t) = \sum_{m=0}^{\infty} a_m t^{2m}, \tag{2.35}$$

$$a_m = \frac{(-1)^m}{[(2m)!!]^2} \int_0^\infty [1 - L(u)] u^{2m} \, du. \tag{2.36}$$

Table 2.2 gives a few values of a_m, found by numerical integration.

Table 2.2: Constants (2.36) for various ν

ν	0.25	0.30	0.35	0.40
a_0	1.3294	1.3769	1.4383	1.5187
a_1	−0.5860	−0.6276	−0.6801	−0.7476
a_2	0.2435	0.2651	0.2922	0.3271
a_3	−0.08890	−0.09759	−0.10853	−0.12261
a_4	0.02987	0.03293	0.03679	0.04177

On the basis of result (2.31) for the function $L(u)$ in the complex plane $z = u + iv$, one can show that, as $|t| \to \infty$, the kernel $K(t)$ vanishes as $\exp(-D|t|)$, therefore, as $|t| \to \infty$, the function $F(t)$ behaves like $|t|^{-1}$.

Let us suppose that the solution $q(\xi, \eta)$ of (2.27) exists in the class $\mathcal{L}(\Omega)$ (absolutely integrable over Ω). Using (2.35), we present the integral equation (2.27) in the form

$$\int_\Omega q(\xi, \eta) \frac{d\Omega}{R} = 2\pi\theta\delta(x, y) + \frac{1}{h} \int_\Omega q(\xi, \eta) F\left(\frac{R}{h}\right) d\Omega \quad ((x, y) \in \Omega). \tag{2.37}$$

Due of the properties of $F(t)$ and of the assumption made with respect to $q(\xi, \eta)$, the function

$$g(x, y) = \int_\Omega q(\xi, \eta) F\left(\frac{R}{h}\right) d\Omega \tag{2.38}$$

is continuous with all its derivatives for $(x, y) \in \Omega$.

This leads to the following important results:

A) as $\lambda \to \infty$, the integral equation (2.37) reduces to the integral equation (1.29) of the contact problem for an elastic half-space;

B) a solution of the integral equation (2.37) has the same character (singularities) as one of the equation (1.29) for all values of λ and μ.

For the case when the contact region is *a priori* known, i.e. the punch is flat, and $\Omega = \Omega^*$, we can prove the following facts:
C) if the function $\delta(x, y)$ (or $f(x, y)$) is such that its first derivatives with respect to x and y satisfy in Ω the Hölder condition with the index $0 < \alpha \leq 1$[1] ($\delta(x, y) \in \mathcal{H}_1^\alpha(\Omega)$), then the solution $q(x, y)$ of the equation (1.29) (or (2.37)) exists and is unique in $\mathcal{L}_p(\Omega)$: it is absolutely integrable over Ω with the power p, where $1 \leq p < 2$, ($\mathcal{L}_1 \equiv \mathcal{L}$), and it can be presented in the form

$$q(x, y) = Q(x, y)[N^*(x, y)]^{-1/2}, \tag{2.39}$$

where $N^*(x, y) = 0$ is the equation of the contour $\partial\Omega^*$ (see 1.22), (1.23), and the function $Q(x, y)$ satisfy in Ω the Hölder condition with the index β ($Q(x, y) \in \mathcal{H}_0^\beta(\Omega)$), and $\beta = \alpha$, if $\alpha < 1$ while $\beta = 1 - \varepsilon$, if $\alpha = 1$; in fact, (2.39) says that the function $q(x, y)$ has a square root singularity as $(x, y) \to \partial\Omega$;
D) if the contour $\partial\Omega$ is as smooth as desired, and the function $\delta(x, y)$ is continuous with all its derivatives in Ω, then the function $Q(x, y)$ in (2.39) is also continuous with all its derivatives in Ω;

If the domain Ω is *a priori* unknown, and $\Omega \subset \Omega^*$, then we will use the complementary condition $q(x, y) = 0$, $(x, y) \in \partial\Omega$ to determine this domain (see a background in Barenblatt (1960)); in this case we obtain from (2.39)

$$q(x, y) = \tilde{Q}(x, y)[N(x, y)]^{1/2},$$

where $N(x, y) = 0$ is the equation of the contour $\partial\Omega$ (see (1.24), (1.25)), and the function $\tilde{Q}(x, y)$ is continuous with all its derivatives in Ω.

2.2.2. INTEGRAL CHARACTERISTICS OF THE SOLUTION

Now, let us obtain formulae for certain integral characteristics of the solution of the equation (2.27)

$$\int_\Omega q(\xi, \eta) K\left(\frac{R}{h}\right) d\Omega = 2\pi\theta h \delta(x, y) \qquad ((x, y) \in \Omega), \tag{2.40}$$

using the symmetry of its kernel $K(t)$ (2.22) with respect to x, y and ξ, η.

Let the region Ω be given, and the function

$$\tilde{q}(x, y) = \delta q_\delta(x, y) + \alpha q_\alpha(x, y) + \beta q_\beta(x, y) \tag{2.41}$$

[1]The Hölder condition with the index $0 < \alpha \leq 1$ for the function $f(x)$ in the domain Ω means that $\forall x_1, x_2 \in \Omega$ we have $|f(x_1) - f(x_2)| \leq M|x_1 - x_2|^\alpha$, where M is a positive constant; this is similar for $f(x, y)$.

be the solution of the integral equation (2.40) for the case

$$\delta(x,y) = \tilde{\delta}(x,y) = \delta + \alpha x + \beta y, \tag{2.42}$$

which corresponds to the action of a punch with a plane sloping face onto the elastic layer. We multiply the both sides of equation (2.40) with an *arbitrary* function $\delta(x,y)$ by $\tilde{q}(x,y)\,d\Omega$ and then integrate over Ω. Changing the order of integration in the left-hand side of the relation (one can prove that it is possible) and using the properties of the solutions and the kernel $K(t)$, we have the formula

$$\int\limits_{\Omega} q(\xi,\eta)\,d\Omega_{\xi\eta} \int\limits_{\Omega} \tilde{q}(x,y) K\left(\frac{R}{h}\right) d\Omega_{xy} = 2\pi h\theta \int\limits_{\Omega} \delta(x,y)\tilde{q}(x,y)\,d\Omega. \tag{2.43}$$

Taking into account that $\tilde{q}(x,y)$ is the solution of the integral equation (2.40) with the right-hand side (2.42), we obtain that

$$\int\limits_{\Omega} (\delta + \alpha\xi + \beta\eta) q(\xi,\eta)\,d\Omega \tag{2.44}$$

$$= \int\limits_{\Omega} \delta(x,y)[\delta q_\delta(x,y) + \alpha q_\alpha(x,y) + \beta q_\beta(x,y)]\,d\Omega.$$

Since δ, α, and β are independent, we obtain from (2.44), (1.30) and (1.31) the following relations

$$P = \int\limits_{\Omega} q_\delta(x,y)\delta(x,y)\,d\Omega, \tag{2.45}$$

$$M_x = \int\limits_{\Omega} q_\beta(x,y)\delta(x,y)\,d\Omega, \qquad M_y = \int\limits_{\Omega} q_\alpha(x,y)\delta(x,y)\,d\Omega. \tag{2.46}$$

Because the function $\delta(x,y)$, which defines the punch base, is arbitrary in (2.45), (2.46), these formulae allow us to find the force P and the moments $M_x = Pm_y$, $M_y = Pm_x$ (see Fig. 2.2) acting on any non-planar punch with the contact region Ω, if we know the solution (2.41) for the plane punch in the same domain Ω. We also give below fairly efficient double-sided estimates for these quantities.

Suppose we know the solution $q(x,y) = \delta q_\delta^\circ(x,y)$ of the integral equation (1.29) for the case $\delta(x,y) \equiv \delta$, i.e. the solution for a punch with horizontal plane base, and shape Ω.

Multiply both sides of (2.27) by $\delta q_\delta^\circ(x,y)$ and then integrate over Ω. Changing the integrals in the left-hand side, we derive a formula which is similar to (2.43). Then using formula (1.30), we find

$$P = J[\delta(x,y)] + \frac{1}{2\pi h\theta} J\left[\int_\Omega q(\xi,\eta) F\left(\frac{R}{h}\right) d\Omega \right], \tag{2.47}$$

$$J[\delta(x,y)] = \int_\Omega q_\delta^\circ(x,y)\delta(x,y)\, d\Omega. \tag{2.48}$$

In fact, J here is a special kind of elastic energy functional in Timoshenko and Goodier (1951). From (2.47), (2.48), supposing for physically real solution that $q(x,y) \geq 0$ in Ω, we obtain double-sided estimate for the quantity of the indenting force, for the general case of $\delta(x,y)$, in the form

$$\frac{J[\delta(x,y)]}{1-(2\pi h\theta)^{-1}J(1)\min\limits_\Omega F(t)} \leq P \leq \frac{J[\delta(x,y)]}{1-(2\pi h\theta)^{-1}J(1)\max\limits_\Omega F(t)}. \tag{2.49}$$

The functional $J(1)$ gives the force for the special case $\delta(x,y) \equiv 1$. Note that the condition $q(x,y) \geq 0$ in Ω always holds if the function $\delta(x,y)$ is positive and convex[2] function in Ω.

This estimate holds for any values of λ, but it is useful only when $\lambda \geq 2$. The values of $\min_\Omega F(t)$ and $\max_\Omega F(t)$ can be easily found for specified value of λ.

2.2.3. ESTIMATES FOR THE FORCE AND MOMENTS

On the basis of the expansion (2.35), one can also obtain the following representation for the function $F(t)$ for sufficiently large values of λ:

$$F(t) = F_0(x^2,y^2,\xi^2,\eta^2) + x\xi h^{-2}F_1(x^2,y^2,\xi^2,\eta^2) \tag{2.50}$$
$$+y\eta h^{-2}F_2(x^2,y^2,\xi^2,\eta^2) + x\xi y\eta h^{-4}F_3(x^2,y^2,\xi^2,\eta^2),$$

where we denote

$$F_0 = a_0 + \frac{a_1}{h^2}\zeta^2 + \frac{a_2}{h^4}[\zeta^4 + 4(x^2\xi^2 + y^2\eta^2)] + \frac{a_3}{h^6}[\zeta^6 \tag{2.51}$$

---

[2]By definition, this means that $\forall(x_1,y_1),(x_2,y_2) \in \Omega$ the inequality

$$\delta\left(\frac{x_1+x_2}{2}, \frac{y_1+y_2}{2}\right) \leq \frac{\delta(x_1,y_1)+\delta(x_2,y_2)}{2}$$

holds; for such a function, we have $\partial^2\delta(x,y)/\partial x^2 < 0$ and $\partial^2\delta(x,y)/\partial y^2 < 0$.

$$+12(x^2\xi^2+y^2\eta^2)\zeta^2]+\frac{a_4}{h^8}[\zeta^8+24(x^2\xi^2+y^2\eta^2)\zeta^4+16(x^2\xi^2$$
$$+y^2\eta^2)^2+64x^2\xi^2y^2\eta^2]+O\left(\frac{1}{\lambda^{10}}\right) \qquad (\zeta^2=x^2+\xi^2+y^2+\eta^2),$$

$$F_1=-2\left\{a_1+\frac{2a_2}{h^2}\zeta^2+\frac{a_3}{h^4}[3\zeta^4+4(x^2\xi^2+3y^2\eta^2)]\right. \tag{2.52}$$
$$\left.+\frac{4a_4}{h^6}[\zeta^6+4(x^2\xi^2+3y^2\eta^2)\zeta^2]+O\left(\frac{1}{\lambda^8}\right)\right\},$$

the expression for F_2 can be obtained from F_1 by replacement of x^2 by y^2 and of ξ^2 by η^2; we do not use F_3.

Now suppose that the contact region Ω has two mutually perpendicular axes of symmetry, and the function $\delta(x,y)$ in (2.37) has the form

$$\delta(x,y)=\delta+\alpha x+\beta y-\varphi(x^2,y^2).$$

Then the solution of equation (2.37) can be presented in the form

$$q(x,y)=q_P(x^2,y^2)+yq_{M_x}(x^2,y^2)+xq_{M_y}(x^2,y^2).$$

Using (1.30) and (1.31), we find

$$P=\int\limits_{\Omega} q_P(x^2,y^2)\,d\Omega,$$

$$M_x=\int\limits_{\Omega} y^2 q_{M_x}(x^2,y^2)\,d\Omega,\quad M_y=\int\limits_{\Omega} x^2 q_{M_y}(x^2,y^2)\,d\Omega.$$

We suppose below that the function $q_P(x^2,y^2)\geq 0$ in Ω; this holds if the function $\delta^*(x,y)=\delta-\varphi(x^2,y^2)$ is positive and convex in Ω. The functions $q_{M_x}(x^2,y^2)$ and $q_{M_y}(x^2,y^2)$ are, apparently, of constant signs in Ω. The quantities M_x and β, M_y and α are positive or negative, if $q_{M_x}(x^2,y^2)$ and $q_{M_y}(x^2,y^2)$ are positive or negative, respectively, and vice versa.

Taking account of what was said above and of (2.50), using the same scheme as for the estimate (2.49), we can obtain other double-sided estimates for P, M_x and M_y, for fairly large λ, in the form

$$\frac{J_0[\delta-\varphi(x^2,y^2)]}{1-(2\pi h\theta)^{-1}J_0(1)\min\limits_{\Omega}F_0}\leq P\leq\frac{J_0[\delta-\varphi(x^2,y^2)]}{1-(2\pi h\theta)^{-1}J_0(1)\max\limits_{\Omega}F_0}, \tag{2.53}$$

$$\frac{|\alpha|J_1(x^2)}{1-(2\pi h^3\theta)^{-1}J_1(x^2)\min\limits_{\Omega}F_1}\leq|M_y|\leq\frac{|\alpha|J_1(x^2)}{1-(2\pi h^3\theta)^{-1}J_1(x^2)\max\limits_{\Omega}F_1}, \tag{2.54}$$

$$\frac{|\beta| J_2(y^2)}{1-(2\pi h^3\theta)^{-1} J_2(y^2) \min\limits_{\Omega} F_2} \le |M_x| \le \frac{|\beta| J_2(y^2)}{1-(2\pi h^3\theta)^{-1} J_2(y^2) \max\limits_{\Omega} F_2}, \tag{2.55}$$

$$J_0[\delta - \varphi(x^2, y^2)] = \int\limits_{\Omega} q_\delta^\circ(x^2, y^2)[\delta - \varphi(x^2, y^2)]\, d\Omega,$$

$$J_1(x^2) = \int\limits_{\Omega} q_\alpha^\circ(x^2, y^2) x^2\, d\Omega, \qquad J_2(y^2) = \int\limits_{\Omega} q_\beta^\circ(x^2, y^2) y^2\, d\Omega.$$

Here, it is supposed that the solution

$$q^\circ(x, y) = \delta q_\delta^\circ(x^2, y^2) + \alpha x q_\alpha^\circ(x^2, y^2) + \beta y q_\beta^\circ(x^2, y^2)$$

of (1.29) is known for $\delta(x, y) = \delta + \alpha x + \beta y$, i.e. for a plane sloping punch with given shape Ω.

The estimates (2.53)–(2.55) are valid for $\lambda \ge 2$.

If, for example, the region Ω is an annulus with internal radius a and external radius b, then in (2.53)–(2.55) we have

$$\min_{\Omega} F_0 = a_0 + \frac{2a_1}{\lambda^2} + \frac{4a_2}{\lambda^4} + \frac{8a_3}{\lambda^6} + \frac{16a_4}{\lambda^8} + O\left(\frac{1}{\lambda^{10}}\right) \quad \left(\lambda = \frac{h}{b}\right),$$

$$\max_{\Omega} F_0 = a_0 + \frac{2a_1}{\mu^2} + \frac{8a_2}{\mu^4} + \frac{32a_3}{\mu^6} + \frac{128a_4}{\mu^8} + O\left(\frac{1}{\mu^{10}}\right) \quad \left(\mu = \frac{h}{a}\right),$$

$$\min_{\Omega} F_1 = \min_{\Omega} F_2 = -2\left\{a_1 + \frac{4a_2}{\lambda^2} + \frac{12a_3}{\lambda^4} + \frac{32a_4}{\lambda^6} + O\left(\frac{1}{\lambda^8}\right)\right\},$$

$$\max_{\Omega} F_1 = \max_{\Omega} F_2 = -2\left\{a_1 + \frac{4a_2}{\mu^2} + \frac{24a_3}{\mu^4} + \frac{128a_4}{\mu^6} + O\left(\frac{1}{\mu^8}\right)\right\},$$

and for the plane sloping annular punch, the integrals $J_0[\delta - \varphi(x^2, y^2)] = \delta J_0(1)$, $J_1(x^2)$ and $J_2(y^2)$ can be found with the help of the results presented in section 5.3.

2.3 A Circular Punch on a Layer

2.3.1. INTEGRAL EQUATIONS

Suppose the contact region Ω is a circle of radius a. Use the variables

$$x = r\cos\varphi, \qquad y = r\sin\varphi, \qquad \xi = \rho\cos\psi, \qquad \eta = \rho\sin\psi,$$

in the integral equation (2.40) and rewrite it in the form

$$\int_0^{2\pi} d\psi \int_0^a q(\rho,\psi)\rho\, d\rho \int_0^\infty L(u)\, \mathcal{J}_0\left(u\frac{R}{h}\right) du = 2\pi\theta h\delta(r,\varphi) \tag{2.56}$$

$$\left(0 \le \varphi < 2\pi, \quad 0 \le r \le a, \quad R = \sqrt{\rho^2 + r^2 - 2r\rho\cos(\varphi - \psi)}\,\right)$$

Expand the functions $\delta(r,\varphi)$ and $q(\rho,\psi)$ in Fourier series in φ and ψ, respectively:

$$\delta(r,\varphi) = \delta_0(r) + 2\sum_{n=1}^\infty [\delta_n(r)\cos(n\varphi) + \tilde{\delta}_n(r)\sin(n\varphi)], \tag{2.57}$$

$$q(\rho,\psi) = q_0(\rho) + 2\sum_{n=1}^\infty [q_n(\rho)\cos(n\psi) + \tilde{q}_n(\rho)\sin(n\psi)], \tag{2.58}$$

and substitute (2.57), (2.58) into the equation (2.56). Using relation 8.531.1 in Gradshteyn and Ryzhik (1994)

$$\mathcal{J}_0\left(u\frac{R}{h}\right) = \mathcal{J}_0\left(u\frac{r}{h}\right)\mathcal{J}_0\left(u\frac{\rho}{h}\right) + 2\sum_{m=1}^\infty \mathcal{J}_m\left(u\frac{r}{h}\right)\mathcal{J}_m\left(u\frac{\rho}{h}\right)\cos[m(\varphi - \psi)],$$

integrating by ψ, we reduce (2.56) to an infinite system of integral equations of the first kind in the harmonics $q_n(\rho)$:

$$\int_0^a q_n(\rho)K_n\left(\frac{\rho}{h},\frac{r}{h}\right)\rho\, d\rho = \theta h\delta_n(r) \tag{2.59}$$

$$(0 \le r \le a; \; n = 0, 1, \ldots),$$

$$K_n(\sigma,\tau) = \int_0^\infty L(u)\,\mathcal{J}_n(\sigma u)\,\mathcal{J}_n(\tau u)\, du. \tag{2.60}$$

There is a similar system for $\tilde{q}_n(\rho)$.

The applied punch force P and moments M_x and M_y are related to δ, α and β by formulae (1.30), (1.31) which now reduce to

$$P = 2\pi\int_0^a q_0(\rho)\rho\, d\rho, \tag{2.61}$$

$$M_x = \pi \int_0^a \tilde{q}_1(\rho)\rho^2 d\rho, \qquad M_y = \pi \int_0^a q_1(\rho)\rho^2 d\rho. \tag{2.62}$$

Clearly, formula (2.59) as $n = 0$ is the integral equation governing the axisymmetric contact for the layer.

We may establish the following results:
A) if the function $\delta_n(r)$ is continuous with all its derivatives in the circle S of radius a, then the solution $q_n(r)$ of the integral equation (2.59) exists for any $\lambda = h/a \in (0, \infty)$; it is unique in $\mathcal{L}_p(S)$ $(1 \leq p < 2)$ and can be presented in the form

$$q_n(r) = Q_n(r)(a^2 - r^2)^{-1/2}, \tag{2.63}$$

where the function $Q_n(r)$ is also continuous with all its derivatives; formula (2.63) shows that $q_n(r)$ may have a square root singularity at the point $r = a$, and certainly has one if $\delta_n(r) = \delta + \alpha r$;
B) if, furthermore, we have the conditions

$$Q_n^{(k)}(a) = 0 \qquad (k = 0, 1, ..., m-1) \tag{2.64}$$

(which, apparently, impose additional restrictions onto the function $\delta_n(r)$), then

$$q_n(r) = Q_n^*(r)(a^2 - r^2)^{m-1/2},$$

and the function $Q_n^*(r)$ is continuous with all its derivatives; conditions (2.64) determine the smoothness of $Q_n(r)$ at $r = a$, for instance $Q_n(r) = (a^2 - r^2)^m$.

On the basis of these results we may prove the following theorem.

Theorem 2.1. *If $q_0(r)$ is the solution of the integral equation* (2.59) *for $n = 0$ for the case*

$$\delta_0(r) = \mathcal{W}^{-m}\{r_1^{-m} g_m(r_1); r\} + \sum_{k=1}^{m} c_k r^{2k-2}, \tag{2.65}$$

$$\mathcal{W}^{-m}\{f(r_1); r\} = \int_0^r r_m dr_m \int_0^{r_m} r_{m-1} dr_{m-1} \cdots \int_0^{r_2} r_1 f(r_1)\, dr_1,$$

where the constants c_k in (2.65) *are defined so that*

$$q_0^{(k)}(a) = 0 \qquad (k = 0, 1, ..., m-1), \tag{2.66}$$

then the function

$$q_m(r) = r^m \mathcal{W}^m\{q_0(r)\} \qquad \left(\mathcal{W} = \frac{1}{r}\frac{d}{dr}\right) \tag{2.67}$$

is the solution of the integral equation (2.59) *for* $n = m$ *for the case* $\delta_n(r) = g_m(r)$.

The meaning of the theorem is the following. Suppose we can solve equation (2.59) for $n = 0$. For equation (2.59) with $n \geq 1$, we must first construct the function (2.65), where $g_m(r) = \delta_n(r)$ and $m = n$, and then solve (2.59) for $n = 0$ and (2.65). The obtained solution will depend linearly on c_k. Then formulae (2.66) give a system of n linear equations for n unknown constants c_k. Solving this system, we obtain the solution of (2.59) for $n \geq 1$ by formula (2.67). Note that the larger is n, the more complicated is the procedure.

Due to this theorem we confine ourselves to the case $n = 0$, omitting the index 0. In this section, we will apply this theorem to obtain an asymptotic solution of (2.59) for $n = 1$ by using an asymptotic solution for $n = 0$.

2.3.2. PROPERTIES OF THE KERNEL FOR AXIAL SYMMETRY

We investigate certain properties of the kernel $K_0(\sigma, \tau) = K(\sigma, \tau)$ (2.60).

Using the value of integral 6.512.1 in Gradshteyn and Ryzhik (1994)

$$I(\sigma, \tau) = \int_0^\infty \mathcal{J}_0(\sigma u)\, \mathcal{J}_0(\tau u)\, du = \frac{2}{\pi(\sigma + \tau)} \mathcal{K}(e) \qquad \left(e = \frac{2\sqrt{\sigma\tau}}{\sigma + \tau}\right),$$

where $\mathcal{K}(e)$ is the complete elliptic integral of the first kind, we rewrite (2.60) in the form

$$K(\sigma, \tau) = I(\sigma, \tau) - F(\sigma, \tau), \tag{2.68}$$

$$F(\sigma, \tau) = \int_0^\infty [1 - L(u)]\, \mathcal{J}_0(\sigma u)\, \mathcal{J}_0(\tau u)\, du. \tag{2.69}$$

Note that $I(\sigma, \tau)$ has a logarithmic singularity at $\sigma = \tau$.

On the basis of the properties (2.21a), (2.21b) of the function $L(u)$, we may prove that the function $F(\sigma, \tau)$ is continuous with all its derivatives, with respect to σ and τ, in the quarter-plane $0 \leq \sigma, \tau < \infty$. In the square $0 \leq \sigma, \tau < 1$, the function $F(\sigma, \tau)$ can be presented by the absolutely and uniformly convergent series

$$F(\sigma, \tau) = \sum_{i=0}^{\infty} \sum_{j=0}^{\infty} b_{ij} \sigma^{2i} \tau^{2j}, \tag{2.70}$$

$$b_{ij} = \frac{(m!)^2}{(i!)^2(j!)^2} a_m \quad (m = i + j),$$

where a_m are defined by (2.36).

On the basis of the behaviour (2.31) of $L(u)$ in the complex plane $z = u + iv$, one may show that $K(\sigma, \tau) \sim \exp(-|\sigma - \tau| D)$ and $F(\sigma, \tau) \sim I(\sigma, \tau)$ as $|\sigma - \tau| \to \infty$.

Using asymptotics 8.451 in Gradshteyn and Ryzhik (1994) for Bessel functions, we have for large σ and τ:

$$K(\sigma, \tau) \sim \frac{1}{\pi\sqrt{\sigma\tau}} K_*(\sigma - \tau), \qquad K_*(\tilde{y}) = \int_0^\infty \frac{L(u)}{u} \cos(u\tilde{y})\, du, \tag{2.71}$$

where, as $\tilde{y} \to 0$, the function $K_*(\tilde{y}) \sim -\ln|\tilde{y}|$.

Using (2.68), (2.69), we present the integral equation (2.59) for $n = 0$ in the form

$$\int_0^a q(\rho)\, \mathcal{K}\left(\frac{2\sqrt{r\rho}}{r+\rho}\right) \frac{\rho\, d\rho}{r+\rho} = \frac{\pi}{2}\theta\delta(r) + \frac{\pi}{2h} \int_0^a q(\rho)\, F\left(\frac{\rho}{h}, \frac{r}{h}\right) \rho\, d\rho \tag{2.72}$$

$$(0 \leq r \leq a).$$

On the basis of these properties of the functions $q(\rho)$ and $F(\sigma, \tau)$, we may show that, as $\lambda \to \infty$, the integral equation (2.72) reduces to the integral equation

$$\int_0^a q(\rho)\, \mathcal{K}\left(\frac{2\sqrt{r\rho}}{r+\rho}\right) \frac{\rho\, d\rho}{r+\rho} = \frac{\pi}{2}\theta\delta(r) \quad (0 \leq r \leq a), \tag{2.73}$$

governing the axisymmetric contact for a half-space. In fact, the solution of (2.73) was given by formulae (1.47) and (1.48) which for $n = 0$ may be written in the form

$$q(r) = \frac{2}{\pi}\left[\frac{p(a)}{\sqrt{a^2 - r^2}} - \int_r^a \frac{p'(\xi)\, d\xi}{\sqrt{\xi^2 - r^2}}\right], \tag{2.74}$$

$$p(x) = \theta\left[\delta(0) + x \int_0^x \frac{\delta'(\rho)\, d\rho}{\sqrt{x^2 - \rho^2}}\right]. \tag{2.75}$$

2.3.3. THE ASYMPTOTIC SOLUTION FOR LARGE λ

By substituting (2.70) in (2.72), we obtain the formula

$$\int_0^a q(\rho)\,\mathcal{K}\left(\frac{2\sqrt{r\rho}}{r+\rho}\right)\frac{\rho\,d\rho}{r+\rho} = \frac{\pi}{2}\theta\delta(r) \tag{2.76}$$

$$+\frac{\pi}{2}\sum_{i=0}^{\infty}\sum_{j=0}^{\infty}\frac{b_{ij}r^{2j}}{h^{2i+2j+1}}\int_0^a q(\rho)\rho^{2i+1}d\rho \quad (0 \le r \le a).$$

Since (2.70) holds for $\sigma = \rho/h < 1$ and $\tau = r/h < 1$, we shall consider equation (2.76) for $\lambda > 1$.

We will seek the asymptotic solution of (2.76) for large λ in the form

$$q(r) = \sum_{n=0}^{\infty}\tilde{q}_n(r)\lambda^{-n}. \tag{2.77}$$

We substitute (2.77) in (2.76) and equate the terms at the same powers of λ^{-1} on both sides; we obtain an infinite system of integral equations for successive determination of the $\tilde{q}_n(r)$. The left-hand side of any such equation is like equation (2.73), and the right-hand side is a polynomial in r whose coefficients are functionals in $\tilde{q}_m(r)$ $(m < n)$. Clearly, the successive determination of the $\tilde{q}_n(r)$ can be easily made with the help of formulae (2.74), (2.75).

Finally, we obtain the solution of the integral equation (2.76), in the form

$$q(r) = q_*(r) + \frac{2}{\pi a\sqrt{a^2-r^2}}\int_0^a p(\xi)\left\{\frac{2a_0}{\pi\lambda}\left[1+\frac{2a_0}{\pi\lambda}+\left(\frac{2a_0}{\pi\lambda}\right)^2\right.\right. \tag{2.78}$$

$$\left.\left.+\left(\frac{2a_0}{\pi\lambda}\right)^3+\frac{8a_1}{3\pi\lambda^3}\right]+\frac{4a_1}{\pi\lambda^3}\left(2\frac{r^2}{a^2}+\frac{\xi^2}{a^2}-1\right)\left(1+\frac{2a_0}{\pi\lambda}\right)\right\}d\xi + O\left(\frac{1}{\lambda^5}\right),$$

where $q_*(r)$ and $p(\xi)$ are defined by formulae (2.74) and (2.75), respectively.

Then we find the acting punch force by (2.61) in the form

$$P = 4\int_0^a p(\xi)\left[1+\frac{2a_0}{\pi\lambda}+\left(\frac{2a_0}{\pi\lambda}\right)^2+\left(\frac{2a_0}{\pi\lambda}\right)^3+\left(\frac{2a_0}{\pi\lambda}\right)^4\right. \tag{2.79}$$

$$\left.+\frac{16a_0a_1}{3\pi^2\lambda^4}+\frac{4a_1}{\pi\lambda^3}\left(\frac{\xi^2}{a^2}+\frac{1}{3}\right)\left(1+\frac{2a_0}{\pi\lambda}\right)\right]d\xi + O\left(\frac{1}{\lambda^5}\right).$$

Using (2.78), we obtain the solution of (2.76) vanishing on the contour of the contact region, i.e. at $r = a$, in the form

$$q(r) = -\frac{2}{\pi}\int_r^a \frac{p'(\xi)\,d\xi}{\sqrt{\xi^2 - r^2}} - \frac{16a_1}{\pi^2 a^3 \lambda^3}\sqrt{a^2 - r^2} \tag{2.80}$$

$$\times\left(1 + \frac{2a_0}{\pi\lambda}\right)\int_0^a p(\xi)\,d\xi + O\left(\frac{1}{\lambda^5}\right),$$

under the condition

$$p(a) + \frac{1}{a}\int_0^a p(\xi)\left\{\frac{2a_0}{\pi\lambda}\left[1 + \frac{2a_0}{\pi\lambda} + \left(\frac{2a_0}{\pi\lambda}\right)^2 + \left(\frac{2a_0}{\pi\lambda}\right)^3 + \frac{8a_1}{3\pi\lambda^3}\right]\right. \tag{2.81}$$

$$\left. + \frac{4a_1}{\pi\lambda^3}\left(\frac{\xi^2}{a^2} + 1\right)\left(1 + \frac{2a_0}{\pi\lambda}\right)\right\} d\xi + O\left(\frac{1}{\lambda^5}\right) = 0,$$

which serves for determination of the radius a of the contact region for specified shape of the punch base, and for specified impressed force P.

We find the force P for the solution (2.80) in the form

$$P = 4\int_0^a p(\xi)\left[1 - \frac{16a_1}{3\pi\lambda^3}\left(1 + \frac{2a_0}{\pi\lambda}\right) + O\left(\frac{1}{\lambda^5}\right)\right] d\xi. \tag{2.82}$$

Now we apply the theorem given above and formula (2.80) to obtain the asymptotic solution of (2.76) for $n = 1$:

$$q_1(r) = -\frac{2}{\pi}\frac{d}{dr}\int_r^a \frac{p_1(\xi)\,d\xi}{\sqrt{\xi^2 - r^2}} - \frac{16a_1}{\pi^2 a^3 \lambda^3}\frac{r}{\sqrt{a^2 - r^2}} \tag{2.83}$$

$$\times\int_0^a \xi p_1(\xi)\,d\xi + O\left(\frac{1}{\lambda^5}\right), \qquad p_1(x) = \theta\frac{d}{dx}x\int_0^x \frac{\delta_1(\rho)\,d\rho}{\sqrt{x^2 - \rho^2}}.$$

Using the second formula of (2.62), we obtain for the moment the formula

$$M = 4\left(1 - \frac{8a_1}{3\pi\lambda^3}\right)\int_0^a \xi p_1(\xi)\,d\xi + O\left(\frac{1}{\lambda^5}\right). \tag{2.84}$$

Note that there are no terms of the orders λ^{-1} and λ^{-2} in (2.83), unlike in (2.78) . Generally, the structure of the asymptotic solution for large λ for the harmonic $q_n(r)$ is

$$q_n(r) = q_n^{\infty}(r) + O(\lambda^{-2n-1}), \tag{2.85}$$

where $q_n^{\infty}(r)$ is the solution for $\lambda = \infty$. We see from (2.85) that, as n increases, the influence of the bottom $z = 0$ of the layer onto the contact pressure decreases; this holds also for small values of λ.

Formulae (2.78)–(2.84) holds for $\lambda \geq 2$.

For the planar punch $\delta(r) \equiv \delta$, we find from (2.78) and (2.79):

$$q(r) = \frac{2\theta\delta}{\pi\sqrt{a^2 - r^2}} \left[1 + \frac{2a_0}{\pi\lambda} + \left(\frac{2a_0}{\pi\lambda}\right)^2 + \left(\frac{2a_0}{\pi\lambda}\right)^3 \right. \tag{2.86}$$

$$\left. + \left(\frac{2a_0}{\pi\lambda}\right)^4 - \frac{8a_1}{3\pi\lambda^3} + \frac{8a_1}{\pi\lambda^3}\left(1 + \frac{2a_0}{\pi\lambda}\right)\frac{r^2}{a^2} + O\left(\frac{1}{\lambda^5}\right) \right],$$

$$P = 4\theta\delta a \left[1 + \frac{2a_0}{\pi\lambda} + \left(\frac{2a_0}{\pi\lambda}\right)^2 + \left(\frac{2a_0}{\pi\lambda}\right)^3 \right. \tag{2.87}$$

$$\left. + \left(\frac{2a_0}{\pi\lambda}\right)^4 + \frac{8a_1}{3\pi\lambda^3}\left(1 + \frac{4a_0}{\pi\lambda}\right) + O\left(\frac{1}{\lambda^5}\right) \right].$$

For the parabolic punch $\delta(r) = \delta - \gamma r^2$, we have from (2.80)–(2.82):

$$q(r) = \frac{8\theta\gamma}{\pi}\sqrt{a^2 - r^2}\left[1 - \frac{8a_1}{3\pi\lambda^3} + O\left(\frac{1}{\lambda^5}\right)\right], \tag{2.88}$$

$$P = \frac{16\theta\gamma a^3}{3}\left[1 - \frac{8a_1}{3\pi\lambda^3} + O\left(\frac{1}{\lambda^5}\right)\right], \tag{2.89}$$

$$P = \frac{8\theta\delta a}{3}\left[1 + \frac{4a_0}{3\pi\lambda} + \left(\frac{4a_0}{3\pi\lambda}\right)^2 + \left(\frac{4a_0}{3\pi\lambda}\right)^3 + \left(\frac{4a_0}{3\pi\lambda}\right)^4 \right. \tag{2.90}$$

$$\left. + \frac{8a_1}{15\pi\lambda^3}\left(1 + \frac{8a_0}{3\pi\lambda}\right) + O\left(\frac{1}{\lambda^5}\right) \right].$$

For specified values of P and γ, we can determine a and δ from (2.89), (2.90). A numerical analysis of formulae (2.88)–(2.90) will be given in subsection 2.9.2 (Table 2.14).

For the planar sloping punch $\delta(r, \varphi) = \delta + \alpha r \cos\varphi$, we find from (2.83) and (2.84):

$$q_1(r) = \frac{4\theta\alpha r}{\pi\sqrt{a^2 - r^2}}\left[1 - \frac{8a_1}{3\pi\lambda^3} + O\left(\frac{1}{\lambda^5}\right)\right], \tag{2.91}$$

$$M = \frac{8\theta\alpha a^3}{3}\left[1 - \frac{8a_1}{3\pi\lambda^3} + O\left(\frac{1}{\lambda^5}\right)\right]. \tag{2.92}$$

2.3.4. THE ASYMPTOTIC SOLUTION FOR SMALL λ

Using asymptotic presentation (2.71) of the kernel, we write the integral equation (2.59), for small λ and $n = 0$, in the form

$$\int_0^1 \varphi(\xi)\, K_*\left(\frac{\xi - x}{\lambda}\right) d\xi = \pi f(x) \qquad (0 \le x \le 1), \tag{2.93}$$

$$\xi = \frac{\rho}{a}, \qquad x = \frac{r}{a}, \qquad \varphi(\xi) = \frac{q(\rho)\sqrt{\rho}}{\theta\sqrt{a}}, \qquad f(x) = \frac{\delta(r)\sqrt{r}}{a^{3/2}}.$$

Introduce the special variables (Aleksandrov, Babeshko and Kucherov (1966))

$$\tau = \frac{f(1) - f(\xi)}{\lambda f'(1)}, \qquad t = \frac{f(1) - f(x)}{\lambda f'(1)}. \tag{2.94}$$

Suppose the function $f(x)$ is strictly monotone and $f'(1) \neq 0$; note that we can present any function of bounded variation as a difference between two strictly monotonic functions (see Natanson (1975)), and use a linearity of equation (2.93). For such a function, we have a one-to-one correspondence between x and t, and formulae (2.63), for small λ, give

$$\xi = 1 - \lambda\tau + O(\lambda^2), \qquad x = 1 - \lambda t + O(\lambda^2). \tag{2.95}$$

Taking account of (2.94) and (2.95), we rewrite (2.93) in the form

$$\int_0^{c/\lambda} \psi(\tau)\, K_*(\tau - t + O(\lambda))\, d\tau = \frac{\pi}{\lambda}[f(1) - \lambda t f'(1)] \quad \left(0 \le t \le \frac{c}{\lambda}\right), \tag{2.96}$$

$$c = f(1)[f'(1)]^{-1}, \qquad \psi(\tau) \equiv \varphi(\xi).$$

We see from (2.96) that the principal term of asymptotics of the solution of equation (2.59), for small λ and $n = 0$, can be found from the integral equation

$$\int_0^{\infty} \psi(\tau)\, K_*(\tau - t)\, d\tau = \frac{\pi}{\lambda}[f(1) - \lambda t f'(1)] \quad (0 \le t < \infty). \tag{2.97}$$

Use approximation (2.32) for $L(u)$ in (2.71) and apply the Wiener–Hopf method (Noble (1958)) to solve (2.97). We obtain the principal term of the asymptotics in the form

$$q(r) = \frac{\theta\sqrt{a}}{\lambda\sqrt{r}}[f(1)\psi_0(t) - \lambda f'(1)\psi_1(t)], \tag{2.98}$$

$$f(1) = \frac{\delta(a)}{a}, \quad f'(1) = \delta'(a) + \frac{\delta(a)}{2a}.$$

Here, t has the form (2.94), $\psi_0(t)$ and $\psi_1(t)$ are given by the formulae

$$\psi_0(t) = \frac{1}{A}\Phi(\sqrt{Et}) + \frac{\exp(-Et)}{\sqrt{\pi At}}, \tag{2.99}$$

$$\psi_1(t) = \frac{t}{A}\Phi(\sqrt{Et}) - \frac{\exp(-Et)}{\sqrt{\pi Et}}\left(1 - \frac{t}{A} - \frac{1}{2\sqrt{AE}}\right), \tag{2.100}$$

where $\Phi(x)$ is the probability integral 8.250.1 in Gradshteyn and Ryzhik (1994)

$$\Phi(x) = \frac{2}{\sqrt{\pi}}\int_0^x \exp(-t^2)\,dt \tag{2.101}$$

Note that this asymptotics has one defect, its singularity at $r = 0$ contradicts what we know about the structure of solution (see (2.63)). Nevertheless, this singularity gives an asymptotically small contribution, as $\lambda \to 0$, because the radius of the neighborhood of the point $r = 0$, where solution (2.98) appreciably deviates from the exact one, tends to zero as $\lambda \to 0$.

For the planar punch $\delta(r) \equiv \delta$, we must set in (2.98)

$$f(1) = 2f'(1) = \delta/a, \qquad t = 2(1-\sqrt{x})/\lambda, \tag{2.102}$$

Using (2.98), (2.102), we find that

$$\lim \frac{1}{\theta\delta}q(r)\sqrt{a^2-r^2} = \sqrt{\frac{s}{\pi}}\left(\frac{1}{\beta} - \frac{1}{2s\beta} + \frac{1}{s}\right) \quad (r \to a), \tag{2.103}$$

$$\frac{P}{\theta\delta a} = 2\pi\left\{\frac{1}{\beta^2}\left[\left(\frac{s}{4} - \frac{1}{2} + \frac{5}{8s} - \frac{3}{8s^2} - \frac{15}{64s^3}\right)D_1 \right.\right. \tag{2.104}$$
$$\left. + \left(\frac{1}{4} - \frac{5}{8s} + \frac{17}{16s^2} + \frac{15}{32s^3}\right)D_2\right]$$

$$+\frac{1}{\beta}\left[\left(1-\frac{3}{2s}+\frac{5}{4s^2}-\frac{3}{8s^3}\right)D_1+\left(\frac{1}{s}-\frac{2}{s^2}+\frac{3}{4s^3}\right)D_2\right]$$

$$+\left(\frac{1}{s}-\frac{1}{s^2}+\frac{3}{4s^3}\right)D_1+\left(\frac{1}{s^2}-\frac{3}{2s^3}\right)D_2\Big\},$$

$$\beta=\sqrt{AE},\qquad s=2E/\lambda,\qquad D_1=\Phi(\sqrt{s}),\qquad D_2=\sqrt{s/\pi}\exp(-s).$$

Formulae (2.103) and (2.104) are valid for $\lambda \leq 2$.

Note that for the parabolic punch $\delta(r) = \delta - \gamma r^2$, we should take a superposition of two asymptotics: for $\delta_1(r) = \delta$ and for $\delta_2(r) = -\gamma r^2$.

2.3.5. ANOTHER SOLUTION FOR SMALL λ

Now we give another approach for construction of asymptotics of equation (2.44) for $n = 0$. We reduce this equation to the equivalent paired integral equations

$$\mathcal{H}_0\{\gamma^{-1}Q(\gamma)L(\gamma h); r\} = \theta\delta(r) \quad (0 \leq r \leq a), \tag{2.105}$$

$$\mathcal{H}_0\{Q(\gamma); r\} = 0 \quad (r > a), \tag{2.106}$$

where the operator $\mathcal{H}_0$ is determined by formula (1.32), $Q(\gamma)$ is the Hankel transform of the discontinuous function $\tilde{q}(r) = q(r)$ $(0 \leq r \leq a)$, $\tilde{q}(r) = 0$ $(a < r < \infty)$:

$$Q(\gamma) = \int_0^a q(\rho)\rho\mathcal{J}_0(\gamma\rho)\,d\rho.$$

Using the Abel operators (1.35), (1.36) and formulae (1.37)–(1.41), we reduce (2.105), (2.106) to the following integral equation for the auxiliary function $p(x)$:

$$\int_0^a p(\xi)\,d\xi \int_0^\infty L(\gamma h)\cos(\gamma\xi)\,\cos(\gamma x)\,d\gamma = \frac{\pi}{2}\theta g(x) \quad (0 \leq x \leq a), \tag{2.107}$$

where $g(x)$ coincides with the expression in the square brackets in (2.75). The contact pressure $q(r)$ may be found by formula (2.74) after solving equation (2.107).

Suppose the functions $p(x)$ and $g(x)$ are even, and extend them to the segment $[-a, 0]$. Introducing the dimensionless quantities

$$x' = \frac{x}{a}, \qquad \xi' = \frac{\xi}{a}, \qquad \varphi(\xi') = \frac{p(\xi' a)}{\theta a}, \qquad f(x') = \frac{g(x' a)}{a},$$

we change (2.107), omitting the primes, to the form

$$\int_{-1}^{1} \varphi(\xi)\, K^*\left(\frac{\xi - x}{\lambda}\right) d\xi = \pi\lambda f(x) \quad (|x| \leq 1), \tag{2.108}$$

$$K^*(\tilde{y}) = \int_0^\infty L(u) \cos(u\tilde{y})\, du.$$

We use the following representation for the kernel $K^*(\tilde{y})$:

$$K^*(\tilde{y}) = \pi\delta_*(\tilde{y}) - \sum_{m=0}^{\infty} \frac{[(2m)!!]^2}{(2m)!} a_m \tilde{y}^{2m},$$

where $\delta_*(\tilde{y})$ is the Dirac δ-function, the quantities a_m are given by formula (2.36); the series converges absolutely as $|\tilde{y}| < 2$. Now seeking the asymptotic solution of equation (2.108) for large λ in the form

$$\varphi(x) = \sum_{n=0}^{\infty} \varphi_n(x)\lambda^{-n}, \tag{2.109}$$

we again arrive at the results (2.78)–(2.92).

Now we consider in detail the construction of the principal asymptotic term for the solution of (2.108) for small λ and for the planar punch, $\delta(r) \equiv \delta$. We use again approximation (2.32) for the function $L(u)$. We will obtain the approximate solution of (2.108) by the formula

$$\varphi(x) = \lim \varphi_\varepsilon(x) \quad (\varepsilon \to 0),$$

where the function $\varphi_\varepsilon(x)$ will be found from the integral equation

$$\int_{-1}^{1} \varphi_\varepsilon(\xi)\, d\xi \int_0^\infty \frac{\sqrt{u^2+\varepsilon^2}\sqrt{u^2+E^2}}{u^2+F^2} \cos\left(\frac{\xi - x}{\lambda} u\right) du = \pi\lambda f \tag{2.110}$$

$$(|x| \leq 1, \qquad f = \delta/a).$$

We seek the principal asymptotic term of the solution of equation (2.110) for small λ in the multiplicative form (Aleksandrov and Babeshko (1965), Aleksandrov and Chebakov (1972))

$$\varphi_\varepsilon(x) = \mu_*\, \omega\left(\frac{1+x}{\lambda}\right) \omega\left(\frac{1-x}{\lambda}\right) \left[v\left(\frac{x}{\lambda}\right)\right]^{-1}, \tag{2.111}$$

where $\omega(t)$ and $v(t)$ satisfy the equations

$$\int_0^\infty \omega(\tau)\,d\tau \int_0^\infty \frac{\sqrt{u^2+\varepsilon^2}\sqrt{u^2+E^2}}{u^2+F^2}\cos[u(\tau-t)]\,du = \pi f \quad (0 \le t < \infty), \tag{2.112}$$

$$\int_{-\infty}^\infty v(\tau)\,d\tau \int_0^\infty \frac{\sqrt{u^2+\varepsilon^2}\sqrt{u^2+E^2}}{u^2+F^2}\cos[u(\tau-t)]\,du = \pi f \quad (-\infty < t < \infty), \tag{2.113}$$

and μ_* is a so far undetermined factor. To solve (2.112), we use the Wiener-Hopf method (Noble (1958)). Equation (2.113) can be easily solved using the theorem on convolution for Fourier transforms (Bochner (1932), Sneddon (1951)). As a result, we derive the formulae

$$\omega(t) = \frac{f}{\sqrt{A\varepsilon}}\left[\exp\left(-\frac{E+\varepsilon}{2}t\right)\mathcal{I}_0\left(\frac{E-\varepsilon}{2}t\right)\right. \tag{2.114}$$

$$\left. + F\int_0^t \exp\left(-\frac{E+\varepsilon}{2}\tau\right)\mathcal{I}_0\left(\frac{E-\varepsilon}{2}\tau\right)d\tau\right], \qquad v(t) = \frac{f}{A\varepsilon},$$

where $\mathcal{I}_0(x)$ is the modified Bessel function.

Using the formulae (2.109), (2.111) and (2.114), we find that

$$\varphi(x) = f\mu_*\chi\left(\frac{1+x}{\lambda}\right)\chi\left(\frac{1-x}{\lambda}\right), \tag{2.115}$$

$$\chi(t) = \exp\left(-\frac{Et}{2}\right)\mathcal{I}_0\left(\frac{Et}{2}\right) + F\int_0^t \exp\left(-\frac{E\tau}{2}\right)\mathcal{I}_0\left(\frac{E\tau}{2}\right)d\tau$$

$$= \frac{1}{\sqrt{AE}}\Phi\left(-\frac{1}{2}, 1; -Et\right) + \left(1 - \frac{1}{\sqrt{AE}}\right)\Phi\left(\frac{1}{2}, 1; -Et\right),$$

where $\Phi(\alpha, \gamma; x)$ is the confluent hypergeometric function (Jahnke, Emde and Lösch (1966)).

To determine the factor μ_*, we substitute (2.115) into equation (2.110) for $\varepsilon = 0$, obtaining the relation

$$\mu_*\int_{-1}^1 \chi\left(\frac{1+\xi}{\lambda}\right)\chi\left(\frac{1-\xi}{\lambda}\right)d\xi \int_0^\infty \frac{\alpha\sqrt{\alpha^2\lambda^2+E^2}}{\alpha^2\lambda^2+F^2} \tag{2.116}$$

$$\times \cos[\alpha(\xi - x)]\,d\alpha = \pi \quad (|x| \le 1).$$

Let in (2.116) $\lambda \to 0$. Taking account of the asymptotic expansion for the confluent hypergeometric function for large x (Jahnke, Emde and Lösch (1966, p.276)), we have

$$\chi(t) \to 2\sqrt{t/(\pi A)} \quad (t \to \infty). \tag{2.117}$$

Using (2.117) and the integral

$$\int_0^\infty \alpha \cos[\alpha(\xi - x)]\, d\alpha = -\frac{d}{dx}\frac{1}{\xi - x},$$

which is a Fourier transform of a generalized function (see Gel'fand and Shilov (1964)), we reduce (2.116) to the formula

$$-\mu_* \frac{4}{\pi}\frac{d}{dx}\int_{-1}^{1} \frac{\sqrt{1-\xi^2}}{\xi - x}\, d\xi = \pi \quad (|x| \le 1). \tag{2.118}$$

Calculating the singular integral in (2.118) (see Muskhelishvili (1953a)), we finally obtain that $\mu_* = \pi/4$.

Thus, the asymptotic solution of the integral equation (2.107), for small λ, has the form

$$p(x) = \frac{\pi}{4}\theta\delta\chi\left(\frac{a+x}{h}\right)\chi\left(\frac{a-x}{h}\right), \tag{2.119}$$

and the corresponding solution of the integral equation (2.59) for $n = 0$ is defined by (2.74). On the basis of the latter, we derive the formulae

$$\lim \frac{1}{\theta\delta} q(r)\sqrt{a^2 - r^2} = \frac{1}{2}\left[\frac{1}{\beta}\Phi\left(-\frac{1}{2}, 1; -s\right)\right. \tag{2.120}$$

$$\left. + \left(1 - \frac{1}{\beta}\right)\Phi\left(\frac{1}{2}, 1; -s\right)\right] \quad (r \to a),$$

$$\frac{P}{\theta\delta a} = \pi\left[\frac{s}{2\beta^2} + \frac{1}{\beta}\left(2 - \frac{1}{\beta}\right) + \left(1 - \frac{1}{\beta}\right)^2 \frac{1}{s}(1 - \exp[-s])\right] \tag{2.121}$$

$$(\beta = \sqrt{AE}, \qquad s = 2E/\lambda),$$

which are similar to (2.103), (2.104), and valid for $\lambda \le 2$.

For a non-planar punch, in order to find an approximate solution of (2.59) for small λ as $n = 0$ (or of the corresponding integral equation (2.107)), we may use Krein's method from subsection 2.4.3. This method allows us to obtain the solution for a non-planar punch ($\delta(r) \neq \text{const}$) in quadratures, if we know the approximate solution for the planar punch $\delta(r) = \text{const}$.

Note that the integral equation (2.59) for $n = 1$ may be reduced to the integral equation (2.108) in which the functions $\varphi(x)$ and $f(x)$ are odd and connected to $q_1(r)$ and $\delta_1(r)$ by the relations

$$q_1(r) = -\frac{2\theta a}{\pi}\frac{d}{dr}\int_r^a \frac{\varphi(\xi)\,d\xi}{\sqrt{\xi^2 - r^2}},$$

$$f(x) = \frac{1}{a}\frac{d}{dx}x\int_0^x \frac{\delta_1(\rho)\,d\rho}{\sqrt{x^2 - \rho^2}}.$$

2.3.6. THE METHOD OF ORTHOGONAL POLYNOMIALS

We briefly describe the method of orthogonal polynomials (Aleksandrov (1967c)) as an alternative to that considered above. This method uses the expansions of the functions $\delta(r)$ and $F(\sigma, \tau)$ in (2.72) as the series

$$\delta(r) = \sum_{i=0}^{\infty} b_i \, \mathrm{P}_{2i}(\sqrt{1 - r^2/a^2}), \tag{2.122}$$

$$F(\rho/h, r/h) = \sum_{i=0}^{\infty}\sum_{j=0}^{\infty} e_{ij}(\lambda)\, \mathrm{P}_{2i}(\sqrt{1 - r^2/a^2})\, \mathrm{P}_{2j}(\sqrt{1 - \rho^2/a^2}), \tag{2.123}$$

where $\mathrm{P}_{2i}(x)$ are the Legendre polynomials (Gradshteyn and Ryzhik (1994)). We seek the function $q(r)$ in the form

$$q(r) = (a^2 - r^2)^{-1/2}\sum_{i=0}^{\infty} a_i \, \mathrm{P}_{2i}(\sqrt{1 - r^2/a^2}), \tag{2.124}$$

and use the spectral relation

$$\int_0^1 \frac{\tau \, \mathrm{P}_{2m}(\sqrt{1 - \tau^2})}{\sqrt{1 - \tau^2}} \mathcal{K}\left(\frac{2\sqrt{\tau t}}{\tau + t}\right)\frac{d\tau}{\tau + t} = \frac{\pi^2[(2m - 1)!!]^2}{4[(2m)!!]^2}\mathrm{P}_{2m}(\sqrt{1 - t^2})$$

$$(m = 0, 1, ...),$$

and the orthogonality condition for the Legendre polynomials

$$\int_0^1 \mathrm{P}_{2i}(\sqrt{1-\tau^2})\,\mathrm{P}_{2j}(\sqrt{1-\tau^2})\frac{\tau\,d\tau}{\sqrt{1-\tau^2}} = \begin{cases} 0 & \text{for } i \neq j \\ (4i+1)^{-1} & \text{for } i = j \end{cases}.$$

Substituting (2.122)–(2.124) into equation (2.72) and equating the coefficients at the same polynomials $\mathrm{P}_{2i}(\sqrt{1-r^2/a^2})$, we obtain an infinite system of algebraic equations for a_i. The system appears to be totally regular, for fairly large λ, and quasiregular[3] for all $\lambda > 0$.

Note also that the coefficients $e_{ij}(\lambda)$ in (2.123) may be presented as the one-dimensional integrals

$$e_{ij}(\lambda) = \pi\lambda(4i+1)(4j+1)\frac{(2i-1)!!(2j-1)!!}{2(2i)!!(2j)!!}$$

$$\times \int_0^\infty [1-L(u)]\,\mathcal{J}_{2i+1/2}\left(\frac{u}{\lambda}\right)\cdot\mathcal{J}_{2j+1/2}\left(\frac{u}{\lambda}\right)\frac{du}{u}.$$

2.4 A Strip Punch on a Layer

2.4.1. ONE-DIMENSIONAL INTEGRAL EQUATION

Note that by using formula (2.18), the integral equation (2.27) of the contact problem for the layer can be also written in the form

$$\int_\Omega q(\xi,\eta)K_*\left(\frac{x-\xi}{h},\frac{y-\eta}{h}\right)d\Omega = \pi^2\theta h\delta(x,y) \quad ((x,y)\in\Omega), \qquad (2.125)$$

[3]The infinite system $\sum_{j=1}^{\infty} a_{ij}x_j = b_i$ $(i = 1, 2, ...)$ is quasiregular, if

(A) $$\sum_{j=1}^{\infty}|a_{ij}| < \infty \quad (i = 1, 2, ..., N);$$

(B) $$\sum_{j=1}^{\infty}|a_{ij}| < 1 \quad (i = N+1, N+2, ...);$$

(C) $$|b_i| \leq M \quad (M = \text{const};\ i = N+1, N+2, ...).$$

The system is totally regular when $N = 0$ (Kantorovich and Krylov (1958)).

where the kernel $K_*(\sigma, \tau)$ is determined by (2.19).

Let us consider the case when the contact region Ω is the infinite strip ($|x| \leq a$, $|y| < \infty$), and introduce the Fourier transforms of the functions $q(\xi, \eta)$ and $\delta(x, y)$, by η and y, by the formulae

$$q(\xi,\eta) = \frac{1}{2\pi} \int_{-\infty}^{\infty} q_\beta(\xi) \exp(-i\beta\eta)\, d\beta, \qquad q_\beta(\xi) = \int_{-\infty}^{\infty} q(\xi,\eta) \exp(i\beta\eta)\, d\eta, \tag{2.126}$$

$$\delta(x,y) = \frac{1}{2\pi} \int_{-\infty}^{\infty} \delta_\beta(x) \exp(-i\beta y)\, d\beta, \qquad \delta_\beta(x) = \int_{-\infty}^{\infty} \delta(x,y) \exp(i\beta y)\, dy. \tag{2.127}$$

With the help of (2.126), (2.127) and (2.19) we rewrite (2.125) as follows

$$\int_{-\infty}^{\infty} \exp(-i\beta y)\, d\beta \left[\int_{-a}^{a} q_\beta(\xi)\, d\xi \int_{0}^{\infty} \frac{L(u)}{u} \cos\left(s\frac{x-\xi}{h}\right) ds \right]$$

$$= \pi\theta \int_{-\infty}^{\infty} \delta_\beta(x) \exp(-i\beta y)\, d\beta.$$

We obtain from here the following one-dimensional integral equation of the first kind and of the convolution type

$$\int_{-a}^{a} q_\beta(\xi) K_\beta\left(\frac{x-\xi}{h}\right) d\xi = \pi\theta\delta_\beta(x) \qquad (|x| \leq a), \tag{2.128}$$

$$K_\beta(\tilde{y}) = \int_{0}^{\infty} \frac{L(u)}{u} \cos(s\tilde{y})\, ds \qquad \left(u = \sqrt{s^2 + (\beta h)^2}\right), \tag{2.129}$$

in the Fourier transform $q_\beta(\xi)$ of the contact pressure $q(\xi, \eta)$. For $\beta = 0$ the kernel $K_\beta(\tilde{y})$ reduces to $K_*(\tilde{y})$ of the form (2.71), and then equation (2.128) corresponds to the problem on the plane deformation of the layer by the punch.

We must replace conditions (1.30), (1.31) for the case under consideration (for the strip punch) by the conditions

$$P(y) = \frac{1}{2\pi} \int_{-\infty}^{\infty} P_\beta \exp(-i\beta y)\, d\beta, \qquad P_\beta = \int_{-a}^{a} q_\beta(\xi)\, d\xi,$$

$$M(y) = \frac{1}{2\pi} \int_{-\infty}^{\infty} M_\beta \exp(-i\beta y)\, d\beta, \qquad M_\beta = \int_{-a}^{a} \xi q_\beta(\xi)\, d\xi,$$

where $P(y)$ and $M(y)$ are the force and the moment acting in the section $y = \text{const}$ of the punch.

The integral equation (2.128) contains two dimensionless geometric parameters $\lambda = h/a$ and βa (or $\beta h = \beta a \lambda$).

Using integral 3.754.1 in Gradshteyn and Ryzhik (1994)

$$\int_0^\infty \frac{\cos(s\tilde{y})\, ds}{\sqrt{s^2 + (\beta h)^2}} = \mathcal{K}_0(\beta h |\tilde{y}|),$$

where $\mathcal{K}_0(x)$ is the modified Bessel function, we present kernel (2.129) as follows

$$K_\beta(\tilde{y}) = \mathcal{K}_0(\beta h |\tilde{y}|) + F(\beta h, \tilde{y}), \tag{2.130}$$

$$F(\beta h, \tilde{y}) = \int_0^\infty \frac{L(u) - 1}{u} \cos(s\tilde{y})\, ds. \tag{2.131}$$

Note that $\mathcal{K}_0(\beta h |\tilde{y}|)$ has a logarithmic singularity as $\tilde{y} = 0$, and vanishes as $|\tilde{y}| \to \infty$ like $|\tilde{y}|^{-1/2} \exp(-\beta h |\tilde{y}|)$. We draw attention to the following properties of the function $F(\beta h, \tilde{y})$ in (2.131):
a) for any $\beta h \neq 0$ the function $F(\beta h, \tilde{y})$ is regular in the complex plane $z = \tilde{y} + i\eta$ in the strip $|\tilde{y}| < \infty$, $|\eta| < 2$ and, therefore, as $0 \leq |\tilde{y}| < \infty$, it is continuous with all its derivatives;
b) for any $|\tilde{y}| \neq \infty$ and large values of βh, we have the following asymptotic estimate, uniform in $\tilde{y}$:

$$F(\beta h, \tilde{y}) = O[(\beta h)^{3/2} \exp(-2\beta h)], \tag{2.132}$$

this shows that the function $F(\beta h, \tilde{y})$ vanishes quickly as $\beta h \to \infty$ ($\beta a \to \infty$ or $\lambda \to \infty$). We also note that the kernel $K_\beta(\tilde{y})$ vanishes as $|\tilde{y}| \to \infty$ like $\exp(-\sqrt{D^2 + \gamma^2}|\tilde{y}|)$, where D is defined by (2.31).

2.4.2. GENERAL RESULTS

Suppose that the solution $q_\beta(\xi)$ of the integral equation (2.128) exists in the class $\mathcal{L}(-a, a)$ (absolutely summable on the segment $[-a, a]$). On the basis of (2.130) we rewrite (2.128) in the form

$$\int_{-a}^{a} q_\beta(\xi)\,\mathcal{K}_0(\beta|x-\xi|)\,d\xi = \pi\theta\delta_\beta(x) - \int_{-a}^{a} q_\beta(\xi)F\left(\beta h, \frac{x-\xi}{h}\right)\,d\xi \quad (|x| \le a). \tag{2.133}$$

In view of the first property (a) of $F(\beta h, \tilde{y})$ and of the assumption made with respect to $q_\beta(\xi)$, the function

$$g_\beta(x) = \int_{-a}^{a} q_\beta(\xi)F\left(\beta h, \frac{x-\xi}{h}\right)\,d\xi$$

is continuous with all its derivatives for $x \in [-a, a]$.

We obtain from this the following important results:
A) because of property (b) above, as $\beta a \to \infty$ or $\lambda \to \infty$, the integral equation (2.133) reduces to the following integral equation for the action of a strip punch onto an elastic half-space

$$\int_{-a}^{a} q_\beta(\xi)\,\mathcal{K}_0(\beta|x-\xi|)\,d\xi = \pi\theta\delta_\beta(x) \quad (|x| \le a); \tag{2.134}$$

B) the solution of the integral equation (2.133) has the same character (singularities) as of the integral equation (2.134), for all values of the dimensionless parameters λ and βa.

For the integral equation (2.128) (or (2.133)) we may prove the following facts:
C) if the function $\delta_\beta(x)$ is such that for $x \in [-a, a]$ its first derivative satisfies the Hölder condition with the index $0 < \alpha \le 1$, ($\delta_\beta(x) \in \mathcal{H}_1^\alpha(-a, a)$), then the solution $q_\beta(x)$ exists and is unique in $\mathcal{L}_p(-a, a)$: it is absolutely summable in the segment $[-a, a]$ with the power p, $1 \le p < 2$, ($\mathcal{L}_1 \equiv \mathcal{L}$), and it can be presented in the form

$$q_\beta(x) = Q(x)(a^2 - x^2)^{-1/2},$$

where $Q(x)$ satisfies on $[-a, a]$ the Hölder condition with the index γ: $Q(x) \in \mathcal{H}_0^\gamma(-a, a)$, where $\gamma = \alpha$, if $\alpha < 1$ while $\gamma = 1 - \varepsilon$, if $\alpha = 1$;
D) if the function $\delta_\beta(x) \in \mathcal{H}_1^\alpha(-a, a)$, where $1/2 < \alpha \le 1$, and

$$q_\beta(\pm a) = 0,$$

then the solution can be presented as follows

$$q_\beta(x) = \tilde{Q}(x)(a^2 - x^2)^{1/2},$$

where $\tilde{Q}(x) \in \mathcal{H}_0^\gamma(-a, a)$ and $\gamma = \alpha - 1/2$.

Note that, if the function $\delta_\beta(x)$ is even ($\delta_\beta^+(x)$) or odd ($\delta_\beta^-(x)$) in x, then the solution $q_\beta(\xi)$ of the integral equation (2.128) is even ($q_\beta^+(\xi)$) or odd ($q_\beta^-(\xi)$) in ξ, respectively. This is due to the evenness of the kernel $K_\beta(y)$ in y and to the symmetry of the integration limits. The even and odd solutions of (2.128) are interconnected by the following theorem (Aleksandrov (1967b)).

Theorem 2.2. *If $\delta_\beta^-(x) \in \mathcal{H}_1^\alpha(-a, a)$, where $1/2 < \alpha \le 1$, and $q_\beta^+(\xi)$ is the solution of equation* (2.128) *for the case*

$$\delta_\beta^+(x) = \int_0^x \delta_\beta^-(x)\, dx + C,$$

vanishing as $\xi = \pm a$, then

$$q_\beta^-(\xi) = [q_\beta^+(\xi)]'$$

is the solution corresponding to $\delta_\beta^-(x)$.

2.4.3. KREIN'S METHOD

Krein obtained general results, published in chapter 4 of Gohberg and Krein (1970), which allows us to confine ourselves, when solving integral equations (2.128) and (2.134), to the case $\delta_\beta(x) \equiv \delta_\beta$, where δ_β is a constant quantity for a specified value of the parameter βa. In other words, his method gives a solution for a non-planar punch, if we know a solution for a planar punch, i.e. for the constant right-hand side of the integral equation. This means that if we know an exact or approximate solution $p(t, \tau) \in \mathcal{L}(a, \tau)$ $(a \le \tau \le b)$ of the integral equation

$$\int_a^\tau p(s, \tau)\, T(t, s)\, ds = 1 \quad (a \le t \le \tau),$$

which has the constant right-hand side, then for the integral equation

$$\int_a^b \psi(s)\, T(t, s)\, ds = g(t) \quad (a \le t \le b),$$

where $T(t,s) \in \mathcal{L}_2(t,s \in [a,b])$ and $g(t) \in C(a,b)$, we obtain the unique solution $\psi(t) \in C(a,b)$ in the quadrature form

$$\psi(t) = \left[\frac{1}{M'(\tau)}\frac{d}{d\tau}\int\limits_a^\tau p(s,\tau)\,g(s)\,ds\right]_{\tau=b} p(t,b) \tag{2.135}$$

$$-\int\limits_t^b p(t,\tau)\frac{d}{d\tau}\left[\frac{1}{M'(\tau)}\frac{d}{d\tau}\int\limits_a^\tau p(s,\tau)\,g(s)\,ds\right]d\tau,$$

$$M(\tau) = \int\limits_a^\tau p(t,\tau)\,dt.$$

We suppose that $M'(\tau) \neq 0$ $(a \le \tau \le b)$.

Example. *Consider the integral equation for the 2-D contact problem for a fairly thick layer* (Aleksandrov and Babeshko (1965))*:*

$$\int\limits_{-1}^{1} \varphi(\xi)[-\ln|\xi - x| + d] = \pi f(x) \quad (|x| \le 1), \quad d = \text{const}. \tag{2.136}$$

The dimensionless constant d depends on the thickness h of the layer, and $d \to \infty$ *as* $h \to \infty$*. For equation* (2.136)*, we have*

$$p(x,\tau) = \frac{1}{\pi[\ln(2/\tau) + d]\sqrt{\tau^2 - x^2}}, \qquad M(\tau) = \frac{1}{\pi[\ln(2/\tau) + d]}.$$

Using (2.135) *and supposing that the function* $f(x)$ *is even, we obtain*

$$\varphi(x) = \frac{\gamma(1)(\ln 2 + d)^{-1} + \gamma'(1)}{\pi\sqrt{1 - x^2}} - \frac{1}{\pi}\int\limits_x^1 \frac{\gamma'(\tau) + \tau\gamma''(\tau)}{\sqrt{\tau^2 - x^2}}\,d\tau, \tag{2.137}$$

$$\gamma(\tau) = 2\int\limits_0^\tau \frac{f(\xi)\,d\xi}{\sqrt{\tau^2 - \xi^2}}.$$

To check solution (2.137)*, we differentiate equation* (2.136) *with respect to* x*; then we obtain equation* (1.18) *on the interval* $(-1,1)$*, which has solution* (1.19)*. To determine the arbitrary constant* C *in* (1.19)*, we must substitute this solution into the original equation* (2.136) *and satisfy it. Finally, we arrive at* (2.137)*.*

2.4.4. ASYMPTOTIC SOLUTIONS

When $\delta_\beta(x) \equiv \delta_\beta$ the solution of (2.128) has the asymptotic expansions:
1) for small values of βa

$$q_\beta(x) = \frac{P_\beta}{\pi\sqrt{a^2-x^2}}\Big[1 + [a_1 + 3b_1/2 - b_1\ln(2\lambda_*)] \tag{2.138}$$

$$\times\left(1 - 2\frac{x^2}{a^2}\right)\frac{1}{\lambda_*^2} + [E_1 + E_2\ln(2\lambda_*)]\left(1 - 2\frac{x^2}{a^2}\right)\frac{1}{\lambda_*^4}$$

$$+[E_3 + E_4\ln(2\lambda_*)]\left(1 + 4\frac{x^2}{a^2} - 8\frac{x^4}{a^4}\right)\frac{1}{\lambda_*^4} + O\left(\frac{\ln^3\lambda_*}{\lambda_*^6}\right)\Big],$$

$$P_\beta = \pi\theta\delta_\beta\Big[\ln(2\lambda_*) + a_0 + [a_1 + b_1 - b_1\ln(2\lambda_*)]\frac{1}{\lambda_*^2} \tag{2.139}$$

$$+[E_5 + E_6\ln(2\lambda_*) - b_1^2\ln^2(2\lambda_*)/4]\frac{1}{\lambda_*^4} + O\left(\frac{\ln^3\lambda_*}{\lambda_*^6}\right)\Big]^{-1},$$

where the notation is as follows:

$$E_1 = -\frac{1}{2}a_1b_1 - \frac{3}{4}b_1^2 + 3a_2 + \frac{13}{4}b_2, \qquad E_2 = \frac{1}{2}b_1^2 - 3b_2,$$

$$E_3 = \frac{1}{12}a_1b_1 + \frac{1}{8}b_1^2 + \frac{1}{2}a_2 + \frac{25}{24}b_2, \qquad E_4 = -\frac{1}{12}b_1^2 - \frac{1}{2}b_2,$$

$$E_5 = -\frac{1}{4}a_1^2 - \frac{3}{4}a_1b_1 - \frac{9}{16}b_1^2 + \frac{9}{4}a_2 + \frac{21}{8}b_2,$$

$$E_6 = \frac{1}{2}a_1b_1 + \frac{3}{4}b_1^2 - \frac{9}{4}b_2, \qquad \lambda_* = \frac{1}{\beta a},$$

and the quantities a_i $(i = 0, 1, 2)$ and b_i $(i = 1, 2)$ are given by the formulae

$$b_1 = -\frac{1}{4}, \qquad b_2 = -\frac{1}{64}, \qquad a_0 = \int_0^\infty [L_*(s) - 1 + \exp(-s)]\frac{ds}{s} \approx 0.116,$$

$$a_1 = \frac{3}{8} + \frac{1}{2}\int_0^\infty \left[s^2 - s^2L_*(s) - \frac{1-\exp(-s)}{2}\right]\frac{ds}{s} \approx 0.279,$$

$$a_2 = \frac{25}{768} - \frac{1}{24}\int_0^\infty \left[s^4\left(1 - \frac{1}{2s^2}\right) - s^4L_*(s) + \frac{3}{8}(1 - \exp(-s))\right]\frac{ds}{s} \approx 0.0252,$$

$$L_*(s) = s(s^2+1)^{-1/2};$$

2) for large values of βa

$$q_\beta(x) = \beta\theta\delta_\beta\,\zeta[\beta(a-x)]\,\zeta[\beta(a+x)], \tag{2.140}$$

$$\zeta(t) = \Phi(\sqrt{t}) + \frac{\exp(-t)}{\sqrt{\pi t}}, \qquad P_\beta = \theta\delta_\beta\left(1+\frac{2}{\lambda_*}\right), \tag{2.141}$$

where $\Phi(x)$ is the probability integral (2.101).

Formulae (2.138), (2.139) and (2.140), (2.141) are valid for $\lambda_* \geq 2$ and $\lambda_* \leq 2$, respectively. In the columns *I*, *II*, and *III* of Table 2.3 we give the quantities

$$\frac{aq_\beta(0)}{\theta\delta_\beta}, \qquad \lim q_\beta(x)\sqrt{a^2-x^2}/(\theta\delta_\beta) \quad (x\to a), \qquad \frac{P_\beta}{\theta\delta_\beta}, \tag{2.142}$$

respectively, calculated for various λ_*. Among the values for $\lambda_* = 2$, *the upper* ones are obtained by (2.138), (2.139), and *the lower* ones by (2.140), (2.141).

Table 2.3: Quantities (2.142)

λ_*	*I*	*II*	*III*
4	0.4639	0.4394	1.419
2	0.6792	0.5898	1.994
2	0.6805	0.5925	2.000
1	1.1030	0.8047	3.000

Note that the closed solution of the integral equation (2.134) for any function $\delta_\beta(x) \in \mathcal{H}_1^\alpha(-a,a)$, $1/2 < \alpha \leq 1$, can be given in the form of a series. To obtain it, we must use the spectral relation (Aleksandrov, Kovalenko and Mkhitarian (1982))

$$\int_{-1}^{1} \frac{\mathrm{ce}_n(\arccos\xi, -q)}{\sqrt{1-\xi^2}}\,\mathcal{K}_0\left(\frac{|\xi - x|}{\lambda_*}\right) d\xi \tag{2.143}$$

$$= \frac{\pi\,\mathrm{Fek}_n(0,-q)}{\mathrm{Fek}_n'(0,-q)}\,\mathrm{ce}_n(\arccos x, -q) \qquad \left(q = \frac{1}{4\lambda_*^2}\right),$$

where $\mathrm{ce}_n(x,-q)$ are the Mathieu functions 8.611 in Gradshteyn and Ryzhik (1994); $\mathrm{Fek}_n(x,-q)$ are the functions 8.664 in Gradshteyn and Ryzhik (1994), corresponding to the second solutions of the modified Mathieu equation (see also McLachlan (1951)).

It follows from (2.143) that if we expand in (2.134) the function $\delta_\beta(x)$ into the series

$$\delta_\beta(x) = \sum_{n=0}^{\infty} \delta_n \operatorname{ce}_n\left(\arccos\frac{x}{a}, -q\right),$$

then the solution $q_\beta(x)$ of equation (2.134) takes the form

$$q_\beta(x) = \frac{\theta}{\sqrt{a^2 - x^2}} \sum_{n=0}^{\infty} \frac{\operatorname{Fek}_n'(0, -q)}{\operatorname{Fek}_n(0, -q)} \delta_n \operatorname{ce}_n\left(\arccos\frac{x}{a}, -q\right).$$

2.4.5. THE METHOD OF ORTHOGONAL FUNCTIONS

Let us return to the integral equation (2.133). Its approximate solution can be obtained by the method of orthogonal functions (Aleksandrov and Kovalenko (1977a)). For instance for the even case, this method is based on the expansions of the functions $\delta_\beta(x)$ and $F(\beta h, \tilde{y})$ into the series

$$\delta_\beta(x) = \sum_{n=0}^{\infty} f_n \delta_n^+ \operatorname{ce}_{2n}\left(\arccos\frac{x}{a}, -q\right),$$

$$F\left(\beta h, \frac{x - \xi}{h}\right) = \sum_{m=0}^{\infty}\sum_{n=0}^{\infty} g_{mn} e_{mn}(\lambda, \lambda_*) \tag{2.144}$$

$$\times \operatorname{ce}_{2m}\left(\arccos\frac{x}{a}, -q\right) \operatorname{ce}_{2n}\left(\arccos\frac{x}{a}, -q\right),$$

where $f_0 = g_{00} = 1$, $f_n = g_{n0} = g_{0n} = 2$ $(n \geq 1)$, $g_{mn} = 4$ $(m, n \geq 1)$, and on the presentation of the solution $q_\beta(x)$ of (2.133) in the form

$$q_\beta(x) = (a^2 - x^2)^{-1/2} \sum_{n=0}^{\infty} a_n \operatorname{ce}_{2n}\left(\arccos\frac{x}{a}, -q\right).$$

Then we use relation (2.143) and the orthogonality condition for the Mathieu functions 6.911.1,2 in Gradshteyn and Ryzhik (1994)

$$\int_{-1}^{1} \operatorname{ce}_{2m}(\arccos\xi, -q) \operatorname{ce}_{2n}(\arccos\xi, -q) \frac{dx}{\sqrt{1 - x^2}} = \begin{cases} 0 & \text{for } m \neq n \\ \pi/2 & \text{for } m = n \end{cases}.$$

After that, as a result of equating the coefficients of the functions

$$\operatorname{ce}_{2n}(\arccos(x/a), -q)$$

in the relation obtained from (2.133), we obtain an infinite system of algebraic equations for a_n, which is totally regular for fairly large λ/λ_* (due of (2.132)) and quasiregular for $\lambda > 0$ and $\lambda_* < \infty$.

The coefficients $e_{mn}(\lambda, \lambda_*)$ in expansion (2.144), for $F(\beta h, y)$, can be obtained as the following one-dimensional integrals:

$$e_{mn}(\lambda, \lambda_*) = (-1)^{m+n} \frac{A_0^{(2m)} A_0^{(2n)}}{\mathrm{ce}_{2m}(0,q)\,\mathrm{ce}_{2n}(0,q)} \int_0^\infty \left[L\left(\frac{\lambda}{\lambda_*}\cosh z\right) - 1\right]$$

$$\times \mathrm{Ce}_{2m}(z,q)\,\mathrm{Ce}_{2n}(z,q)\,dz,$$

where $\mathrm{Ce}_{2n}(z,q)$ are the associated Mathieu functions 8.630 in Gradshteyn and Ryzhik (1994), and $A_0^{(2n)}$ are the first coefficients in representation 8.611 of Gradshteyn and Ryzhik (1994)

$$\mathrm{ce}_{2n}(z,q) = \sum_{r=0}^{\infty} A_r^{(2n)} \cos(2rz).$$

To determine $A_r^{(2n)}$, we use recurrence relations 8.621 in Gradshteyn and Ryzhik (1994).

2.4.6. AN ANNULAR PUNCH ON A HALF-SPACE

Now consider the impression of an annular punch. The contact region Ω is an annulus with the internal radius a and external radius b. This problem is reduced to the integral equation (2.59) with the limits of integration a and b. Kernel (2.60) can be represented as follows:

$$K_n(\sigma,\tau) = \frac{1}{\pi}\int_0^\pi \cos(n\psi)\,K(t)\,d\psi \quad \left(t = \sqrt{\sigma^2 + \tau^2 - 2\sigma\tau\cos\psi}\right), \tag{2.145}$$

where $K(t)$ is defined by (2.22). We will use the fact that kernel (2.129) of the integral equation (2.128) for the strip punch can be rewritten in the form

$$K_\beta(\tilde{y}) = \int_0^\infty \cos(\beta h\vartheta)\,K(t)\,d\vartheta \qquad \left(t = \sqrt{\tilde{y}^2 + \vartheta^2}\right), \tag{2.146}$$

where $K(t)$ is also defined by (2.22).

Let us now show that, for a relatively narrow annular punch ($\varepsilon = a/b \sim 1$) and for a relatively small thickness of the layer ($\lambda = h/b \ll 1$), kernel

(2.145) is asymptotically connected to kernel (2.146). To do this, we rewrite formula (2.145) in the form

$$K_n(\sigma,\tau) = \frac{1}{\pi e^2} \int_c^d \frac{\cos(n\psi)}{\sin\psi} N(t)\,dt, \tag{2.147}$$

$$N(t) = tK(t), \qquad c = |\sigma - \tau|, \qquad d = \sigma + \tau, \qquad e = \sqrt{\sigma\tau},$$

$$\sin\psi = (2e^2)^{-1} \times \sqrt{(d^2 - t^2)(t^2 - c^2)}. \tag{2.148}$$

If λ is small, then σ and τ are large, therefore, d and e are large too. If ε is close to 1, then σ and τ are close to each other, therefore, $d \approx 2e$. Thus, $\sin\psi$ is small and

$$\sin\psi \approx \psi \approx e^{-1}\sqrt{t^2 - c^2}.$$

By substituting (2.148) into (2.147) and taking into account that d is large, we obtain

$$K_n(\sigma,\tau) \approx \frac{h}{\pi\sqrt{r\rho}} \int_c^\infty \frac{\cos(ne^{-1}\sqrt{t^2 - c^2})}{\sqrt{t^2 - c^2}} N(t)\,dt. \tag{2.149}$$

The kernel $K_\beta(\tilde{y})$ of the form (2.146) can be represented in the similar form

$$K_\beta(\tilde{y}) = \int_{\tilde{y}}^\infty \frac{\cos(\beta h\sqrt{t^2 - \tilde{y}^2})}{\sqrt{t^2 - \tilde{y}^2}} N(t)\,dt. \tag{2.150}$$

Putting $e \approx \sqrt{\varepsilon}\lambda^{-1}$ and comparing then (2.149) with (2.150), we find that

$$K_n(\sigma,\tau) \approx \frac{h}{\pi\sqrt{r\rho}} K_\beta(c), \qquad \beta a = \frac{n}{\sqrt{\varepsilon}}, \tag{2.151}$$

because of $\lambda \approx h/a$. Using (2.151), we write the integral equation (2.59), with the limits of integration from a to b, in the form

$$\int_a^b q_n(\rho)\sqrt{\rho}\, K_\beta\left(\frac{r-\rho}{h}\right) d\rho = \pi\theta\delta_n(r)\sqrt{r} \quad (a \le r \le b).$$

By introducing the new variables and notation by the formulae

$$x = \frac{a_*(2r - b - a)}{b - a}, \qquad \xi = \frac{a_*(2\rho - b - a)}{b - a}, \qquad h_* = \frac{2a_* h}{b - a},$$

$$q_\beta(\xi) = q_n(\rho)\sqrt{\rho}, \qquad \delta_\beta(x) = 2a_*(b-a)^{-1}\delta_n(r)\sqrt{r},$$

we finally come to the integral equation (2.128) in which a and h have been replaced by a_* and h_*, respectively.

Thus, the problem for a relatively narrow annular punch impressing a relatively thin layer has been reduced to the integral equation for a strip punch on the layer.

2.5 Contact Problems for a Thick Layer

2.5.1. THE ASYMPTOTIC SOLUTION

Let us substitute (2.35) into the integral equation (2.37):

$$\int_\Omega q(\xi,\eta)\frac{d\Omega}{R} = 2\pi\theta\delta(x,y) + \sum_{m=0}^{\infty} \frac{a_m}{h^{2m+1}} \int_\Omega q(\xi,\eta)R^{2m}\, d\Omega \quad ((x,y)\in\Omega). \tag{2.152}$$

Equation (2.151) and all the results, obtained from it, are true for $\lambda = h/a > 1$, since formula (2.35) holds if $t = R/h < 2$ and $\max_\Omega R = 2a$. So we will obtain an asymptotic solution for a thick layer: its thickness h must be larger than the parameter a, describing the size of the contact region Ω. This solution was first derived in Aleksandrov and Vorovich (1960) and Aleksandrov (1963).

The right-hand side of (2.152) suggests that we should seek the solution in the form of a series

$$q(\xi,\eta) = \sum_{k=0}^{\infty} q_k(\xi,\eta)h^{-k}. \tag{2.153}$$

By substituting (2.153) into (2.152) and equating the terms at the same powers of $\lambda^{-1} = a/h$, we obtain the following infinite system of integral equations for $q_k(x,y)$:

$$\int_\Omega q_0(\xi,\eta)\frac{d\Omega}{R} = 2\pi\theta\delta(x,y) \qquad ((x,y)\in\Omega), \tag{2.154}$$

$$\int_\Omega q_1(\xi,\eta)\frac{d\Omega}{R} = a_0\int_\Omega q_0(\xi,\eta)\, d\Omega, \tag{2.155}$$

$$\int_\Omega q_2(\xi,\eta)\frac{d\Omega}{R} = a_0\int_\Omega q_1(\xi,\eta)\, d\Omega, \tag{2.156}$$

$$\int\limits_{\Omega} q_3(\xi,\eta)\frac{d\Omega}{R} = \int\limits_{\Omega} [a_0 q_2(\xi,\eta) + a_1 q_0(\xi,\eta)R^2]\, d\Omega, \tag{2.157}$$

$$\int\limits_{\Omega} q_4(\xi,\eta)\frac{d\Omega}{R} = \int\limits_{\Omega} [a_0 q_3(\xi,\eta) + a_1 q_1(\xi,\eta)R^2]\, d\Omega, \tag{2.158}$$

etc.

Equation (2.154) coincides with (1.29) and corresponds the impression of a punch into a half-space. Let us suppose that we know the solution of this integral equation for an arbitrary right-hand side, i.e., we know the inverse integral linear operator $\mathcal{A}$ which transforms the function $\delta(x,y)$ into the function $q_0(x,y)$. Precisely, we have

$$q_0(x,y) = \theta\mathcal{A}(\tilde{\delta}) \qquad (\tilde{\delta} = \delta(x,y)/a);$$

then we successively determine q_1,..., q_4 from the integral equations (2.155)–(2.158). As a result, the asymptotic solution for large λ of equation (2.152) can be presented in the form

$$\begin{aligned}\frac{q(x,y)}{\theta} &= \mathcal{A}(\tilde{\delta}) + \frac{\mathcal{A}(1)}{2\pi\lambda}\Bigg\langle a_0\mathcal{F}_{00}(\tilde{\delta})\Bigg[1 \\ &+\frac{a_0\mathcal{F}_{00}(1)}{2\pi\lambda} + \left(\frac{a_0\mathcal{F}_{00}(1)}{2\pi\lambda}\right)^2 + \left(\frac{a_0\mathcal{F}_{00}(1)}{2\pi\lambda}\right)^3\Bigg] \\ &+\frac{a_1}{\lambda^2}\left\{[\mathcal{F}_{20}(\tilde{\delta}) + \mathcal{F}_{02}(\tilde{\delta})]\left(1+\frac{a_0\mathcal{F}_{00}(1)}{2\pi\lambda}\right) + \frac{a_0\mathcal{F}_{00}(\tilde{\delta})}{2\pi\lambda}[\mathcal{F}_{20}(1)+\mathcal{F}_{02}(1)]\right\} \\ &+\frac{a_0 a_1}{2\pi\lambda^3}[-2\mathcal{F}_{10}(\tilde{\delta})\mathcal{F}_{00}(\tilde{x}) - 2\mathcal{F}_{01}(\tilde{\delta})\mathcal{F}_{00}(\tilde{y}) + \mathcal{F}_{00}(\tilde{\delta})\mathcal{F}_{00}(\tilde{x}^2+\tilde{y}^2)]\Bigg\rangle \\ &-\frac{a_1}{\pi\lambda^3}\left\{\mathcal{F}_{10}(\tilde{\delta})\mathcal{A}(\tilde{x}) + \mathcal{F}_{01}(\tilde{\delta})\mathcal{A}(\tilde{y}) + \frac{a_0\mathcal{F}_{00}(\tilde{\delta})}{2\pi\lambda}[\mathcal{F}_{10}(1)\mathcal{A}(\tilde{x}) + \mathcal{F}_{01}(1)\mathcal{A}(\tilde{y})]\right\} \\ &+\frac{a_1\mathcal{F}_{00}(\tilde{\delta})\mathcal{A}(\tilde{x}^2+\tilde{y}^2)}{2\pi\lambda^3}\left(1+\frac{a_0\mathcal{F}_{00}(1)}{2\pi\lambda}\right) + O\left(\frac{1}{\lambda^5}\right).\end{aligned} \tag{2.159}$$

Here, we denote $\tilde{x} = x/a$, $\tilde{y} = y/a$,

$$\mathcal{F}_{ij}(f) = \frac{1}{a^{i+j+2}}\int\limits_{\Omega} \mathcal{A}(f)\xi^i\eta^j\, d\Omega.$$

2.5.2. AN ELLIPTIC PLANAR SLOPING PUNCH

Let us consider the particular case when the contact region Ω is an ellipse with semiaxes a and b, and the function $\delta(x,y) = \delta + \alpha x + \beta y$ (the case of a planar sloping punch). On the basis of formula (1.56), we have

$$\mathcal{A}(1) = \frac{a}{\mathcal{K}b}\left(1 - \frac{x^2}{a^2} - \frac{y^2}{b^2}\right)^{-1/2}, \tag{2.160}$$

$$\mathcal{A}(\tilde{x}) = \frac{e^2 x}{\psi_\alpha b}\left(1-\frac{x^2}{a^2}-\frac{y^2}{b^2}\right)^{-1/2}, \quad \mathcal{A}(\tilde{y}) = \frac{e^2 y}{\psi_\beta b}\left(1-\frac{x^2}{a^2}-\frac{y^2}{b^2}\right)^{-1/2}, \tag{2.161}$$

$$\mathcal{A}(\tilde{x}^2 + \tilde{y}^2) = \frac{a}{\psi \mathcal{K} b}\left(1 - \frac{x^2}{a^2} - \frac{y^2}{b^2}\right)^{-1/2}\biggl\{-2[\psi_1 + (1-e^2)\psi_2] \tag{2.162}$$

$$+\frac{4\mathcal{K}x^2}{a^2}[\psi_{11} - (1-e^2)\psi_{10}] + \frac{4\mathcal{K}y^2}{a^2}(\psi_{00} - \psi_{01})\biggr\},$$

where $\mathcal{K} = \mathcal{K}(e)$ is the complete elliptic integral of the first kind. By using integral (1.55), we find

$$\mathcal{F}_{00}(1) = \frac{2\pi}{\mathcal{K}}, \qquad \mathcal{F}_{00}(\tilde{x}^2 + \tilde{y}^2) = \frac{2\pi(2-e^2)}{3\mathcal{K}}, \tag{2.163}$$

$$\mathcal{F}_{10}(\tilde{x}) = \frac{2\pi e^2}{3\psi_\alpha}, \qquad \mathcal{F}_{01}(\tilde{y}) = \frac{2\pi e^2(1-e^2)}{3\psi_\beta}, \tag{2.164}$$

$$\mathcal{F}_{20}(1) = \frac{2\pi}{3\mathcal{K}}, \qquad \mathcal{F}_{02}(1) = \frac{2\pi(1-e^2)}{3\mathcal{K}}, \tag{2.165}$$

moreover,

$$\mathcal{F}_{10}(1) = \mathcal{F}_{01}(1) = 0, \tag{2.166}$$

$$\mathcal{F}_{00}(\tilde{x}) = \mathcal{F}_{01}(\tilde{x}) = \mathcal{F}_{20}(\tilde{x}) = \mathcal{F}_{02}(\tilde{x}) = 0, \tag{2.167}$$

$$\mathcal{F}_{00}(\tilde{y}) = \mathcal{F}_{10}(\tilde{y}) = \mathcal{F}_{20}(\tilde{y}) = \mathcal{F}_{02}(\tilde{y}) = 0, \tag{2.168}$$

where the expression for $\mathcal{F}_{00}(\tilde{x}^2 + \tilde{y}^2)$ has been simplified with help of the identity

$$-2[\psi_1 + (1-e^2)\psi_2] + \frac{4}{3}\mathcal{K}\{[\psi_{11} - (1-e^2)\psi_{10}] \tag{2.169}$$

$$+(1-e^2)(\psi_{00} - \psi_{01})\} = \frac{1}{3}(2-e^2)\psi.$$

By substituting (2.160)–(2.168) into (2.159), we obtain the following asymptotic solution, for large λ, of the problem on impression of the elliptic planar sloping punch into the layer:

$$q(x,y) = \frac{\theta}{b}\left(1 - \frac{x^2}{a^2} - \frac{y^2}{b^2}\right)^{-1/2} \left\{\frac{\delta}{\mathcal{K}}\left[R(\lambda) + S\left(\lambda, \frac{x^2}{a^2}, \frac{y^2}{b^2}\right)\right]\right. \tag{2.170}$$

$$\left. + \frac{\alpha x e^2}{\psi_\alpha} T_\alpha(\lambda) + \frac{\beta y e^2}{\psi_\beta} T_\beta(\lambda)\right\},$$

$$R(\lambda) = 1 + \frac{a_0}{\mathcal{K}\lambda} + \left(\frac{a_0}{\mathcal{K}\lambda}\right)^2 + \left(\frac{a_0}{\mathcal{K}\lambda}\right)^3 + \left(\frac{a_0}{\mathcal{K}\lambda}\right)^4$$

$$+ \frac{2a_1}{3\mathcal{K}\lambda^3}(2 - e^2)\left(1 + \frac{2a_0}{\mathcal{K}\lambda}\right) + O\left(\frac{1}{\lambda^5}\right),$$

$$S\left(\lambda, \frac{x^2}{a^2}, \frac{y^2}{b^2}\right) = -\frac{a_1}{3\mathcal{K}\lambda^3}(2 - e^2)\left(1 + \frac{a_0}{\mathcal{K}\lambda}\right) + \frac{2a_1}{\psi\mathcal{K}\lambda^3}\left(1 + \frac{a_0}{\mathcal{K}\lambda}\right)$$

$$\times\left\{-[\psi_1 + (1 - e^2)\psi_2] + \frac{2\mathcal{K}x^2}{a^2}[\psi_{11} + (1 - e^2)\psi_{10}]\right.$$

$$\left. + \frac{2\mathcal{K}y^2}{b^2}(1 - e^2)(\psi_{00} - \psi_{01})\right\} + O\left(\frac{1}{\lambda^5}\right),$$

$$T_\alpha(\lambda) = 1 - \frac{2e^2 a_1}{3\psi_\alpha\lambda^3} + O\left(\frac{1}{\lambda^5}\right),$$

$$T_\beta(\lambda) = 1 - \frac{2e^2(1 - e^2)a_1}{3\psi_\beta\lambda^3} + O\left(\frac{1}{\lambda^5}\right).$$

Using formulae (1.30), (1.31) and the integral (1.55), we find that

$$P = \frac{2\pi\theta\delta a}{\mathcal{K}}R(\lambda) = \theta\delta a\left[A + \frac{B}{\lambda} + \frac{C}{\lambda^2} + \frac{D}{\lambda^3} + \frac{E}{\lambda^4} + O\left(\frac{1}{\lambda^5}\right)\right], \tag{2.171}$$

$$M_x = Pm_x = \frac{2\pi\theta\beta a^3 e^2(1 - e^2)}{3\psi_\beta}T_\beta(\lambda) = \theta\beta a^3\left[F + \frac{G}{\lambda^3} + O\left(\frac{1}{\lambda^5}\right)\right], \tag{2.172}$$

$$M_y = Pm_y = \frac{2\pi\theta\alpha a^3 e^2}{3\psi_\alpha}T_\alpha(\lambda) = \theta\alpha a^3\left[H + \frac{I}{\lambda^3} + O\left(\frac{1}{\lambda^5}\right)\right]. \tag{2.173}$$

Note that the formula for P has been also simplified by using the identity (2.169). Note also that, as $e \to 0$, formulae (2.170)–(2.173) reduce to formulae (2.87), (2.91) and (2.92).

The values of A, B, C, D, E, F, G, H, and I are given in Tables 2.4 and 2.5 for $\nu = 0.3$ and various values of e^2.

Table 2.4: Constants in (2.171)

e^2	A	B	C	D	E
0.0	4.0000	3.5062	3.0734	0.5632	−1.3742
0.1	3.8967	3.3275	2.8414	0.5052	−1.2091
0.2	3.7859	3.1410	2.6059	0.4440	−1.0570
0.3	3.6660	2.9452	2.3661	0.3794	−0.9175
0.4	3.5348	2.7381	2.1210	0.3117	−0.7898
0.5	3.3889	2.5167	1.8690	0.2408	−0.6730
0.6	3.2229	2.2762	1.6076	0.1670	−0.5659
0.7	3.0275	2.0086	1.3326	0.0907	−0.4663
0.8	2.7836	1.6980	1.0358	0.0127	−0.3700
0.85	2.6300	1.5158	0.8736	−0.0262	−0.3204
0.9	2.4371	1.3016	0.6952	−0.0638	−0.2664
0.95	2.1604	1.0228	0.4842	−0.0971	−0.2005
0.99	1.7002	0.6334	0.2360	−0.1065	−0.1121

The limits of λ for practical use of the *asymptotic* formulae (2.170)–(2.173) are determined from the condition that dropping the term $O(\lambda^{-4})$ in (2.171) changes the value of P not more than by 3%. So we find that formulae (2.170)–(2.173) can be used for $\lambda \geq \lambda^*$. The values of λ^* are given in the last column in Table 2.5 for different e^2.

Table 2.5: Constants in (2.172) and (2.173)

e^2	F	G	H	I	λ^*
0.0	2.6667	1.4206	2.6667	1.4206	1.58
0.1	2.3692	1.1214	2.5640	1.3133	1.54
0.2	2.0770	0.8618	2.4555	1.2045	1.50
0.3	1.7905	0.6405	2.3399	1.0938	1.47
0.4	1.5100	0.4555	2.2155	0.9806	1.43
0.5	1.2360	0.3052	2.0801	0.8644	1.39
0.6	0.9692	0.1877	1.9299	0.7441	1.35
0.7	0.7105	0.1008	1.7585	0.6178	1.32
0.8	0.4609	0.0424	1.5533	0.4820	1.28
0.85	0.3402	0.0231	1.4292	0.4081	1.26
0.9	0.2226	0.0099	1.2794	0.3270	1.24
0.95	0.1087	0.0024	1.0767	0.2316	1.21
0.99	0.0212	0.00014	0.7738	0.1196	1.13

Note that one can likewise obtain the asymptotic solution for large λ for an elliptic punch with parabolic base ($\delta(x, y) = \delta - x^2/(2R_1) - y^2/(2R_2)$). In this case, if we demand that the contact pressure $q(x, y)$ vanishes on the contour of the contact region Ω, then the contact region will be elliptic only with an accuracy up to the terms $O(\lambda^{-5})$, and we can obtain two relations which are similar to (1.66) and serve for determination of the semiaxes of Ω.

2.5.3. THE FORCE AND MOMENTS

For the case when the region Ω is an ellipse, relations (2.45), (2.46), due to formula (2.170), take the form

$$P = \frac{\theta}{\mathcal{K}b}\int\limits_{\Omega}\left[R(\lambda) + S\left(\lambda, \frac{\xi^2}{a^2}, \frac{\eta^2}{b^2}\right)\right]\left(1 - \frac{\xi^2}{a^2} - \frac{\eta^2}{b^2}\right)^{-1/2}\delta(\xi,\eta)\,d\Omega,$$

$$M_x = \frac{\theta e^2}{\psi_\beta}T_\beta(\lambda)\int\limits_{\Omega}\eta\left(1 - \frac{\xi^2}{a^2} - \frac{\eta^2}{b^2}\right)^{-1/2}\delta(\xi,\eta)\,d\Omega,$$

$$M_y = \frac{\theta e^2}{\psi_\alpha}T_\alpha(\lambda)\int\limits_{\Omega}\xi\left(1 - \frac{\xi^2}{a^2} - \frac{\eta^2}{b^2}\right)^{-1/2}\delta(\xi,\eta)\,d\Omega,$$

and allow us to determine the force P and the moments M_x, M_y acting onto an arbitrary non-planar elliptic punch embedded into the layer with relative thickness λ.

Let us also consider the question on accuracy of bilateral estimates (2.53)–(2.55) for the case of the elliptic planar sloping punch. On the basis of (1.58), we find that

$$J_0(1) = \frac{2\pi a\theta}{\mathcal{K}}, \tag{2.174}$$

$$J_1(x^2) = \frac{2\pi a^3 e^2\theta}{3\psi_\alpha}, \qquad J_2(y^2) = \frac{2\pi ab^2 e^2\theta}{3\psi_\beta}. \tag{2.175}$$

By substituting (2.174), (2.175) into formulae (2.53)–(2.55), we will have the estimates

$$\frac{2\pi a\theta\delta}{\mathcal{K} - \lambda^{-1}\min\limits_{\Omega} F_0} \le P \le \frac{2\pi a\theta\delta}{\mathcal{K} - \lambda^{-1}\max\limits_{\Omega} F_0}, \tag{2.176}$$

$$\frac{2\pi a^3\theta|\alpha|}{3\psi_\alpha e^{-2} - \lambda^{-3}\min\limits_{\Omega} F_1} \le |M_y| \le \frac{2\pi a^3\theta|\alpha|}{3\psi_\alpha e^{-2} - \lambda^{-3}\max\limits_{\Omega} F_1}, \tag{2.177}$$

$$\frac{2\pi(1-e^2)a^3\theta|\beta|}{3\psi_\beta e^{-2} - \lambda^{-3}(1-e^2)\min\limits_{\Omega} F_2} \le |M_x| \le \frac{2\pi(1-e^2)a^3\theta|\beta|}{3\psi_\beta e^{-2} - \lambda^{-3}(1-e^2)\max\limits_{\Omega} F_2}. \tag{2.178}$$

It is easy to check using (2.51), (2.52) that

$$\min\limits_{\Omega} F_0 = a_0 + \frac{a_1(2-e^2)}{\lambda^2} + \frac{a_2(2-e^2)^2}{\lambda^4} + \frac{a_3(2-e^2)^3}{\lambda^6}$$

$$+\frac{a_4(2-e^2)^4}{\lambda^8}+O\left(\frac{1}{\lambda^{10}}\right), \qquad \max_{\Omega} F_0 = a_0,$$

$$\min_{\Omega} F_1 = \min_{\Omega} F_2 = -2\left[a_1 + \frac{2a_2(2-e^2)}{\lambda^2} + \frac{3a_3(2-e^2)^2}{\lambda^4}\right.$$

$$\left.+\frac{4a^4(2-e^2)^3}{\lambda^6}+O\left(\frac{1}{\lambda^8}\right)\right], \qquad \max_{\Omega} F_1 = \max_{\Omega} F_2 = -2a_1,$$

where the constants a_i must be taken from Table 2.2.

Table 2.6: Constants in (2.179)

e^2	0.2	0.2	0.7	0.7	0.95	0.95
λ	2	4	2	4	2	4
κ_δ	6.12	4.75	4.39	3.62	2.78	2.45
κ_δ	6.00	4.74	4.35	3.61	2.77	2.44
κ_β	2.17	2.09	0.722	0.712	0.109	0.109
κ_β	2.18	2.09	0.723	0.712	0.109	0.109
κ_α	2.59	2.47	1.83	1.77	1.10	1.08
κ_α	2.61	2.47	1.84	1.77	1.11	1.08

On the basis of (2.176)–(2.178), mean values of the force P and of the moments M_x, M_y can be presented in the form

$$P = a\theta\delta\kappa_\delta, \quad M_y = a^3\theta\alpha\kappa_\alpha, \quad M_x = a^3\theta\beta\kappa_\beta. \tag{2.179}$$

The quantities κ_δ, κ_α, and κ_β are given in Table 2.6 for $\nu = 0.3$ and certain values of e^2. The second lines of the table for κ_δ, κ_α, and κ_β correspond to the almost exact values obtained by formulae (2.171)–(2.173). One can see that, as $\lambda \geq 2$, the maximal difference does not exceed 2%. The calculation of the quantity κ_δ with the help of the estimate (2.49) gives a little worse result.

2.6 Contact Problems for a Thin Layer

2.6.1. THE CONFLUENT SOLUTION

The case in which the value of λ is small ($\lambda < 1$), does not generally correspond to small relative thickness of the layer. One must suppose that parameter μ, which has been introduced in subsection 2.2.1, is small. Let us assume for simplicity that the contact region Ω is convex (the non-convex case is more complicated and was considered in detail in Aleksandrov (1969)); then we have $\mu = h/\rho_0$ where ρ_0 is the minimal radius of curvature of the contour $\partial\Omega$[4], and h is the thickness of the layer.

Let us introduce the auxiliary domains Ω_0 and Ω_ε whose points lie inside the domain Ω and are distant from $\partial\Omega$ along the normal not less than by $\rho_0(1-0)$ and $\rho_0(1-\varepsilon)$, $\varepsilon > 0$, respectively (see Fig. 2.5).

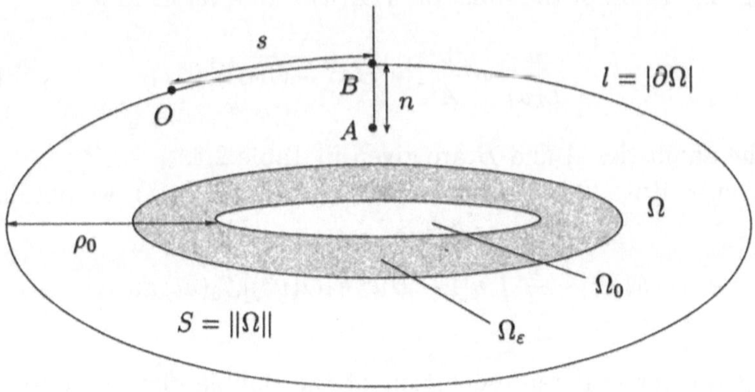

Figure 2.5: Domains $\Omega_0 \subset \Omega_\varepsilon \subset \Omega$

Let us consider the integral equation (2.27) and note that for the all points $(x, y) \in \Omega_\varepsilon$, as $\mu \to 0$ while h is fixed, the boundary of the domain Ω goes to infinity, i.e., under the indicated conditions one can reckon that the domain Ω expands into the whole plane. This fact allows us, if the value of μ is small, to construct the following internal (or *confluent*) solution of equation (2.27) (Aleksandrov (1963,1964,1969,1993)):

$$q(x,y) = \frac{\theta}{2\pi h^3}\left\{\int_\Omega \delta(\xi,\eta) M\left(\frac{R}{h}\right) d\Omega + O\left[\frac{1}{\sqrt{\mu}} \exp\left(-\frac{1-\varepsilon}{\mu}C\right)\right]\right\}, \tag{2.180}$$

$$(x,y) \in \Omega_\varepsilon, \qquad M(t) = \int_0^\infty \frac{u^2}{L(u)} \mathcal{J}_0(ut)\, du, \tag{2.181}$$

[4]For example, if $\partial\Omega$ is an ellipse $x^2/a^2 + y^2/b^2 = 1$, $a > b$, then $\rho_0 = b^2/a$ at the points $(\pm a; 0)$.

where the values of C are presented in Table 2.1, $\mathcal{J}_0(x)$ is the Bessel function, and the integral in the expression for $M(t)$ must be understood as in the impulse function theory.

The internal solution (2.180) gives a more accurate result for $q(x, y)$, if the point (x, y) is distant from the boundary of Ω. Let us represent this solution in another form which is more convenient than (2.180).

By substituting the variable $u = \beta/t$ into the integral in (2.180), we see that to determine the asymptotics of $M(t)$ for large t, one must use the following expansion of the function $u/L(u)$ into a series in u^2:

$$\frac{u}{L(u)} = \frac{1}{A}[1 - Bu^2 + O(u^4)], \tag{2.182}$$

where the quantities A and B are given in Table 2.1.

If we substitute (2.182) into formula (2.181) for $M(t)$, we obtain

$$M(t) \sim \frac{1}{A}\int_0^\infty u[1 - Bu^2 + O(u^4)]\mathcal{J}_0(ut)\,du.$$

By using the integral representation of the double delta-function in the form[5]

$$\int_0^\infty u\mathcal{J}_0(ut)\,du = 2\pi\delta_*(t), \tag{2.183}$$

and the differential identity in the form

$$\Delta_{s,\sigma}\mathcal{J}_0(ut) = -u^2\mathcal{J}_0(ut)$$

$$\left(t = \sqrt{s^2+\sigma^2}, \quad s = \frac{x-\xi}{h}, \quad \sigma = \frac{y-\eta}{h}, \quad \Delta_{s,\sigma} = \frac{\partial^2}{\partial s^2} + \frac{\partial^2}{\partial \sigma^2}\right),$$

[5]To prove (2.183), consider the integral 6.623.2 in Gradshteyn and Ryzhik (1994)

$$\int_0^\infty u\exp(-au)\mathcal{J}_0(ut)\,du = \frac{a}{(a^2+t^2)^{3/2}} = 2\pi\delta_*(t,a). \tag{$*$}$$

Integrating the function $\delta_*(t,a)$ over the whole plane by cylindrical coordinates t and φ, we get

$$\int_0^{2\pi} d\varphi \int_0^\infty t\delta_*(t,a)dt = \int_0^\infty \frac{at\,dt}{(a^2+t^2)^{3/2}} = 1. \tag{$**$}$$

By passing to the limit $a \to 0$ in $(*)$ and $(**)$, we get formula (2.183).

we find

$$M(t) \sim \frac{2\pi}{A}[\delta_*(t) + B\Delta_{s,\sigma}\delta_*(t) - ...]. \tag{2.184}$$

Now, by substituting expression (2.184) into (2.180) and taking account of the following well-known property of the double delta-function

$$\int_{-\infty}^{\infty}\int_{-\infty}^{\infty} \varphi(\xi', \eta')\delta_*(t)\, d\xi'\, d\eta' = \varphi(x', y')$$

$$(t = \sqrt{(x' - \xi')^2 + (y' - \eta')^2}),$$

we obtain the new form of the internal solution

$$q(x, y) \sim \theta(hA)^{-1}[1 + Bh^2\Delta_{x,y} - ...]\delta(x, y). \tag{2.185}$$

For the plane inclined punch ($\delta(x, y) = \delta + \alpha x + \beta y$), we find from (2.185) that

$$q(x, y) \sim \theta(hA)^{-1}(\delta + \alpha x + \beta y), \tag{2.186}$$

i.e., there is a direct proportional dependence between the settlement of the punch and the contact pressure as for the Winkler foundation (Timoshenko (1951)). For the parabolic punch ($\delta(x, y) = \delta - x^2/(2R_1) - y^2/(2R_2)$), we obtain from (2.185) that

$$q(x, y) \sim \frac{\theta}{hA}\left(\delta_0 - \frac{x^2}{2R_1} - \frac{y^2}{2R_2}\right), \quad \delta_0 = \delta - Bh^2\left(\frac{1}{R_1} + \frac{1}{R_2}\right). \tag{2.187}$$

If the contact domain Ω is unknown, then one should set the supplementary condition that $q(x, y) = 0$ on $\partial\Omega$. Then on the basis of (2.187) we find that the contour of the domain Ω is an ellipse with the semiaxes $a = \sqrt{2R_1\delta_0}$ and $b = \sqrt{2R_2\delta_0}$. By using formulae (1.30) and (2.187), we find the following dependence between the impressed force P and the settlement δ of the punch:

$$P \sim \frac{\pi ab\theta}{hA}\left(\delta_0 - \frac{a^2}{8R_1} - \frac{b^2}{8R_2}\right). \tag{2.188}$$

Formulae (2.186)–(2.188) are derived neglecting the boundary layer solution of the integral equation (2.27) in the domain $\Omega\backslash\Omega_0$, therefore, these formulae can be used only for very small values of μ ($\mu < 1/4$).

2.6.2. THE BOUNDARY LAYER SOLUTION

To construct the *boundary layer solution*, we rewrite equation (2.27) in the form

$$\int_{\Omega\backslash\Omega_0} q(\xi,\eta)K\left(\frac{R}{h}\right)\,d\Omega + \int_{\Omega_0} q(\xi,\eta)K\left(\frac{R}{h}\right)\,d\Omega = 2\pi\theta h\delta(x,y), \qquad (2.189)$$

and consider the points $(x,y) \in \Omega\backslash\Omega_\varepsilon$. Then for the last integral in (2.189) one can obtain the following estimate:

$$\int_{\Omega_0} q(\xi,\eta)K\left(\frac{R}{h}\right)\,d\Omega = O\left[\sqrt{\mu}\exp\left(-\frac{\varepsilon}{\mu}D\right)\right], \qquad (2.190)$$

where the values of D are presented in Table 2.1.

Let us draw the normal to the contour $\partial\Omega$ from the point $A(x,y) \in \Omega\backslash\Omega_0$, let n be the length of this normal, and $B(x_1,y_1)$ be its point of intersection with the contour as shown in Fig 2.6. We choose the certain point $O(x_0,y_0)$ on the contour $\partial\Omega$ as the origin, and measure the distance s between the points O and B along the contour. We take the quantities n and s as the new coordinates of the point A in the curvilinear coordinate system (n,s). Under the condition $-l/2 < s \leq l/2$ (l is the perimeter of the contour $\partial\Omega$), there will be the one-to-one correspondence between the pair (x,y) in $\Omega\backslash\Omega_0$ and (n,s).

The integral equation (2.189), taking account of (2.190), can be written as follows

$$\int_0^{\rho_0} d\nu \int_{-l/2}^{l/2} \tilde{q}(\nu,\tau)K\left(\frac{R}{h}\right)\,d\tau + O\left[\sqrt{\mu}\exp\left(-\frac{\varepsilon}{\mu}D\right)\right] \qquad (2.191)$$

$$= 2\pi\theta h\tilde{\delta}(n,s), \qquad 0 \leq n < \rho_0, \quad |s| \leq l/2,$$

$$\tilde{q}(\nu,\tau) \equiv q(\xi,\eta), \qquad \tilde{\delta}(n,s) \equiv \delta(x,y).$$

We change the variables in equation (2.191) and introduce the notation

$$b = n/h, \qquad c = s/h, \qquad \beta = \nu/h, \qquad \gamma = \tau/h,$$

$$\tilde{q}(\nu,\tau) \equiv \varphi(\beta,\gamma), \qquad \tilde{\delta}(n,s) \equiv f(b,c), \qquad k = l/(2\rho_0).$$

Then we let μ tend to zero. Note that as μ tends to zero, the contour $\partial\Omega$ straightens into an infinite straight line in the coordinate system (n,s); at the same time, evidently, the belt domain $0 \le b < 1/\mu$, $|c| \le k/\mu$ expands into a plane and the coordinate system (n, s) degenerates into a rectangular one. Thus, as $\mu \to 0$, the integral equation (2.191) takes the form

$$\int_0^\infty d\beta \int_{-\infty}^\infty \varphi(\beta,\gamma) K\left(\sqrt{(\beta - b)^2 + (\gamma - c)^2}\right) d\gamma = \frac{2\pi\theta}{h} f(b, c), \qquad (2.192)$$

$$0 \le b < \infty, \qquad |c| < \infty.$$

One can see from here that for small μ the principal term in the asymptotics of the solution of the integral equation (2.191) does not depend on the curvature of the contour $\partial\Omega$.

The exact solution of (2.192) can be found by using the Fourier transform with respect to c with the following use of the Wiener–Hopf technique as in section 2.4 for the problem on the strip punch.

2.6.3. AN IMPORTANT SPECIAL CASE

Let us now confine ourselves by a consideration of the important particular case when $\tilde{\delta}(n, s) \equiv \tilde{\delta}(n)$ or $f(b, c) \equiv f(b)$. This includes the case of axial symmetry and also any punch with planar base ($\delta(x, y) \equiv \delta = \text{const}$). Now $\varphi(\beta, \gamma) \equiv \varphi(\beta)$ and the internal integral by γ in (2.192) can be calculated. As a result, we come to the one-dimensional integral equation

$$\int_0^\infty \varphi(\beta) K_*(\beta - b)\, d\beta = \frac{\pi\theta f(b)}{h} \qquad (0 \le b < \infty), \qquad (2.193)$$

where the function $K_*(\tilde{y})$ has the form

$$K_*(\tilde{y}) = \int_0^\infty \frac{L(u)}{u} \cos(u\tilde{y})\, du \qquad (2.194)$$

(cf. (2.193) and (2.194) with (2.97) and (2.71)).

Equations (2.193) and (2.194) correspond to the plane problem of a semi-infinite punch onto the strip as shown in Fig. 2.6 (Aleksandrov and Babeshko (1965)). Thus, for small μ and under the condition $\delta(x, y) \equiv f(b)$, the principal term in the asymptotics of the solution of integral equation (2.191) (or, equivalently, of integral equation (2.27) in the domain $\Omega \backslash \Omega_0$), takes the form of a plane boundary layer determined by the Wiener–Hopf equation (2.193). For small μ, the boundaries of applicability of the asymptotic relation

$$q(x,y) \sim \varphi(b), \qquad (x,y) \in \Omega\backslash\Omega_0 \tag{2.195}$$

depend on the thickness of the boundary layer. The function $\varphi(b)$ always has the form $\varphi(b) = \varphi^*(b) + \varphi_*(b)$, where the first term does not decrease when $b \to 0$ while the second term exponentially decreases when $b \to 0$. For instance, if the function $f(b)$ in (2.193) is a polynomial, then the function $\varphi^*(b)$ is a polynomial of the same degree. In other words, $\varphi^*(b)$ is the principal part of the function $\varphi(b)$ as $b \to \infty$.

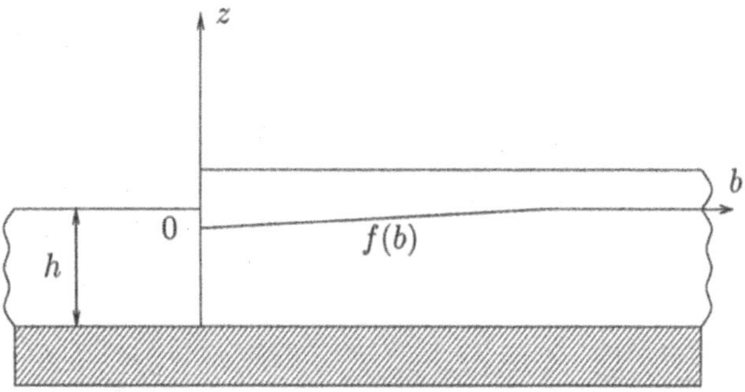

Figure 2.6: A semi-infinite punch on an elastic strip in 2-D

The relative thickness H of the boundary layer is determined by the expression

$$H = b_0, \qquad |\varphi(b_0) - \varphi^*(b_0)| \cdot |\varphi^*(b_0)|^{-1} = 0.03. \tag{2.196}$$

Formula (2.195) holds for the boundary layer thickness determined by (2.196). Clearly, if the boundary layer is correctly constructed, then the value $\varphi^*(b_0)$ must asymptotically coincide with the corresponding value of the internal solution determined by formula (2.180) or (2.185).

Now, we can easily outline the bound of applicability of the boundary layer solution $\varphi(b)$ of the integral equation (2.193). Precisely, in order that the boundary layers, which adjoin each point of the contour $\partial\Omega$, not intersect or lie on each other, it is necessary and sufficient for these layers to be packed by "thickness" in the domain $\Omega\backslash\Omega_0$, where we have the one-to-one correspondence between the coordinates x, y and n, s. The boundary layer solution is a line orthogonal to the boundary $\partial\Omega$ reaching the boundary

$\partial\Omega_0$ as shown in Fig. 2.7. We obtain from this the following restriction for the range of μ: $b_0 h < \rho_0$ or $\mu < H^{-1}$.

Thus, the function $\varphi(b)$ gives the principal part in the asymptotics of the solution, for small μ, of the integral equation (2.27) in the domain $\Omega\backslash\Omega_0$. Formula (2.180) or (2.185) gives the solution of (2.27), for small μ, in $\Omega_\varepsilon \supset \Omega_0$. On the boundary of the domain Ω_ε both the solutions come asymptotically close. Thus, for small μ, we have a complex of approximate solutions of the integral equation (2.27) in the whole domain Ω.

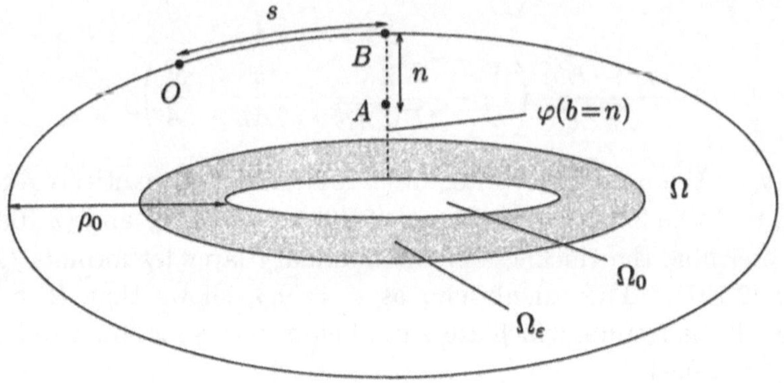

Figure 2.7: The boundary layer solution $\varphi(b)$

For small μ, calculation of the contact pressure at a fixed point $A(x,y)$ in the domain Ω can be reduced to the following procedure. If the point $A \in \Omega_\varepsilon$, then the calculation must be made by the internal (confluent) solution; if $A \in \Omega\backslash\Omega_\varepsilon$, then it is necessary to draw the perpendicular from this point to the contour $\partial\Omega$ (Fig. 2.7), to measure the length n of the normal, and then to calculate the contact stress in accordance with (2.195). Note that for axial symmetry, the domain Ω_ε is a circle of radius ε around 0.

2.6.4. EXAMPLES

To construct the boundary layer solution, we use the approximation (2.32) for the function $L(u)$. By solving the integral equation (2.193) with (2.194) using the Wiener–Hopf technique, we find that

1) if $\theta f(b)h^{-1} = 1$, then

$$\varphi(b) = \frac{1}{A}\Phi(\sqrt{Eb}) + \frac{\exp(-Eb)}{\sqrt{\pi A b}}, \tag{2.197}$$

2) if $\theta f(b)h^{-1} = b$, then

$$\varphi(b) = \frac{b}{A}\Phi(\sqrt{Eb}) - \frac{\exp(-Eb)}{\sqrt{\pi Eb}}\left(1 - \frac{b}{A} - \frac{1}{2\sqrt{AE}}\right), \tag{2.198}$$

3) if $\theta f(b)h^{-1} = b^2$, then

$$\varphi(b) = \left(\frac{b^2}{A} + \frac{1}{AE^2} - \frac{2}{E}\right)\Phi(\sqrt{Eb}) \tag{2.199}$$

$$-\frac{\exp(-Eb)}{\sqrt{\pi Eb}}\left(\frac{1}{E} - \frac{3}{4E\sqrt{AE}} + \frac{b}{2AE} - \frac{b^2}{A}\right),$$

where $\Phi(x)$ is the probability integral (2.101), and the quantities A and E are given in Table 2.1 (cf. (2.197) and (2.198) with (2.99) and (2.100)).

We determine the thickness of the boundary layer by formula (2.196) by using (2.197). The calculation, as $\nu = 0.3$, shows that $H \approx 1.06$, therefore, all the results, which are given below in this section, can be used as $\mu < H^{-1} \approx 0.94$.

Let us consider in detail the case of a planar punch with arbitrary shape. The internal solution in the domain Ω_ε, according to (2.186), takes the form

$$q(x,y) \sim \theta\delta(Ah)^{-1}. \tag{2.200}$$

The boundary layer solution in the domain $\Omega\backslash\Omega_0$ is presented by the relation

$$q(x,y) = \theta\delta h^{-1}\varphi(b), \tag{2.201}$$

where $\varphi(b)$ has the form (2.197), and it can be represented on the internal boundary of this domain as follows

$$q(x,y) = \theta\delta(Ah)^{-1} + O(\mu^{-1/2}\exp(-E\mu)). \tag{2.202}$$

By comparing formulae (2.200) and (2.202), one can see that on the boundary of Ω_0, asymptotically for small μ, the internal solution comes close to the boundary layer solution, providing the approximate determination of the contact stress in the whole domain Ω.

If Ω is an ellipse with the semiaxes a and b, then, evidently, as $y = 0$, the coordinate $n = a - x$, and as $x = 0$, the coordinate $n = b - y$. Thus, using formulae (2.200) and (2.201), we obtain the following approximate formulae for the contact pressures on the x- and y-axes in the contact region for a planar elliptic punch, for small μ:

$$q(x,0) = \frac{\theta\delta}{Ah}\left[\Phi\left(\sqrt{\frac{E(a-|x|)}{h}}\right) + \sqrt{\frac{Ah}{\pi(a-|x|)}}\exp\left(-\frac{E(a-|x|)}{h}\right)\right], \tag{2.203}$$

$$q(0,y) = \frac{\theta\delta}{Ah}\left[\Phi\left(\sqrt{\frac{E(b-|y|)}{h}}\right) + \sqrt{\frac{Ah}{\pi(b-|y|)}}\exp\left(-\frac{E(b-|y|)}{h}\right)\right]. \tag{2.204}$$

Taking into account that the minimal radius of curvature of the ellipse is $a(1-e^2)$ when $a > b$ (e is the eccentricity), we obtain the following relation between the parameters μ and λ:

$$\mu = \lambda(1-e^2)^{-1}. \tag{2.205}$$

Now, we can draw on the plane (e^2, λ) the boundaries of the domains in which the asymptotic solutions of the integral equation (2.27) are applicable for small λ (formulae (2.203), (2.204)) and for large λ (formula (2.170)), for the planar elliptic punch (see Fig. 2.8). We use the last column from Table 2.5 and relation (2.205) taking into account that $\mu < 0.94$.

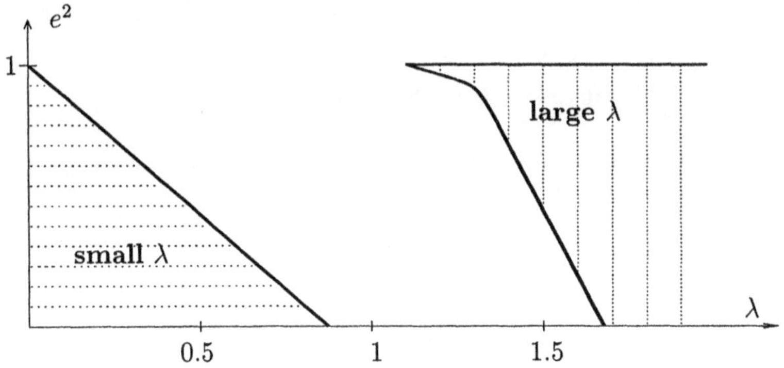

Figure 2.8: Ranges of applicability for the asymptotic solutions for the planar elliptic punch

2.6.5. DETERMINATION OF THE FORCE

We present the approximate determination of the force P acting on an arbitrary planar punch for small μ.

By using formulae (2.200) and (2.201), we present the asymptotic solution for small μ in the form

$$q(x,y) = \begin{cases} \theta\delta/(Ah) & \text{for } (x,y) \in \Omega_0 \\ \theta\delta(Ah)^{-1}[1+\psi(En/h)] & \text{for } (x,y) \in \Omega\backslash\Omega_0 \end{cases},$$

$$\psi(t) = \Phi(\sqrt{t}) - 1 + \sqrt{AE/(\pi t)}\,\exp(-t).$$

Then we find the following approximate formula for the force P with the help of formula (1.30):

$$P = \frac{\theta\delta}{Ah}\left[S + lh\int_0^{1/\mu} \psi(E\beta)\,d\beta\right], \tag{2.206}$$

where S is the area of Ω, and l is the perimeter of $\partial\Omega$ (Fig. 2.6). If we assume that asymptotically for small μ in (2.206) the upper limit of integration is equal to ∞, then, calculating this integral, we obtain the formula

$$P = \frac{\theta\delta}{Ah}(S + Dhl), \qquad D = \sqrt{\frac{A}{E}} - \frac{1}{2E}. \tag{2.207}$$

On the basis of Table 2.1 we have the following values of the constant D:

ν	0.25	0.30	0.35	0.40
D	0.1841	0.1238	0.006041	−0.2341

For a planar elliptic punch, the approximate expression (2.207) takes the form

$$P = \frac{\theta\delta a}{Ah}[\pi\sqrt{1-e^2} + 4D\lambda\,\mathcal{E}(e)],$$

where $\mathcal{E}(e)$ is the complete elliptic integral of the second kind.

We note that for a noncircular contact domain Ω, this asymptotic solution for small μ does not join that for large λ (see Fig. 2.8). Only for a circular contact domain do we have $\mu = \lambda$, and can show the closeness of the asymptotic solutions. We will do it in the following section.

2.7 A Circular Punch on a Layer (Examples)

2.7.1. A PLANAR PUNCH

Consider a planar punch: the function $\delta(x,y) = \delta(r) \equiv \delta = \text{const}$. For this case we have the asymptotic solution (2.86), (2.87) as $\lambda \geq 2$ and also two variants of the asypmtotic solutions as $\lambda \leq 2$, which are determined by formulae (2.98)–(2.104) and (2.74), (2.119)–(2.121), respectively. Moreover, for small λ we have on the basis of (2.197), (2.201) and (2.202), the formula

$$q(r) = \frac{\theta\delta}{Ah}\left[\Phi\left(\sqrt{\frac{E(a-r)}{h}}\right) + \sqrt{\frac{Ah}{\pi(a-r)}}\exp\left(-\frac{E(a-r)}{h}\right)\right]. \tag{2.208}$$

In accordance with the estimate obtained in the previous section, formula (2.208) can be used for $\lambda < 0.94$, although, calculations show that it is useful provided $\lambda \leq 1$.

From formula (2.208) we arrive at the relations

$$\lim \frac{1}{\theta\delta} q(r)\sqrt{a^2 - r^2} = \sqrt{\frac{2}{\pi A\lambda}} \qquad (r \to a), \tag{2.209}$$

$$\frac{P}{\theta\delta a} = \frac{\pi}{A\lambda}\left[\Phi\left(\sqrt{\frac{E}{\lambda}}\right)\left(1 - \frac{\lambda}{E} + \frac{3\lambda^2}{4E^2} + 2\lambda\sqrt{\frac{A}{E}} - \frac{\lambda^2}{E}\sqrt{\frac{A}{E}}\right)\right. \tag{2.210}$$
$$\left. + \sqrt{\frac{\lambda}{\pi E}}\exp(-E/\lambda)\left(1 - \frac{3\lambda}{2E} + 2\lambda\sqrt{\frac{A}{E}}\right)\right].$$

From the first formula of (2.207) for the circular punch we obtain the formula

$$\frac{P}{\theta\delta a} = \frac{\pi}{A\lambda}\left[1 + \lambda\left(2\sqrt{\frac{A}{E}} - \frac{1}{E}\right)\right]. \tag{2.211}$$

It is easy to see that the approximate formula (2.211) gives the first two terms of the asymptotics for small λ in (2.210). It confirms the good accuracy of the approximate formula (2.207) for P. Calculations show that the difference between the values obtained by formulae (2.211) and (2.210) as $\lambda \leq 1$ does not exceed 5%.

By using the method of orthogonal polynomials, which was briefly considered at the end of section 2.3, as $\lambda = 1$ and $\nu = 0.3$, we find for the case under consideration the following approximate solution:

$$q(r) = \frac{\theta\delta}{\sqrt{a^2 - r^2}}\left(2.183 - 1.048\frac{r^2}{a^2} + 0.181\frac{r^4}{a^4}\right). \tag{2.212}$$

Formula (2.212) is derived by means of reduction of the infinite algebraic system for a_i from expression (2.124), to three equations. Increasing the number of equations in the reduced system does not essentially improve the solution (2.212), therefore, this solution can be treated practically as the exact one.

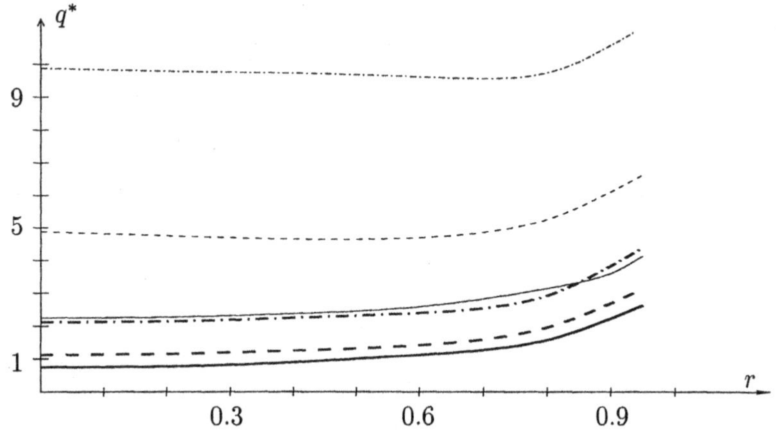

Figure 2.9: The plots $q^*(r)$ for $\lambda = 1/4$ (dotted-dashed, (2.208)); 1/2 (dashed, (2.208)); 1 (2.208); 1 (bold dotted-dashed, (2.212)); 2 (bold dashed, (2.86)); 4 (bold (2.86))

The values of the quantities

$$\lim \frac{1}{\theta\delta} q(r)\sqrt{a^2 - r^2} \quad (r \to a), \qquad \frac{P}{\theta\delta a}, \tag{2.213}$$

calculated by various formulae in the frames of their applicability for different λ and $\nu = 0.3$, are given in Table 2.7 and are numbered by I and II, respectively. The values from Ufliand (1965) are given in the last line of this table. The plots of $q^*(r) = q(r)a/\theta\delta$, calculated by various formulae for different λ and $\nu = 0.3$, are shown in Fig 2.9. One can see that, as λ diminishes, the contact pressure in the central part of the contact region becomes more plane. At the same time, as $\lambda > 1$, this pressure increases monotonically when r grows from 0 up to a. The nonmonotonicity appears as $\lambda \leq 1$ amplifying itself as λ decreases; to be exact, the stress at first decreases as r grows, and then it begins to increase. The point $r = r_*$, from which the stress begins to grow, shifts, as λ diminishes, to the point $r = a$.

Table 2.7: Characteristics (2.213)

	λ	4	2	1	1/2	1/4
I	(2.86)	0.801	0.974			
I	(2.103)		0.992	1.326	1.821	2.536
I	(2.120)		0.944	1.281	1.790	2.516
I	(2.209)			1.249	1.766	2.498
I	(2.212)			1.316		
II	(2.87)	5.07	6.51			
II	(2.104)		6.43	10.12	17.63	32.87
II	(2.121)		6.56	10.17	17.62	32.86
II	(2.210)			10.21	17.79	32.99
II	(2.212)			9.93		
II	Ufliand (1965)		6.48	9.92	17.36	

2.7.2. A PARABOLIC PUNCH

Consider the impression of a parabolic punch; the function $\delta(x, y) = \delta(r) = \delta - r^2/(2R)$ where R is the radius of curvature at the apex. Here, we have the asymptotic solution for $\lambda \geq 2$ defined by formulae (2.88)–(2.90).

From the previous section we obtain the asymptotic solution valid for $\lambda < 0.94$. The internal solution, in accordance with formulae (2.187), has the form

$$q(r) \sim \frac{\theta}{Ah}\left(\delta - \frac{2Bh^2}{R} - \frac{r^2}{2R}\right), \qquad r \in \Omega_\varepsilon, \tag{2.214}$$

and, taking into account that

$$\delta - \frac{r^2}{2R} = \delta - \frac{a^2}{2R} + \frac{ahb}{2R} - \frac{h^2b^2}{2R} \qquad \left(b = \frac{a-r}{h}\right),$$

we can present the solution of the boundary layer type, in accordance with formulae (2.195) and (2.197)–(2.199), in the form

$$q(r) \sim \frac{\theta}{Ah}\left\langle \Phi\left(\sqrt{\frac{E(a-r)}{h}}\right)\left[\delta - \frac{1}{2R}\left(r^2 + \frac{h^2}{E^2} - \frac{2Ah^2}{E}\right)\right]\right. \tag{2.215}$$

$$+\sqrt{\frac{Ah}{\pi(a-r)}}\exp\left(-E\,\frac{a-r}{h}\right)\left\{\delta - \frac{1}{2R}\left[a^2 - k(a^2-r^2)\right.\right.$$

$$\left.\left.\left. -\frac{hk}{2E}(a-r) + ha\left(2kA - \frac{1}{E}\right) + h^2\left(\frac{3}{4E^2} - \frac{kA}{E}\right)\right]\right\}\right\rangle,$$

$$k = (AE)^{-1/2}, \qquad r \in \Omega\backslash\Omega_0.$$

Note that due to relations (2.33), the solutions (2.214) and (2.215) converge asymptotically on the boundary of the domain Ω_ε when λ is small. Therefore, the asymptotic solution for the parabolic punch, as $\lambda < 0.94$, can be presented by the single analytical expression (2.215) valid for the whole contact domain Ω.

After some reduction, on the basis of (2.215), we find the asymptotic solution, vanishing on the contour $r = a$, for the parabolic punch for small λ in the form

$$q(r) = \frac{\theta}{2RAh}\left\langle \Phi\left(\sqrt{\frac{E(a-r)}{h}}\right)\left[a^2 - r^2 + h(a + hkA)\right.\right. \tag{2.216}$$

$$\left.\left. \times\left(2kA - \frac{1}{E}\right) - \frac{h^2}{4E^2}\right] + \sqrt{\frac{h(a-r)}{\pi E}}\exp\left(-E\frac{a-r}{h}\right)\left(a + r + \frac{h}{2E}\right)\right\rangle$$

under the condition

$$\delta = \frac{1}{2R}\left(a^2 + 2hakA - \frac{ha}{E} + \frac{3h^2}{4E^2} - \frac{h^2kA}{E}\right). \tag{2.217}$$

The force P for the case (2.216) is determined by formula (1.30):

$$P = \frac{\pi\theta a^3}{RA}\left\{\lambda kA\left(1 - \frac{\lambda}{2E} + \lambda kA\right)\left[\Phi\left(\sqrt{\frac{E}{\lambda}}\right)\left(\frac{1}{\lambda} - \frac{1}{E} + \frac{3\lambda}{4E^2}\right)\right.\right. \tag{2.218}$$

$$\left. + \frac{1}{\sqrt{\pi E\lambda}}\exp(-E/\lambda)\left(1 - \frac{3\lambda}{2E}\right)\right] + \Phi\left(\sqrt{\frac{E}{\lambda}}\right)\left(\frac{1}{4\lambda} - \frac{1}{2E} + \frac{5\lambda}{8E^2}\right.$$

$$\left.\left. - \frac{3\lambda^2}{8E^3} - \frac{15\lambda^3}{64E^4}\right) + \frac{1}{\sqrt{\pi E\lambda}}\exp(-E/\lambda)\left(\frac{1}{4} - \frac{5\lambda}{8E} + \frac{17\lambda^2}{16E^2} + \frac{15\lambda^3}{32E^3}\right)\right\},$$

and we also find the quantity $(r \to a)$

$$\psi = \frac{2R}{\theta}\lim q(r)(a^2 - r^2)^{-1/2} = 2\sqrt{\frac{2}{\pi\lambda A}}\left(1 + \lambda kA - \frac{\lambda}{2E}\right). \tag{2.219}$$

By using the method of orthogonal polynomials, we obtain the following approximate solution for the parabolic punch as $\lambda = 1$ and $\nu = 0.3$:

$$q(r) = \frac{\theta}{\sqrt{a^2 - r^2}}\left[\delta\left(2.183 - 1.048\frac{r^2}{a^2} + 0.181\frac{r^4}{a^4}\right)\right. \tag{2.220}$$

$$\left. + \frac{a^2}{2R}\left(0.455 - 2.062\frac{r^2}{a^2} - 0.112\frac{r^4}{a^4}\right)\right].$$

Because of (2.220), we find the solution bounded on $r = a$, in the form

$$q(r) = \frac{\theta}{2R}\sqrt{a^2 - r^2}\left(3.305 - 0.124\frac{r^2}{a^2}\right), \qquad \delta = 1.306\frac{a^2}{2R}. \tag{2.221}$$

The quantities

$$\frac{2Rq(0)}{\theta a}, \qquad \psi, \qquad \frac{2RP}{\theta a^3}, \qquad \frac{P}{\theta\delta a}, \tag{2.222}$$

calculated using the various formulae in the ranges of their applicability for different values of λ as $\nu = 0.3$, are given in the columns in Table 2.8 numbered by I, II, III and IV, respectively.

Table 2.8: Characteristics (2.222)

		I	II	III	IV
(2.88)–(2.90)	$\lambda = 4$	2.568	2.568	5.38	3.12
(2.88)–(2.90)	$\lambda = 2$	2.716	2.716	5.69	3.70
(2.221)	$\lambda = 1$	3.305	3.181	6.82	5.22
(2.216)–(2.219)	$\lambda = 1/2$	5.465	3.751	10.25	8.81
(2.216)–(2.219)	$\lambda = 1/4$	10.342	5.150	17.66	16.48

2.7.3. A PLANAR SLOPING PUNCH

Now consider the planar inclined (sloping) punch; the function $\delta(x, y) = \delta + \alpha r \cos\varphi$. By using the superposition principle and taking into account that for $\delta(x, y) = \delta$ the problem was investigated above, we consider the additional term $\delta(x, y) = \delta(r, \varphi) = \alpha r \cos\varphi$. Here, we have the asymptotic solution determined by formulae (2.91)–(2.92), valid for $\lambda \geq 2$.

Because of the theorem on the connection between the harmonics of the solution for the circular punch, which was presented in subsection 2.3.1, and due to formula (2.216), we can write the asymptotic solution, for $\lambda < 0.94$, in the form

$$q(r, \varphi) = \frac{\theta\alpha\cos\varphi}{A\lambda}\left\{r\Phi\left(\sqrt{\frac{E(a-r)}{h}}\right) + \sqrt{\frac{h}{\pi E(a-r)}}\right. \tag{2.223}$$

$$\left.\times \exp\left(-E\frac{a-r}{h}\right)\left[\frac{1}{k}\left(a - \frac{h}{2E} + hkA\right) - a + r\right]\right\}.$$

We find from (2.223) that

$$\psi = \frac{1}{\theta\alpha a\cos\varphi}\lim q(r, \varphi)\sqrt{a^2 - r^2} \tag{2.224}$$

$$= \sqrt{\frac{2}{\pi A\lambda}}\left(1 - \frac{\lambda}{2E} + \lambda k A\right) \quad (r \to a),$$

$$\chi = \frac{1}{\theta\alpha\cos\varphi}\lim q'_r(r,\varphi) = \frac{1}{A\lambda}\left[\Phi\left(\sqrt{\frac{E}{\lambda}}\right)\right. \tag{2.225}$$

$$\left. + \sqrt{\frac{\lambda}{\pi E}}\exp(-E/\lambda)\left(\frac{1}{2} - \frac{E}{\lambda} + \frac{E}{\lambda k} - \frac{1}{2k} + EA\right)\right] \quad (r \to 0).$$

The expression for the moment M coincides with (2.218), in which one should replace $1/R$ by α.

Note that the approximate solution (2.223) is defective at $r = 0$; the function $q(r,\varphi)$ is not equal to zero at $r = 0$ (cf. with (2.91)). This is the implication of the fact that the derivative at zero of the function $q(r)$ of the form (2.216) is not equal to zero. The point $r = 0$ is a conic point[6] of the function $q(r)$, but, according to the general information about the solution given in section 2.2, this solution should be as smooth as desired in the vicinity of $r = 0$.

This defect of approximate solutions (2.216) and (2.223) disappears asymptotically as $\lambda \to 0$ and it does not practically influence the correctness of the determination of the main characteristics of the solution.

By using the theorem from section 2.3 and the first formula of (2.221), we find the following approximate solution for the term $\delta(r,\varphi) = \alpha r\cos\varphi$ as $\lambda = 1$ and $\nu = 0.3$:

$$q(r,\varphi) = \frac{\theta\alpha r\cos\varphi}{\sqrt{a^2 - r^2}}\left(1.776 - 0.186\frac{r^2}{a^2}\right). \tag{2.226}$$

Table 2.9: Characteristics (2.227)

		I	II	III	IV
(2.91),(2.92)	$\lambda = 4$	1.284	1.284	2.69	
(2.91),(2.92)	$\lambda = 2$	1.358	1.358	2.84	2.82
(2.226)	$\lambda = 1$	1.776	1.590	3.41	3.38
(2.218),(2.223)–(2.225)	$\lambda = 1/2$	4.588	1.876	5.12	5.01
(2.218),(2.223)–(2.225)	$\lambda = 1/4$	9.676	2.575	8.83	

The quantities

$$\chi, \qquad \psi, \qquad M/(\theta\alpha), \tag{2.227}$$

[6]The point $r = 0$ is conic for the function $z = q(r,\phi)$ (r, ϕ, z are cylindrical coordinates, the origin is at the conic point), if in the small vicinity of this point, $q(r,\phi) \approx c_1(\phi)r + c_2(\phi)$.

calculated using the various formulae in the ranges of their applicability for different values of λ as $\nu = 0.3$, are given in the columns in Table 2.9 numbered by I, II and III, respectively. The values of $M/(\theta\alpha)$ from Ufliand (1965) are presented in the last column numbered by IV.

2.8 A Narrow Rectangular Punch on a Half-Space

2.8.1. BASIC ASSUMPTIONS

Suppose the contact region Ω is a rectangle: $|x| \leq a$, $|y| \leq \delta$. We use the integral equation (2.125) with kernel (2.19). For a half-space, we must set $L(u) = 1$ in (2.19), then equation (2.125) for the rectangular punch takes the form

$$\int_{-a}^{a} d\xi \int_{-\delta}^{\delta} q(\xi,\eta)\, d\eta \int_0^\infty \int_0^\infty \frac{\cos[\alpha(x-\xi)]\,\cos[\beta(y-\eta)]}{\sqrt{\alpha^2+\beta^2}}\, d\alpha\, d\beta \tag{2.228}$$

$$= \pi^2\theta\delta(x,y), \qquad |x| \leq a, \quad |y| \leq \delta.$$

For $a \gg \delta$ and $\delta(x,y) \equiv \delta_0(x)$, it was suggested by Galin (1961) to seek the solution of the integral equation (2.228) in the form

$$q(x,y) = \frac{p(x)\delta}{\sqrt{\delta^2 - y^2}}, \tag{2.229}$$

i.e., to reckon that the distribution of the contact stresses across the contact area is approximately that for the corresponding two-dimensional problem.

Substituting (2.229) into (2.228) and setting for simplicity $y = 0$ in the left-hand side of (2.228) (because now the right-hand side of (2.228) does not depend on y), we successively calculate the integrals in η and β (Gradshteyn and Ryzhik (1994, formulae 3.753.2, 6.552.1)), and obtain the following integral equation for the function $p(x)$:

$$\int_{-a}^{a} p(\xi)\, K\left(\frac{x-\xi}{\delta}\right) d\xi = \pi\theta\delta_0(x) \qquad (|x| \leq a), \tag{2.230}$$

$$K(t) = \int_0^\infty \frac{L(u)}{u} \cos(ut)\, du, \quad L(u) = u\,\mathcal{I}_0\left(\frac{u}{2}\right)\mathcal{K}_0\left(\frac{u}{2}\right). \tag{2.231}$$

The kernel $K(t)$ of equation (2.230) can also be presented in the form (Prudnikov, Brychkov and Marichev (1992, Vol. 1, formula 2.16.30.3))

$$K(t) = \frac{1}{\sqrt{t^2+1}} \mathcal{K}\left(\frac{1}{\sqrt{t^2+1}}\right), \tag{2.232}$$

where $\mathcal{K}(e)$ is the complete elliptic integral of the first kind. Due to the second formula of (2.231) and to formula (2.232), we find that the basic properties of the function $L(u)$ and of the kernel $K(t)$ are defined by the relations

$$L(u) = 1 + O(u^{-2}) \qquad (u \to \infty), \tag{2.233}$$

$$L(u) = u[-\ln u + \ln 4 - \mathrm{C} + O(u^2 \ln u)] \qquad (u \to 0), \tag{2.234}$$

$$K(t) = \ln|t|[-1 + O(t^2)] \qquad (t \to 0),$$

$$K(t) = |t|^{-1}[\pi/2 + O(t^{-2})] \qquad (|t| \to \infty), \tag{2.235}$$

where C is Euler's constant.

For the further investigation of the problem it is convenient to rewrite the integral equation (2.230) in the dimensionless form

$$\int_{-1}^{1} \varphi(\xi') \, K\left(\frac{x'-\xi'}{\lambda}\right) d\xi' = \pi f(x') \qquad (|x'| \le 1), \tag{2.236}$$

$$x' = xa^{-1}, \qquad \xi' = \xi a^{-1}, \qquad \lambda = \delta a^{-1},$$

$$\varphi(\xi') = p(a\xi')\theta^{-1}, \qquad f(x') = \delta_0(ax')a^{-1}$$

(we will omit the primes). Note that, in view of the statement of the problem, the dimensionless geometric parameter $\lambda \ll 1$.

Borodachev (1970) and Borodachev and Galin (1974) sought the solution of equation (2.236) for the planar punch ($f(x) \equiv \text{const}$) in the form

$$\varphi(x) = (1-x^2)^{-1/2} \sum_{n=0}^{\infty} a_n \, \mathrm{T}_{2n}(x),$$

where $\mathrm{T}_{2n}(x)$ are the even Chebyshev polynomials of the first kind. The infinite algebraic system in the unknown coefficients a_n was obtained by using the method of orthogonal polynomials. However, it is known that the classical method of orthogonal polynomials is effective only for fairly large values of λ, therefore, the use of this method is not correct for this problem. We introduce below, in accordance with Aleksandrov and Sumbatyan (1980), an asymptotic method which is effective for small λ.

2.8.2. THE INTEGRAL EQUATION

We rewrite the integral equation (2.236) in the form

$$\int_{-1}^{1} \varphi(\xi)\,d\xi \int_{0}^{\infty} \frac{L(\lambda v)}{v} \cos[v(x-\xi)]\,dv = \pi f(x) \quad (|x| \le 1),$$

and, taking account of the parameter λ is small, change the function $L(u)$ to its asymptotic form (2.234) as $u \to 0$. We then obtain ($|x| \le 1$)

$$\int_{-1}^{1} \varphi(\xi)\,d\xi \int_{0}^{\infty} \left(-\ln\left[\frac{\lambda v}{4}\right] - \mathrm{C}\right) \cos[v(x-\xi)]\,dv = \frac{\pi f(x)}{\lambda}. \qquad (2.237)$$

The solution $\varphi(x) = \varphi_0(x)$ of the integral equation (2.237) is known as a *confluent* solution for small λ.

By using the following integrals:

$$\int_{0}^{\infty} \cos(v\tilde{y})\,dv = \pi\delta_*(\tilde{y}),$$

$$\int_{0}^{\infty} \ln v \cos(v\tilde{y})\,dv = \frac{d}{d\tilde{y}} \int_{0}^{\infty} \ln v \,\frac{\sin(v\tilde{y})}{v}\,dv = -\frac{\pi}{2}\frac{d}{d\tilde{y}}[(\ln|\tilde{y}| + \mathrm{C})\,\mathrm{sign}\,\tilde{y}],$$

which are understood in the generalized sense, $\delta_*(\tilde{y})$ is the delta-function, we reduce equation (2.237) to the form

$$\left(-\ln\frac{\lambda}{4} - \mathrm{C}\right) \int_{-1}^{1} \varphi(\xi)\,\delta(x-\xi)\,d\xi + \frac{1}{2}[\ln(1-x^2) + 2\mathrm{C}]\varphi(x)$$

$$+\frac{1}{2}\int_{-1}^{1} [\varphi(\xi) - \varphi(x)] \frac{d}{d\xi}[(\mathrm{C} + \ln|\xi - x|)\,\mathrm{sign}\,(\xi - x)]\,d\xi = \frac{f(x)}{\lambda}.$$

Using the relation $(\mathrm{sign}\,\tilde{y})' = 2\delta(\tilde{y})$ and the main property of the delta-function, we finally obtain the integral equation

$$\varphi(x) \ln\frac{16(1-x^2)}{\lambda^2} + \int_{-1}^{1} \frac{\varphi(\xi) - \varphi(x)}{|\xi - x|}\,d\xi = \frac{f(x)}{\lambda}, \qquad (2.238)$$

which completely coincides with the equation considered in Kalker (1972), Sivashinsky (1975), and Panek and Kalker (1977).

It can be shown that the solution $\varphi_0(x)$ of equation (2.238) is continuous as $|x| < 1$, and vanishes at its endpoints as $[\ln(1 \pm x)]^{-1/2}$. However, it is known (see, e.g., formula (2.39)) that in general the contact pressure has the typical square root singularity near any smooth part of the boundary $\partial\Omega$. Therefore, when λ is small, the *confluent* solution $\varphi_0(x)$ is not correct close to the small sides of the rectangular punch. Nevertheless, the confluent solution gives quite satisfactory results for the dimensionless ratio between the settlement of the punch and impressed force, for fairly small λ. We will now obtain for small λ the asymptotic solution of the integral equation (2.236) which will be free of this defect arising in the confluent one.

2.8.3. PERTURBATION OF THE KERNEL

Let us perturb the kernel of the equation (2.236) so that the perturbed kernel $K_\varepsilon(t)$ exponentially decreases at infinity (note that the original kernel (2.232) decreases as $|t|^{-1}$ according to (2.235)). We choose $K_\varepsilon(t)$ so that

$$\lim K_\varepsilon(t) = K(t) \quad (\varepsilon \to 0).$$

In this case the corresponding Fourier transform $M_\varepsilon(u) = L_\varepsilon(u)/u$ of the kernel, which is defined by the first formula of (2.231), will be analytic in a certain strip of the complex u-plane containing the real axis.

Now, the asymptotics of the solution of the perturbed equation for small λ can be constructed in the well-known way (Aleksandrov and Babeshko (1965)). In particular, the first term of the asymptotics can be written in the multiplicative form

$$\varphi_\varepsilon(x) = \frac{\omega_1[(1+x)/\lambda]\,\omega_2[(1-x)/\lambda]}{v(x/\lambda)} \qquad (|x| \le 1), \tag{2.239}$$

where $\omega_n(x)$ $(n = 1, 2)$ and $v(x)$ can be found from the following integral equations:

$$\int_0^\infty \omega_n(t)\, K_\varepsilon(x-t)\, dt = \frac{\pi}{\lambda} f(\pm\lambda x \mp 1) \quad (0 \le x < \infty;\ n = 1, 2), \tag{2.240}$$

$$\int_{-\infty}^\infty v(t)\, K_\varepsilon(x-t)\, dt = \frac{\pi}{\lambda} f(\lambda x) \quad (|x| < \infty). \tag{2.241}$$

We will seek solutions of equations (2.240) by using the Wiener–Hopf technique. The main difficulty is the factorization of the transform $M_\varepsilon(u)$. It can be successfully solved if the function $M_\varepsilon(u)$ is approximated by a function $N_\varepsilon(u)$ which is analytic in a certain vicinity of the real axis, is near $M_\varepsilon(u)$, in a certain sense, on this axis and easily factorizable. Such an approximation should be uniform in ε, because, as $\varepsilon \to 0$, we must obtain a function near the function $L(u)u^{-1}$ of (2.231).

As the function $N_\varepsilon(u)$, we can take the following function [7]:

$$N_\varepsilon(u) = \left[\frac{\ln(\varepsilon - iuA)\ln(\varepsilon + iuA)(B - iuA - 1)(B + iuA - 1)}{\ln(B - iuA)\ln(B + iuA)(\varepsilon - iuA - 1)(\varepsilon + iuA - 1)}\right]^{1/2}$$

$$\times \frac{u^2 + D^2}{\sqrt{(u^2 + E^2)(u^2 + \alpha^2)(u^2 + \beta^2)}}.$$

As $A = \exp(\mathrm{C} - \ln 4)$, $B = 0.4$, $D = 0.2$, $E = (B-1)/(\ln B)$, $\alpha = 0.365$, and $D^2 = \alpha\beta$, for the case when $\varepsilon = 0$ the behaviour of the function $N_\varepsilon(u)$ near zero and infinity is the same as $L(u)/u$ of (2.231), i.e., it is defined by formulae (2.233) and (2.234), and the relative error of this approximation on the whole real axis does not exceed 4.5%.

Now, we can easily factorize the function $N_\varepsilon(u)$ in the form

$$N_\varepsilon(u) = N_\varepsilon^+(u)\, N_\varepsilon^-(u), \qquad N_\varepsilon^-(u) = N_\varepsilon^+(-u),$$

$$N_\varepsilon^+(u) = \left[\frac{\ln(\varepsilon - iuA)(B - iuA - 1)}{\ln(B - iuA)(\varepsilon - iuA - 1)}\right]^{1/2} \frac{D - iu}{\sqrt{(E - iu)(\alpha - iu)(\beta - iu)}}.$$

2.8.4. A PLANAR PUNCH

For simplicity, we confine ourselves to the planar punch, i.e., $\delta(x) \equiv w$ so that $f(x) \equiv f = w/a$. Now, equations (2.240) for $\omega_n(x)$, $n = 1, 2$, coincide, and we omit the index n. Solving this equation using the Wiener–Hopf technique, we obtain the following functional equation for the Fourier transform:

$$\Omega^+(u)N_\varepsilon(u) = f\,(-iu\lambda)^{-1} + G^-(u),$$

where $\Omega^+(u)$ is the transform of $\omega(x)$, and $G^-(u)$ is the transform of the function

$$g(x) = \int_0^\infty \omega(t)\, K_\varepsilon(x - t)\, dt \qquad (-\infty < x < 0).$$

[7] This approximation was suggested by R.A.Gruntfest.

The sign "+"or "−" denotes that the function is analytic in the complex variable u in the upper or lower half-plane, respectively. After factorization of the function $N_\varepsilon(u)$ we obtain

$$\Omega^+(u)\, N_\varepsilon^+(u) = f\,[-iu\lambda\, N_\varepsilon^-(u)]^{-1} + H^-(u).$$

The expression of the function $[-iu\, N_\varepsilon^-(u)]^{-1}$ as the sum of two functions, one of which is regular in the upper half-space and another in the lower one, can be presented in the form

$$\frac{1}{-iu\, N_\varepsilon^-(u)} = \Psi^+(u) + \Psi^-(u), \qquad \Psi^+(u) = \frac{1}{-iu\, N_\varepsilon^-(0)},$$

$$\Psi^-(u) = \frac{1}{-iu}\left[\frac{1}{N_\varepsilon^-(u)} - \frac{1}{N_\varepsilon^-(0)}\right].$$

As a result, we obtain the relation

$$\Omega^+(u)\left[\frac{\ln(\varepsilon - iuA)(B - iuA - 1)}{\ln(B - iuA)(\varepsilon - iuA - 1)}\right]^{1/2} \tag{2.242}$$

$$\times \frac{D - iu}{\sqrt{(E - iu)(\alpha - iu)(\beta - iu)}} - \frac{f}{-iu\lambda}\left(\frac{\varepsilon - 1}{\ln\varepsilon}\right)^{1/2} \equiv 0.$$

The vanishing of the last expression follows from the condition that the singularity of the function $\omega(x)$ in zero must be integrable.

Finding the function $\Omega^+(u)$ from (2.242) and changing the variable $-iu = p$ to pass from the Fourier transforms to the Laplace transforms, for the transform of $\omega(x)$ we obtain the expression

$$\Omega(p) = \frac{f}{\lambda}\left(\frac{\varepsilon - 1}{\ln\varepsilon}\right)^{1/2} \frac{1}{p(p + D)}$$

$$\times \left[\frac{(\varepsilon + pA - 1)(E + p)(\alpha + p)(\beta + p)\ln(B + pA)}{\ln(\varepsilon + pA)\,(B + pA - 1)}\right]^{1/2}.$$

The solution of equation (2.241) can be easily found by using the Fourier transformation as follows

$$v(x) = f/[\lambda M_\varepsilon(0)], \qquad M_\varepsilon(0) = N_\varepsilon(0) = (\ln\varepsilon)/(\varepsilon - 1).$$

So after the substitution of $\omega(x)$ and $v(x)$ into (2.239) and passage to the limit $\varepsilon \to 0$, the main term in the asymptotics of the solution of equation (2.236) for small λ takes the form

$$\varphi(x) = \frac{f}{\lambda}\, r\left(\frac{1 + x}{\lambda}\right) r\left(\frac{1 - x}{\lambda}\right) \qquad (|x| \le 1),$$

where $r(x)$ is the inverse transform of

$$R(p) = \frac{1}{p(p+D)} \left[\frac{(pA-1)(E+p)(\alpha+p)(\beta+p)\ln(B+pA)}{\ln(pA)\,(B+pA-1)} \right]^{1/2}. \tag{2.243}$$

We have the following relation for the applied force:

$$P = \int_{-a}^{a} d\xi \int_{-\delta}^{\delta} q(\xi,\eta)\,d\eta = \pi\theta\delta w \int_{0}^{2/\lambda} r(\xi)\, r\left(\frac{2}{\lambda}-\xi\right) d\xi = \pi\theta\delta w\, r_1\left(\frac{2}{\lambda}\right). \tag{2.244}$$

By using the theorem on convolution, we obtain the Laplace transform of the function $r_1(x)$ in the form

$$R_1(p) = \frac{1}{p^2(p+D)^2} \frac{(pA-1)(E+p)(\alpha+p)(\beta+p)\ln(B+pA)}{\ln(pA)\,(B+pA-1)}. \tag{2.245}$$

2.8.5. NUMERICAL RESULTS

Let us invert the transforms (2.243) and (2.245).

Since the function $r_1(\tilde{x})$ in (2.244) is required only for large values of $\tilde{x} = 2/\lambda$, we construct its asymptotics. This is determined by the behaviour of the function $R_1(p)$ near zero (see Bleistein and Handelsman (1975)). Expanding the latter function for small p and confining ourselves only to the first and second terms

$$R_1(p) \sim -\frac{1}{p^2\ln(pA)} - \frac{N}{p\ln(pA)} \qquad (p \to 0, \quad N = 2.474),$$

we obtain the following power-logarithmic asymptotics for $r_1(\tilde{x})$:

$$r_1(\tilde{x}) \sim \tilde{x} \sum_{k=0}^{\infty} (-1)^k \left(\frac{1}{\Gamma(2)}\right)^{(k)} \left(\ln\frac{\tilde{x}}{A}\right)^{-k-1} \tag{2.246}$$

$$+ \sum_{k=0}^{\infty} a_k \left(\ln\frac{\tilde{x}}{A}\right)^{-k-1} \qquad (\tilde{x} \to \infty),$$

$$a_k = (A+N)\frac{b_k}{k+1} - A\left(\frac{1}{\Gamma(1)}\right)^{(k)},$$

$$b_0 = 1, \quad b_k = \sum_{j=0}^{[k/2]} (-1)^j A^{2j} \binom{k+1}{2j+1} \Gamma^{(k-2j)}(1) \quad (k \geq 1),$$

where $\Gamma(x)$ is the Γ-function, the superscript (k) means k-derivative, and $[k/2]$ means the integer part.

Table 2.10 presents the comparison of the dimensionless coefficient

$$\gamma = \frac{2aw\theta}{P} = \frac{2/\lambda}{\pi r_1(2/\lambda)}, \tag{2.247}$$

corresponding to the compliance of the punch, where the value $r_1(2/\lambda)$ is determined by formula (2.246), with the corresponding values from Borodachev and Galin (1974) (line B. and G.), and from Panek and Kalker (1977) (line P. and K.)

Table 2.10: The values of γ (2.247)

λ	0.2	0.15	0.1	0.05	0.02
B. and G.	0.835	0.922	1.04	1.26	1.53
P. and K.		0.927	1.05	1.27	1.57
(2.247)	0.817	0.911	1.05	1.29	1.60

One can see that the results, based on the asymptotic method of *small* λ, are close to those from Borodachev and Galin (1974) when λ is not very small, and are close to those from Panek and Kalker (1977) as $\lambda \leq 0.1$. This is to be expected, since the method of orthogonal polynomials, applied by Borodachev and Galin (1974), is efficient only for large and average λ, and is unstable for small λ. At the same time, the *confluent* solution in Panek and Kalker (1977) is efficient only for very small λ.

To calculate the function $r(\tilde{x})$ for large $\tilde{x}$, it is also convenient to use its asymptotic representation which is similar to (2.246). For values of $\tilde{x}$ which are not large, it is necessary to apply a numerical inversion of the function (2.243). As is known, the function $r(\tilde{x})$ has a square root singularity at zero, i.e.,

$$r(\tilde{x}) \sim (\pi\tilde{x})^{-1/2} \qquad (\tilde{x} \to 0).$$

Table 2.11 presents the results of calculation of the dimensionless quantity $\pi a\delta p(ax)/P$ for various x and λ. If the contact pressures were uniformly distributed along the larger side of the contact region, as in the two-dimensional problem, then this quantity would be equal to 0.5. One can see from the table that, as the ratio δ/a diminishes, this quantity tends to 0.5 at points distant from the boundary (in particular, at $x = 0$): this demonstrates a gradual approach to the two-dimensional problem. At the same time, this tendency is slow because of the logarithmic terms in the asypmtotic solution.

Table 2.11: The values of $\pi a \delta p(ax)/P$

x	$\lambda = 0.2$	$\lambda = 0.15$	$\lambda = 0.1$	$\lambda = 0.05$	$\lambda = 0.02$
0.0	0.4410	0.4558	0.4662	0.4670	0.4690
0.1	0.4413	0.4561	0.4668	0.4674	0.4695
0.2	0.4426	0.4575	0.4688	0.4704	0.4711
0.3	0.4448	0.4597	0.4722	0.4756	0.4738
0.4	0.4485	0.4633	0.4770	0.4828	0.4821
0.5	0.4543	0.4686	0.4836	0.4930	0.4881
0.6	0.4645	0.4768	0.4924	0.5022	0.4975
0.7	0.4827	0.4914	0.5055	0.5196	0.5186
0.8	0.5218	0.5226	0.5296	0.5437	0.5483
0.85	0.5612	0.5547	0.5534	0.5613	0.5709
0.9	0.6366	0.6182	0.6013	0.5908	0.6016
0.95	0.8305	0.7880	0.7372	0.6738	0.6532
0.975	1.130	1.057	0.9642	0.8278	0.7318
0.99	1.748	1.623	1.457	1.193	0.9460

Note that the approximate solution for the narrow rectangular punch does not pretend to correctly mirror the behaviour of the contact pressure near the angular points of the punch. The behaviour of the contact pressure near the angular points will be considered in section 4.3.

2.9 Galanov–Newton Method

2.9.1. NONLINEAR BOUNDARY EQUATIONS

An integral equation of a 3-D frictionless contact problem can generally be written in the form

$$\theta^{-1} \int\limits_{\Omega} K(N,M)\, q(N)\, d\Omega_N = g(M), \qquad M \in \Omega, \tag{2.248}$$

$$K(N,M) = K(M,N),$$

$$M = (x,y), \quad N = (r,z), \quad g(x,y) = 2\pi[\delta - f(x,y)],$$

where $q(x,y)$ is an unknown contact pressure in a contact zone Ω, δ is a settlement of a punch and the function $f(x,y)$ describes the base of the punch. We suppose in (2.248) that the punch presses without rotation. For an elastic half-space the kernel of equation (2.248) has the form

$$K(N,M) = 1/R = 1/\sqrt{(r-x)^2 + (z-y)^2}$$

(see equation (1.29)). For many elastic solids such as layers, multilayers, cylinders, wedges, cones, spheres and so on, we are able to find the kernel $K(N, M)$ in an analytic form from the problem for a concentrated force. The principal parts of these kernels will always be $1/R$, so that the kernels will have the structure

$$K(N, M) = 1/R + \text{ a smooth function.} \tag{2.249}$$

The integral equation (2.248) has exact solutions for a very few special cases. We need a numerical method for solving this general equation. For a smooth punch when the contact pressure is zero on the boundary of the contact domain, such a method was suggested by Galanov (1985), who reduced equation (2.248) to a nonlinear integral equation with a special type of nonlinearity and then used Newton's method. We describe this method of nonlinear boundary integral equations of the Hammerstein type, which allows the normal contact pressure, the unknown contact domain and the normal displacement outside the contact domain to be determined simultaneously. Suppose the punch is elongated along the y-axis and symmetric with respect to y, and the domain Ω is completely contained within a rectangle S, two sides of which are parallel to the coordinate axes, with centre at the x-axis and semiaxes b, along the y-axis, and c ($b \geq c$, see Fig 2.12 below). The contact problem can be reduced to the following integral problem:

$$\theta^{-1} \int_S K(N, M) q(N)\, d\Omega_N = g(M), \quad q(M) \geq 0, \quad M \in \Omega, \tag{2.250}$$

$$\theta^{-1} \int_S K(N, M) q(N)\, d\Omega_N > g(M), \quad q(M) = 0, \quad M \in (S \setminus \Omega), \tag{2.251}$$

where equation (2.250) follows from (2.248) and inequality (2.251) shows the absence of contact outside the contact zone, where the elastic displacement on the left-hand side is greater than the displacement of the punch.

It is also assumed that there exists a bounded region $S_0 = \{M : g(M) > 0\}$ such that $\Omega \subset \overline{S_0} \subset S$. Introduce the nonlinear operators

$$v^+(M) = \sup\{v(M), 0\}, \quad v^-(M) = \inf\{v(M), 0\}. \tag{2.252}$$

Any function $v(M)$ can be presented as

$$v(M) = v^+(M) + v^-(M). \tag{2.253}$$

The principal idea of Galanov's method consists in the presentation of the normal contact pressure $q(M)$ in the form (2.253) and the automatic satisfaction of the integral inequality (2.251) by considering the nonlinear operator equation

$$\mathcal{T}v = 0 \ \ (M \in \Omega), \quad \mathcal{T}v \equiv \eta v^- + \theta^{-1}Kv^+ - g, \ \ \eta = \text{const}, \tag{2.254}$$

where $v^\pm = v^\pm(M)$, $g = g(M)$ and K is an integral operator of the form

$$Kv^+ = \int_S K(N,M)v^+(N)\,dS_N. \tag{2.255}$$

The parameter $\eta > 0$ can be choosen arbitrarily (we took $\eta = 1$ in calculations); the dependence of the solution of equation (2.254) on η will be shown in Theorem 2.5.

Theorem 2.3. *If $v_* = v_*(M)$ is a solution of equation* (2.254), *then $q = q(M) = v_*^+$ with $\Omega = \{M : v_* \geq 0\}$ is a solution of system* (2.250), (2.251) *and $\Omega \neq \emptyset$ if $S_0 \neq \emptyset$; vice versa, if $q = q(M)$, $M \in \Omega$ is a solution of system* (2.250), (2.251), *then*

$$v_* = \eta^{-1}g + q - (\theta\eta)^{-1}Kq, \quad M \in \Omega \tag{2.256}$$

is a solution of equation (2.254).

Proof. First we show that $\Omega \neq \emptyset$ if $S_0 \neq \emptyset$. Let us assume the opposite. Then (2.254) implies the inequality $g < 0$ which contradicts the existence of $S_0 \neq \emptyset$.

Let v_* be a solution of (2.254). If $M \in \Omega$, (2.252)–(2.254) imply the relations $q = v_*^+ > 0$, $\theta^{-1}Kq = g$. If $M \notin \Omega$, then $v_* < 0$, $\eta v^- + \theta^{-1}Kv^+ = g$ and $\theta^{-1}Kv^+ > g$, and this proves the right-hand side of the theorem.

Now let q and Ω be a solution of system (2.250), (2.251). If $M \in \Omega$, the equality $v_* = q$ follows from (2.250) and (2.256). If $M \in (S \setminus \Omega)$, then we have $v_* = \eta^{-1}g - (\theta\eta)^{-1}Kq < 0$ and, using formula (2.253), we can rewrite (2.256) as:

$$v_* = \eta^{-1}g + v_*^+ - (\theta\eta)^{-1}Kv_*^+, \quad M \in \Omega, \tag{2.257}$$

i.e. v_* is a solution of (2.254). •

The mechanical meaning of the solution $v = v_*$ of equation (2.254) is following: this function is the contact pressure in the domain where $v_* > 0$ (this inequality determines the contact domain Ω), and in the complement $S\backslash\Omega$, where $v_* < 0$ and $v_*^+ = 0$, this function gives the normal displacement u_z of an elastic solid in the form (see formula (2.257) and Fig. 2.10)

$$u_z = -\theta^{-1}Kv_*^+ = \eta v_* - g, \quad M \in (S \setminus \Omega). \tag{2.258}$$

Theorem 2.4. *Equation* (2.254) *has a unique solution* $v_* \in \mathcal{L}_2(S)$ *if and only if the function* $v_0 = v_0(M)$, *which serves as the solution of the equation* $(\varepsilon_* > 0)$

$$\varepsilon_* v^+ + \eta v^- + \theta^{-1} K v^+ = g \quad (M \in \Omega), \tag{2.259}$$

satisfies the condition

$$\|v_0\|_{\mathcal{L}_2} \leq C, \quad \varepsilon_* \in (0, \varepsilon_0], \quad \varepsilon_0 = const > 0,$$

where the constant C *is independent of* ε_*.

Equation (2.259) *has a solution by virtue of the principle of contractive mappings for sufficiently large values of* η.

Theorem 2.5. *Suppose* $v_1(M)$ *and* $v_2(M)$ *are the solutions of equation* (2.254) *for* $\eta = \eta_1$ *and* $\eta = \eta_2$, *respectively* $(\eta_1 \neq \eta_2)$. *Then* $v_1^+(M) = v_2^+(M)$.

The proofs of Theorems 2.4 and 2.5 repeat the proofs of the corresponding theorems for a half-space in Galanov (1985), because the main properties of the integral operator K (2.255) (strict positiveness and complete continuity) are preserved for the general case (2.249).

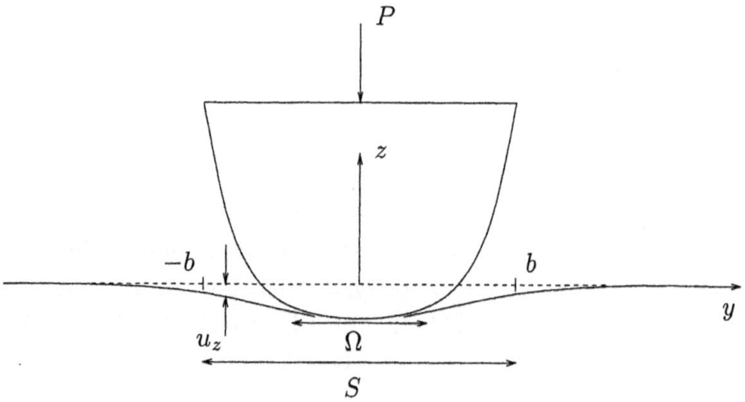

Figure 2.10: Normal displacement u_z outside the contact zone (the y, z-axes section)

For a numerical solution of equation (2.254) we will use a modified Newton method proposed in Krasnosel'skii, Vainikko, Zabreiko, et al (1972), which is based on the construction of successive approximations using the formulae

$$v_{n+1} = v_n - (Q' v_n)^{-1} \mathcal{T} v_n, \tag{2.260}$$

$$v_n = v_n(M), \quad n = 0, 1, ..., \quad v_0 = g. \tag{2.261}$$

Here, Q is a differentiable operator which approximates the operator $\mathcal{T}$ in (2.254) in a uniform metric, and has the form

$$Qv = \eta(v - Q_1 v) + \theta^{-1} K Q_1 v - g;$$

$$Q_1 v = 0 \quad (v < -\varepsilon_1), \qquad Q_1 v = v \quad (v > \varepsilon_1 > 0),$$

$$Q_1 v = {}^1\!/\!_2 \, [v - v^2/(2\varepsilon_1)] + 3\varepsilon_1/4 \quad (|v| \leq \varepsilon_1).$$

In dependence on the choice of the constant ε_1 in these formulae, the operator $\mathcal{T}$ can be approximated with any specified accuracy.

In view of the symmetry of the problem with respect to y it is sufficient to consider only the upper half of the rectangle S, which we will cover with a rectangular net of m nodes with spacing h_1 along the x-axis and h_2 along the y-axis (see Fig. 2.11, where $m = 81$). When calculating the values of the function $K(N, M)$ of the form (2.249) at these nodes we smooth its singularity using the formula

$$1/R \to 1/R_*, \quad R_* = \sqrt{(r - x)^2 + (z - y)^2 + \delta_*}. \tag{2.262}$$

It can be shown that the regularizing parameter δ_* in formula (2.262) should be related to the net spacings h_1 and h_2 (we took $\delta_* = h_1 h_2/16$ in calculations).

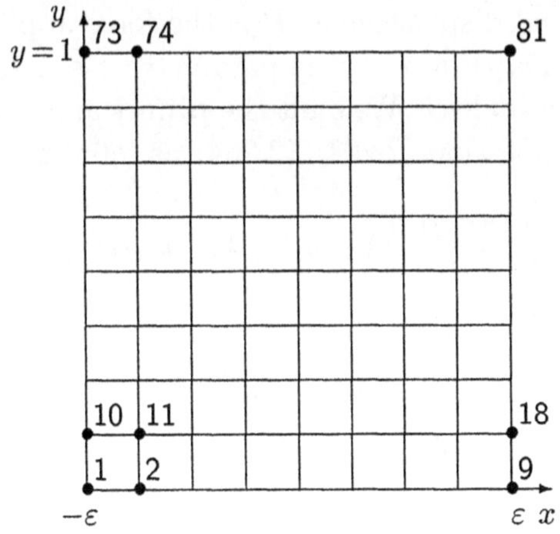

Figure 2.11: The rectangular net

We note that the Galanov–Newton method can be sucessfully modified for 3-D contact problems with friction (Pozharskii (2000)) as well as for 2-D contact problems.

2.9.2. TESTING THE COMPUTER PROGRAM

In appendix we give Galanov's computer program for a half-space when the kernel is equal to $1/R$. For a general case, one should just change the kernel of the integral equation in this program. To test the program, we consider below some known solutions for an elliptic paraboloid when $f(x,y) = x^2/(2R_1) + y^2/(2R_2)$, $R_1 \leq R_2$.

A) For the *axisymmetric contact problem* for a half-space we have a circular paraboloid with $R_1 = R_2$; the contact domain, shown in Fig. 1.3, must be circular too. The solution of this problem can be obtained as the limiting case in which the ellipse eccentricity $e_1 \to 0$ in formulae (1.64)–(1.69). Due to formulae 8.113.1 and 8.114.1 in Gradshteyn and Ryzhik (1994) we have

$$\mathcal{K}(0) = \mathcal{E}(0) = \frac{\pi}{2}, \quad \lim_{e_1 \to 0} \frac{\mathcal{K}(e_1) - \mathcal{E}(e_1)}{e_1^2} = \frac{\pi}{4}.$$

We introduce the following dimensionless quantities and notation:

$$x = x'b, \quad r = r'b, \quad y = y'b, \quad z = z'b, \quad \delta = \delta'b, \tag{2.263}$$

$$A = b/(2R_1), \quad B = b/(2R_2), \quad \varepsilon = c/b, \quad u_z = -u_z'b, \tag{2.264}$$

$$q(x,y)/\theta = 2\pi q'(x',y'), \quad P/\theta = 2\pi b^2 P', \quad S' \to S, \quad \Omega' \to \Omega, \tag{2.265}$$

where u_z is the normal displacement, P is the force applied to the punch, i.e., the integral of $q(x,y)$ over Ω; the parameter $\varepsilon \leq 1$ is the ratio of the dimensions of the rectangle S. We omit the primes and rewrite the integral equation (2.248) for notation (2.263)–(2.264) as follows

$$\int_{\Omega} \frac{q(r,z)}{R} \, d\Omega = \delta - (Ax^2 + By^2),$$

$$R = \sqrt{(r-x)^2 + (z-y)^2}, \quad (x,y) \in \Omega.$$

For $\delta = A = B = 1$ the solution of the axisymmetric problem takes the form

$$q(x,y) = (2\sqrt{2}/\pi^2)\sqrt{1 - 2x^2 - 2y^2}. \tag{2.266}$$

A comparison between exact and calculated solution, for $\varepsilon = 1$ and $9 \times 9 = 81$ nodes, is given in Table 2.12, where m is the node number in Fig. 2.11, and the sign "$-$" means that the node does not belong to the contact region Ω. The exact value of the force is $P = \int_{\Omega} q(x,y)dxdy = 2\sqrt{2}/(3\pi) \approx 0.300$, while in the program it is calculated as $P = 0.299$.

Table 2.12: Comparison between exact and numerical (prog.) solution

m	(2.266)	prog.	m	(2.266)	prog.	m	(2.266)	prog.
1	–	–	10	–	–	19	–	–
2	–	0.00435	11	–	–	20	–	–
3	0.203	0.204	12	0.196	0.196	21	0.176	0.172
4	0.268	0.269	13	0.263	0.264	22	0.248	0.249
5	0.286	0.287	14	0.282	0.283	23	0.268	0.269
6	0.268	0.269	15	0.263	0.264	24	0.248	0.249
7	0.203	0.204	16	0.196	0.196	25	0.176	0.172
8	–	0.00435	17	–	–	26	–	–
9	–	–	18	–	–	27	–	–

Similar agreements were checked out for $15 \times 15 = 225$ and $21 \times 41 = 841$ nodes.

The normal displacement of the half-space surface outside the contact has the form

$$u_z = u_z(x,y) = \int\limits_{\Omega} \frac{q(r,z)}{R}\, dr dz, \quad (x,y) \notin \Omega. \tag{2.267}$$

By substituting expression (2.266) into (2.267), using polar coordinates, $\rho = \sqrt{x^2+y^2}$, and integral 3.674.1 in Gradshteyn and Ryzhik (1994), we obtain

$$u_z = u_z(\rho) = \frac{8\sqrt{2}}{\pi^2 \rho} \int\limits_0^{1/\sqrt{2}} t\sqrt{1-2t^2}\, \mathcal{K}\left(\frac{t}{\rho}\right) dt, \quad \rho > \frac{1}{\sqrt{2}}. \tag{2.268}$$

A comparison between calculated by formula (2.268) and by the program (formula (2.258)) normal displacement, for $\varepsilon = 1$ and $15 \times 15 = 225$ nodes, is given in Table 2.13, where ρ is the polar radius.

Table 2.13: Analytical and numerical (prog.) values of displacement

m	1	2	16	17	31	32	225
ρ	1	6/7	$\sqrt{196}/14$	$\sqrt{125}/14$	$\sqrt{50}/7$	$\sqrt{37}/7$	$\sqrt{2}$
(2.268)	0.318	0.382	0.317	0.380	0.315	0.376	0.218
prog.	0.318	0.381	0.317	0.379	0.314	0.374	0.218

The conclusion is that even for 81 nodes the program gives quite satisfactory results.

B) The program was tested against the axisymmetric asymptotic solution (2.88)–(2.90) of the contact problem for a layer, where $\gamma = 1/(2R_1)$, $\lambda = h/a$, h is a thickness of the layer and a is a radius of the contact circle.

In the notation (2.263)–(2.265), $c_* = a/b$, $\lambda_* = h/b$ (we recall that b is the major semiaxis of the contact rectangle), we rewrite formulae (2.88)–(2.90) in the form

$$q(r) = \frac{4A}{\pi^2}\sqrt{c_*^2 - r^2}\left[1 - \frac{8a_1c_*^3}{3\pi\lambda_*^3}\right], \tag{2.269}$$

$$P = \frac{8Ac_*^3}{3\pi}\left[1 - \frac{8a_1c_*^3}{3\pi\lambda_*^3}\right], \tag{2.270}$$

$$P = \frac{4\delta c_*}{3\pi}\left[1 + \frac{4a_0c_*}{3\pi\lambda_*} + \left(\frac{4a_0c_*}{3\pi\lambda_*}\right)^2 + \left(\frac{4a_0c_*}{3\pi\lambda_*}\right)^3\right. \tag{2.271}$$

$$\left. + \left(\frac{4a_0c_*}{3\pi\lambda_*}\right)^4 + \frac{8a_1c_*^3}{15\pi\lambda_*^3}\left(1 + \frac{8a_0c_*}{3\pi\lambda_*}\right)\right].$$

We take $A = \delta = 1$ in (2.269)–(2.271) and determine P and c from equations (2.270), (2.271); the values of a_0 and a_1 should be taken from Table 2.2, where we choose $\nu = 0.3$.

We present kernel (2.22) of the integral equation (2.27) for the layer, using our notation and integral (1.15), as follows:

$$K\left(\frac{R}{\lambda_*}\right) = \frac{1}{R} + \frac{1}{\lambda_*}\int_0^\infty [L(u) - 1]\mathcal{J}_0\left(u\frac{R}{\lambda_*}\right) du. \tag{2.272}$$

The principal part, $1/R$, is separated in (2.272). Calculating the integral in (2.272), we use Gaussian quadrature formula 25.4.30 in Abramowitz and Stegun (1972) with 32 nodes. Due to formulae (2.15), $(2.17)_2$ and (2.20), we have $L(u) - 1 \sim \exp(-2u)$ as $u \to \infty$. Appling the quadrature formula on the interval (0;10) to integral 2.12.8.3 in Prudnikov, Brychkov and Marichev (1992, Vol. 2)

$$\int_0^\infty \exp(-2u)\mathcal{J}_0(up)\, du = \frac{1}{\sqrt{4 + p^2}},$$

which is similar to the integral in (2.272), we can conclude that for $\lambda_* \geq 1/4$ the quadrature formula gives 7 correct significant digits for the integral in (2.272) in the contact rectangle, $|x| \leq 1$, $0 \leq y \leq 1$.

In columns 2-4 of Table 2.14 we give values, calculated by formulae (2.269)–(2.271); the corresponding values, calculated by the program ($\varepsilon = 1$ and $9 \times 9 = 81$ nodes), are given in the last two columns.

Table 2.14: Analytical and numerical solution for a layer

λ_*	c_*	P	$q(0)$	P (prog.)	$q(0)$ (prog.)
10	0.722	0.320	0.293	0.319	0.294
5	0.739	0.343	0.300	0.342	0.301
2	0.790	0.432	0.331	0.434	0.331

We can see from this table that the thinner the layer (the smaller the λ_*), the greater is the force P and the contact area (the larger is its dimensionless radius c), for a fixed value of the settlement δ. This is due to the rigid fixation of the lower surface of the layer.

2.9.3. CONTACT PROBLEM FOR A TWO-LAYER FOUNDATION

We consider now a contact problem of a punch whose shape is defined by a function $f(x, y)$, pressed by a force P into a layer of thickness h with elastic characteristics G_1, ν_1. The layer is glued completely to a half-space with elastic characteristics G_2, ν_2. For simplicity, we suppose again that the punch is an elliptic paraboloid

$$f(x, y) = x^2/(2R_1) + y^2/(2R_2), \tag{2.273}$$

and the force P is applied centrally, so that the punch settles by an amount δ without any rotation.

Using the double integral Fourier transformation and the scheme described in section 2.1, we obtain the following integral equation (Aleksandrova (1973); Aleksandrov, Kalker and Pozharskii (1999)) for the unknown contact pressure $q(x, y)$ under the punch:

$$\int_\Omega q(r, z) K_* \left(\frac{R}{h}\right) d\Omega = 2\pi h \theta_1 (\delta - f(x, y)) \qquad ((x, y) \in \Omega), \tag{2.274}$$

$$K_*(t) = \int_0^\infty N_*(u) \mathcal{J}_0(ut)\, du,$$

$$N_*(u) = \frac{M_* + 4u \exp(-2u) - L_* \exp(-4u)}{M_* - (1 + 4u^2 + L_* M_*) \exp(-2u) + L_* \exp(-4u)},$$

$$M_* = \frac{G_2 \kappa_1 + G_1}{G_1 - G_2}, \qquad L_* = \frac{G_1 \kappa_2 - G_2 \kappa_1}{G_1 \kappa_2 + G_2}, \qquad \theta_n = \frac{G_n}{1 - \nu_n},$$

$$R = \sqrt{(r - x)^2 + (z - y)^2}, \qquad \kappa_n = 3 - 4\nu_n \quad (n = 1, 2),$$

where Ω is an unknown contact domain.

Note that ($h \to \infty$)

$$\lim K_*\left(\frac{R}{h}\right) = \lim \int_0^\infty [N_*(u) - 1 + 1]\mathcal{J}_0\left(\frac{R}{h}\right) du = \frac{h}{R}, \tag{2.275}$$

where we have used the last integral of (1.15). This means that, as $h \to \infty$, equation (2.274) with (2.273) reduces to the equation

$$\int_\Omega \frac{q(r,z)}{R}\, d\Omega = 2\pi\theta_1\left(\delta - \frac{x^2}{2R_1} - \frac{y^2}{2R_2}\right) \quad ((x,y) \in \Omega), \tag{2.276}$$

which corresponds to an elastic half-space (see (1.30)), and has an exact solution because of its special right-hand side. Suppose $R_2 \geq R_1$ in (2.276); then the contact region Ω in (2.276) is an ellipse

$$\frac{x^2}{a_1^2(1-e_1^2)} + \frac{y^2}{a_1^2} = 1,$$

and the contact pressure has the form (cf. ((1.64) and (1.65))

$$q(x,y) = \frac{2\theta_1\delta}{a_1\sqrt{1-e_1^2}\mathcal{K}(e_1)}\sqrt{1 - \frac{x^2}{a_1^2(1-e_1^2)} - \frac{y^2}{a_1^2}}. \tag{2.277}$$

The eccentricity e_1 should be obtained first from equation (1.68), in which we must interchange R_1 and R_2. We recall that, in (1.68), $\mathcal{K} = \mathcal{K}(e_1)$ and $\mathcal{E} = \mathcal{E}(e_1)$ are the complete elliptic integrals. Afterwards, the value of a_1 is found by formula (1.69), in which we must change R_1 to R_2 and θ to θ_1; the value of δ is found, using formulae (1.67) and (1.69), in the form

$$\delta = \frac{3P^{2/3}}{4\pi\theta_1^{2/3}}\mathcal{K}(e_1)\left[\frac{2\pi e_1^2}{3R_2(\mathcal{K}(e_1) - \mathcal{E}(e_1))}\right]^{1/3}. \tag{2.278}$$

For equation (2.274), we assume that the area Ω is completely contained within a rectangle $S: \{|x| \leq c, |y| \leq b\}$, $b \geq c$ (Fig. 2.12), and use the Galanov–Newton method. We introduce dimensionless notation (2.263)–(2.265), $\lambda = h/b$, $R = R'b$, omit the primes and, using formula (2.275), separate out the principal part of the kernel in the form

$$K_*(R) = \frac{1}{R} + \frac{1}{\lambda}\int_0^\infty [N_*(u) - 1]\, \mathcal{J}_0\left(u\frac{R}{\lambda}\right) du.$$

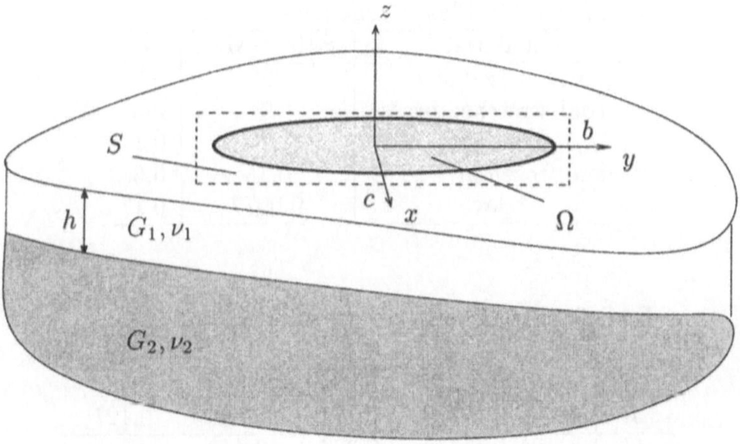

Figure 2.12: The contact region Ω inside a rectangle S

To determine the connection between the quantities P and δ, we use the equilibrium condition

$$\int_{\Omega} q(\xi, \eta)\, d\Omega = P.$$

We give the numerical results for the case when $A = 0.002$, $B = 0.001$ and $\varepsilon = 0.8$. To check these results, we have used the exact solutions (formulae (2.277) and (2.278)), as $\delta = 0.001$ and $\lambda \to \infty$:

$$q(x,y)10^3 \approx 0.348\sqrt{1 - \left(\frac{x}{0.469}\right)^2 - \left(\frac{y}{0.744}\right)^2}, \qquad P10^3 \approx 0.254.$$

The following four cases of the two-layer foundation are considered: (I) steel over brass; (II) brass over steel; (III) concrete over frozen sand; (IV) concrete over clay. The values of the elastic characteristics for these materials (Young's modulus E (MPa) and Poisson's ratio ν) are given in Table 2.15 (see Nordling and Österman (1980)). We note that the shear modulus $G = E/[2(1+\nu)]$. For these four cases, the values of the force P and the contact pressure $q_0 = q(0,0)$ at the point where the punch and the foundation initially touch are given in Table 2.16 depending on λ when $\delta = 0.001$; the quarter-boundaries of the contact area Ω for $\lambda = 1/4$ and the same value of δ are shown in Fig. 2.13. The dependences of the indenting force P on the settling δ of the punch for different values of λ are shown in Fig. 2.14a,b,c,d for the cases I–IV, respectively.

Table 2.15: Elastic constants

material	$E10^{-4}$ MPa	ν
carburized steel	20	0.28
cold-stretched brass	9	0.35
concrete	2	0.17
sand frozen at $-10°$C	0.15	0.30
clay	0.0077	0.17

Table 2.16: Contact pressure and force

case	$\lambda =$	2	1	1/2	1/4
I	$q_0 10^3$	0.313	0.289	0.243	0.186
I	$P10^3$	0.200	0.171	0.143	0.127
II	$q_0 10^3$	0.358	0.384	0.448	0.541
II	$P10^3$	0.288	0.328	0.386	0.438
III	$q_0 10^3$	0.243	0.202	0.143	0.0626
III	$P10^3$	0.0964	0.0627	0.0392	0.0263
IV	$q_0 10^3$	0.145	0.116	0.0857	0.0512
IV	$P10^3$	0.0211	0.0114	0.00603	0.00316

If the contact stiffness θ_1 of the layer is greater than the stiffness θ_2 of the half-space, we see that the contact area and the force decrease when the relative thickness λ of the layer decreases. If the half-space stiffness exceeds that of the layer, the contact area and the force increase as the parameter λ decreases.

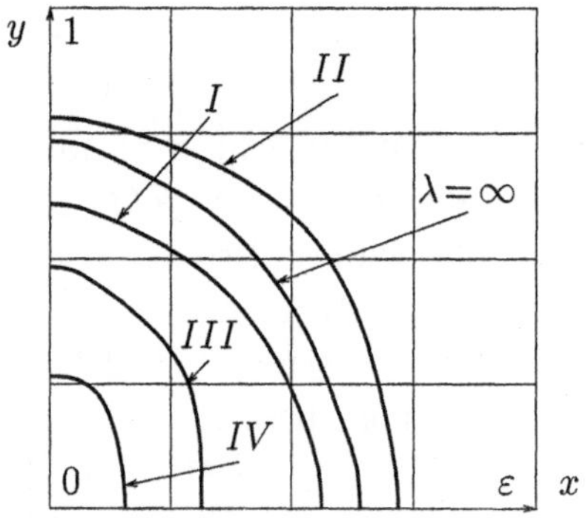

Figure 2.13: The quarters of $\partial\Omega$ for cases I–IV

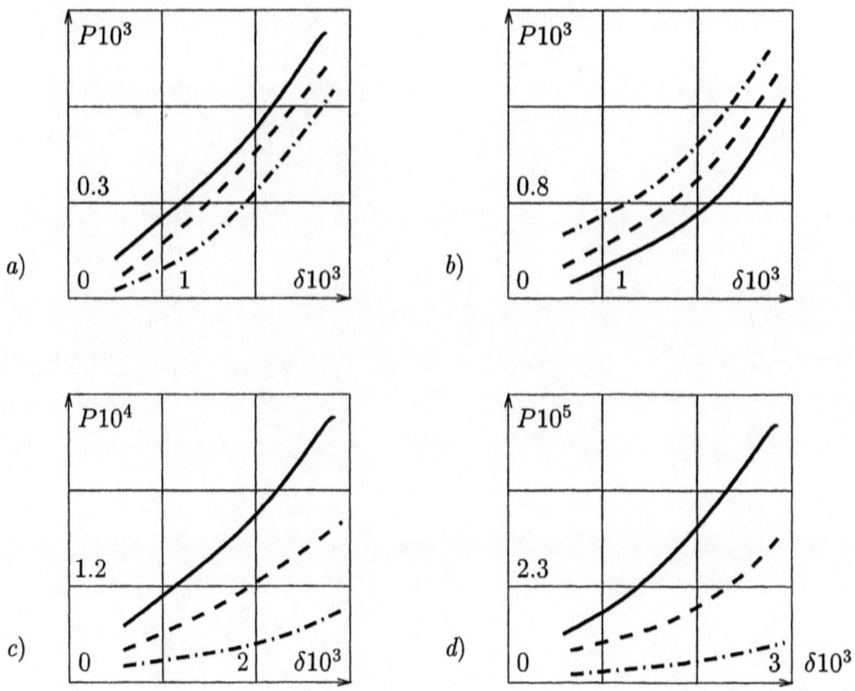

Figure 2.14: The plots $P(\delta)$ for $\lambda = 2, 1$ (dashed), 1/4 (dotted-dashed); *a)*–*d)* for cases *I*–*IV*

Chapter 3

CONTACT PROBLEMS FOR A CYLINDER

In this chapter we present the methods and results for certain static contact problems for an infinite elastic circular cylinder, an elastic space with an infinite cylindrical circular cavity, and also for an elastic circular plate and a finite cylinder.

The equilibrium of an infinite circular cylinder and circular plate under the action of normal loads has been considered, e.g., in Lur'e (1964). The contact problem for an infinite circular cylinder was formulated for the first time, apparently, by Voronin (1957). The development and investigation of this problem has been made in Aleksandrov (1962), and Aleksandrov and Belokon' (1967,1968). The contact problem for an elastic space with an infinite circular cylindrical cavity was investigated in Aleksandrov and Belokon' (1967,1968). Efficient solutions for the contact problems for finite solids, in particular, for a circular plate, have been suggested in Aleksandrov (1973,1975), Aleksandrov and Chebakov (1973,1977), and Gladwell (1980).

In this chapter we consider axisymmetric problems, although the methods can be used also for investigation of non-axisymmetric problems.

3.1 Equilibrium under Given Normal Loads

3.1.1. AN INFINITE CYLINDER

Using cylindrical coordinates r, φ, and z, we consider the static axisymmetric problem of an infinite linear-elastic, isotropic and homogeneous cylinder

($r \leq R$, $0 \leq \varphi < 2\pi$, $|z| < \infty$), and also the static axisymmetric problem of a space with an infinite cylindrical cavity.

We use the general solution of axisymmetric problems through the biharmonic Love function $\Phi(r, z)$:

$$\Delta^2 \Phi = 0, \qquad \Delta = \frac{\partial^2}{\partial r^2} + \frac{1}{r}\frac{\partial}{\partial r} + \frac{\partial^2}{\partial z^2}. \tag{3.1}$$

The fields of displacements and stresses are determined via this function as follows:

$$2Gu_r = -\frac{\partial^2 \Phi}{\partial r \partial z}, \qquad 2Gu_z = \left[2(1-\nu)\Delta - \frac{\partial^2}{\partial z^2}\right]\Phi, \tag{3.2}$$

$$\sigma_r = \frac{\partial}{\partial z}\left(\nu\Delta - \frac{\partial^2}{\partial r^2}\right)\Phi, \qquad \sigma_\varphi = \frac{\partial}{\partial z}\left(\nu\Delta - \frac{1}{r}\frac{\partial}{\partial r}\right)\Phi, \tag{3.3}$$

$$\sigma_z = \frac{\partial}{\partial z}\left[(2-\nu)\Delta - \frac{\partial^2}{\partial z^2}\right]\Phi, \qquad \tau_{rz} = \frac{\partial}{\partial r}\left[(1-\nu)\Delta - \frac{\partial^2}{\partial z^2}\right]\Phi, \tag{3.4}$$

where G is the shear modulus and ν is Poisson's ratio.

We will seek the Love function as a Fourier integral by z, namely,

$$\Phi = \frac{1}{2\pi}\int_{-\infty}^{\infty} \Phi_*(r, \alpha)\exp(-i\alpha z)\, d\alpha. \tag{3.5}$$

Then to determine the Fourier transform Φ_*, we obtain the following ordinary differential equation:

$$\left(\frac{d^2}{dr^2} + \frac{1}{r}\frac{d}{dr} - \alpha^2\right)^2 \Phi_* = 0,$$

whose general solution has the form

$$\Phi_*(r, \alpha) = a_1\mathcal{I}_0(|\alpha|r) + a_2|\alpha|r\mathcal{I}_1(|\alpha|r) + b_1\mathcal{K}_0(|\alpha|r) + b_2|\alpha|r\mathcal{K}_1(|\alpha|r). \tag{3.6}$$

Here, a_n and b_n ($n = 1, 2$) are functions of α, $\mathcal{I}_n(x)$ and $\mathcal{K}_n(x)$ are the modified Bessel functions in Gradshteyn and Rhyzhik (1994), often called cylindrical functions (Lur'e (1964)).

Substituting (3.5) and (3.6) into (3.2)–(3.4), we get the following expressions for the displacement u_r and stresses σ_r, τ_{rz}:

$$u_r = \frac{i}{4\pi G}\int_{-\infty}^{\infty} \alpha|\alpha|[a_1\mathcal{I}_1(|\alpha|r) + a_2|\alpha|r\mathcal{I}_0(|\alpha|r) \tag{3.7}$$

$$-b_1\mathcal{K}_1(|\alpha|r) - b_2|\alpha|r\mathcal{K}_0(|\alpha|r)]\exp(-i\alpha z)\,d\alpha,$$

$$\sigma_r = \frac{i}{2\pi}\int_{-\infty}^{\infty} \alpha^3\{a_1[\mathcal{I}_0(|\alpha|r) - (|\alpha|r)^{-1}\mathcal{I}_1(|\alpha|r)] \tag{3.8}$$

$$+a_2[(1-2\nu)\mathcal{I}_0(|\alpha|r) + |\alpha|r\mathcal{I}_1(|\alpha|r)] + b_1[\mathcal{K}_0(|\alpha|r) + (|\alpha|r)^{-1}\mathcal{K}_1(|\alpha|r)]$$

$$+b_2[-(1-2\nu)\mathcal{K}_0(|\alpha|r) + |\alpha|r\mathcal{K}_1(|\alpha|r)]\}\exp(-i\alpha z)\,d\alpha,$$

$$\tau_{rz} = \frac{1}{2\pi}\int_{-\infty}^{\infty} |\alpha|^3\{a_1\mathcal{I}_1(|\alpha|r) + a_2[|\alpha|r\mathcal{I}_0(|\alpha|r) + 2(1-\nu)\mathcal{I}_1(|\alpha|r)] \tag{3.9}$$

$$-b_1\mathcal{K}_1(|\alpha|r) - b_2[|\alpha|r\mathcal{K}_0(|\alpha|r) - 2(1-\nu)\mathcal{K}_1(|\alpha|r)]\}\exp(-i\alpha z)\,d\alpha.$$

The components u_z, σ_φ and σ_z can be obtained similarly.

We consider the static problem for an infinite cylinder of radius R under the boundary conditions

$$\sigma_r(R,z) = -\tilde{q}(z), \qquad \tau_{rz}(R,z) = 0, \tag{3.10}$$

$$\tilde{q}(z) = q(z) \quad (|z| \le a), \qquad \tilde{q}(z) = 0 \quad (|z| > a),$$

and under the condition of boundedness of stresses at $r = 0$.

We suppose that the function $q(z) \in \mathcal{L}(-a,a)$ (absolutely integrable on the segment $|z| \le a$) and $q(z) \in \mathcal{C}(-a+\varepsilon, a-\varepsilon)$ (continuous on the segment $|z| \le a - \varepsilon$). For this case, we have the relations

$$\tilde{q}(z) = \frac{1}{2\pi}\int_{-\infty}^{\infty} Q(\alpha)\exp(-i\alpha z)\,d\alpha,$$

$$Q(\alpha) = \int_{-a}^{a} q(\zeta)\exp(i\alpha\zeta)\,d\zeta, \tag{3.11}$$

and, to satisfy the boundary conditions (3.10), we can use formulae (3.7)–(3.9).

The condition of boundedness of the stresses at $r = 0$ will be satisfied if in (3.7) we set $b_1 = b_2 = 0$; then the boundary conditions (3.10) give the relations

$$a_1(\mathcal{I}_0 - u^{-1}\mathcal{I}_1) + a_2[(1-2\nu)\mathcal{I}_0 + u\mathcal{I}_1] = Q(\alpha)/(i\alpha^3), \tag{3.12}$$

$$a_1\mathcal{I}_1 + a_2[u\mathcal{I}_0 + 2(1-\nu)\mathcal{I}_1] = 0 \qquad (u = |\alpha|R). \tag{3.13}$$

Here and later we omit the argument u in the functions $\mathcal{I}_0(u)$ and $\mathcal{I}_1(u)$ for brevity. Solving equations (3.12), (3.13) in a_1 and a_2, we obtain

$$a_1 = -\frac{Q(\alpha)[u\mathcal{I}_0 + 2(1-\nu)\mathcal{I}_1]u}{i\alpha^3 D}, \qquad a_2 = \frac{Q(\alpha)\mathcal{I}_1 u}{i\alpha^3 D}, \tag{3.14}$$

$$D(u) = u^2(\mathcal{I}_0^2 - \mathcal{I}_1^2) - 2(1-\nu)\mathcal{I}_1^2. \tag{3.15}$$

By substituting (3.14) into (3.7)–(3.9) and using formula (3.11), we find

$$u_r(R,z) = -\frac{1}{\pi\theta}\int_{-a}^{a} q(\zeta)K_1\left(\frac{\zeta - z}{R}\right)d\zeta, \tag{3.16}$$

$$K_1(t) = \int_0^\infty \frac{L_1(u)}{u}\cos(ut)\,du, \qquad \theta = \frac{G}{1-\nu}, \tag{3.17}$$

$$L_1(u) = \frac{u}{u^2(\omega_0^2 - 1) - 2(1-\nu)}, \qquad \omega_0(u) = \frac{\mathcal{I}_0}{\mathcal{I}_1}. \tag{3.18}$$

Now consider the static problem for a space with an infinite cylindrical cavity of radius R under the boundary conditions (3.10) and the condition of boundedness of the stresses when $r \to \infty$.

The condition of boundedness of the stresses when $r \to \infty$ will be satisfied if in (3.7)–(3.9) we set $a_1 = a_2 = 0$; then the boundary conditions (3.10) give the relations

$$b_1(\mathcal{K}_0 + u^{-1}\mathcal{K}_1) + b_2[-(1-2\nu)\mathcal{K}_0 + u\mathcal{K}_1] = -Q(\alpha)/(i\alpha^3), \tag{3.19}$$

$$b_1\mathcal{K}_1 + b_2[u\mathcal{K}_0 - 2(1-\nu)\mathcal{K}_1] = 0 \qquad (u = |\alpha|R). \tag{3.20}$$

Here and later we also omit the argument u in the functions $\mathcal{K}_0(u)$ and $\mathcal{K}_1(u)$ for brevity. Solving equations (3.19), (3.20) in b_1 and b_2, we obtain

$$b_1 = \frac{Q(\alpha)[u\mathcal{K}_0 - 2(1-\nu)\mathcal{K}_1]u}{i\alpha^3 D}, \qquad b_2 = \frac{Q(\alpha)\mathcal{K}_1 u}{i\alpha^3 D}, \tag{3.21}$$

$$D(u) = 2(1-\nu)\mathcal{K}_1^2 - u^2(\mathcal{K}_0^2 - \mathcal{K}_1^2).$$

By substituting (3.21) into (3.7)–(3.9) and using formula (3.11), we find

$$u_r(R,z) = \frac{1}{\pi\theta}\int_{-a}^{a} q(\zeta)K_2\left(\frac{\zeta - z}{R}\right)d\zeta, \tag{3.22}$$

$$K_2(t) = \int_0^\infty \frac{L_2(u)}{u} \cos(ut)\, du, \tag{3.23}$$

$$L_2(u) = \frac{u}{2(1-\nu) - u^2(\Omega_0^2 - 1)}, \qquad \Omega_0(u) = \frac{\mathcal{K}_0}{\mathcal{K}_1}. \tag{3.24}$$

3.1.2. A FINITE CYLINDER UNDER SPECIAL CONDITIONS

Now consider the static problem for a cylinder of radius R and finite length $2h$ ($|z| \leq h$), which is symmetrically loaded by normal forces on the end-walls and is subjected to a sliding support along the cylindrical surface (Fig. 3.1). The boundary conditions have the form

$$\sigma_z(r, \pm h) = -\tilde{q}(r), \qquad \tau_{rz}(r, \pm h) = 0, \tag{3.25}$$

$$\tilde{q}(r) = q(r) \quad (0 \leq r \leq a < R), \qquad \tilde{q}(r) = 0 \quad (a < r \leq R), \tag{3.26}$$

$$u_r(R, z) = \tau_{rz}(R, z) = 0. \tag{3.27}$$

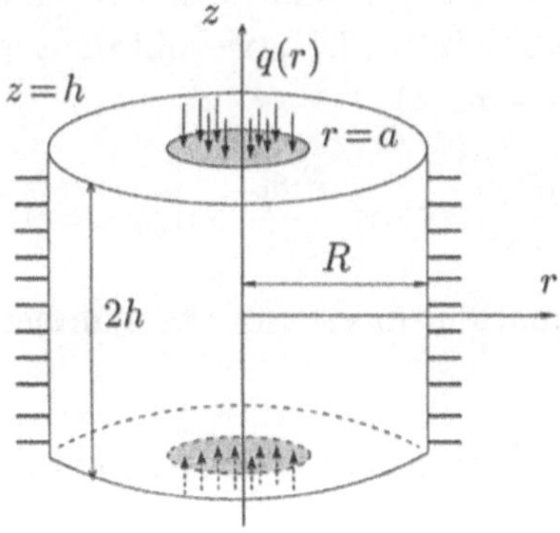

Figure 3.1: A finite cylinder under special conditions

We will seek the solution of equation (3.1) in the form

$$\Phi(r, z) = \sum_{k=0}^{\infty} \varphi_k(z) \mathcal{J}_0(\alpha_k r), \tag{3.28}$$

where α_k are the roots of the transcendental equation

$$\mathcal{J}_1(\alpha R) = 0 \tag{3.29}$$

($\mathcal{J}_0(x)$ and $\mathcal{J}_1(x)$ are the Bessel functions).

We note that all the roots of equation (3.29) are real and simple; $\alpha_0 = 0$, $\alpha_1 = 3.832/R$, $\alpha_2 = 7.016/R$, $\alpha_3 = 10.17/R$, $\alpha_4 = 13.32/R$, $\alpha_5 = 16.47/R$, and the large zeros have the asymptotic form

$$\alpha_k \sim \pi(k + 1/4)/R.$$

It may be verified that the system $\mathcal{J}_0(\alpha_k r)$ is a complete system of linearly independent functions as $0 \leq r \leq R$, and that any solution of equation (3.1) in the cylinder can be presented by the expansion (3.28). The system of functions $\mathcal{J}_0(\alpha_k r)$ satisfies the following orthogonality condition (Lebedev (1965)):

$$\int_0^R r\mathcal{J}_0(\alpha_k r)\mathcal{J}_0(\alpha_l r)\,dr = \begin{cases} 0 & \text{for } k \neq l \\ 2^{-1}R^2\mathcal{J}_0^2(\alpha_k R) & \text{for } k = l \end{cases}. \tag{3.30}$$

Using the first formula of (3.2) and the second formula of (3.4), we see that, when seeking the function $\Phi(r, z)$ in the form (3.28), we can satisfy the boundary conditions (3.27) automatically in view of (3.29).

By substituting (3.28) into (3.1), we obtain the ordinary differential equation for the function $\varphi_k(z)$

$$\frac{d^4\varphi_k}{dz^4} - 2\alpha_k^2\frac{d^2\varphi_k}{dz^2} + \alpha_k^4\varphi_k = 0. \tag{3.31}$$

The solution of this equation, in view of the symmetry of the loading, can be taken in the form

$$\varphi_k = a_1^{(k)}\sinh(\alpha_k z) + a_2^{(k)}\alpha_k z\cosh(\alpha_k z), \tag{3.32}$$

where $a_1^{(k)}$ and $a_2^{(k)}$ are[1] the functions in α_k which can be determined from the boundary conditions (3.25).

On the basis of formulae (3.2)–(3.4), (3.28) and (3.32), we obtain for displacements u_r and u_z and for stresses σ_r, σ_z and τ_{rz} the formulae

$$u_r = \frac{1}{2G}\sum_{k=0}^{\infty}\alpha_k^2[(a_1^{(k)} + a_2^{(k)})\cosh(\alpha_k z) + a_2^{(k)}\alpha_k z\sinh(\alpha_k z)]\mathcal{J}_1(\alpha_k r), \tag{3.33}$$

$$u_z = \frac{1}{2G}\sum_{k=0}^{\infty}\alpha_k^2\{[-a_1^{(k)} + 2(1-2\nu)a_2^{(k)}]\sinh(\alpha_k z) \tag{3.34}$$

[1]In this section the subscript (k) does not mean a differentiation.

$$-a_2^{(k)}\alpha_k z\,\cosh(\alpha_k z)\}\mathcal{J}_0(\alpha_k r),$$

$$\sigma_r = \sum_{k=0}^{\infty}\alpha_k^3\{2\nu a_2^{(k)}\cosh(\alpha_k z)\,\mathcal{J}_0(\alpha_k r) + [\mathcal{J}_0(\alpha_k r) - (\alpha_k r)^{-1}\mathcal{J}_1(\alpha_k r)]$$

$$\times[(a_1^{(k)} + a_2^{(k)})\cosh(\alpha_k z) + a_2^{(k)}\alpha_k z\,\sinh(\alpha_k z)]\},$$

$$\sigma_z = \sum_{k=0}^{\infty}\alpha_k^3\{[-a_1^{(k)} + (1-2\nu)a_2^{(k)}]\cosh(\alpha_k z) \tag{3.35}$$

$$-a_2^{(k)}\alpha_k z\,\sinh(\alpha_k z)\}\mathcal{J}_0(\alpha_k r),$$

$$\tau_{rz} = \sum_{k=o}^{\infty}\alpha_k^3[(a_1^{(k)} + 2\nu a_2^{(k)})\sinh(\alpha_k z) + a_2^{(k)}\alpha_k z\,\cosh(\alpha_k z)]\mathcal{J}_1(\alpha_k r). \tag{3.36}$$

We will not need the expression for σ_φ.

We assume that the function $q(r)$ in (3.26) satisties $q(r) \in \mathcal{L}(\Omega_a)$ (absolutely summable in the circle $r \le a$) and $q(r) \in C(\Omega_{a-\varepsilon})$ (continuous in the circle $r \le a - \varepsilon$). For this case we have with regard for (3.30) the relations

$$\tilde{q}(r) = \sum_{k=0}^{\infty} Q_k\mathcal{J}_0(\alpha_k r), \tag{3.37}$$

$$Q_k = \frac{2}{R^2\mathcal{J}_0^2(\alpha_k R)}\int_0^a \rho q(\rho)\mathcal{J}_0(\alpha_k\rho)\,d\rho.$$

By using formulae (3.35), (3.36) and (3.37), we see that the conditions (3.25) lead us to the following equations in $a_1^{(k)}$ and $a_2^{(k)}$:

$$-a_1^{(k)}\cosh(\alpha_k h) + a_2^{(k)}[(1-2\nu)\cosh(\alpha_k h) \tag{3.38}$$

$$-\alpha_k h\,\sinh(\alpha_k h)] = -Q_k/(\alpha_k^3),$$

$$a_1^{(k)}\sinh(\alpha_k h) + a_2^{(k)}[2\nu\,\sinh(\alpha_k h) + \alpha_k h\,\cosh(\alpha_k h)] = 0. \tag{3.39}$$

From here we find

$$a_1^{(k)} = \frac{2Q_k}{\alpha_k^3 D^{(k)}}[2\nu\,\sinh(\alpha_k h) + \alpha_k h\,\cosh(\alpha_k h)], \tag{3.40}$$

$$a_2^{(k)} = -\frac{2Q_k}{\alpha_k^3 D^{(k)}}\sinh(\alpha_k h), \quad D^{(k)} = \sinh(2\alpha_k h) + 2\alpha_k h. \tag{3.41}$$

By substituting (3.40), (3.41) into (3.34), we specifically obtain that

$$u_z(r, \pm h) = \mp\frac{1}{\theta}\sum_{k=0}^{\infty}\frac{L(\alpha_k h)}{\alpha_k}Q_k\mathcal{J}_0(\alpha_k r), \tag{3.42}$$

$$L(u) = \frac{\cosh(2u) - 1}{\sinh(2u) + 2u}. \tag{3.43}$$

3.1.3. A LAYER UNDER SPECIAL CONDITIONS

Let us now consider the problem of equilibrium of a layer $(0 \leq r < \infty)$ of thickness $2h$ $(|z| \leq h)$, which is symmetrically loaded by the normal forces in the end-walls (Fig 3.2). The boundary conditions have the form

$$\sigma_z(r, \pm h) = -\tilde{q}(r), \qquad \tau_{rz}(r, \pm h) = 0, \tag{3.44}$$

$$\tilde{q}(r) = q(r) \quad (0 \leq r \leq a), \qquad \tilde{q}(r) = 0 \quad (a < r < \infty).$$

We will seek the solution of equation (3.1) as a Hankel integral

$$\Phi(r, z) = \int_0^\infty \alpha\,\varphi(\alpha, z)\mathcal{J}_0(\alpha r)\,d\alpha. \tag{3.45}$$

If we substitute (3.45) into (3.1), we find that the function $\varphi(\alpha, z)$ satisfies equation (3.31) in which we should omit the index k. The solution of this equation in view of the symmetry of the loading should be taken, as before, in the form

$$\varphi = a_1\sinh(\alpha z) + a_2\alpha z\,\cosh(\alpha z), \tag{3.46}$$

where a_1 and a_2 are functions of α which should be determined from the boundary conditions (3.44).

By comparing (3.28), (3.32) with (3.45), (3.46), we see that we have the formal relation

$$\sum_{k=0}^{\infty}\langle\ldots\rangle \sim \int_0^\infty \alpha\langle\ldots\rangle d\alpha,$$

in which one should omit the index k in the angular brackets under the integral sign. On the basis of this formalism we can rewrite formulae (3.33)–(3.36) and obtain the expressions for the displacements and stresses in the layer.

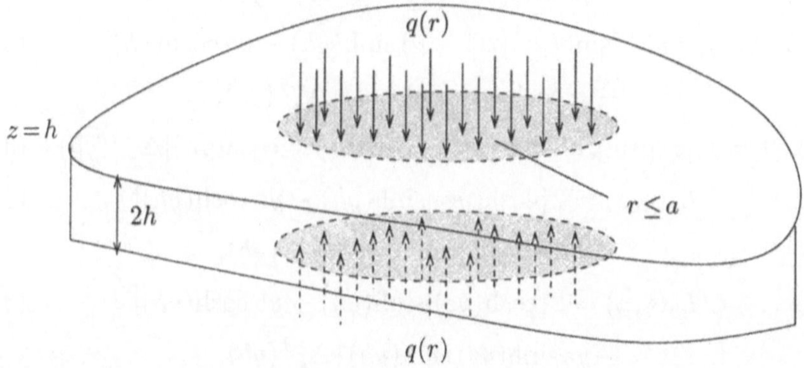

Figure 3.2: A layer under special conditions

By introducing the Hankel transform $Q(\alpha)$ for the function $\tilde{q}(r)$ by the formulae

$$\tilde{q}(r) = \int_0^\infty \alpha Q(\alpha)\mathcal{J}_0(\alpha r)\,d\alpha, \qquad Q(\alpha) = \int_0^a \rho q(\rho)\mathcal{J}_0(\alpha\rho)\,d\rho,$$

and then satisfying, as in subsection 2, the boundary conditions (3.44), we obtain

$$u_r = \frac{1}{G}\int_0^\infty Q(y)L_1(y,z)\mathcal{J}_1(yr)y\,dy, \tag{3.47}$$

$$u_z = \frac{1}{G}\int_0^\infty Q(y)L_2(y,z)\mathcal{J}_0(yr)y\,dy, \tag{3.48}$$

$$\tau_{rz} = \int_0^\infty Q(y)L_3(y,z)\mathcal{J}_1(yr)y\,dy, \tag{3.49}$$

$$\sigma_z(r,z) = \int_0^\infty Q(y)L_6(y,z)\mathcal{J}_0(yr)y\,dy, \tag{3.50}$$

$$\sigma_r(r,z) = \int_0^\infty Q(y)\left[L_4(y,z)\mathcal{J}_0(yr) + \frac{L_5(y,z)}{yr}\mathcal{J}_1(yr)\right]y\,dy, \tag{3.51}$$

$$L_1(y,z) = y^{-1}\{\cosh(yz)[(1-2\nu)\sinh(yh) - yh\cosh(yh)] \tag{3.52}$$

$$+yz\sinh(yz)\sinh(yh)\}\Delta_*^{-1}(yh),$$

$$L_2(y,z) = y^{-1}\{\sinh(yz)[2(1-\nu)\sinh(yh) + yh\cosh(yh)] \tag{3.53}$$

$$-yz\cosh(yz)\sinh(yh)\}\Delta_*^{-1}(yh),$$

$$L_3(y,z) = 2y[z\cosh(yz)\sinh(yh) - h\sinh(yz)\cosh(yh)]\Delta_*^{-1}(yh), \tag{3.54}$$

$$L_4(y,z) = 2\{\cosh(yz)[\sinh(yh) - yh\cosh(yh)] \tag{3.55}$$

$$+yz\sinh(yz)\sinh(yh)\}\Delta_*^{-1}(yh),$$

$$L_6(y,z) = 2\{\cosh(yz)[\sinh(yh) + yh\cosh(yh)] \tag{3.56}$$

$$-yz\sinh(yz)\sinh(yh)\}\Delta_*^{-1}(yh),$$

$$L_5(y,z) = -2yL_1(y,z), \qquad \Delta_*(v) = \sinh(2v) + 2v. \tag{3.57}$$

From these formulae we specifically find

$$u_z(r,\pm h) = \mp\frac{1}{\theta}\int_0^a \rho q(\rho)K(r,\rho)\,d\rho,$$

$$K(r,\rho) = \int_0^\infty L(uh)\mathcal{J}_0(ru)\mathcal{J}_0(\rho u)\,du, \tag{3.58}$$

where $L(u)$ is given by formula (3.43).

3.1.4. HOMOGENEOUS SOLUTIONS

Let us find the system of homogeneous solutions of the static problem (3.44) for the layer, i.e., the system of nontrivial solutions of the equation (3.1) in the layer for the case when $\tilde{q}(r) \equiv 0$. We will seek the functions $\Phi^{(k)}(r,z)$ $(k = 0,1,2,...)$ in the form

$$\Phi^{(k)}(r,z) = [a_1^{(k)}\sinh(\alpha_k z) + a_2^{(k)}\alpha_k z\cosh(\alpha_k z)]\mathcal{J}_0(\alpha_k r).$$

As a result of satisfaction of the homogeneous boundary conditions (3.44), we come to the homogeneous system (3.38), (3.39) in $a_1^{(k)}$ and $a_2^{(k)}$ from which we obtain the relation

$$a_1^{(k)} = -[2\nu + \alpha_k h\coth(\alpha_k h)]a_2^{(k)}, \tag{3.59}$$

where now $\alpha_k = v_k/h$, and v_k are the roots of the transcendental equation

$$\Delta_*(v) = 0. \tag{3.60}$$

Note that all the roots of this equation are simple and complex excepting the trivial root $v_0 = 0$. Let $v_n = \gamma_n + i\delta_n$ $(n = 1, 2, ...)$ be the set of the complex roots lying in the first quadrant of the complex variable v, with successively increasing imaginary parts. Then we can easily check that $\gamma_n - i\delta_n$, $-\gamma_n + i\delta_n$, $-\gamma_n - i\delta_n$ are roots of (3.60) also. The values of the first five complex roots v_n, lying in the first quadrant, are given in Table 3.1. The large zeros have the asymptotic form (Vorovich, Aleksandrov and Babeshko (1974))

$$v_n \sim \frac{1}{2}\ln\left[4\pi\left(n - \frac{1}{4}\right)\right] + i\pi\left(n - \frac{1}{4}\right). \tag{3.61}$$

Table 3.1: First roots for (3.60)

v_1	$1.125 + i\,2.106$
v_2	$1.552 + i\,5.356$
v_3	$1.776 + i\,8.537$
v_4	$1.929 + i\,11,70$
v_5	$2.047 + i\,14.85$

Substituting $a_1^{(k)}$ from (3.59) into expressions (3.33)–(3.36) under the Σ signs and setting $a_2^{(k)} = \alpha_k^{-3}\cosh(\alpha_k h)$, we arrive at the following formulae for the system of nontrivial homogeneous solutions:

$$u_r^{(k)} = L_{1k}^0(z)\mathcal{J}_1(\alpha_k r), \tag{3.62}$$

$$u_z^{(k)} = L_{2k}^0(z)\mathcal{J}_0(\alpha_k r), \tag{3.63}$$

$$\tau_{rz}^{(k)} = GL_{3k}^0(z)\mathcal{J}_1(\alpha_k r), \tag{3.64}$$

$$\sigma_r^{(k)} = G[L_{4k}^0(z)\mathcal{J}_0(\alpha_k r) + (\alpha_k r)^{-1}L_{5k}^0(z)\mathcal{J}_1(\alpha_k r)], \tag{3.65}$$

$$\sigma_z^{(k)} = GL_{6k}^0(z)\mathcal{J}_0(\alpha_k r), \tag{3.66}$$

where the functions $L_{nk}^0(z)$ $(n = 1, 2, ..., 6)$ can be expressed via the functions $L_n(y, z)$ with the help of the relation

$$L_{nk}^0(z) = L_n(\alpha_k, z)\alpha_k\Delta_*(\alpha_k h).$$

We will not need the expression for $\sigma_\varphi^{(k)}$. In section 3.6 we will use only such homogeneous solutions (3.62)–(3.66) for which $\operatorname{Im}\alpha_k > 0$ $(\alpha_{2k} = \bar{\alpha}_{2k-1})$.

For the homogeneous solutions (3.62)–(3.66), we have the relation (Chebakov (1980))

$$W_{kl} = \int_{-h}^{h} [L_{4k}^0(z)L_{1l}^0(z) - L_{2k}^0(z)L_{3l}^0(z)]\,dz = 0 \qquad (k \neq l), \tag{3.67}$$

which is known as a condition of generalised orthogonality, and

$$W_{kk} = 8h^3\alpha_k^3(1-\nu). \tag{3.68}$$

The homogeneous solutions for a layer will be used in contact problems for a plate (section 3.6). Lur'e (1964) defined a plate as a part of a layer bounded by a cylindrical surface C with generators perpendicular to the plane faces of the layer. Note that this lateral surface C may be non-cylindrical also. For the plate, we must not only satisfy the boundary conditions on the faces, but also the conditions on the surface C. Therefore, apart from solutions which satisfy the boundary conditions on the faces, we must make use of a class of solutions of the equations of elasticity theory which leave the faces free from loading; these solutions are called homogeneous, and the solutions satisfying loading conditions on the faces non-homogeneous. The homogeneous solutions (3.62)–(3.66) determine the state of stress of a plate produced by loads distributed over its lateral surface. The sum of the homogeneous and non-homogeneous solutions obviously satisfies the loading conditions on the faces of the plate, and the arbitrariness of the selection of the homogeneous solutions must be used to satisfy the boundary conditions on the surface C.

3.2 Contact Problems for a Cylinder, and a Space with a Cylindrical Cavity

3.2.1. INTEGRAL EQUATIONS

We consider the contact between an infinite elastic cylinder of radius R and a rigid sleeve of the width $2a$ having the base $r = R - \delta(z)$ (Fig. 3.3). We suppose there are no friction forces in the domain of contact, and no load outside the contact domain; then the boundary conditions can be written in the form

$$\sigma_r(R,z) = 0 \quad (|z| > a), \qquad \tau_{rz}(R,z) = 0 \quad (|z| < \infty), \tag{3.69}$$

$$u_r(R,z) = -\delta(z) \quad (|z| \le a). \tag{3.70}$$

We assume that the stresses are bounded at $r = 0$.

To satisfy the boundary conditions (3.69), (3.70), one can use solution (3.16)–(3.18) of the auxiliary problem (3.10) for the cylinder. As a result, we obtain for the contact pressure $q(z) = -\sigma_r(R, z)$ ($|z| \le a$) the integral equation

$$\int_{-a}^{a} q(\zeta) K_1\left(\frac{\zeta - z}{R}\right) d\zeta = \pi\theta\delta(z) \qquad (|z| \le a), \tag{3.71}$$

in which the kernel $K_1(t)$ is determined by formulae (3.17).

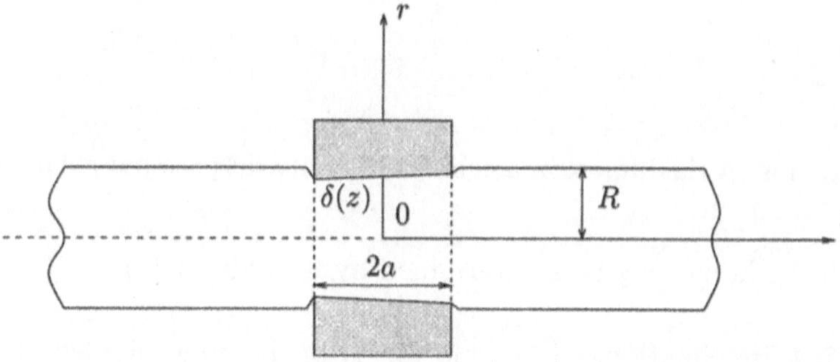

Figure 3.3: Contact between a sleeve and an infinite cylinder (an axial section)

We also consider the interaction between an elastic space, weakened by an infinite cylindrical cavity of radius R, and a rigid insert of width $2a$ having base radius $r = R + \delta(z)$ (Fig. 3.4). We again assume that there are no friction forces in the domain of contact between the insert and the space with the cavity, and no load outside the contact domain; then the boundary conditions on the cavity surface can be written in the form

$$\sigma_r(R, z) = 0 \quad (|z| > a), \qquad \tau_{rz}(R, z) = 0 \quad (|z| < \infty), \tag{3.72}$$

$$u_r(R, z) = \delta(z) \quad (|z| \le a). \tag{3.73}$$

We assume that the stresses are bounded when $r \to \infty$.

To satisfy the boundary conditions (3.72), (3.73), one can use solution (3.22)–(3.24) of the auxiliary problem (3.5) for the space with a cylindrical cavity; we obtain for the contact pressure $q(z) = -\sigma_r(R, z)$ ($|z| \le a$) the integral equation

$$\int_{-a}^{a} q(\zeta) K_2\left(\frac{\zeta - z}{R}\right) d\zeta = \pi\theta\delta(z) \qquad (|z| \le a), \tag{3.74}$$

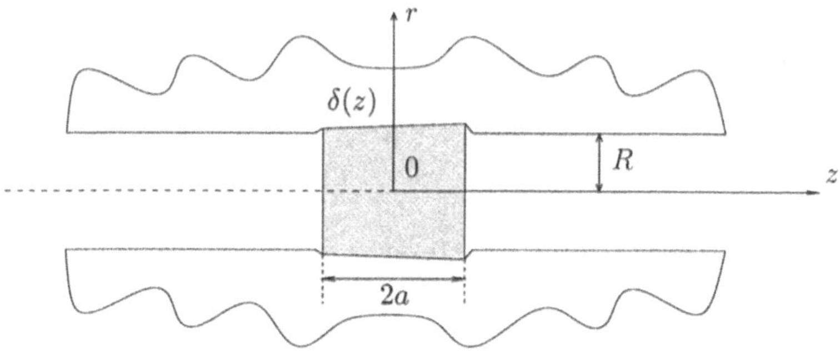

Figure 3.4: A rigid insert in a cylindrical cavity in space (an axial section)

in which the kernel $K_2(t)$ is determined by formula (3.23).

We rewrite equations (3.71) and (3.74) in the dimensionless form

$$\int_{-1}^{1} \varphi(\xi) K_n\left(\frac{\xi - x}{\lambda}\right) d\xi = \pi f(x) \qquad (|x| \le 1;\ n = 1, 2), \tag{3.75}$$

where we have introduced the notation

$$x = \frac{z}{a}, \quad \xi = \frac{\zeta}{a}, \quad \lambda = \frac{R}{a}, \quad \varphi(x) = \frac{q(ax)}{\theta}, \quad f(x) = \frac{\delta(ax)}{a}. \tag{3.76}$$

We will now omit the index n in the kernel $K_n(t)$ saying, if necessary, that it is problem 1 ($n = 1$) or problem 2 ($n = 2$).

We recall that

$$K(t) = \int_0^\infty \frac{L(u)}{u} \cos(ut)\, du, \tag{3.77}$$

where the $L(u)$ for problems 1 and 2 are given by formulae (3.18) and (3.24), respectively. The function $L(u)$ in (3.77) possesses the following properties: for large values of u

$$L(u) = 1 + c_1 u^{-1} + c_2 u^{-2} + c_3 u^{-3} + c_4 u^{-4} + O(u^{-5}), \tag{3.78}$$

where for problem 1, and $\nu = 0.3$,

$$c_1 = 0.4, \qquad c_2 = -0.965, \qquad c_3 = -2.336, \qquad c_4 = -2.9097125,$$

and for problem 2, and $\nu = 0.3$,

$$c_1 = -0.4, \qquad c_2 = -0.965, \qquad c_3 = 2.336, \qquad c_4 = -2.9097125;$$

for small values of u we have

$$L(u) = u[2(1+\nu)]^{-1}[1 + O(u^4)] \tag{3.79}$$

for problem 1, and

$$L(u) = u[2(1-\nu)]^{-1}\{1 - u^2[2(1-\nu)]^{-1} + O(u^4 \ln^2 u)\} \tag{3.80}$$

for problem 2.

Now if the integrals 3.951.8 and 3.741.3 in Gradshteyn and Ryzhik (1994)

$$\int_0^\infty \frac{\cos(ut) - \exp(-u)}{u}\, du = -\ln|t|, \quad \int_0^\infty \frac{1-\cos(ut)}{u^2}\, du = \frac{\pi}{2}\,|t| \tag{3.81}$$

are used together with properties (3.78)–(3.80), then the kernel $K(t)$ may be presented as:

$$K(t) = -\ln|t| - \frac{\pi c_1}{2}\,|t| + \frac{c_2}{2}\,t^2 \ln|t| - \frac{3c_2}{4}\,t^2 + \frac{\pi c_3}{12}\,|t|^3 \tag{3.82}$$

$$+ \int_0^\infty \Big[(u^3 L(u) - u^3 - c_1 u^2 - c_2 u - c_3)\cos(ut) + u^3 \exp(-u) + c_1 u^2$$

$$-c_2 u \left(\frac{1}{2} u^2 t^2 \exp(-u) - 1 \right) - c_3 \left(\frac{1}{2} u^2 t^2 - 1 \right) \Big] \frac{du}{u^4}.$$

We hence obtain the following asymptotic representation for the kernel for small t (or equivalently, for large $\lambda = R/a$):

$$K(t) = -\ln|t| + F(t) \qquad (|t| < \infty), \tag{3.83}$$

where the function $F(t)$ is such that its first derivative satisfies a Lipschitz condition and has the form[2]

$$F(t) = a_{30} + a_{20}|t| + a_{11}t^2 \ln|t| + a_{31}t^2 + a_{21}|t|^3 + O(t^4 \ln|t|), \tag{3.84}$$

---

[2]For problem 1 the process of calculation of the function $F(t)$ for any values of $|t| < \infty$ will be described in detail in section 3.5.

$$a_{30} = \int_0^\infty \frac{L(u) - 1 + \exp(-u)}{u}\, du, \quad a_{20} = -\frac{\pi c_1}{2}, \quad a_{11} = \frac{c_2}{2}, \quad a_{21} = -\frac{\pi c_3}{12},$$

$$a_{31} = -\frac{3c_2}{4} + \frac{1}{2}\int_0^\infty [u^2 - u^2 L(u) + c_1 u + c_2(1 - \exp(-u))]\, \frac{du}{u}.$$

We calculate

$$a_{30} = -0.556, \qquad a_{20} = -0.628, \qquad a_{11} = -0.482,$$

$$a_{31} = 1.427, \qquad a_{21} = -0.612$$

for problem 1 and $\nu = 0.3$, and

$$a_{30} = -0.455, \qquad a_{20} = 0.628, \qquad a_{11} = -0.482,$$

$$a_{31} = -0.390, \qquad a_{21} = 0.612$$

for problem 2 and $\nu = 0.3$.

Below we will need the approximation of $L(u)u^{-1}$ by the easily factorizable function (Aleksandrov (1986), Aleksandrov and Pozharskii (1999b))

$$K^*(u) = \frac{\sqrt{u^2 + B^2}}{u^2 + C^2} \exp\left(\frac{D}{\sqrt{u^2 + E^2}}\right). \tag{3.85}$$

We choose the constants B, C, D, and E in this approximation so that the function $K^*(u)$ accurately reflects the behaviour of the function $L(u)u^{-1}$ on the real axis, especially as $u \to 0$ and $u \to \infty$. We choose

$$D = c_1, \qquad BC^{-2}\exp(DE^{-1}) = A,$$

where $A = [2(1 + \nu)]^{-1}$ for problem 1, and $A = [2(1 - \nu)]^{-1}$ for problem 2. Besides, it will be necessary later to satisfy the approximate equality (3.129), restricting the value of E: for the both problems under consideration one should take $E > 1$. As a result, for problem 1, and $\nu = 0.3$, we choose

$$B = 1.4442, \qquad C = 2.0929, \qquad D = 0.4, \qquad E = 2.5968, \tag{3.86}$$

then the error of the approximation (3.85) for all $0 \leq u < \infty$ does not exceed 8.7%; for problem 2, and $\nu = 0.3$, we choose

$$B = 0.8353, \qquad C = 0.8970, \qquad D = -0.4, \qquad E = 1.0699, \tag{3.87}$$

then the error of approximation (3.85) for all $0 \le u < \infty$ does not exceed 5%. Moreover, for problem 1 the asymptotic behaviour (3.79) has been exactly taken into account in (3.85).

3.2.2. IMPORTANT SINGULAR INTEGRALS

We will also need some singular integrals. Consider the singular integral

$$I(x) = \int_{-1}^{1} \frac{d\tau}{\sqrt{1-\tau^2}\,(\tau - x)} \quad (|x| \le 1). \tag{3.88}$$

Let us show that $I(x) \equiv 0$. We change the integration variable in (3.88) to

$$\tau = (1-s^2)/(1+s^2).$$

Doing some obvious transformations, we have

$$\int_{-1}^{1} \frac{d\tau}{\sqrt{1-\tau^2}\,(\tau - x)} = 2\int_{0}^{\infty} \frac{ds}{1 - x - (1+x)s^2} \tag{3.89}$$

$$= \frac{1}{\sqrt{1-x^2}} \ln \frac{\sqrt{1-x} + s\sqrt{1+x}}{\sqrt{1-x} + s\sqrt{1+x}}\Bigg|_0^{\infty} = 0.$$

The integral $I(x)$ plays a key role for calculation of a number of singular integrals. Namely, by using (3.89), the identity

$$\frac{\tau^m}{\tau - x} = \frac{x^m}{\tau - x} + \tau^{m-1} + \tau^{m-2}x + \ldots + \tau x^{m-2} + x^{m-1},$$

and the integral 3.226.2 in Gradshteyn and Ryzhik (1994)

$$\int_{-1}^{1} \frac{\tau^{2n}\,d\tau}{\sqrt{1-\tau^2}} = \frac{\pi(2n-1)!!}{(2n)!!},$$

we get

$$\int_{-1}^{1} \frac{\tau^m\,d\tau}{\sqrt{1-\tau^2}\,(\tau - x)} = \begin{cases} \pi P_n(x) & \text{for } m = 2n+1 \\ \pi x P_n(x) & \text{for } m = 2n+2 \end{cases}, \tag{3.90}$$

$$P_n(x) = \sum_{k=0}^{n} \frac{(2k-1)!!}{(2k)!!} x^{2n-2k}.$$

On the basis of (3.90) we can also derive the formulae ($|x| \leq 1$)

$$\int_{-1}^{1} \frac{\tau^m \sqrt{1-\tau^2}}{\tau - x} d\tau = \begin{cases} -\pi x & \text{for } m = 0 \\ \pi Q_n(x) & \text{for } m = 2n+1 \\ \pi x Q_n(x) & \text{for } m = 2n+2 \end{cases}, \tag{3.91}$$

$$Q_n(x) = -x^{2n+2} \sum_{k=0}^{n} \frac{(2k-1)!!}{(2k+2)!!} x^{2n-2k},$$

and obtain the integrals ($|x| \leq 1$)

$$-\int_{-1}^{1} \frac{\tau^m \ln|\tau - x|}{\sqrt{1-\tau^2}} d\tau = \mu_m(x), \tag{3.92}$$

$$\mu_0(x) = \pi \ln 2, \qquad \mu_{2n+1}(x) = \pi \sum_{k=0}^{n} \frac{(2k-1)!!}{(2k)!!} \frac{x^{2n-2k+1}}{2n-2k+1},$$

$$\mu_{2n+2}(x) = \pi \sum_{k=0}^{n} \frac{(2k-1)!!}{(2k)!!} \frac{x^{2n-2k+2}}{2n-2k+2}$$

$$+\pi \left[\ln 2 + \sum_{k=1}^{2n+2} \frac{(-1)^k}{k} \right] \frac{(2n+1)!!}{(2n+2)!!} \qquad (n = 0, 1, \ldots).$$

3.2.3. SOLUTION FOR A THICK CYLINDER

Suppose radius R of the cylinder (cylindrical cavity) is considerably larger than the half-length a of the contact zone, i.e. the parameter $\lambda \gg 1$ in (3.76). We substitute kernel (3.83), with (3.84), into the integral equation (3.75) and apply the asymptotic method of *large* λ. According to the structure (3.83), (3.84) of the kernel for large λ (or small t), we seek the asymptotic solution of (3.75) in the form

$$\varphi(x) = \frac{1}{\sqrt{1-x^2}} \left[\omega_{00}(x) + \frac{1}{\lambda}\omega_{10}(x) + \frac{1}{\lambda^2}\omega_{20}(x) \right. \tag{3.93}$$

$$\left. + \frac{\ln\lambda}{\lambda^2}\omega_{21}(x) + \frac{1}{\lambda^3}\omega_{30}(x) + \frac{\ln\lambda}{\lambda^2}\omega_{31}(x) + O\left(\frac{\ln^2\lambda}{\lambda^4}\right) \right].$$

Differentiating (3.75) with respect to x and equating the terms of the same powers of $\lambda^{-n} \ln^m \lambda$, we get a sequence of singular integral equations (1.18) on the interval $x \in (-1, 1)$, which have solution (1.19). So we obtain

$$\omega_{00}(x) = \frac{1}{\pi}\left[N_0 - \int_{-1}^{1} \frac{f'(\xi)\sqrt{1-\xi^2}}{\xi - x}\, d\xi\right], \qquad N_0 = \int_{-1}^{1} \varphi(x)\, dx, \tag{3.94}$$

$$\omega_{10}(x) = -\frac{a_{20}}{\pi^2}\int_{-1}^{1} \frac{\sqrt{1-\xi^2}}{\xi - x}\, d\xi \int_{-1}^{1} \frac{\omega_{00}(t)\,\mathrm{sign}(t-\xi)}{\sqrt{1-t^2}}\, dt, \tag{3.95}$$

$$\omega_{20}(x) = -\frac{1}{\pi^2}\int_{-1}^{1} \frac{\sqrt{1-\xi^2}}{\xi - x}\, d\xi \int_{-1}^{1} [(t-\xi)(2a_{11}\ln|t-\xi| \tag{3.96}$$

$$+a_{11} + 2a_{31})\omega_{00}(t) + a_{20}\,\mathrm{sign}(t-\xi)\omega_{10}(t)]\, \frac{dt}{\sqrt{1-t^2}},$$

$$\omega_{21}(x) = \frac{2a_{11}}{\pi^2}\int_{-1}^{1} \frac{\sqrt{1-\xi^2}}{\xi - x}\, d\xi \int_{-1}^{1} \frac{(t-\xi)\omega_{00}(t)\,\mathrm{sign}(t-\xi)}{\sqrt{1-t^2}}\, dt, \tag{3.97}$$

$$\omega_{30}(x) = -\frac{1}{\pi^2}\int_{-1}^{1} \frac{\sqrt{1-\xi^2}}{\xi - x}\, d\xi \int_{-1}^{1} \{(t-\xi)[2a_{11}\ln|t-\xi| + a_{11} \tag{3.98}$$

$$+2a_{31}]\omega_{10}(t) + a_{20}\,\mathrm{sign}(t-\xi)\omega_{20}(t)$$

$$+3a_{21}(t-\xi)^2\,\mathrm{sign}(t-\xi)\omega_{00}(t)\}\, \frac{dt}{\sqrt{1-t^2}},$$

$$\omega_{31}(x) = -\frac{1}{\pi^2}\int_{-1}^{1} \frac{\sqrt{1-\xi^2}}{\xi - x}\, d\xi \int_{-1}^{1} [-2a_{11}(t-\xi)\omega_{10}(t) \tag{3.99}$$

$$+a_{20}\,\mathrm{sign}(t-\xi)\omega_{21}(t)]\, \frac{dt}{\sqrt{1-t^2}}.$$

As a result, for the case of the cylindrical sleeve (insert) when $\delta(x) \equiv \delta$, or $f(x) \equiv f$, by using formulae (3.94)–(3.99) and (3.89)–(3.92), we get

$$\omega_{00}(x) = \frac{N_0}{\pi}, \qquad \omega_{10}(x) = \frac{4}{\pi^3} a_{20} N_0 S_1(x), \tag{3.100}$$

$$\omega_{21}(x) = -\frac{N_0}{\pi} a_{11}(1 - 2x^2), \tag{3.101}$$

$$\omega_{20}(x) = \frac{N_0}{\pi}\Big[(a_{11}(1.5 - \ln 2) + a_{31})(1 - 2x^2) \tag{3.102}$$

$$+\frac{32}{\pi^4}a_{20}^2(S_2(x) - S_0)\bigg],$$

$$\omega_{31}(x) = -\frac{2}{\pi^3}a_{11}a_{20}N_0 S_4(x), \quad \omega_{30}(x) = \frac{N_0}{\pi^3}\bigg[\frac{8}{9}a_{11}a_{20}S_3(x) \tag{3.103}$$

$$+\left(6a_{21}(1+2x^2) - \frac{128}{\pi^4}a_{20}^3 S_0\right)S_1(x) + [9a_{21}$$

$$+2a_{20}(a_{11}(1.5 - \ln 2) + a_{31})]S_4(x) + \frac{8}{3}a_{21} + \frac{64}{\pi^4}a_{20}^3 S_5(x)\bigg],$$

where

$$S_0 = \sum_{m=0}^{\infty} \frac{4m}{(4m^2-1)^3} \approx 0.1508, \tag{3.104}$$

$$S_1(x) = (1-2x^2) + 2\sqrt{1-x^2}\sum_{m=1}^{\infty}\frac{\sin[(2m+1)\arccos x]}{(2m+1)^2}, \tag{3.105}$$

$$S_2(x) = (1-x^2)\sum_{m=1}^{\infty}\frac{\nu_{2m}(x)}{(4m^2-1)^2}, \tag{3.106}$$

$$\nu_0(x) = 0, \quad \nu_2(x) = 4 + 2x\ln\frac{1-x}{1+x}, \tag{3.107}$$

$$\nu_{2m+2}(x) = -2(1-2x^2)\nu_{2m}(x) \tag{3.108}$$

$$+\frac{16m}{4m^2-1} - \nu_{2m-2}(x) \quad (m = 1, 2, ...),$$

$$S_3(x) = -(1-2x^2) + 144\sqrt{1-x^2} \tag{3.109}$$

$$\times\sum_{m=0}^{\infty}\frac{\sin[(2m+3)\arccos x]}{(2m+1)^2\,(2m+3)^2\,(2m+5)^2},$$

$$S_4(x) = \frac{1}{3} + (1-2x^2) + x(1-x^2)\ln\frac{1-x}{1+x}, \tag{3.110}$$

$$S_5(x) = \int_{-1}^{1}\frac{\sqrt{1-\xi^2}}{\xi - x}\,d\xi\int_{0}^{\xi}\frac{S_2(t)\,dt}{\sqrt{1-t^2}}. \tag{3.111}$$

Here, the function $S_2(x)$ for all $|x| \leq 1$ can be approximated by the expression

$$S_2(x) \approx \left[0.4356 + 0.1321x^2 + 0.2494x\ln\frac{1-x}{1+x}\right]\sqrt{1-x^2}, \tag{3.112}$$

with the error less than 0.7%. Using this approximation, we can also approximate $S_5(x)$:

$$S_5(x) \approx 0.3547 - 8.463x^2 + 0.3442x^4 \tag{3.113}$$

$$+x(1-x^2)\ln\left(\frac{1-x}{1+x}\right)(0.1180 + 0.03305x^2)$$

$$-0.4156(1-x^2)^2\ln^2\left(\frac{1-x}{1+x}\right) + 0.3026S_1(x).$$

For the integral characteristic N_0 given by the second formula of (3.94) we obtain, using (3.93) and (3.100)–(3.113), the approximation

$$N_0 = \pi f[a_{30} + 0.8106a_{20}\lambda^{-1} + (a_{11} + a_{31} - 0.03287a_{20}^2)\lambda^{-2} \tag{3.114}$$

$$+(1.442a_{21} - 0.2762a_{11}a_{20} - 0.1807a_{20}a_{31} - 0.02450a_{20}^3)\lambda^{-3}$$

$$+\ln(2\lambda)(1 - a_{11}\lambda^{-2} + 0.1801a_{11}a_{20}\lambda^{-3}) + O(\lambda^{-4}\ln^2(2\lambda))]^{-1}.$$

3.2.4. SOLUTION FOR A THIN CYLINDER

Suppose radius R of the cylinder (cylindrical cavity) is considerably less than the half-length a of the contact zone, i.e. the parameter $\lambda \ll 1$ in (3.76). We now consider the asymptotic method of *small* λ based on approximation (3.85).

For simplicity we confine ourselves to a cylindrical sleeve (insert) $f(x) \equiv f$. We represent the integral equation (3.75) as an equivalent system of three integral equations (Aleksandrov (1968))

$$\int_{-1}^{\infty} \omega\left(\frac{1+\xi}{\lambda}\right) K\left(\frac{\xi - x}{\lambda}\right) d\xi = \pi\lambda \tag{3.115}$$

$$+\int_{-\infty}^{-1}\left[\omega\left(\frac{1-\xi}{\lambda}\right) - v\left(\frac{\xi}{\lambda}\right)\right] K\left(\frac{\xi - x}{\lambda}\right) d\xi \quad (-1 \le x < \infty),$$

$$\int_{-\infty}^{1} \omega\left(\frac{1-\xi}{\lambda}\right) K\left(\frac{\xi - x}{\lambda}\right) d\xi = \pi\lambda \tag{3.116}$$

$$+\int_{-1}^{\infty}\left[\omega\left(\frac{1+\xi}{\lambda}\right) - v\left(\frac{\xi}{\lambda}\right)\right] K\left(\frac{\xi - x}{\lambda}\right) d\xi \quad (-\infty < x \le 1),$$

$$\int_{-\infty}^{\infty} v\left(\frac{\xi}{\lambda}\right) K\left(\frac{\xi - x}{\lambda}\right) d\xi = \pi\lambda \quad (-\infty < x < \infty), \tag{3.117}$$

under the condition

$$\varphi(x) = \frac{f}{\lambda}\left[\omega\left(\frac{1+x}{\lambda}\right) + \omega\left(\frac{1-x}{\lambda}\right) - v\left(\frac{x}{\lambda}\right)\right] \quad (|x| \leq 1). \tag{3.118}$$

The sum of equations (3.115)–(3.117) gives equation (3.75).

We emphasize that the function v gives a *confluent* solution, corresponding to an infinite sleeve (insert) and having no singularities. This function properly represents the contact stresses far from the edges of a sleeve (insert), i.e. deep inside the contact region, whereas the function ω brings a *boundary layer solution*, having the well-known square root singularities, and properly represents the contact stresses near the points $x = \pm 1$.

The solution of (3.117) can be found by using the convolution theorem for Fourier transforms. As a result, we obtain

$$v\left(\frac{\xi}{\lambda}\right) = \frac{1}{A}, \qquad A = \lim_{u \to 0} \frac{L(u)}{u}, \tag{3.119}$$

see (3.79) and (3.80) to determine the constant A.

By a change of variable the integral equations (3.115), (3.116) are reduced to just the one $(0 \leq t < \infty)$

$$\int_0^{\infty} \omega(\tau) K(\tau - t)\, d\tau = \pi + \int_{2/\lambda}^{\infty} [\omega(\tau) - A^{-1}] K\left(\frac{2}{\lambda} - \tau - t\right) d\tau. \tag{3.120}$$

It can be shown that the integral on the right-hand side of (3.120) has an exponential decay if $\lambda \to 0$. For small λ the asymptotic solution of (3.120) can be obtained by succesive approximations. At each step it is necessary to find the solution of the same Wiener–Hopf equation, but with different right sides. It means that for small λ the principal term (zero approximation) in the asymptotic solution of (3.75) with the kernel in the form (3.77), (3.85) is given by formulae (3.118), (3.119), where the function $\omega(s)$ is determined from the Wiener–Hopf equation

$$\int_0^{\infty} \omega(\tau)\, d\tau \int_0^{\infty} K^*(u) \cos u(\tau - s)\, du = \pi \quad (0 \leq s < \infty), \tag{3.121}$$

To solve (3.121), we apply the well-known technique (Noble (1958)). As a result, we will have

$$\omega(s) = \frac{1}{2\pi} \int\limits_{\Gamma} \frac{\exp(-ius)\, du}{(-iu) K_+^*(0) K_+^*(u)}, \tag{3.122}$$

where the contour Γ is a straight line lying just above the real axis in the complex u-plane (Fig. 3.5a), and the functions $K_\pm^*(u)$ are found from the factorization

$$K^*(u) = K_+^*(u) K_-^*(u), \tag{3.123}$$

where $K_+^*(u)$ is regular in the half-plane $\operatorname{Im} u > -\varepsilon$ and has no zeros there, and $K_-^*(u)$ is regular in the half-plane $\operatorname{Im} u < \varepsilon$ and it too has no zeros there; $\varepsilon = \min(B, C, E)$.

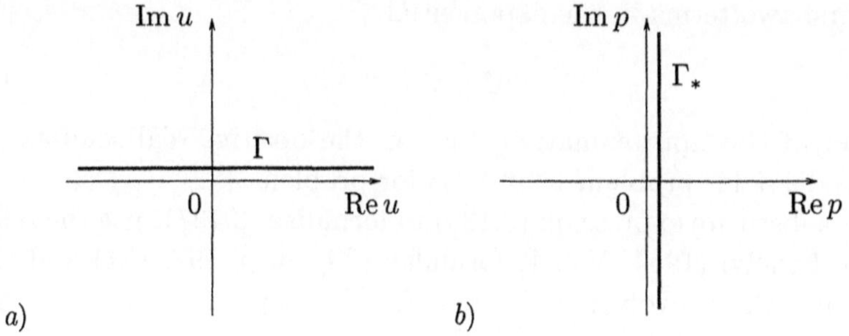

Figure 3.5: Contours $a)$ Γ and $b)$ Γ_*

The factorization (3.123) can be carried out in the form (Noble (1958, p. 91))

$$K_\pm^*(u) = \frac{\sqrt{B \mp iu}}{C \mp iu} \exp[D f_\pm(u)], \tag{3.124}$$

$$\frac{1}{\sqrt{u^2 + E^2}} = f_+(u) + f_-(u), \tag{3.125}$$

$$f_\pm(u) = \frac{\pm i}{\pi \sqrt{u^2 + E^2}} \ln \frac{u + \sqrt{u^2 + E^2}}{\pm iE}. \tag{3.126}$$

Here, the functions $f_+(u)$ and $f_-(u)$ are regular in the half-planes $\operatorname{Im} u > -\varepsilon$ and $\operatorname{Im} u < \varepsilon$, respectively.

We now find that $K_+^*(0) = \sqrt{A}$. For convenience we set $u = ip$ in (3.122), then with the help of formulae (3.124)–(3.126) we will have

$$\omega(s) = \frac{1}{2\pi i} \int\limits_{\Gamma_*} \frac{\Omega(p)}{p} \exp(ps)\, dp, \quad \Omega(p) = \frac{p + C}{\sqrt{A(p+B)}} \exp[g(p)], \tag{3.127}$$

$$g(p) = \frac{-D}{\pi\sqrt{p^2 - E^2}} \ln \frac{p + \sqrt{p^2 - E^2}}{E}, \tag{3.128}$$

where the contour Γ_* is a straight line lying just to the right of the imaginary axis in the complex p-plane (Fig. 3.5*b*).

One can see from (3.127) that the function $\omega(s)$ is the inverse Laplace transform of $\Omega(p)p^{-1}$ (or it is the inverse Laplace-Carson transform of $\Omega(p)$). Such a connection will be denoted by $\Omega(p) :\rightarrow \omega(s)$. Note that the function $\exp[g(p)]$, where $g(p)$ has the form (3.128), (3.86), (3.87), can be approximated in the half-plane $\mathrm{Re}\, p > 0$ with a high degree of accuracy by the first two terms in the exponential:

$$\exp[g(p)] \approx 1 + g(p). \tag{3.129}$$

The error of this approximation, e.g., on the positive real semi-axis, does not exceed 1% for problem 1, and 2% for problem 2.

If we substitute expression (3.129) in formulae (3.127), use the reference tables in Erdelyi (1954, Vol. 1, formulae (23) on p. 197, (21) and (23) on p. 235) and the convolution theorem for the Laplace transform, then we find an approximation for $\omega(s)$ in the form

$$\omega(s) = \frac{1}{\sqrt{A}} \left(\frac{\exp(-Bs)}{\sqrt{\pi s}} + \frac{C}{\sqrt{B}} \Phi(\sqrt{Bs}) + I(s) \right), \tag{3.130}$$

$$I(s) = \frac{-D}{\pi} \int\limits_0^s \left[\frac{\exp[-B(s-\tau)]}{\sqrt{\pi(s-\tau)}} + \frac{C}{\sqrt{B}} \Phi(\sqrt{B(s-\tau)}) \right]$$

$$\times\, \mathcal{K}_0(E\tau)\, d\tau \leftarrow: \frac{p + C}{\sqrt{p+B}} g(p).$$

Here $\Phi(x)$ is the probability integral (error function) (2.101), $\mathcal{K}_0(x)$ is the modified Bessel function.

On the basis of (3.118), for the integral characteristic $(3.94)_2$ for the cylindrical sleeve (insert) we obtain the formula

$$\frac{N_0}{f} = 2 \int\limits_0^\zeta \omega(\tau)\, d\tau - \frac{\zeta}{A} \qquad (\zeta = 2/\lambda). \tag{3.131}$$

By substituting expression (3.130) into (3.131) and taking into account that[3]

$$\int_0^s I(\tau)\,d\tau \leftarrow: \frac{p+C}{p\sqrt{p+B}}\,g(p) :\rightarrow \tag{3.132}$$

$$J(s) = \frac{-D}{\pi}\int_0^s \left\{\frac{C}{B}\sqrt{\frac{s-\tau}{\pi}}\exp[-B(s-\tau)]\right.$$

$$\left.+\frac{\Phi(\sqrt{B(s-\tau)})}{\sqrt{B}}\left[1-\frac{C}{2B}+C(s-\tau)\right]\right\}\mathcal{K}_0(E\tau)\,d\tau,$$

we find

$$\frac{N_0}{f} = \frac{2C}{\sqrt{AB}}\left[\left(\zeta-\frac{1}{2B}+\frac{1}{C}\right)\Phi(\sqrt{B\zeta})\right. \tag{3.133}$$

$$\left.+\sqrt{\frac{\zeta}{\pi B}}\exp(-B\zeta)\right]-\frac{\zeta}{A}+\frac{2}{\sqrt{A}}J(\zeta).$$

The expression (3.133) can be considerably simplified for $\lambda \leq 1/4$, if we note that from the relation

$$\frac{p+C}{p\sqrt{p+B}}\,g(p) = \frac{-CD}{2\sqrt{B}Ep}\left[1-\left(\frac{1}{2B}-\frac{1}{C}+\frac{2}{\pi E}\right)p+O(p^2)\right] \quad (p\to 0)$$

we obtain the asymptotic equality (Bleistein and Handelsman (1975))

$$J(s) = \frac{-CD}{2\sqrt{B}E}\left(s-\frac{1}{2B}+\frac{1}{C}-\frac{2}{\pi E}+O(s^{-1})\right) \quad (s\to+\infty). \tag{3.134}$$

If we substitute presentation (3.134) into formula (3.133), then we obtain, for $\lambda \leq 1/4$, that

$$\frac{N_0}{f} = \zeta\left[\frac{C}{\sqrt{AB}}\left(2-\frac{D}{E}\right)-\frac{1}{A}\right] \tag{3.135}$$

$$+\frac{C}{\sqrt{AB}}\left[\frac{2}{C}-\frac{1}{B}-\frac{D}{E}\left(\frac{1}{C}-\frac{1}{2B}-\frac{2}{\pi E}\right)\right].$$

Setting the values of constants (3.86), (3.87) in (3.135), we have

$$N_0 f^{-1} = 2.5838\zeta + 0.7882 \tag{3.136}$$

[3]To derive (3.132), we use the same formulae in Erdelyi (1954, Vol. 1), then use formula (6) for $a = 0$ in Erdelyi (1954, Vol. 1, p. 227) and formula 1.5.1.10 in Prudnikov, Brychkov and Marichev (1992, Vol. 2) to calculate the integral of the error function $\Phi(\sqrt{s})$.

for problem 1, and

$$N_0 f^{-1} = 1.3567\zeta + 1.1648 \tag{3.137}$$

for problem 2. When $\lambda \leq 1/4$, the values calculated by formulae (3.136), (3.137) differ less than by 1% from the corresponding values obtained from formula (3.133).

We note that instead of using the additive form (3.118) to approximate $\varphi(x)$ for small λ we can use an equivalent multiplicative form (Vorovich, Aleksandrov and Babeshko (1974)) which for the cylindrical sleeve (insert) has the form

$$\varphi(x) = \frac{fA}{\lambda}\,\omega\left(\frac{1+x}{\lambda}\right)\omega\left(\frac{1-x}{\lambda}\right). \tag{3.138}$$

Table 3.2: Quantities (3.139)

λ	8	4	2	1	1/2	1/4
χ_1^1 (3.114)	1.44	2.09	3.55	—	—	—
χ_1^1 (3.133)	—	—	3.42	5.97	11.1	21.5
χ_2^1 (3.93)	0.459	0.685	1.20	—	—	—
χ_2^1 (3.118)	—	—	1.29	2.52	5.14	10.3
χ_3^1 (3.93)	0.459	0.651	1.00	—	—	—
χ_3^1 (3.138)	—	—	0.896	1.28	1.81	2.57
χ_1^2 (3.114)	1.32	1.77	2.52	—	—	—
χ_1^2 (3.133)	—	—	2.45	3.86	6.59	12.0
χ_2^2 (3.93)	0.432	0.602	0.924	—	—	—
χ_2^2 (3.118)	—	—	0.872	1.48	2.76	5.43
χ_3^2 (3.93)	0.402	0.518	0.640	—	—	—
χ_3^2 (3.138)	—	—	0.686	0.937	1.32	1.86

In Table 3.2 we present the values of the quantities ($\nu = 0.3$)

$$\chi_1^n = N_0 f^{-1}, \quad \chi_2^n = \varphi(0) f^{-1}, \quad \chi_3^n = \lim \varphi(x)\sqrt{1-x^2}\,f^{-1} \;\; (x \to 1), \tag{3.139}$$

calculated by the formulae from the previous subsection for large λ and by formulae from this subsection for small λ. The upper index $n = 1, 2$ corresponds to problems 1 and 2, respectively. The calculated values of χ_2^n, χ_3^n, for $\lambda = 2$, approach the corresponding values from Aleksandrov and Belokon' (1968) found for small λ on the basis of another approximation. Near $\lambda = 2$, the two asymptotic solutions are close, suggesting that the two together give an adequate solution for the whole range $(0, \infty)$ of the parameter λ. One can see from Table 3.2 that, for a fixed length $2a$ of the contact zone, the smaller the radius R of the cylinder (cylindrical cavity), the larger are the values of the integral characteristic N_0[4] (χ_1^n), of the contact stress (χ_2^n) and of its intensity factor (χ_3^n). The comparison of formulae (3.136) and (3.137) shows that for $\lambda = R/a \to 0$ the normal load N_0 for the sleeve is almost twice greater than for the insert. In practice (in metallurgy), it may lead to collapse of a steel sleeve with a small λ, especially, for a large interference f.

3.3 Interaction between a Cylinder and Deformable Sleeves

3.3.1. FORMULATION AND ASYMPTOTIC SOLUTIONS

Consider the axisymmetric interaction between an elastic sleeve (G_1, ν_1) of radius $r = R - \delta(z)$ ($\delta(z) \ll R$) and an infinite elastic circular cylinder (G, ν) of radius R ($\delta(z)$ is the interference, Fig. 3.6). We assume that there are no friction forces between the sleeve and cylinder in the contact domain $|z| \leq a$, and there is no load outside the contact domain. We write the condition of the contact between the sleeve and cylinder in the form

$$u_r(R, z) - u_r^1(R, z) = -\delta(z) \qquad (|z| \leq a), \tag{3.140}$$

where $u_r(R, z)$ and $u_r^1(R, z)$ are the radial displacements of the surface points of the cylinder and sleeve, respectively.

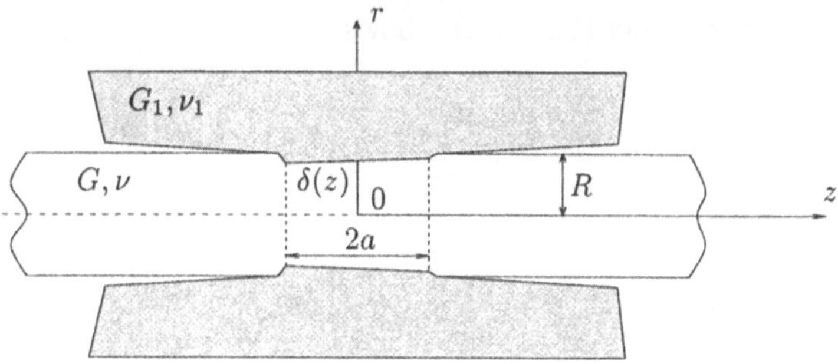

Figure 3.6: Contact between a cylinder and deformable sleeve (an axial section)

[4]The quantity $2\pi N_0 = 2\pi \int_{-1}^{1} \varphi(x)dx$ equals the total circular normal load arising between the cylinder and the internal side of a sleeve or between the cylindrical cavity and the external side of an insert.

We also suppose that the elastic sleeve is a sufficiently massive solid that its dimensions considerably exceed the length $2a$ of the contact domain. Then according to the Hertz theory, the radial displacements of the sleeve surface due to the pressure $q(z)$ can be approximated with a fair degree of accuracy by the radial displacements of a surface of an infinite cylindrical cavity of radius R in an elastic space due to the same pressure. Thus, on the basis of the condition (3.140) and formulae (3.16), (3.22), we reduce the problem to the following integral equation for the unknown pressure $\sigma_r(R, z) = -q(z)$ $(\theta_1 = G_1/(1 - \nu_1))$:

$$\int_{-a}^{a} q(\xi)\, K\left(\frac{\xi - z}{R}\right) d\xi = \frac{\pi}{2}\theta\delta(z) \qquad (|z| \leq a), \tag{3.141}$$

$$K(t) = \int_{0}^{\infty} \frac{L(s)}{s}\cos(ts)\, ds, \quad L(s) = \frac{L_1(s) + mL_2(s)}{2}, \quad m = \frac{\theta}{\theta_1}, \tag{3.142}$$

where the functions $L_n(s)$, $n = 1, 2$, are determined by formulae (3.18), (3.24), where, in expression (3.24), ν is replaced by ν_1.

We suppose for simplicity that the elastic characteristics of the cylinder and sleeves are identical ($G_1 = G$, $\nu_1 = \nu$), i.e. $m = 1$. Using the dimensionless notation

$$\varphi(x) = q(z)\theta^{-1}, \quad x = za^{-1}, \quad \lambda = Ra^{-1}, \quad f(x) = \delta(z)(2a)^{-1}, \tag{3.143}$$

we can write equation (3.141) as follows

$$\int_{-1}^{1} \varphi(\xi)\, K\left(\frac{\xi - x}{\lambda}\right) d\xi = \pi f(x) \qquad (|x| \leq 1). \tag{3.144}$$

Let us obtain an explicit solution of the equation (3.144) for a special approximation of its kernel. One can easily see that, when $m = 1$, the function $L(s)$ of the form $(3.142)_2$ has the following properties:

$$L(s) = As + O(s^3) \quad (s \to 0), \qquad A = \frac{1}{2}(1 - \nu^2)^{-1}, \tag{3.145}$$

$$L(s) = 1 + c_1 s^{-2} + c_2 s^{-4} + O(s^{-6}) \quad (s \to \infty), \tag{3.146}$$

where, for $\nu = 0.3$, $c_1 = -0.965$, $c_2 = -2.910$.

Using these properties, we approximate the function $L(s)$ by the expression $\tanh(As)$. The error of this approximation for $\nu = 0.3$ for all $s \in [0, \infty)$

does not exceed 13%. It is known (see, e.g. Noble (1958)) that the error of the approximate solution, based on such an approximation, does not exceed the error of the approximation. By using the value of integral 4.116.2 in Gradshteyn and Ryzhik (1994)

$$\int_0^\infty \frac{\tanh(As)}{s} \cos(ts)\, ds = -\ln\left|\tanh\frac{\pi t}{4A}\right|, \tag{3.147}$$

and changing the variables in (3.144) by the formulae

$$\eta = \exp(-B\xi), \quad y = \exp(-Bx), \quad B = \pi(2A\lambda)^{-1}, \quad c = \exp(-B),$$

$$d = \exp B, \quad \varphi_*(\eta) = 2A\pi^{-1}\varphi(\xi)\exp(B\xi), \quad f_*(y) = f(x),$$

we arrive at the equation

$$-\int_c^d \varphi_*(\eta) \ln\frac{|\eta - y|}{\eta + y}\, d\eta = \pi f_*(y) \quad (c \le y \le d). \tag{3.148}$$

Differentiating equation (3.148) with respect to y, we can reduce it to the well-known singular integral equation (1.18) on an interval, having the exact solution (1.19). For a cylindrical sleeve $f(x) \equiv f$ we obtain

$$\varphi(x) = \frac{\pi f d}{2A\lambda\mathcal{K}(e)}\left[2\cosh\left(\frac{\pi}{A\lambda}\right) - 2\cosh\left(\frac{\pi x}{A\lambda}\right)\right]^{-1/2}, \quad e = \frac{c}{d}, \tag{3.149}$$

$$N_0 = \int_{-1}^{1} \varphi(x)\, dx = \frac{f\mathcal{K}'(e)}{\mathcal{K}(e)}, \quad \mathcal{K}'(e) = \mathcal{K}(\sqrt{1-e^2}), \tag{3.150}$$

where $\mathcal{K}(e)$ is the complete elliptic integral of the first kind.

Let us find a solution of equation (3.144) for large values of the relative thickness λ of the cylinder. In view of properties (3.145), (3.146) of the function $L(s)$, the kernel $K(t)$ of the form (3.142) can be presented as follows (cf. with (3.82)–(3.84))

$$K(t) = \ln|t|\, F_1(t) + F_2(t), \tag{3.151}$$

$$F_1(t) = \sum_{k=0}^{\infty} b_k t^{2k}, \qquad F_2(t) = \sum_{k=0}^{\infty} a_k t^{2k}, \tag{3.152}$$

where the few first coefficients in the series (3.152) are calculated by the formulae

$$b_0 = -1, \quad b_1 = \frac{c_1}{2}, \quad b_2 = -\frac{c_2}{24}, \quad a_0 = \int_0^\infty \frac{L(s) - 1 + \exp(-s)}{s} ds, \quad (3.153)$$

$$a_1 = -\frac{3}{4}c_1 + \frac{1}{2}\int_0^\infty [s^2 - s^2 L(s) + c_1(1 - \exp(-s))]\frac{ds}{s}, \quad (3.154)$$

$$a_2 = \frac{25}{288}c_2 - \frac{1}{24}\int_0^\infty [s^4(1 + c_1 s^{-2}) - s^4 L(s) \quad (3.155)$$

$$+ c_2(1 - \exp(-s))]\frac{ds}{s},$$

and, for $\nu = 0.3$, we have $b_1 = -0.482$, $b_2 = 0.121$, $a_0 = -0.506$, $a_1 = 0.682$, $a_2 = -0.00920$.

Using the asymptotic method of *large* λ, we will seek the solution of the integral equation (3.144) with kernel (3.151) in the form

$$\varphi(x) = \sum_{n=0}^{N}\sum_{j=0}^{n} \varphi_{2n,j}(x)\lambda^{-2n}\ln^j \lambda + O(\lambda^{-2N-2}\ln^{N+1}\lambda). \quad (3.156)$$

Differentiating (3.144) with respect to x and equating the coefficients of the powers of $\lambda^{-2n}\ln^j \lambda$, we get a sequence of singular integral equations (1.18) on the interval $x \in (-1, 1)$, which have solution (1.19). If we confine ourselves to the accuracy $O(\lambda^{-6}\ln^3 \lambda)$, then for $f(x) \equiv f$ we have

$$\varphi(x) = \frac{N_0}{\pi\sqrt{1-x^2}}\left[1 + \left(a_1 + \frac{3}{2}b_1 - b_1\ln(2\lambda)\right)(1 - 2x^2)\frac{1}{\lambda^2}\right. \quad (3.157)$$

$$+[E_1 + E_2\ln(2\lambda)](1 - 2x^2)\frac{1}{\lambda^4}$$

$$\left. +[E_3 + E_4\ln(2\lambda)](1 + 4x^2 - 8x^4)\frac{1}{\lambda^4} + O\left(\frac{\ln^3\lambda}{\lambda^6}\right)\right],$$

$$N_0 = \pi f\left[\ln(2\lambda) + a_0 + [a_1 + b_1 - b_1\ln(2\lambda)]\frac{1}{\lambda^2}\right. \quad (3.158)$$

$$\left. + \left(E_5 + E_6\ln(2\lambda) - \frac{1}{4}b_1^2\ln^2(2\lambda)\right)\frac{1}{\lambda^4} + O\left(\frac{\ln^3\lambda}{\lambda^6}\right)\right]^{-1},$$

$$E_1 = -\frac{1}{2}a_1b_1 - \frac{3}{4}b_1^2 + 3a_2 + \frac{13}{4}b_2, \qquad E_2 = \frac{1}{2}b_1^2 - 3b_2,$$

$$E_3 = \frac{1}{12}a_1b_1 + \frac{1}{8}b_1^2 + \frac{1}{2}a_2 + \frac{25}{24}b_2, \qquad E_4 = -\frac{1}{12}b_1^2 - \frac{1}{2}b_2,$$

$$E_5 = -\frac{1}{4}a_1^2 - \frac{3}{4}a_1b_1 - \frac{9}{16}b_1^2 + \frac{9}{4}a_2 + \frac{21}{8}b_2,$$

$$E_6 = \frac{1}{2}a_1b_1 + \frac{3}{4}b_1^2 - \frac{9}{4}b_2.$$

For $\lambda \geq 2$, formulae (3.156)–(3.158) give an error not more than 5%.

Now we construct a solution of equation (3.144) with the help of the asymptotic method of *small* λ. This method is based on the following subdivision of equation (3.144):

$$\int_{-1}^{\infty} \varphi_1(\xi) K\left(\frac{\xi - x}{\lambda}\right) d\xi = \pi f_1(x) \tag{3.159}$$

$$+ \int_{-\infty}^{-1} \varphi_2(\xi) K\left(\frac{\xi - x}{\lambda}\right) d\xi \quad (-1 \leq x < \infty),$$

$$\int_{-\infty}^{1} \varphi_2(\xi) K\left(\frac{\xi - x}{\lambda}\right) d\xi = \pi f_2(x) \tag{3.160}$$

$$+ \int_{1}^{\infty} \varphi_1(\xi) K\left(\frac{\xi - x}{\lambda}\right) d\xi \quad (-\infty < x \leq 1).$$

Here, the functions $f_n(x)$, $n = 1, 2$, are such that

$$f_1(x) = O(\exp[-\alpha_1 x]) \quad (x \to \infty), \tag{3.161}$$

$$f_2(x) = O(\exp[\alpha_2 x]) \quad (x \to -\infty), \tag{3.162}$$

$$f_1(x) + f_2(x) = f(x) \quad (|x| \leq 1), \qquad \alpha_{1,2} > 0. \tag{3.163}$$

The representation (3.161)–(3.163) is always available, e.g., if $f(x) \equiv f$, then after solving equation (3.144) with the right-hand side

$$f_\varepsilon(x) = f \exp(-\varepsilon/\lambda) \cosh(\varepsilon x/\lambda),$$

$$f_1(x) = \frac{1}{2} f \exp[-\varepsilon(1 + x)/\lambda],$$

$$f_2(x) = \frac{1}{2} f \exp[-\varepsilon(1-x)/\lambda],$$

we can obtain the solution for the original case $f(x) \equiv f$ by passing to the limit $\varepsilon \to 0$.

If the first condition of (3.163) holds, then the solution of equation (3.144) can be found as a sum of the solutions of equations (3.159) and (3.160), i.e.

$$\varphi(x) = \varphi_1(x) + \varphi_2(x).$$

Separating the problem into the even and odd ones, after evident changes of variables, we can find the solution of equations (3.159), (3.160) by the method of successive approximations by the scheme[5]

$$\int_0^\infty \zeta_n(\tau)K(t-\tau)\,d\tau = \pi h(t) \pm \int_{2/\lambda}^\infty \zeta_{n-1}(\tau)K\left(\frac{2}{\lambda} - t - \tau\right) d\tau \qquad (3.164)$$

$$(0 \le t < \infty; \quad n \ge 0; \quad \zeta_{-1}(t) \equiv 0),$$

$$\zeta(t) = \varphi_1(\lambda t - 1), \qquad h(t) = \lambda^{-1} f_1(\lambda t - 1).$$

For the even case $f(-x) = f(x)$, we choose the sign "+" in the right-hand side of (3.164) and have

$$f_1(x) = f_2(-x), \qquad \varphi_1(x) = \varphi_2(-x).$$

For the odd case $f(-x) = -f(x)$, we choose the sign "−" in (3.164) and have

$$f_1(x) = -f_2(-x), \qquad \varphi_1(x) = -\varphi_2(-x).$$

For each approximation we solve a Wiener–Hopf integral equation of the form (3.164) whose right-hand side is a known function. To solve such an equation, we apply the approximation

$$L(s) \approx s\frac{\sqrt{s^2 + D^2}}{s^2 + C^2} \qquad \left(\frac{D}{C^2} = A\right), \qquad (3.165)$$

[5]This scheme converges for small λ since we can prove that, for $\lambda \to 0$,

$$\int_{2/\lambda}^\infty \zeta_{n-1}(\tau)K\left(\frac{2}{\lambda} - t - \tau\right) d\tau \sim O(\exp[-2\beta_1/\lambda]),$$

if we suppose that $\varphi_1(x) = O(\exp[-\beta_1 x])$ $(x \to \infty)$, where β_1 is a positive constant.

whose error, for $C = 2.335$, $D = 2.996$, does not exceed 8.5%. For the cylindrical sleeve ($f(x) \equiv f$) for the approximation (3.165), the approximate solution of the problem, when λ is small, takes the form

$$\varphi(x) = \frac{f}{2\lambda}\left[\zeta^0\left(\frac{1-x}{\lambda}\right) + \zeta^0\left(\frac{1+x}{\lambda}\right)\right], \tag{3.166}$$

$$\zeta^0(t) = \frac{1}{A}\Phi(\sqrt{Dt}) + \frac{\exp(-Dt)}{\sqrt{\pi At}},$$

where $\Phi(x)$ is the probability integral (2.101). For $\lambda \le 2$, solution (3.166) gives an error not more than 13.5%.

We note that the solution of the problem for small λ can be also written in the multiplicative form

$$\varphi(x) = \frac{fA}{\lambda}\zeta^0\left(\frac{1-x}{\lambda}\right)\zeta^0\left(\frac{1+x}{\lambda}\right), \tag{3.167}$$

which is equivalent to (3.166). On the basis of (3.167) for the integral characteristic of the solution we obtain the expression

$$N_0 = \int_{-1}^{1} \varphi(x)\,dx = f\left[\frac{2}{A\lambda} + \frac{1}{\sqrt{AD}}\left(2 - \frac{1}{\sqrt{AD}}\right)\right. \tag{3.168}$$

$$\left. + \left(1 - \frac{1}{\sqrt{AD}}\right)^2 \exp\left(-\frac{2D}{\lambda}\right)\right].$$

3.3.2. TWO IDENTICAL SLEEVES

Let us consider the interaction between two identical symmetrically situated elastic sleeves and an infinite elastic cylinder (Fig. 3.7). The sleeves are fitted with the interference $\delta(z) = \delta(-z)$ ($b \le |z| \le a$). Again we take $m = 1$.

The integral equation of the problem in the dimensionless notation (3.143) and when $k = ba^{-1}$ can be written in the form

$$\left(\int_{-1}^{-k} + \int_{k}^{1}\right)\varphi(\xi)\,K\left(\frac{\xi - x}{\lambda}\right)d\xi = \pi f(x) \quad (k \le |x| \le 1). \tag{3.169}$$

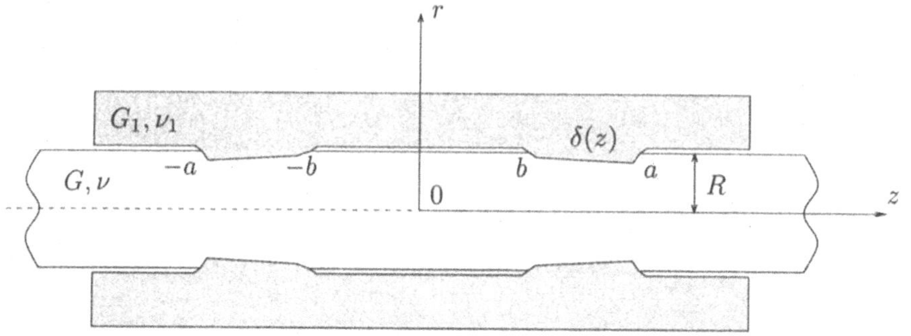

Figure 3.7: Contact between a cylinder and two deformable sleeves (an axial section)

In view of the evenness of the functions $\varphi(x)$, $f(x)$ the integral equation (3.169) can be reduced to the equation on a single segment

$$\int_k^1 \varphi(\xi)\left[K\left(\frac{\xi-x}{\lambda}\right)+K\left(\frac{\xi+x}{\lambda}\right)\right]d\xi=\pi f(x) \quad (k\le x\le 1). \tag{3.170}$$

We obtain an explicit solution of equation (3.170) when its kernel is approximated by expression (3.147). Changinig the variables in (3.170) by the formulae

$$\eta=\cosh(B\xi),\quad y=\cosh(Bx),\quad B=\pi(2A\lambda)^{-1},\quad c=\cosh(Bk),$$
$$d=\cosh B,\quad \varphi_*(\eta)=\varphi(\xi)[B\sinh(B\xi)]^{-1},\quad f_*(y)=f(x),$$

we reduce it to equation (3.148) whose solution, returning to the old notation, can be written as follows

$$\varphi(x)=\frac{2|\sinh(Bx)|}{A\lambda s(x)}\left\{B\int_0^1\left[2\cosh^2 B\,\frac{\mathcal{E}(e)}{\mathcal{K}(e)}-[\cosh(2B)-\cosh(2B\xi)]\right]\right. \tag{3.171}$$
$$\left.\times\frac{f(\xi)\sinh(B\xi)}{s(\xi)}\,d\xi-\int_k^1\frac{s(\xi)f'(\xi)\cosh(B\xi)}{\cosh(2B\xi)-\cosh(2Bx)}\,d\xi\right\},$$

$$s(x)=\left[\left(\cosh\frac{\pi}{A\lambda}-\cosh\frac{\pi x}{A\lambda}\right)\left(\cosh\frac{\pi x}{A\lambda}-\cosh\frac{\pi k}{A\lambda}\right)\right]^{1/2},\quad e=\frac{c}{d}$$

where $\mathcal{E}(e)$ is the complete elliptic integral of the second kind. For the integral characteristic, which is equivalent to the external circular normal load applied to one sleeve, we obtain the expression

$$N_0 = \int_k^1 \varphi(x)\,dx = \frac{2B\cosh B}{\mathcal{K}(e)} \int_k^1 \frac{f(x)\sinh(Bx)}{s(x)}\,dx. \tag{3.172}$$

Analysis of formulae (3.171) and (3.172) shows that the interaction between the sleeves weakens when the relative thickness λ of the cylinder diminishes for the fixed value of k.

For cylindrical sleeves, $f(x) \equiv f$, formulae (3.171) and (3.172) can be considerably simplified to the form

$$\varphi(x) = 2fB\cosh B\,\mathcal{K}^{-1}(e)|\sinh(Bx)|\,[s(x)]^{-1}, \quad N_0 = f\mathcal{K}'(e)\mathcal{K}^{-1}(e), \tag{3.173}$$

$$N_1 = \int_k^1 \varphi(x)x\,dx = \frac{2Bf\cosh B}{\mathcal{K}(e)} \int_k^1 \frac{x\sinh(Bx)}{s(x)}\,dx.$$

Let us find a solution of equation (3.170) for large values of the relative thickness λ of the cylinder seeking it in the form (3.156). Neglecting terms of the order $O(\lambda^{-4}\ln^2\lambda)$, we represent equation (3.170) as follows

$$\int_k^1 \varphi(\xi)\left[-\ln\frac{|\xi^2 - x^2|}{\lambda^2} + 2a_0\right] d\xi = \pi\psi(x) \quad (k \le x \le 1), \tag{3.174}$$

$$\psi(x) = f(x) - \frac{1}{\pi\lambda^2}\int_k^1 \varphi(\xi)[(\xi^2 + x^2)(2a_1 + b_1\ln|\xi^2 - x^2|)$$

$$-2b_1\xi x\ln\frac{|\xi - x|}{\xi + x} - 2b_1(\xi^2 + x^2)\ln\lambda + O(\lambda^{-4}\ln^2\lambda)]d\xi.$$

The inversion of the integral operator on the left-hand side of equation (3.174) is given by the formulae (Aleksandrov and Kucherov (1968))

$$\varphi(x) = \frac{2|x|}{\pi g(x)}\left[N_0 - \int_k^1 \frac{g(\xi)\psi'(\xi)}{\xi^2 - x^2}d\xi\right], \quad g(x) = \sqrt{(1 - x^2)(x^2 - k^2)}, \tag{3.175}$$

$$N_0 = \int_k^1 \varphi(x)\,dx = \left[\ln\frac{2\lambda}{\sqrt{1 - k^2}} + a_0\right]^{-1} \int_k^1 \frac{\psi(x)x}{g(x)}dx, \tag{3.176}$$

$$N_1 = \int_k^1 \varphi(x) x\, dx = \frac{2}{\pi}\left[N_0 \mathcal{E}'(k) - \int_k^1 \frac{x^2 dx}{g(x)} \int_k^1 \frac{g(\xi)\psi'(\xi)}{\xi^2 - x^2} d\xi\right], \tag{3.177}$$

where $\mathcal{E}'(k) = \mathcal{E}(\sqrt{1-k^2})$. Substituting expression (3.156) into equation (3.174), equating the terms at the same powers of $\lambda^{-2n} \ln^j \lambda$ and using formulae (3.175)–(3.177), we successively find that

$$\varphi_{00}(x) = \frac{2|x|}{\pi g(x)}\left[N_0 - \int_k^1 \frac{g(\xi) f'(\xi)}{\xi^2 - x^2} d\xi\right], \tag{3.178}$$

$$\varphi_{20}(x) = \frac{4|x|}{\pi^2 g(x)} \int_k^1 \frac{g(\xi) d\xi}{\xi^2 - x^2} \int_k^1 \varphi_{00}(t)[\xi(2a_1 + b_1 \tag{3.179}$$

$$+ b_1 \ln|t^2 - \xi^2|) + b_1 t \ln|(t+\xi)/(t-\xi)|]\, dt,$$

$$\varphi_{21}(x) = -\frac{8|x| b_1}{\pi g(x)} \int_k^1 \frac{g(\xi)\xi\, d\xi}{\xi^2 - x^2} \int_k^1 \varphi_{00}(t)\, dt \tag{3.180}$$

$$= 4 b_1 N_0 \pi^{-1} g^{-1}(x) |x| [x^2 - (1+k^2)/2].$$

To calculate the integrals occuring in formulae (3.175)–(3.180), we use the integrals 3.152.9, 3.153.7, 3.154.7, 3.248.2,3, 3.157.10 in Gradshteyn and Ryzhik (1994):

$$\mu_n = \int_k^1 \frac{x^n\, dx}{g(x)}, \quad \mu_0 = \mathcal{K}'(k), \quad \mu_1 = \frac{\pi}{2}, \quad \mu_2 = \mathcal{E}'(k), \tag{3.181}$$

$$\mu_3 = \frac{\pi}{4}(1+k^2), \quad \mu_4 = \frac{1}{3}[(2(1+k^2)\mathcal{E}'(k) - k^2 \mathcal{K}'(k)], \tag{3.182}$$

$$\mu_5 = \pi(3 + 2k^2 + 3k^4)/16, \tag{3.183}$$

$$\int_k^1 \frac{g(x)\, dx}{x^2 - \xi^2} = (\xi^2 - k^2)\, \Pi\left(\frac{\pi}{2}, \frac{1-k^2}{1-\xi^2}, \sqrt{1-k^2}\right) \tag{3.184}$$

$$+ (1 + k^2 - \xi^2)\mathcal{K}'(k) - \mathcal{E}'(k) = w(\xi, k),$$

where $\Pi(\varphi, n, k)$ is the elliptic integral of the third kind, the integral

$$\int_k^1 \frac{t^2}{g(t)} \ln\left|\frac{t+\xi}{t-\xi}\right| dt = \pi[\xi + \mathcal{K}(k) - \mathcal{E}(k)], \tag{3.185}$$

which is a particular case of formula (1.33) in Aleksandrov, Kovalenko and Mkhitarian (1982), the integral

$$\int_k^1 \frac{g(x)x}{x^2-\xi^2}\,dx = -\frac{\pi}{2}\left(\xi^2 - \frac{1+k^2}{2}\right), \tag{3.186}$$

which can be reduced to (3.91), and the integral

$$\nu_n(\xi) = \int_k^1 \frac{t^{2n+1}}{g(t)} \ln|t^2-\xi^2|\,dt, \quad \nu_0(\xi) = \frac{\pi}{2}\ln\frac{1-k^2}{4}, \tag{3.187}$$

$$\nu_1(\xi) = \frac{\pi}{4}\left[(1+k^2)\left(1+\ln\frac{1-k^2}{4}\right) - 2\xi^2\right], \tag{3.188}$$

which can be reduced to (3.90), if we change the variable and then differentiate $(3.187)_1$, and also, to derive (3.187), (3.188), we should use integrals 4.241.1,2 in Gradshteyn and Ryzhik (1994). In formulae $(3.181)_1$, (3.185)–$(3.187)_1$ $g(x)$ has the form $(3.175)_2$ and $k < \xi < 1$.

Taking account of formulae (3.178)–(3.188), for the special case when $f(x) \equiv f$ we obtain the solution of equation (3.170) in the form

$$\varphi(x) = \frac{2N_0|x|}{\pi g(x)}\left[1 + \left(x^2 - \frac{1+k^2}{2}\right)\left(-\frac{\varepsilon_0}{\lambda^2} + 2b_1\frac{\ln\lambda}{\lambda^2}\right)\right. \tag{3.189}$$

$$\left. + \frac{4b_1[\mathcal{K}(k) - \mathcal{E}(k)]}{\pi\lambda^2} w(x,k) + O\left(\frac{\ln^2\lambda}{\lambda^4}\right)\right],$$

$$\varepsilon_0 = 2a_1 + 3b_1 + b_1\ln\frac{1-k^2}{4},$$

$$N_0 = \int_k^1 \varphi(x)\,dx = \pi f\left[-\ln\frac{1-k^2}{4} + 2(a_0 + \ln\lambda)\right. \tag{3.190}$$

$$\left. + \frac{1+k^2}{\lambda^2}\left(2(a_1+b_1) + b_1\ln\frac{1-k^2}{4}\right) + O\left(\frac{\ln^2\lambda}{\lambda^4}\right)\right]^{-1},$$

$$N_1 = \int_k^1 \varphi(x)x\,dx = \frac{2N_0}{\pi}\left[\mathcal{E}'(k) + \frac{1}{3\lambda^2}\left(k^2\mathcal{K}'(k) - \frac{1+k^2}{2}\mathcal{E}'(k)\right)\right. \tag{3.191}$$

$$\times(\varepsilon_0 - 2b_1 \ln\lambda) + O\left(\frac{\ln^2\lambda}{\lambda^4}\right)\Bigg].$$

The approximate solutions (3.178)–(3.180), (3.189)–(3.191) can be used for $\lambda \geq 3$. The numerical analysis of these formulae shows that the interaction between the sleeves weakens as the value of λ decreases for a fixed value of k. In Table 3.3 we present the values of the quantity N_0 calculated by formulae $(3.173)_2$, (3.190) for $f(x) \equiv 1$, $\nu = 0.3$, and $k = 0.5$ versus the various values of λ.

Table 3.3: Values of N_0

λ	5	4	3	2
N_0 $(3.173)_2$	0.740	0.820	0.946	1.175
N_0 (3.190)	0.797	0.890	1.038	1.295

One can see that the smaller the radius R of the cylinder for a fixed value of a, the larger is the internal normal circular load, applied to the sleeves, caused by the given interference $f(x)$.

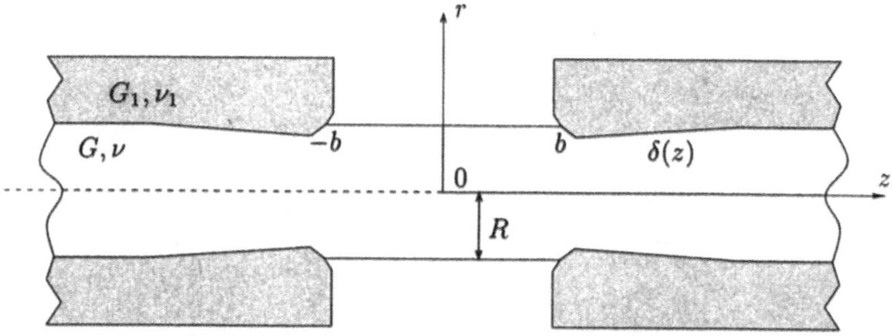

Figure 3.8: Contact between a cylinder and two semi-infinite sleeves (an axial section)

We now find a solution of equation (3.170) for relatively wide sleeves. Let us consider the auxiliary integral equation corresponding to the even problem for two semi-infinite sleeves on an infinite cylinder (Fig. 3.8):

$$\int_k^\infty q_+(\xi)\left[K\left(\frac{\xi - x}{\lambda}\right) + K\left(\frac{\xi + x}{\lambda}\right)\right] d\xi = \pi f_+(x) \quad (k \leq x < \infty). \quad (3.192)$$

Assuming that the stresses and displacements vanish at infinity, we confine ourselves by the case when $f_+(x) = \exp(-\mu|x|)$ $(\mathrm{Re}\,\mu > 0)$. We will seek the solution of equation (3.192) for the small cylinder thickness R with respect to the one-half distance b between the sleeves, i.e. when the parameter $\rho = R/b = \lambda/k$ is small. Then, changing the variables in equation (3.192) and introducing the notation by the formulae

$$|\xi| = \lambda\tau + k, \quad |x| = \lambda t + k, \quad \rho = \lambda/k,$$
$$\varphi^+(\tau) = \lambda \exp(\mu k) q_+(\xi), \quad \varepsilon = \lambda\mu,$$

we present it in the form

$$\int_0^\infty \varphi^+(\tau) K(\tau - t)\, d\tau = \pi \exp(-\varepsilon t) \tag{3.193}$$
$$- \int_0^\infty \varphi^+(\tau) K\left(\tau + t + \frac{2}{\rho}\right) d\tau \quad (0 \le t < \infty),$$

which is convenient for an investigation for small ρ.

Since the integral on the right-hand side of equation (3.193) has an exponential decay for $\rho \to 0$ and $t \to \infty$ (see the footnote below, Kucherov (1973)), to obtain the solution $\varphi^+(\tau)$ of (3.193), we can apply the method of successive approximations by the scheme

$$\varphi^+(\tau) = \lim_{n\to\infty} \varphi_n(\tau), \quad \varphi_n(\tau) = \varphi_0(\tau) + \sum_{k=1}^{n} (-1)^k \varphi_k^*(\tau),$$

$$\int_0^\infty \varphi_n(\tau) K(\tau - t)\, d\tau = \pi \exp(-\varepsilon t) - \int_0^\infty \varphi_{n-1}(\tau) K\left(\tau + t + \frac{2}{\rho}\right) d\tau,$$

$$\int_0^\infty \varphi_0(\tau) K(\tau - t)\, d\tau = \pi \exp(-\varepsilon t) \quad (0 \le t < \infty). \tag{3.194}$$

The solution of Wiener–Hopf equation (3.194), for the zeroth approximation $\varphi_0(\tau)$, corresponds to one semi-infinite sleeve. By using the exact factorization of the function $L(s)/s$ in the kernel $K(t)$ of the form (3.142), we can obtain $\varphi_0(\tau)$ in a complicated form and prove that, as $\tau \to \infty$, the function $\varphi_0(\tau)$ behaves as $\exp(-\chi\tau)$, $\chi > 0$ (see also (3.196)). Let us obtain for small ρ the asymptotic estimate for the first approximation $\varphi_1^*(\tau)$. The function $\varphi_1^*(\tau)$ is determined from the equation

$$\int_0^\infty \varphi_1^*(\tau) K(\tau - t)\, d\tau = \int_0^\infty \varphi_0(\tau) K\left(\tau + t + \frac{2}{\rho}\right) d\tau \quad (0 \le t < \infty). \tag{3.195}$$

In view of properties of the function $K(t)$[6] we conclude that, for small ρ and $t \to \infty$, the right-hand side of equation (3.195) behaves as $\exp[-\kappa(t + 2/\rho)]$, $\kappa > 0$, therefore,

$$\varphi_1^*(\tau) \sim \exp(-\mu\tau - 2\kappa/\rho), \quad \mu > 0, \quad \tau \to \infty.$$

Reasoning in a similar way, we obtain the asymptotic estimates for each approximation $\varphi_n(\tau)$. Using these estimates, we can present the exact solution $\varphi^+(\tau)$ of equation (3.193) in the form of an asymptotically convergent series for small ρ:

$$\varphi^+(\tau) = \sum_{n=0}^{\infty}(-1)^n \tilde{\varphi}_n(\tau)\exp(-2n\kappa/\rho), \quad \tilde{\varphi}_0(\tau) = \varphi_0(\tau),$$

$$\int_0^{\infty} \tilde{\varphi}_n(\tau)K(\tau - t)\, d\tau = \exp(2\kappa/\rho)\int_0^{\infty} \tilde{\varphi}_{n-1}(\tau)K\left(\tau + t + \frac{2}{\rho}\right) d\tau$$

$$(n = 1, 2, ..., \quad 0 \le t < \infty).$$

To obtain the approximate solution $\varphi^+(\tau)$ in the form, which is convenient for engineering purposes, we use approximation (3.165). The zeroth approximation $\varphi_0(\tau)$ for (3.165) is given by the formula ($-C < \mathrm{Re}\,\varepsilon \le D$)

$$\varphi_0(\tau) = \frac{C^2 - \varepsilon^2}{\sqrt{D^2 - \varepsilon^2}}\Phi(\sqrt{(D - \varepsilon)\tau})\exp(-\varepsilon\tau) \tag{3.196}$$

[6]The function $L(s)/s$ in (3.142) is meromorphic and can be represented as the infinite product

$$\frac{L(s)}{s} = A\prod_{n=1}^{\infty}\left(1 + \frac{s^2}{\delta_n^2}\right)\left(1 + \frac{s^2}{\gamma_n^2}\right)^{-1}$$

in the complex s-plane, which converges uniformly in s outside ε-vicinities of the poles $i\gamma_n$ (by the way, this representation allows us to factorize $L(s)/s$ exactly). On the other hand, this infinite product can be decomposed as a sum (Lawrentjew and Schabat (1967))

$$\frac{L(s)}{s} = \frac{2}{\pi}\sum_{n=1}^{\infty}\frac{\gamma_n b_n}{s^2 + \gamma_n^2}, \quad \lim_{s\to 0}\frac{L(s)}{s} = A = \frac{2}{\pi}\sum_{n=1}^{\infty}\frac{b_n}{\gamma_n}, \tag{*}$$

$$b_n = \pi i\{[i\gamma_n/L(i\gamma_n)]'\}^{-1}, \quad \mathrm{Re}\, b_n \sim n^{-1} \ (n \to \infty).$$

The expansion $(*)_1$ converges uniformly in s in the same domain. By substituting $(*)_1$ into formula $(3.142)_1$ and taking the integral in the complex s-plane, we obtain for the kernel $K(t)$ the formula

$$K(t) = \sum_{n=1}^{\infty} b_n \exp(-|t|\gamma_n).$$

This series converges uniformly for all $0 < \varepsilon_1 \le |t| < \infty$.

$$+\frac{C+\varepsilon}{\sqrt{D+\varepsilon}}\frac{\exp(-D\tau)}{\sqrt{\pi\tau}},$$

where $\Phi(x)$ has the form (2.101). The $\varphi_n(\tau)$, $n = 1, 2, ...$, can be constructed similarly. The integral characteristics N_0^0 and N_1^0 for the zeroth approximation $q_+^0(x)$ of the solution $q_+(x)$ of equation (3.192), for $f_+(x) = \exp(-\mu|x|)$, are determined by the formulae

$$N_0^0 = \int_k^\infty q_+^0(x)\,dx = \exp(-\mu k)\int_0^\infty \varphi_0(\tau)\,d\tau = \frac{C(C+\varepsilon)}{\varepsilon\sqrt{D(D+\varepsilon)}}\exp(-\mu k),$$

$$N_1^0 = \int_k^\infty xq_+^0(x)\,dx = \exp(-\mu k)\int_0^\infty (\lambda\tau + k)\varphi_0(\tau)\,d\tau$$

$$= kN_0^0[1 + \rho(\varepsilon^{-1} + (2D)^{-1} - C^{-1})].$$

Having the approximate solution $q_+(x)$ of the integral equation (3.192) for small ρ for the case when $f_+(x) = \exp(-\mu|x|)$ and μ is close to zero, one can obtain the solution $\varphi(x)$ of equation (3.170) for two finite sleeves of width $1-k$ when the quantity $(1-k)\lambda^{-1}$ is fairly large, i.e. $\lambda_k = \lambda(1-k)^{-1}$ is fairly small ($\lambda_k \leq 1$). To be exact, if we know the solutions $\omega_\pm(x)$, $v(x)$ of the integral equations

$$\int_{-1}^\infty \omega_-(\xi)\,K\left(\frac{\xi - x}{\lambda}\right)\,d\xi = \pi\exp(-\mu|x|) \quad (-1 \leq x < \infty), \tag{3.197}$$

$$\int_{-\infty}^1 \omega_+(\xi)\,K\left(\frac{\xi - x}{\lambda}\right)\,d\xi = \pi\exp(-\mu|x|) \quad (-\infty < x \leq 1), \tag{3.198}$$

$$\int_{-\infty}^\infty v(\xi)\,K\left(\frac{\xi - x}{\lambda}\right)\,d\xi = \pi\exp(-\mu|x|) \quad (-\infty < x < \infty), \tag{3.199}$$

then the solution $\varphi(x)$ of equation (3.170) for small values of the parameter λ_k can be presented as the following combination (cf. with (3.118))

$$\varphi(x) = q_+(x) + \omega_+(x) + \omega_-(x) - 2v(x). \tag{3.200}$$

The integral equations (3.197)–(3.197) can be reduced to Wiener–Hopf equations and solved using the approximation (3.165). Solving these equations and taking the zeroth approximation for $q_+(x)$ in formula (3.200), we obtain the approximate expression $\varphi(x) \approx \varphi_0(x)$, where

$$\varphi_0(x) = \frac{1}{A\lambda}\Phi\sqrt{\frac{D}{\lambda}}(|x|-k) + \frac{\exp[-D(|x|-k)/\lambda]}{\sqrt{\pi A\lambda(|x|-k)}} \tag{3.201}$$

$$+\frac{1}{A\lambda}\left(\Phi\sqrt{\frac{D}{2\lambda}}(1-x^2)-1\right) + \frac{\sqrt{2}\exp[-D(1-x^2)/(2\lambda)]}{\sqrt{\pi A\lambda(1-x^2)}},$$

where $\Phi(x)$ is defined by formula (2.101).

In Table 3.4 we give the values of the function $\varphi(x)$ (the solution of (3.170)) for $x = (1+k)/2$, $\nu = 0.3$ for various values of the parameters λ, k, calculated by formulae $(3.173)_1$ and (3.201).

Table 2.4: The solution of (3.170) at the middle point

λ, k	1/2, 1/2	1/2, 1/4	1/4, 1/2	1/4, 1/4
$(3.173)_1$	3.858	3.688	7.304	7.281
(3.201)	4.054	3.812	7.403	7.309

These data mean that the smaller the distance $2b$ between the sleeves for fixed values of R and a (R/b should be small), the smaller is the contact pressure at the middle contact points.

3.3.3. THE METHOD OF ORTHOGONAL POLYNOMIALS

We now use the method of orthogonal polynomials to solve equation (3.170). We rewrite (3.170) in the form

$$-\int_k^1 \varphi(\xi)\ln\frac{|\xi^2-x^2|}{\lambda^2}d\xi = \pi f(x) \tag{3.202}$$

$$+\int_k^1 \varphi(\xi)G\left(\frac{\xi}{\lambda},\frac{x}{\lambda},\frac{k}{\lambda}\right)d\xi \quad (k \le x \le 1),$$

where the function $G(\xi, x, k)$ is symmetric with respect to ξ, x and has no singularities for $\xi = x$. We change the variables in equation (3.202) and introduce the notation

$$x = \sqrt{k'^2u^2+k^2}, \quad \xi = \sqrt{k'^2v^2+k^2}, \quad k' = \sqrt{1-k^2},$$

$$\varphi_+(v) = k'^2|v|(k'^2v^2+k^2)^{-1/2}\varphi(\sqrt{k'^2v^2+k^2}), \quad f_+(u) = f(\sqrt{k'^2u^2+k^2}),$$

$$G_+\left(\frac{v}{\mu}, \frac{u}{\mu}, \varepsilon\right) = G(\sqrt{\mu^{-2}v^2+\varepsilon}, \sqrt{\mu^{-2}u^2+\varepsilon}, \varepsilon), \quad \varepsilon = \frac{k}{\lambda}, \quad \mu = \frac{\lambda}{k'}.$$

We then get

$$-\int_0^1 \varphi_+(v) \ln \frac{|v^2-u^2|}{\mu^2}\, dv = \pi f_+(x) \tag{3.203}$$

$$+\int_0^1 \varphi_+(v) G_+\left(\frac{v}{\mu}, \frac{u}{\mu}, \varepsilon\right) dv \qquad (0 \le u \le 1).$$

We seek the solution of (3.203) in the form $\varphi_+(v) = \varphi_0(v) + \varphi_1(v)$, where $\varphi_0(v)$ is determined from the equation

$$-\int_{-1}^1 \varphi_0(u) \ln \frac{|u-v|}{\mu}\, du = \pi f_+(v) + G_+(0) N_0 \qquad (|u| \le 1),$$

$$G_+(0) = G_+(0,0,\varepsilon), \qquad N_0 = \int_{-1}^1 \varphi_0(u)\, du$$

and has the form

$$\varphi_0(v) = \frac{1}{\pi\sqrt{1-v^2}}\left[N_0 - \int_{-1}^1 \frac{f'_+(u)\sqrt{1-u^2}}{u-v}\, du\right],$$

$$N_0 = \frac{1}{\ln(2\mu) - G_+(0)} \int_{-1}^1 \frac{f_+(u)\, du}{\sqrt{1-u^2}}.$$

Then the function $\varphi_1(v)$ can be found from equation (3.203) in which, instead of $f_+(u)$, one should take

$$f_1(u) = \frac{1}{\pi}\int_0^1 \varphi_0(v)\left[2G_+(0) - G_+\left(\frac{v}{\mu}, \frac{u}{\mu}, \varepsilon\right)\right] dv.$$

We seek $\varphi_1(v)$ in the form

$$\varphi_1(v) = \Phi_1(v)(1-v^2)^{-1/2}, \quad \Phi_1(v) = \sum_{k=0}^{\infty} q_k \mathrm{T}_{2k}(v), \tag{3.204}$$

where $\mathrm{T}_{2k}(x)$ are the Chebyshev polynomials of the first kind.

We also expand the functions $G_+(v/\mu, u/\mu, \varepsilon)$ and $f_1(u)$ into the series

$$G_+\left(\frac{v}{\mu}, \frac{u}{\mu}, \varepsilon\right) = \sum_{m=0}^{\infty}\sum_{n=0}^{\infty} e_{mn}(\mu, \varepsilon)\mathrm{T}_{2m}(u)\mathrm{T}_{2n}(v), \tag{3.205}$$

$$f_1(u) = \sum_{k=0}^{\infty} r_k \mathrm{T}_{2k}(u), \tag{3.206}$$

$$e_{mn}(\mu, \varepsilon) = \frac{\beta_{mn}}{\pi^2}\int_0^{\pi}\int_0^{\pi} G_+\left(\frac{\cos\psi}{\mu}, \frac{\cos\phi}{\mu}, \varepsilon\right)$$

$$\times \cos(2n\psi)\cos(2m\phi)\, d\phi\, d\psi,$$

$$\beta_{00} = 1, \qquad \beta_{m0} = \beta_{0n} = 2, \qquad \beta_{mn} = 4 \quad (m, n \geq 1),$$

$$r_k = \frac{\beta_k}{\pi}\int_{-1}^{1}\frac{f_1(u)}{\sqrt{1-u^2}}\mathrm{T}_{2k}(u)\, du, \qquad \beta_0 = 1, \qquad \beta_k = 2 \quad (k \geq 1).$$

Substituting relations (3.204)–(3.206) into the integral equation (3.203) for $\varphi_1(v)$, after some reductions, we obtain the following infinite algebraic system for unknown coefficients q_k:

$$r_k + q_0 e_{k0}(\mu, \varepsilon) + \frac{1}{2}\sum_{n=1}^{\infty} q_n e_{kn}(\mu, \varepsilon) = \begin{cases} q_0 \ln(2\mu) & \text{for } k = 0 \\ q_k/(2k) & \text{for } k = 1, 2, \ldots \end{cases}. \tag{3.207}$$

The approximate solution of system (3.207) can be found by a reduction of its matrix (Kantorovich and Krylov (1958), Aleksandrov and Kucherov (1970)) for $\mu = \lambda/k' \geq \mu_0$. If the value of k increases, then the ranges of applicability of this method extend for the parameter λ.

Remark. *We note that these solutions are locally incorrect near the endpoints of the contact domains because the sleeves are deformable. For such sleeves (punches), having sharp edges, there is no square root singularity. The exact singularity at the endpoints is determined by the elastic characteristics of the interacting solids (see* Theocaris and Gdoutos (1976), Aleksandrov, Grishin and Kovalenko (1985) *and* subsection 3.5.1*). Nevertheless, we can use the constructed asymptotics everywhere outside the small vicinities. To ensure correctness, one should consider sleeves with a smooth base and construct solutions bounded at the endpoints of contact domains, taking the above formulae as a first stage of solving the problem.*

3.4 Periodic System of Sleeves on a Cylidner

3.4.1. FORMULATION AND REDUCTION TO INFINITE SYSTEMS

Consider the axisymmetric periodic contact problem for an infinite elastic circular cylinder (G, ν) of radius R. The infinite periodic (with period $l = a + b$) system of identical elastic sleeves (G_1, ν_1) is situated as shown in Fig. 3.9 (as in section 3.3, we assume that the sleeves are fairly massive, i.e. their dimensions considerably exceed dimensions of contact domains). This problem, evidently, is equivalent to the contact problem for interaction between one sleeve and a finite cylinder of length l whose end-walls $z = 0$ and $z = l$ are in a state of sliding support: $\tau_{rz} = u_z = 0$. The stress $\sigma_r(R, z) = -q(z)$, in the contact domain, and the function $R - \delta(z)$, describing the shape of the base of the sleeve, are periodic functions with period l.

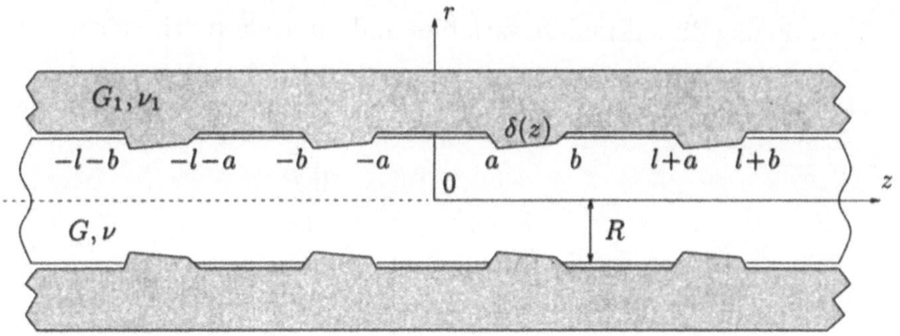

Figure 3.9: A periodic system of sleeves (an axial section)

For the periodic contact zone, shown in Fig 3.9, integral equation (3.141) with kernel (3.142) can be rewritten in the form

$$\int_a^b q(\xi)\, K_*\left(\frac{\xi - z}{R}\right) d\xi = \frac{\pi}{2}\theta\delta(z) \quad (a \le z \le b), \tag{3.208}$$

$$K_*(t) = \sum_{k=-\infty}^{\infty} \int_0^{\infty} \frac{L(s)}{s} \cos\left(ts + \frac{kls}{R}\right) ds, \tag{3.209}$$

where the function $L(s)$ is determined by formulae $(3.142)_2$ with $m = 1$ (we suppose $G = G_1$ and $\nu = \nu_1$), and has properties (3.145), (3.146). Since $L(-s) = -L(s)$, we can also represent (3.209) as follows:

$$K_*(t) = \frac{1}{2} \int_{-\infty}^{\infty} \frac{L(s)}{s} \cos(ts) \left[1 + 2 \sum_{k=1}^{\infty} \cos \frac{kls}{R}\right] ds. \tag{3.210}$$

Using the relation (Gel'fand and Shilov (1964, formula (1) on p. 31))

$$1 + 2 \sum_{k=1}^{\infty} \cos(ky) = 2\pi \sum_{k=-\infty}^{\infty} \delta_*(y - 2\pi k),$$

where $\delta_*(y)$ is the delta function, we reduce (3.210) to the form

$$K_*(t) = \frac{1}{2} \sum_{k=-\infty}^{\infty} \frac{L(2\pi Rk/l)}{k} \cos \frac{2\pi Rkt}{l}. \tag{3.211}$$

Introducing dimensionless variables and notation by the formulae

$$x = \frac{2z - b - a}{b - a}, \qquad \varphi(x) = \frac{q(z)}{\theta}, \qquad f(x) = \frac{\delta(z)}{b - a}, \tag{3.212}$$

$$\lambda = \frac{l}{b - a}, \qquad \mu = \frac{2\pi R}{l}, \tag{3.213}$$

from (3.208), (3.211) we get the integral equation

$$\int_{-1}^{1} \varphi(\xi) K\left(\frac{\xi - x}{\lambda}\right) d\xi = \pi f(x) \qquad (|x| \le 1), \tag{3.214}$$

with the kernel

$$K(t) = \frac{1}{2} \sum_{k=-\infty}^{\infty} \frac{L(\mu k)}{k} \exp(i\pi kt).$$

Note that the parameter $\lambda > 1$. We also suppose $f'(x)$ satisfies a Hölder condition for $|x| \le 1$.

Historically, the first approach to the solution of equation (3.214) was the method of orthogonal polynomials, as suggested and proved in (Aleksandrov and Kovalenko (1977b)). Using the relation 1.441.2 in Gradshteyn and Ryzhik (1994)

$$\sum_{k=1}^{\infty} \frac{\cos(kx)}{k} = -\ln\left|2 \sin \frac{x}{2}\right|,$$

we separate the principal part of the kernel of equation (3.214) and then write this equation as follows

$$-\int_{-1}^{1} \varphi(\xi) \ln\left|2\sin\frac{\pi(\xi - x)}{2\lambda}\right| d\xi = \pi f^*(x) \tag{3.215}$$

$$-\int_{-1}^{1} \varphi(\xi)\, N\left(\mu, \frac{\xi - x}{\lambda}\right) d\xi,$$

$$N(\mu, t) = \sum_{k=1}^{\infty} \frac{L(\mu k) - 1}{k} \cos(\pi k t), \qquad |x| \leq 1,$$

$$f^*(x) = f(x) - N_0 A\mu/(2\pi), \qquad L(s) \sim As \quad (s \to 0).$$

The integral operator on the left-hand side of (3.215) can be inverted exactly with the help of the spectral relation for the Chebyshev polynomials $\mathrm{T}_n(x)$ (see below). We divide the problem into even and odd ones (lower indices "+" and "−", respectively) and we analyze the even ($\mathcal{A}$) and odd ($\mathcal{B}$) variants of equation (3.215) for the functions $\varphi_\pm(x)$ ($\varphi(x) = \varphi_+(x) + \varphi_-(x)$), in the form

$$\varphi_\pm(x) = \frac{\omega_\pm(x)\cos^{\pm 1}(px)}{\sqrt{\cos(2px) - \cos(2p)}}, \qquad p = \frac{\pi}{2\lambda}, \tag{3.216}$$

$$\omega_+(x) = \sum_{k=0}^{\infty} a_k \mathrm{T}_{2k}\left(\frac{\sin(px)}{\sin p}\right), \tag{3.217}$$

$$\omega_-(x) = \sum_{k=0}^{\infty} a_k \mathrm{T}_{2k+1}\left(\frac{\tan(px)}{\tan p}\right). \tag{3.218}$$

We also expand the functions $f^*_\pm(x)$ ($f^*(x) = f^*_+(x) + f^*_-(x)$), $N(\mu, t)$ into series in the Chebyshev polynomials:

$$f^*_+(x) = \sum_{k=0}^{\infty} f_k \mathrm{T}_{2k}\left(\frac{\sin(px)}{\sin p}\right), \tag{3.219}$$

$$f^*_-(x) = \sum_{k=0}^{\infty} f_k \mathrm{T}_{2k+1}\left(\frac{\tan(px)}{\tan p}\right), \tag{3.220}$$

$\mathcal{A}$) :

$$N(\mu, t) = \sum_{m=0}^{\infty}\sum_{n=0}^{\infty} e_{mn}(\mu, \lambda)$$

$$\times \mathrm{T}_{2m}\left(\frac{\sin(p\xi)}{\sin p}\right)\mathrm{T}_{2n}\left(\frac{\sin(px)}{\sin p}\right)+\ldots, \tag{3.221}$$

$\mathcal{B})$:

$$N(\mu,t)=\sum_{m=0}^{\infty}\sum_{n=0}^{\infty}e_{mn}(\mu,\lambda)$$

$$\mathrm{T}_{2m+1}\left(\frac{\tan(p\xi)}{\tan p}\right)\mathrm{T}_{2n+1}\left(\frac{\tan(px)}{\tan p}\right)+\ldots, \tag{3.222}$$

where

$\mathcal{A})$:

$$e_{mn}(\mu,\lambda)=\frac{2p^2\beta_{mn}}{\pi^2}\int_{-1}^{1}\int_{-1}^{1}\frac{N(\mu,t)\mathrm{T}_{2m}\left(\frac{\sin(p\xi)}{\sin p}\right)\mathrm{T}_{2n}\left(\frac{\sin(px)}{\sin p}\right)}{\sqrt{\cos(2p\xi)-\cos(2p)}}$$

$$\times\frac{\cos(p\xi)\cos(px)\,d\xi\,dx}{\sqrt{\cos(2px)-\cos(2p)}},$$

$\mathcal{B})$:

$$e_{mn}(\mu,\lambda)=\frac{8p^2\cos^2 p}{\pi^2}\int_{-1}^{1}\int_{-1}^{1}\frac{N(\mu,t)\mathrm{T}_{2m+1}\left(\frac{\tan(p\xi)}{\tan p}\right)}{\sqrt{\cos(2p\xi)-\cos(2p)}}$$

$$\times\frac{\mathrm{T}_{2n+1}\left(\frac{\tan(px)}{\tan p}\right)\,d\xi\,dx}{\sqrt{\cos(2px)-\cos(2p)}\,\cos(p\xi)\cos(px)},$$

and $\beta_{00}=1$, $\beta_{m0}=\beta_{0n}=2$, $\beta_{mn}=4$ $(m,n\geq 1)$. The double and single series (3.219)–(3.222) are uniformly convergent for all $|x|\leq 1$, $|\xi|\leq 1$, $\lambda>1$, $\mu\geq\mu_0>0$.

Substituting expansions (3.216)–(3.222) into the even or odd variant of equation (3.215) and using the spectral relations[7]

$\mathcal{A})$:

$$-\int_{-1}^{1}\frac{\mathrm{T}_{2n}\left(\frac{\sin(p\xi)}{\sin p}\right)}{\sqrt{\cos(2p\xi)-\cos(2p)}}\ln\left|2\sin\frac{\pi(\xi-x)}{2\lambda}\right|\cos(p\xi)\,d\xi$$

[7]These spectral relations are obtained by changing variables in the well-known spectral relation (6.11.2) in Gladwell (1980), for $|x|\leq 1$:

$$-\frac{1}{\pi}\int_{-1}^{1}\frac{\mathrm{T}_n(\xi)}{\sqrt{1-\xi^2}}\ln|\xi-x|d\xi=\begin{cases}\mathrm{T}_n(x)/n & \text{for } n\geq 1\\ \ln 2\,\mathrm{T}_0(x) & \text{for } n=0\end{cases}.$$

$$= \pi \lambda_n \, \mathrm{T}_{2n} \left(\frac{\sin(px)}{\sin p} \right),$$

$\mathcal{B})$:

$$- \int_{-1}^{1} \frac{\mathrm{T}_{2n+1}\left(\frac{\tan(p\xi)}{\tan p}\right)}{\sqrt{\cos(2p\xi) - \cos(2p)}} \ln \left| 2 \sin \frac{\pi(\xi - x)}{2\lambda} \right| \frac{d\xi}{\cos(p\xi)}$$

$$= \pi \lambda_n \, \mathrm{T}_{2n+1} \left(\frac{\tan(px)}{\tan p} \right),$$

where

$\mathcal{A})$: $\lambda_0 = -(\sqrt{2}p)^{-1} \ln|\sin p|, \quad \lambda_n = (2\sqrt{2}pn)^{-1} \quad (n \geq 1);$

$\mathcal{B})$: $\lambda_n = [\sqrt{2}p \cos p \, (2n+1)]^{-1} \quad (n \geq 0),$

we obtain an infinite system of algebraic equations in the unknown coefficients a_k, occurring in (3.217), in the form

$$a_n = r_n - \sum_{m=0}^{\infty} a_m c_{mn} \qquad (n = 0, 1, ...). \tag{3.223}$$

Here, we denote $r_n = f_n \lambda_n^{-1}$ and for the variants $\mathcal{A}$ and $\mathcal{B}$ we have

$\mathcal{A})$: $c_{m0} = -(\ln|\sin p|)^{-1} e_{m0}, \qquad c_{mn} = n e_{mn};$

$\mathcal{B})$: $c_{mn} = (n + 1/2) e_{mn}.$

The system (3.223) can be solved by reduction for $\lambda > 1$, $\mu \geq \mu_0 > 0$.

We suggest another method of solving the problem which is suitable for any μ, and $\lambda > 1$.

3.4.2. REDUCTION TO SINGULAR INTEGRAL EQUATIONS

For $m = 1$ the function $L(s)$ can be presented in the form

$$L(s) = \tanh(As) + g(s), \qquad |g(s)| \leq \delta \quad (0 \leq |s| < \infty), \tag{3.224}$$

$$g(|s|) = O(s^{-2}) \quad (|s| \to \infty), \qquad g(|s|) = O(s^3) \quad (|s| \to 0),$$

where the quantity δ is fairly small.

We consider the series 1.441.2, 8.146.10 in Gradshteyn and Ryzhik (1994)

$$M(t) = \sum_{k=1}^{\infty} \tanh(\gamma k) \sin(kt) = \frac{1}{2} \cot \frac{t}{2} \tag{3.225}$$

$$-2 \sum_{k=1}^{\infty} \frac{q^{2k}}{1+q^{2k}} \sin(kt) = \frac{\mathcal{K}(e) \operatorname{cn} u}{\pi \operatorname{sn} u} \qquad (q = e^{\gamma}, \gamma = \mu A).$$

Here and later, $u = \pi^{-1}\mathcal{K}(e)t$, and the quantity $e < 1$ is determined from the transcendental equation

$$\pi\mathcal{K}(\sqrt{1-e^2})\,[\mathcal{K}(e)]^{-1} = \gamma,$$

$\mathcal{K}(e)$ is the complete elliptic integral of the first kind; $\operatorname{sn} u$, $\operatorname{cn} u$, $\operatorname{dn} u$ are the Jacobi elliptic functions.

We differentiate the integral equation (3.214) with respect to x and, on the basis of formulae (3.224), (3.225), write it as follows

$$\beta \int_{-1}^{1} \varphi(\xi) \frac{\operatorname{cn}[\beta(\xi - x)]}{\operatorname{sn}[\beta(\xi - x)]} d\xi = \pi f'(x) - \alpha \int_{-1}^{1} \varphi(\xi) G[\alpha(\xi - x)] d\xi, \tag{3.226}$$

$$G(t) = \sum_{k=1}^{\infty} g(\mu k) \sin(kt), \qquad \beta = \pi^{-1}\mathcal{K}(e)\alpha, \qquad \alpha = \pi\lambda^{-1}.$$

It can be shown using the properties of $g(s)$, that the function $G(t)$ in (3.226) satisfies a Hölder condition for $|t| \leq 2\alpha$.

We consider the even ($\mathcal{A}$: $\varphi(x)$ and $f(x)$ are the even functions) and odd ($\mathcal{B}$: $\varphi(x)$ and $f(x)$ are the odd functions) variants of equation (3.226), separately. We note that (Gradshteyn and Ryzhik (1994, formulae 8.156))

$$\frac{\operatorname{cn}[\beta(\xi - x)]}{\operatorname{sn}[\beta(\xi - x)]} - \frac{\operatorname{cn}[\beta(\xi + x)]}{\operatorname{sn}[\beta(\xi + x)]} = \frac{2 \operatorname{sn}(\beta x) \operatorname{cn}(\beta x) \operatorname{dn}(\beta\xi)}{\operatorname{sn}^2(\beta\xi) - \operatorname{sn}^2(\beta x)},$$

$$\frac{\operatorname{cn}[\beta(\xi - x)]}{\operatorname{sn}[\beta(\xi - x)]} + \frac{\operatorname{cn}[\beta(\xi + x)]}{\operatorname{sn}[\beta(\xi + x)]} = \frac{2 \operatorname{sn}(\beta\xi) \operatorname{cn}(\beta\xi) \operatorname{dn}(\beta x)}{\operatorname{sn}^2(\beta\xi) - \operatorname{sn}^2(\beta x)}.$$

On the basis of these formulae and taking into account that $\beta < \mathcal{K}(e)$, and the functions $\operatorname{cn}(\mathcal{K}x)$ and $\operatorname{dn}(\mathcal{K}x)$ are monotonically decreasing, if x increases from 0 to 1 (Jahnke, Emde and Lösch (1966, Figs. 48 and 49)), we reduce (3.226) to the equations

$$\mathcal{A}): \qquad \beta \int_{-1}^{1} \frac{\varphi(\xi) \operatorname{dn}(\beta\xi)}{\operatorname{sn}(\beta\xi) - \operatorname{sn}(\beta x)} d\xi = \frac{\pi f'(x)}{\operatorname{cn}(\beta x)}$$

$$-\frac{\alpha}{\mathrm{cn}\,(\beta x)}\int\limits_{-1}^{1}\varphi(\xi)G[\alpha(\xi-x)]d\xi, \tag{3.227}$$

$\mathcal{B})$:
$$\beta\int\limits_{-1}^{1}\frac{\varphi(\xi)\,\mathrm{cn}\,(\beta\xi)}{\mathrm{sn}\,(\beta\xi)-\mathrm{sn}\,(\beta x)}d\xi=\frac{\pi f'(x)}{\mathrm{dn}\,(\beta x)}$$

$$-\frac{\alpha}{\mathrm{dn}\,(\beta x)}\int\limits_{-1}^{1}\varphi(\xi)G[\alpha(\xi-x)]d\xi, \tag{3.228}$$

Since $\beta < \mathcal{K}(e)$ and the function $\mathrm{sn}\,(\mathcal{K}x)$ is monotonically increasing from 0 to 1, if x increases from 0 to 1 (Jahnke, Emde and Lösch (1966, Fig. 47)), we introduce the new variables

$$\tau=\mathrm{sn}\,(\beta\xi),\quad t=\mathrm{sn}\,(\beta x),\quad c=\mathrm{sn}\,\beta. \tag{3.229}$$

We also note that (Gradshteyn and Ryzhik (1994, formulae 8.154.4,5 and 8.158.1))

$$\tau=\beta\,\mathrm{cn}\,(\beta\xi)\;\mathrm{dn}\,(\beta\xi)\,d\xi,\quad \mathrm{cn}\,(\beta x)=\sqrt{1-t^2},\quad \mathrm{dn}\,(\beta x)=\sqrt{1-e^2t^2}, \tag{3.230}$$

and introduce the inverse to $\mathrm{sn}\,u$ (Gradshteyn and Ryzhik (1994, formula 8.144.1))

$$\xi=\frac{\mathrm{asn}\,\tau}{\beta},\quad x=\frac{\mathrm{asn}\,t}{\beta},\quad \mathrm{asn}\,t=\mathrm{F}(\arcsin t,e)=\int\limits_{0}^{t}\frac{dx}{\sqrt{(1-x^2)(1-e^2x^2)}}. \tag{3.231}$$

Here, $\mathrm{F}(x,e)$ is the elliptic integral of the first kind.

On the basis of formulae (3.229)–(3.231) we reduce equations (3.227), (3.228) to the form

$$\int\limits_{-c}^{c}\frac{\psi_i(\tau)}{\tau-t}d\tau=\pi h_i(t)-\int\limits_{-c}^{c}\psi_i(\tau)\,H_i(\tau,t)\,d\tau\qquad(|t|\le c), \tag{3.232}$$

where $i=1$ and $i=2$ correspond to variants $\mathcal{A}$ and $\mathcal{B}$, respectively, and we have introduced the notation

$$\frac{\varphi(\xi)}{\mathrm{cn}\,(\beta\xi)}=\psi_1(\tau),\qquad \frac{f'(x)}{\mathrm{cn}\,(\beta x)}=h_1(t), \tag{3.233}$$

$$\frac{\varphi(\xi)}{\mathrm{dn}\,(\beta\xi)} = \psi_2(\tau), \qquad \frac{f'(x)}{\mathrm{dn}\,(\beta x)} = h_2(t), \tag{3.234}$$

$$\mathcal{A}): \qquad H_1(\tau,t) = \frac{\pi}{\mathcal{K}(e)\sqrt{1-e^2\tau^2}\sqrt{1-t^2}}\, G\left[\frac{\pi}{\mathcal{K}(e)}(\mathrm{asn}\,\tau - \mathrm{asn}\,t)\right],$$

$$\mathcal{B}): \qquad H_2(\tau,t) = \frac{\pi}{\mathcal{K}(e)\sqrt{1-e^2t^2}\sqrt{1-\tau^2}}\, G\left[\frac{\pi}{\mathcal{K}(e)}(\mathrm{asn}\,\tau - \mathrm{asn}\,t)\right].$$

It is important that $c < 1 < 1/e$ and the square root singularities in the denominators in the expressions $H_i(\tau,t)$ lie outside the intervals of definition and integration of (3.232).

To solve the singular integral equation of the first kind (3.232), one can use the various well-known approximate methods (Erdogan (1969), Erdogan, Gupta and Cook (1973), Ivanov (1976), Belotserkovsky and Lifanov (1993)) We choose the method of collocation by Chebyshev nodes, which is also known as the Multopp–Kalandiya method (Kalandiya (1973)).

Using the properties of the functions $f(x)$ and $G(y)$, it can be proved that, if, for specified values of α and μ, the solution of equation (3.232) exists for the class of functions for which the integral

$$\int_{-c}^{c} |\psi_i(\tau)|^p \, d\tau \qquad (0 < p < 2)$$

converges, then this solution can be presented in general in the form

$$\psi_i(t) = \Psi_i(t)(c^2 - t^2)^{-1/2}, \tag{3.235}$$

where the function $\Psi_i(t)$ satisfies a Hölder condition for $|t| < c$.

We substitute expression (3.235) into (3.232), pass to the new variables

$$\tau = c\cos\omega, \qquad t = c\cos\vartheta,$$

and present it in the form

$$\frac{1}{c}\int_0^\pi \frac{\tilde{\Psi}_i(\omega)}{\cos\omega - \cos\vartheta}\, d\omega = \pi\tilde{h}_i(\vartheta) - \int_0^\pi \tilde{\Psi}_i(\omega) H_i(c\cos\omega, c\cos\vartheta)\, d\omega, \tag{3.236}$$

where $0 \le \vartheta \le \pi$, $\tilde{\Psi}_i(\vartheta) = \Psi_i(c\cos\vartheta)$ and $\tilde{h}_i(\vartheta) = h_i(c\cos\vartheta)$. We will omit the tildes over $\tilde{\Psi}_i(\vartheta)$ and $\tilde{h}_i(\vartheta)$.

We construct for $\Psi_i(t)$ the Lagrange interpolational polynomial using the nodes

$$t_n = c\cos\vartheta_n, \quad \vartheta_n = \pi(2n-1)/(2N) \quad (n = 1, 2, ..., N),$$

which are the zeros of the Chebyshev polynomial $\mathrm{T}_N(t/c)$. In the special cases when $\Psi_i(t)$ is an even or odd function and $N = 2r+1$ $(r \geq 1)$, such polynomials have the form

$$\mathcal{A}): \quad \Psi_1(\vartheta) \approx \frac{1}{r+1/2} \sum_{n=1}^{r+1} \Psi_1(\vartheta_n)\delta_n$$

$$\times \left(1 + 2\sum_{m=1}^{r} \cos(2m\vartheta_n)\cos(2m\vartheta)\right), \tag{3.237}$$

$$\mathcal{B}): \quad \Psi_2(\vartheta) \approx \frac{2}{r+1/2} \sum_{n=1}^{r} \Psi_2(\vartheta_n)$$

$$\times \sum_{m=1}^{r} \cos[(2m-1)\vartheta_n]\cos[(2m-1)\vartheta], \tag{3.238}$$

respectively, where $\delta_n = 1$ $(n \neq r+1)$, $\delta_n = 1/2$ $(n = r+1)$.

We substitute the approximate expression $\Psi_i(\vartheta)$ in one of the forms (3.237), or (3.238), into equation (3.236) and use the relation (Gradshteyn and Ryzhik (1994, formula 7.344.1)), often called Glauert's integral (Gladwell (1980, formula (2.3.37)):

$$\int_0^\pi \frac{\cos(l\omega)}{\cos\omega - \cos\vartheta}\, d\omega = \pi\frac{\sin(l\vartheta)}{\sin\vartheta},$$

where $0 \leq \vartheta \leq \pi$ and $l = 0, 1, \ldots$. Then we exactly calculate the integral on the left-hand side of (3.236). To calculate approximately the integral on the right-hand side of (3.236), we use the Gauss quadrature formula 25.4.30 in Abramowitz and Stegun (1972)

$$\int_0^\pi f(\omega)\, d\omega \approx \frac{\pi}{N}\sum_{n=1}^{N} f(\vartheta_n). \tag{3.239}$$

After the calculation of the integrals in (3.236), we set $\vartheta = \vartheta_s$ in the relation and come to a system of r linear algebraic equations in the values $\Psi_i(\vartheta_n)$. To be exact we have

$$\mathcal{A}): \quad \sum_{n=1}^{r+1} \Psi_1(\vartheta_n)\delta_n \left\{ \frac{1}{c\sin\vartheta_s}\chi_r^+(\vartheta_n,\vartheta_s) + \left[\frac{1}{2}H_1(c\cos\vartheta_n, c\cos\vartheta_s)\right.\right.$$

$$\left.\left. - \frac{1}{2}H_1(c\cos\vartheta_n, -c\cos\vartheta_s)\right]\right\} = \left(r+\frac{1}{2}\right) h_1(\vartheta_s) \quad (s = 1,2,...,r), \quad (3.240)$$

and for the odd one we have

$$\mathcal{B}): \quad \sum_{n=1}^{r} \Psi_2(\vartheta_n) \left\{ \frac{1}{c\sin\vartheta_s}\chi_r^-(\vartheta_n,\vartheta_s) + \left[\frac{1}{2}H_2(c\cos\vartheta_n, c\cos\vartheta_s)\right.\right.$$

$$\left.\left. + \frac{1}{2}H_2(c\cos\vartheta_n, -c\cos\vartheta_s)\right]\right\} = \left(r+\frac{1}{2}\right) h_2(\vartheta_s) \quad (s = 1,2,...,r). \quad (3.241)$$

In (3.240), (3.241) we used the notation

$$\chi_r^+(\omega,\vartheta) = 2\sum_{m=1}^{r} \cos(2m\omega)\,\sin(2m\vartheta),$$

$$\chi_r^-(\omega,\vartheta) = 2\sum_{m=1}^{r} \cos[(2m-1)\omega]\,\sin[(2m-1)\vartheta].$$

To close the system of equations (3.240), it is necessary to have one more relation. This is obtained from the condition that the solution in the form (3.235), $(3.233)_1$ for variant $\mathcal{A}$ should also satisfy the original (not differentiated by x) equation (3.214), e.g. at the point $x = 0$. Note that the point has been chosen only for simplicity. We could take any other point on the segment $|x| \leq 1$, because the first equation of (3.227), and, therefore, equation (3.232), differs from (3.214) by only one operation of differentiation.

Setting $x = 0$ in (3.214) and using formula (2.5) from Kovalenko, Tarasov and Chebakov (1990), which in our notation takes the form

$$\frac{1}{2}\gamma + \sum_{k=1}^{\infty} \frac{\tanh(\gamma k)}{k}\cos(kt) = \frac{1}{2}\ln\frac{1+\mathrm{dn}\,u}{1-\mathrm{dn}\,u},$$

after some reduction, taking account of (3.233)–(3.235), we get the relation

$$I_1 + I_2 = \pi\beta f(0), \quad (3.242)$$

where

$$I_1 = -\int_{-c}^{c} \frac{\Psi_1(\tau)}{\sqrt{c^2-\tau^2}} \ln\frac{e|\tau|}{2}\, d\tau,$$

$$I_2 = \int_{-c}^{c} \frac{\Psi_1(\tau)}{\sqrt{c^2-\tau^2}} [P(\tau)+R(\tau)]\, d\tau,$$

$$P(\tau) = \frac{1}{2\sqrt{1-e^2\tau^2}} \ln\frac{1+\sqrt{1-e^2\tau^2}}{1-\sqrt{1-e^2\tau^2}} + \ln\frac{e|\tau|}{2},$$

$$R(\tau) = \frac{1}{2\sqrt{1-e^2\tau^2}} Q\left(\frac{\pi}{\mathcal{K}(e)} \operatorname{asn}\tau\right), \quad Q(t) = \sum_{k=1}^{\infty} \frac{g(\mu k)}{k} \cos(kt).$$

To calculate the integral I_1, we use the first representation of (3.237) and the value of integral

$$-\frac{1}{\pi}\int_{-c}^{c} \frac{\mathrm{T}_{2m}(\tau/c)}{\sqrt{c^2-\tau^2}} \ln\frac{e|\tau|}{2}\, d\tau = A_m, \tag{3.243}$$

$$A_0 = \ln\frac{4}{ec}, \qquad A_m = \frac{(-1)^m}{2m} \quad (m \geq 1),$$

where the $\mathrm{T}_n(x)$ are the Chebyshev polynomials of the first kind. Integral (3.243) can be easily derived from spectral relation (6.11.2) in Gladwell (1980). The integral I_2 can be calculated by means of the quadrature formula (3.239). As a result, we get the following supplementary relation for the system (3.240):

$$\sum_{n=1}^{r+1} \Psi_1(\vartheta_n)\delta_n[\chi_r^0(\vartheta) + P(c\cos\vartheta_n) + R(c\cos\vartheta_n)] \tag{3.244}$$

$$= \left(r+\frac{1}{2}\right)\beta f(0), \qquad \chi_r^0(\omega) = A_0 + 2\sum_{m=1}^{r} A_m \cos(2m\omega).$$

Determining the values $\Psi_i(\theta_n)$ from the system of algebraic equations (3.240), (3.244), we then find the corresponding approximate solution of equation (3.214) for problem $\mathcal{A}$ by the formula

$$\varphi(x) = \operatorname{cn}(\beta x)\, \psi_1(\operatorname{sn}\beta x), \tag{3.245}$$

$$\psi_1(t) = \frac{1}{(r+1/2)\sqrt{c^2-t^2}} \sum_{n=1}^{r+1} \Psi_1(\vartheta_n)\delta_n \tag{3.246}$$

$$\times\left[1+2\sum_{m=1}^{r}\cos(2m\vartheta_n)\,\mathrm{T}_{2m}\left(\frac{t}{c}\right)\right].$$

If, for specified values of α and μ in the even variant $\mathcal{A}$, the function $f(x)$ in (3.214) is defined with an accuracy up to the constant term c_0, then, to find c_0, we need the additional condition $\varphi(1)=0$, giving the equation

$$\sum_{n=1}^{r+1}\Psi_1(\vartheta_n)\delta_n\left[1+2\sum_{m=1}^{r}\cos(2m\vartheta_n)\right]=0. \tag{3.247}$$

By using (3.247) and the relation

$$\mathrm{T}_{2m}(x)-1=-2(1-x^2)\sum_{s=0}^{m-1}\mathrm{U}_{2s}(x) \qquad (m\geq 1),$$

where $\mathrm{U}_n(x)$ are the Chebyshev polynomials of the second kind, we reduce formula (3.246) to the form

$$\psi_1(t)\equiv\psi_+(t)=-\frac{4\sqrt{c^2-t^2}}{c^2(r+1/2)}\sum_{n=1}^{r+1}\Psi_1(\vartheta_n)\delta_n\sum_{m=1}^{r}\cos(2m\vartheta_n)\sum_{s=1}^{m-1}\mathrm{U}_{2s}\left(\frac{t}{c}\right).$$

Unlike (3.240), the system (3.241) is already closed and gives the approximate solution of equation (3.214) for the odd variant $\mathcal{B}$ in the form

$$\varphi(x)=\mathrm{dn}\,(\beta x)\,\psi_2(\mathrm{sn}\,\beta x), \tag{3.248}$$

$$\psi_2(t)=\frac{2}{(r+1/2)\sqrt{c^2-t^2}}\sum_{n=1}^{r}\Psi_2(\vartheta_n) \tag{3.249}$$

$$\times\sum_{m=1}^{r}\cos[(2m-1)\vartheta_n]\,\mathrm{T}_{2m-1}\left(\frac{t}{c}\right).$$

If, for specified values of α and μ in the odd variant $\mathcal{B}$, the function $f(x)$ in (3.214) is defined with an accuracy up to the term c_1x, then, to find c_1, we need the additional condition $\varphi(1)=0$, giving the equation

$$\sum_{n=1}^{r}\Psi_2(\vartheta_n)\sum_{m=1}^{r}\cos[(2m-1)\vartheta_n]=0. \tag{3.250}$$

By using (3.247) and the relation

$$\mathrm{T}_{2m-1}(x)-x=-2(1-x^2)\sum_{s=1}^{m-1}\mathrm{U}_{2s-1}(x) \qquad (m\geq 2),$$

we reduce formula (3.249) to the form

$$\psi_2(t) \equiv \psi_-(t) = -\frac{4\sqrt{c^2-t^2}}{c^2(r+1/2)} \sum_{n=1}^{r} \Psi_2(\vartheta_n)$$

$$\times \sum_{m=2}^{r} \cos[(2m-1)\vartheta_n] \sum_{s=1}^{m-1} \mathrm{U}_{2s-1}\left(\frac{t}{c}\right).$$

The mathematical proof of convergence, when r increases, of these approximate solutions to the exact ones is given in Kalandiya (1973). If the quantity δ in $(3.224)_2$ is small, then, to reach an acceptable accuracy in practice for any values of α and μ, it is sufficient to solve system (3.240), (3.244) or (3.241) for five to six equations. For a numerical realization of the method one should have a routine for calculation of the functions $G(y)$ and $Q(y)$ for a specified value of μ. We note that these functions are connected by the relation $Q'(y) = -G(y)$.

In the general case the function $f(x)$ and, therefore, the function $\varphi(x)$ in equation (3.214) has as an even part as an odd one, i.e.

$$f(x) = f_+(x) + f_-(x), \qquad \varphi(x) = \varphi_+(x) + \varphi_-(x),$$

where the sign "+" indicates the even parts, and the sign "−" indicates the odd ones. Combining (3.245) and (3.248) for this case, we get

$$\varphi(x) = \mathrm{cn}\,(\beta x)\,\psi_+(\mathrm{sn}\,\beta x) + \mathrm{dn}\,(\beta x)\,\psi_-(\mathrm{sn}\,\beta x). \tag{3.251}$$

The integral characteristics

$$N_0 = \int_{-1}^{1} \varphi_+(\xi)\,d\xi, \qquad N_1 = \int_{-1}^{1} \xi\varphi_-(\xi)\,d\xi, \tag{3.252}$$

after some reduction, using formulae (3.235), (3.239) and (3.251) can be obtained in the form

$$N_0 = \frac{\pi}{\beta(r+1/2)} \sum_{n=1}^{r+1} \Psi_1(\vartheta_n) B_0(c\cos\vartheta_n), \qquad B_0(\tau) = \frac{1}{\sqrt{1-e^2\tau^2}},$$

$$N_1 = \frac{\pi}{\beta^2(r+1/2)} \sum_{n=1}^{r} \Psi_2(\vartheta_n) C(c\cos\vartheta_n), \qquad C(\tau) = \frac{\mathrm{asn}\,\tau}{\sqrt{1-\tau^2}}.$$

If, for specified values of α and μ for the general case, the function $f(x)$ in (3.214) is defined with an accuracy up to the linear term $c_0 + c_1 x$, then, to find c_0 and c_1, we need the additional condition $\varphi(\pm 1) = 0$, giving the equation

$$\sqrt{1-c^2}\,\psi_+(c) \pm \sqrt{1-e^2c^2}\,\psi_-(c) = 0.$$

Consider the case when the function $g(s) \equiv 0$ in (3.224). For this case equation (3.232) degenerates into the classical singular integral equation (1.18) of the first kind with the Cauchy kernel, which can be solved in the closed form (see (1.19)). Then for variant $\mathcal{A}$ we get

$$\varphi_+(x) = \frac{\operatorname{cn}(\beta x)}{\pi\sqrt{c^2 - \operatorname{sn}^2(\beta x)}}\left[P \right. \tag{3.253}$$

$$\left. -\beta \int_{-1}^{1} \frac{\sqrt{c^2 - \operatorname{sn}^2(\beta\xi)}\, f_+'(\xi)\operatorname{dn}(\beta\xi)}{\operatorname{sn}(\beta\xi) - \operatorname{sn}(\beta x)} d\xi\right],$$

whereas the additional relation (3.242) takes the form

$$\int_{-1}^{1} \frac{\varphi_+(\xi)\operatorname{dn}(\beta\xi)}{\sqrt{1-e^2\operatorname{sn}^2(\beta\xi)}} \ln\frac{1+\sqrt{1-e^2\operatorname{sn}^2(\beta\xi)}}{1-\sqrt{1-e^2\operatorname{sn}^2(\beta\xi)}} d\xi = \pi\beta f_+(0). \tag{3.254}$$

Under the condition $\varphi(1) = 0$ we obtain

$$\varphi_+(x) = -\frac{\beta\operatorname{cn}(\beta x)}{\pi}\sqrt{c^2 - \operatorname{sn}^2(\beta x)} \int_{-1}^{1} \frac{f_+'(\xi)\operatorname{dn}(\beta\xi)\, d\xi}{\sqrt{c^2 - \operatorname{sn}^2(\beta\xi)}\,[\operatorname{sn}(\beta\xi) - \operatorname{sn}(\beta x)]}.$$

The relation (53) still holds, and the additional condition (3.247) takes the form

$$P + \beta\int_{-1}^{1} \frac{f_+'(\xi)\operatorname{sn}(\beta\xi)\operatorname{dn}(\beta\xi)}{\sqrt{c^2 - \operatorname{sn}^2(\beta\xi)}} d\xi = 0.$$

Note that if $f_+(x) \equiv f_+ = \text{const}$, then, using formulae (3.253), (3.254), $(3.252)_1$ and relations 3.152.7, 4.317.10 in Gradshteyn and Ryzhik (1994), we derive the solution

$$\varphi_+(x) = \frac{\beta f_+ \operatorname{cn}(\beta x)}{\mathcal{K}(\sqrt{1-e^2c^2})\sqrt{c^2 - \operatorname{sn}^2(\beta x)}}, \quad N_0 = \frac{2 f_+ \mathcal{K}(ec)}{\mathcal{K}(\sqrt{1-e^2c^2})}. \tag{3.255}$$

The error of solution (3.255) does not exceed 13% because the $\tanh(As)$ approximates the $L(s)$ in (3.224) with this error.

For variant $\mathcal{B}$ for the case when $g(s) \equiv 0$ we similarly obtain

$$\varphi_-(x) = -\frac{\beta \operatorname{dn}(\beta x)}{\pi\sqrt{c^2 - \operatorname{sn}^2(\beta x)}} \int_{-1}^{1} \frac{\sqrt{c^2 - \operatorname{sn}^2(\beta\xi)}\, f'_-(\xi) \operatorname{cn}(\beta\xi)}{\operatorname{sn}(\beta\xi) - \operatorname{sn}(\beta x)} d\xi.$$

The condition $\varphi(1) = 0$ gives

$$\varphi_-(x) = -\frac{\beta \operatorname{dn}(\beta x)}{\pi}\sqrt{c^2 - \operatorname{sn}^2(\beta x)} \int_{-1}^{1} \frac{f'_-(\xi) \operatorname{cn}(\beta\xi)\, d\xi}{\sqrt{c^2 - \operatorname{sn}^2(\beta\xi)}[\operatorname{sn}(\beta\xi) - \operatorname{sn}(\beta x)]},$$

and thc additional condition (3.250) takes the form

$$\int_{-1}^{1} \frac{f'_-(\xi) \operatorname{cn}(\beta\xi)}{\sqrt{c^2 - \operatorname{sn}^2(\beta\xi)}} d\xi = 0.$$

3.4.3. RIGID SLEEVES

For nondeformable (absolutely rigid) sleeves, the kernel symbol $L_1(s)$ (3.18) of the integral equation of the periodic contact problem cannot be presented in the form (3.224), because in its asymptotic expansion as $s \to \infty$ there is a term of the order s^{-1}. For this case we apply the method of dual equations (Sneddon (1966), Filippova, Tsvetkov and Chebakov (1991)). Introducing new notation by the formulae

$$a_0 = \frac{b-a}{2R}, \quad b_0 = \frac{l}{2R}, \quad \tilde{x} = \frac{2z-l}{2R}, \quad \varphi(\tilde{x}) = \frac{q(z)}{\theta}, \quad f_0(\tilde{x}) = \frac{\delta(z)}{R}, \tag{3.256}$$

and using the Fourier series (we assume that $f_0(-\tilde{x}) = f_0(\tilde{x})$)

$$\sum_{k=0}^{\infty} Q_k \cos(\beta_k \tilde{x}) = \begin{cases} \varphi(\tilde{x}) & \text{for } |\tilde{x}| \le a_0 \\ 0 & \text{for } a_0 < |\tilde{x}| < b_0 \end{cases},$$

where $\beta_k = \pi k / b_0$, we reduce the problem to the dual series-equations in the coefficients Q_k

$$\sum_{k=1}^{\infty} Q_k L_1(\beta_k) \cos(\beta_k \tilde{x}) + Q_0 k_0 = f_0(\tilde{x}) \quad (|\tilde{x}| \le a_0), \tag{3.257}$$

$$\sum_{k=0}^{\infty} Q_k \cos(\beta_k \tilde{x}) = 0 \quad (a_0 < |\tilde{x}| < b_0). \tag{3.258}$$

Here,

$$Q_0 = \frac{1}{b_0}\int\limits_0^{a_0} \varphi(\tilde{x})\,d\tilde{x} = \frac{N_0}{2b_0}, \qquad k_0 = \frac{1-2\nu}{2(1-\nu)}.$$

Taking into account that $\lim L_1(s) = A = [2(1+\nu)]^{-1}$ $(s \to 0)$, we transform equations (3.257), (3.258) to the form $(n_0 = A + k_0)$

$$\sum_{k=0}^{\infty} Q_k L_1(\beta_k)\cos(\beta_k\tilde{x}) = f_0(\tilde{x}) + \frac{n_0 N_0}{2b_0} \qquad (|\tilde{x}| \le a_0), \tag{3.259}$$

$$\sum_{k=0}^{\infty} Q_k \cos(\beta_k\tilde{x}) = 0 \qquad (a_0 < |\tilde{x}| < b_0). \tag{3.260}$$

Let the function $f_0(\tilde{x})$ also be presented by a Fourier series. It is sufficient then to consider the dual integral equations (3.259), (3.260) for the case when the right-hand side of (3.259) is $\cos(i\varepsilon\tilde{x})$ where ε is a known constant. Such dual equations can be reduced to an infinite system of linear algebraic equations

$$(B_1 + B_2)X(\varepsilon) = D_1(\varepsilon) + D_2(\varepsilon), \tag{3.261}$$

where the matrices B_i and the vector columns D_i $(i = 1, 2)$ have the elements

$$B_1 = \left\{\frac{\gamma_m\cosh(\delta_n a_0) + \delta_n\sinh(\delta_n a_0)}{(\delta_n^2 - \gamma_m^2)\cosh(\delta_n a_0)}\right\}, \tag{3.262}$$

$$B_2 = -\left\{\frac{\gamma_m\cosh(\delta_n a_0)\sinh(\gamma_m a_0) - \delta_n\sinh(\delta_n a_0)\cosh(\gamma_m a_0)}{(\delta_n^2 - \gamma_m^2)\cosh(\delta_n a_0)\sinh(\gamma_m b_0)} \times \exp(-\gamma_m b_0)\right\}, \tag{3.263}$$

$$D_1(\varepsilon) = \left\{\frac{\gamma_m\cosh(\varepsilon a_0) + \varepsilon\sinh(\varepsilon a_0)}{(\gamma_m^2 - \varepsilon^2)L_1(i\varepsilon)}\right\}, \tag{3.264}$$

$$D_2(\varepsilon) = -\left\{\frac{\gamma_m\cosh(\varepsilon a_0)\sinh(\gamma_m a_0) - \varepsilon\sinh(\varepsilon a_0)\cosh(\gamma_m a_0)}{(\gamma_m^2 - \varepsilon^2)L_1(i\varepsilon)\sinh(\gamma_m b_0)} \times \exp(-\gamma_m b_0)\right\}. \tag{3.265}$$

Here, $i\delta_n$, $i\gamma_m$ are the denumerable sets of the complex zeros and poles, respectively, of the meromorphic function $L_1(s)$ lying in the upper half-space. The matrices B_n, $n = 1, 2$, are singular because of δ_n and γ_m have the same asymptotics at infinity. The method of solving the system will be given in subsection 3.6.1. We confine ourselves to the case when $\delta(z) \equiv \delta$, $f_0(\tilde{x}) \equiv f$, and set $\varepsilon = 0$ in system (3.261)–(3.265). Then the required function $\varphi(\tilde{x})$ and the integral characteristic N_0 can be expressed via the solution $x_n(0) = x_n$ of system (3.315), (3.262)–(3.265) by the relations

$$\varphi(\tilde{x}) = \left(f + \frac{n_0 N_0}{2b_0}\right)\left[\frac{1}{A} + \sum_{n=1}^{\infty} x_n \frac{\cosh(\delta_n x)}{\cosh(\delta_n a_0)}\right], \tag{3.266}$$

$$N_0 = 2f\left(\frac{a_0}{A} + S\right)\left[\frac{b_0 - a_0}{b_0} + \frac{a_0 k_0}{b_0 A} - \frac{n_0}{b_0} S\right]^{-1}, \tag{3.267}$$

$$S = \sum_{n=1}^{\infty} x_n \frac{\sinh(\delta_n a_0)}{\delta_n \cosh(\delta_n a_0)}. \tag{3.268}$$

To solve system (3.315), (3.262)–(3.265) by using reduction to an infinite system, we approximate the function $L_1(s)s^{-1}$ with $\nu = 0.3$ by expression (3.316) with $B = 0.3846$, $N = 2$, $p_1 = 2.1586$, $p_2 = 1.2245$, $q_1 = 2.0540$ and $q_2 = 1.2868$; the error of this approximation does not exceed 1.1%. We then use formulae (3.317)–(3.320). In the upper half of Table 3.5, for the rigid sleeves, we give the values of the function $q(z)R(\theta\delta)^{-1} = \varphi(\tilde{x})f^{-1}$ (see (3.256)) for certain values of $\tilde{x}$, and of the load

$$\frac{N_0}{f} = \frac{1}{\theta\delta}\int_a^b q(z)\,dz, \tag{3.269}$$

calculated by formulae (3.266)–(3.268) for $\nu = 0.3$, $a_0 = 2$ and for various values of b_0/a_0. The number of equation in the finite system (3.320) was $M = 40$.

In the lower half of Table 3.5, for the same deformable sleeves, we give the values of the function $q(z)R(\theta\delta)^{-1} = \varphi(x)\lambda\mu(\pi f)^{-1}$ (see (3.212), (3.213)) at the same points $\tilde{x} = x\pi(\lambda\mu)^{-1}$, and of the load (3.269), calculated by formulae (3.255) for the same values of ν, $a_0 = \pi(\lambda\mu)^{-1}$ and $b_0/a_0 = \lambda$.

To estimate the interaction between the sleeves, we should calculate also the corresponding values of the integral characteristics of the contact pressures in the problems considered in sections 3.2 and 3.3 for a rigid or deformable sleeve. Note that $a_0 = (b-a)/(2R) = 2$ corresponds to $\lambda = R/a \sim a_0^{-1} = 1/2$ from formulae $(3.76)_3$, $(3.143)_3$. For one rigid sleeve with a cylindrical base ($\delta(z) \equiv \delta$) for the $\lambda = 1/2$ we get, using formula (3.136), $(\theta\delta)^{-1}\int_{-a}^{a} q(z)dz = N_0 f^{-1} \approx 11.1$. And for one deformable sleeve with a cylindrical base for the $\lambda = 1/2$ we get, using formula $(3.150)_1$,

$(\theta\delta)^{-1}\int_{-a}^{a} q(z)dz = N_0(2f)^{-1} \approx 4.08$[8]. One can see from Table 3.5 that the larger the value of $b_0/a_0 = l/(b-a)$, which characterizes the relative distance between the sleeves, the closer is the value of $N_0 f^{-1}$ to the corresponding value for one sleeve. For $b - a = 4R$ ($a_0 = 2$) and $b_0/a_0 \geq 2$ ($l = b + a \geq 2(b-a)$ or $2a \geq b - a$, i.e., the distance $2a$ between the sleeves is greater than the length $b - a$ of a sleeve, see Fig. 3.9) we can say that there is no interaction between deformable sleeves, but there is a little interaction between rigid sleeves for $b_0/a_0 = 2$.

Table 3.5: The values of $\varphi(\tilde{x})f^{-1}$ and $N_0 f^{-1}$

$\tilde{x} =$	0	1.2	1.6	1.9	$N_0 f^{-1}$
$b_0/a_0 = 2$	2.75	2.99	3.04	4.01	12.9
$b_0/a_0 = 1.5$	3.17	3.16	3.22	4.25	13.6
$b_0/a_0 = 1.1$	3.45	3.45	3.50	4.36	14.6
$b_0/a_0 = 1.05$	3.49	3.49	3.52	4.06	14.5
$b_0/a_0 = 2$	0.912	0.916	0.962	1.38	4.09
$b_0/a_0 = 1.5$	0.907	0.912	0.957	1.37	4.07
$b_0/a_0 = 1.1$	0.909	0.911	0.928	1.13	3.92
$b_0/a_0 = 1.05$	0.910	0.909	0.905	0.939	3.84

For the system of deformable sleeves with the same elastic constants as the cylinder, due to a compliance of elastic material, we have a fall of contact pressure and total load roughly $3 - 3.5$ times that with the system of absolutely rigid sleeves. The risk for the rigid sleeves to collapse is much more higher than for the elastic ones. Calculations show that for a system of rigid sleeves, the quantity N_0 has a maximum as the function of the parameter b_0/a_0 (when the other parameters are fixed) for the value $b_0/a_0 \approx 1.1$ (Table 3.5). There is no such a maximum for elastic sleeves. Closeness of these sleeves brings a smaller value of total load $N_0 f^{-1}$ than for one elastic sleeve. This is a result of interaction. For close elastic sleeves ($b_0/a_0 = 1.05$ in the table) we have a local minimum of the function $\varphi(\tilde{x})f^{-1}$ ($|\tilde{x}| \leq a_0 = 2$) not far from the edges of the sleeves; inside the contact zone this function is almost constant. We have already seen a similar result for small λ in Fig. 2.9 in chapter 2 ($\lambda = 1/2$, the dashed line).

In conclusion we note that the contact problem for a finite cylinder with end-walls free of stresses or rigidly fixed, will be solved in section 3.7 by means of the method of homogeneous solutions.

[8]There is a difference between the last formulae of (3.76) and (3.143).

3.5 Interaction between a Cylinder and a Wheel or Cylindrical Shell

3.5.1. INTERACTION BETWEEN A CYLINDER AND A WHEEL

Consider the axisymmetric contact of an infinite circular elastic cylinder compressed by a circular elastic wheel (ring) fitted onto the cylinder with an interference. Solving this problem, we will follow Arutyunyan (1994), who, under certain assumptions, gave a simple and elegant approximate solution for this complicated problem.

Let $C = \{0 \le r \le R,\ 0 \le \varphi < 2\pi,\ |z| < \infty\}$ be the cylinder having elastic moduli G and ν, and $W = \{R_1 \le r \le R_2,\ 0 \le \varphi < 2\pi,\ |z| \le h\}$ be the ring having elastic moduli G_1 and ν_1, and let h be small with respect to the cylinder radius R and to the value of $R_2 - R_1$, and let the quantity $\delta = R - R_1$ be positive, and have the order of an elastic displacement in the linear theory (Fig. 3.10*a*,*b*).

a) b)

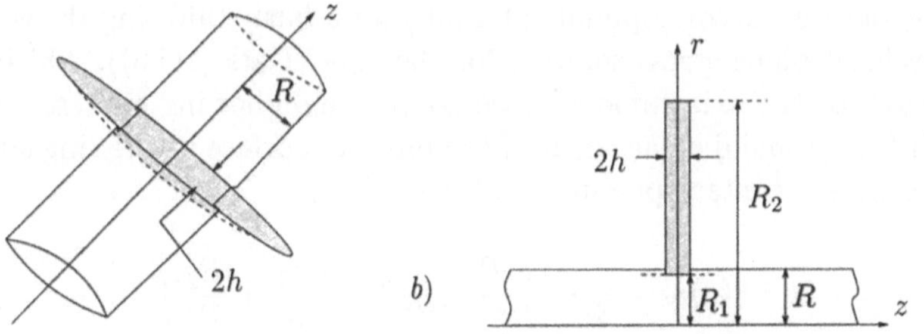

Figure 3.10: A wheel on a cylinder *a*) in 3-D, *b*) a one-half of axial section

In the contact interaction between the cylinder C and the ring W, the quantity δ is known as the interference. We suppose that the contact is smooth, i.e., there is only the normal contact pressure $q(z)$ between the elastic solids. We suppose also that in the contact region $\{r = R,\ 0 \le \varphi < \pi,\ |z| \le h\}$ the difference of the mean radial displacements of the ring and cylinder points is δ, i.e., the condition of the contact between the ring and cylinder has the form

$$\langle u \rangle_W - \langle u \rangle_C = \delta \qquad \left(\langle u \rangle = \frac{1}{2h} \int_{-h}^{h} u(z)\, dz \right). \tag{3.270}$$

The approximate contact condition (3.270) holds for $h \ll R$, when there are no large gradients of the displacement field in the very thin contact zone.

The boundary conditions for the cylinder have the form

$$\sigma_r = -q(z)\,\mathrm{H}(h-z)\,\mathrm{H}(h+z), \quad \tau_{rz} = 0 \quad (r = R, \ \ |z| < \infty), \tag{3.271}$$

($\mathrm{H}(x)$ is the Heaviside function) with the additional condition of boundedness of stresses at $r = 0$. Equations (3.271) are similar to the boundary conditions (3.10): since $q(z)$ is even in z, formulae (3.16)–(3.18) give:

$$\langle u \rangle_C = -\frac{1}{\pi\theta h}\int_0^\infty \frac{L_1(\alpha R)}{\alpha^2} Q(\alpha)\sin(\alpha h)\,d\alpha, \tag{3.272}$$

$$Q(\alpha) = \int_{-h}^{h} q(\zeta)\cos(\alpha\zeta)\,d\zeta.$$

To obtain the corresponding formula for a fairly thin ring W, we use the generalized plane stress solution for the ring (Kats (1956)), which can be changed for the axisymmetric case, giving the following expression for the mean (z) normal displacement of the internal surface of the ring subject to the mean (z) contact pressure $\langle \sigma_r \rangle = -\langle q \rangle$:

$$\langle u \rangle_W = \frac{\langle q \rangle R^3}{2G_1(R_2^2 - R^2)}\left(\frac{1-\nu_1}{1+\nu_1} + \frac{R_2^2}{R^2}\right). \tag{3.273}$$

To solve the problem finally, it is necessary to determine the character of the contact stresses. Since both bodies in contact are elastic, and the ring has sharp edges near the points $z = \pm h$, the contact pressure takes the form

$$q(z) \sim A_i r_i^{\kappa-1} \quad (A_i = \text{const}, \ \ i = 1,2), \tag{3.274}$$

$$r_1 = h - z, \quad r_2 = h + z,$$

near the edges. The quantity κ can be determined from the solution of the plane problem of interaction without friction between an elastic quarter-plane with the constants G_1, ν_1 and a half-plane with the constants G, ν (Fig. 3.11a). The more general problem of the asymptotic behaviour of the stress field in a multi-wedge corner of a composite body consisting of a number of homogeneous and isotropic wedges with rough interfaces in linear elasticity was studied by Theocaris and Gdoutos (1976). They used the two complex functions of the Muskhelishvili formulation of the plane stress problem, expressed as sums of powers of complex exponents and their conjugates, and they obtained the characteristic equations for the evaluation of the singularity in a bi-wedge. For the elastic half- and quarter-plane this characteristic equation has the form (Aleksandrov, Grishin and Kovalenko (1985))

$$\tan(\pi s)\sin(\pi s) = -n[(1 - 2s^2 - \cos(\pi s)], \tag{3.275}$$

where $n = \theta_1/\theta$, $\theta_1 = G_1(1-\nu_1)^{-1}$ and $\theta = G(1-\nu)^{-1}$. The quantity κ is the solution of this equation in the complex s-plane with the least real part. The value of κ must also ensure the finiteness of the integral of $q(r)$ over the contact zone. In the strip $0 < \mathrm{Re}\, s < 1$ equation (3.275) has a unique solution κ on the real axis. The form of the dependence between κ and n is shown in Fig. 3.11*b*: $\kappa = 1$ when $n = 0$, and $\kappa \to 1/2$ as $n \to \infty$. When cylinder stiffness θ_1 is much greater than ring stiffness θ ($n = 0$), then $q(z) = B = \text{const}$; and when the ring stiffness is much greater than the cylinder stiffness ($n \to \infty$), then for fairly small values of h/R we have $q(z) = C(r_1 r_2)^{-1/2}$, $C = \text{const}$ (Aleksandrov (1962)). The latter is the well-known solution of the 2-D contact problem for a punch with a plane base.

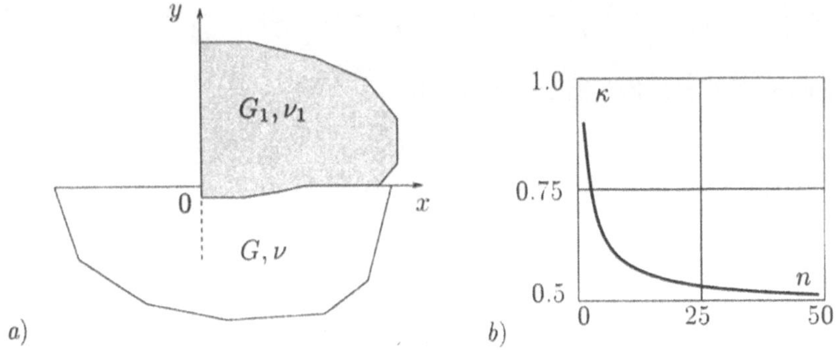

Figure 3.11: *a*) Contact between a quarter- and half-plane; *b*) the function $\kappa(n)$

Taking this into account, we suppose for arbitrary n that

$$q(z) = D(r_1 r_2)^{\kappa-1} = D(h^2 - z^2)^{\kappa-1} \qquad (D = \text{const}), \tag{3.276}$$

and we then find

$$Q(\alpha) = D\sqrt{\pi}\left(\frac{2h}{\alpha}\right)^{\kappa-1/2}\Gamma(\kappa)\,\mathcal{J}_{\kappa-1/2}(\alpha h) = Df_1(\alpha, h, \kappa), \tag{3.277}$$

$$\langle q\rangle = D\frac{\sqrt{\pi}}{2}h^{2(\kappa-1)}\frac{\Gamma(\kappa)}{\Gamma(\kappa+1/2)} = Df_2(h,\kappa), \tag{3.278}$$

where $\Gamma(x)$ is the gamma-function and $\mathcal{J}_\mu(x)$ is the Bessel function of order μ.

Substituting (3.272), (3.273), (3.277) and (3.278), into the contact condition (3.270), we obtain the following expression for the constant D from (3.276):

$$D = \delta\left[\frac{f_2(h,\kappa)R^3}{2G_1(R_2^2 - R^2)}\left(\frac{1-\nu_1}{1+\nu_1} + \frac{R_2^2}{R^2}\right)\right. \tag{3.279}$$

$$\left. + \frac{1}{\pi\theta h}\int_0^\infty \frac{L_1(\alpha R)}{\alpha^2} f_1(\alpha, h, \kappa)\sin(\alpha h)\, d\alpha\right]^{-1}.$$

Thus, for a specified interference δ, the contact pressure can be calculated by formulae (3.276), (3.279), and then the stress-strain state of the cylinder and ring can be determined. To apply these result, we introduce the following dimensionless notation

$$x = \frac{z}{h}, \quad q_*(x) = \frac{q(z)R}{\theta\delta}, \quad \beta_1 = \frac{R_2}{R} > 1, \quad \beta_2 = \frac{h}{R} \ll 1, \tag{3.280}$$

and rewrite (3.276), (3.279) in the form

$$q_*(x) = \frac{D_*}{(1-x^2)^{1-\kappa}},$$

$$D_* = \left[\frac{\sqrt{\pi}\,\Gamma(\kappa)}{4n\,\Gamma(\kappa+1/2)(\beta_1^2-1)}\left(\frac{1}{1+\nu_1} + \frac{\beta_1^2}{1-\nu_1}\right)\right. \tag{3.281}$$

$$\left. + \frac{\Gamma(\kappa)}{\sqrt{\pi}}\left(\frac{2}{\beta_2}\right)^{\kappa-1/2}\int_0^\infty \frac{L_1(u)}{u^{\kappa+3/2}}\mathcal{J}_{\kappa-1/2}(u\beta_2)\sin(u\beta_2)\, du\right]^{-1}.$$

For numerical examples we take $\beta_1 = 2$, $\beta_2 = 0.01$ and $\nu = \nu_1 = 0.3$. In Fig. 3.12 we give the function $q_*(x)$ for $n = 1/2$ ($\kappa = 0.854$), $n = 1$ ($\kappa = 0.774$) and $n = 2$ ($\kappa = 0.690$). In design of such a system one should choose the elastic materials of the ring and cylinder with the ratio $n \ll 1$ to diminish the risk of dangerous contact pressures. Since $1 - \kappa \to 0$ as

$n \to 0$, we also will have a decrease of singularity in the distribution (3.280) together with a decreasing stress intensity factor (the coefficient at this singularity). Due to the asymptotics of the Bessel function $\mathcal{J}_\mu(u) = O(u^\mu)$, $u \to 0$, the second term vanishes in formula (3.281) for $h \to 0$. This means that the very small thickness $2h$ of the ring gives only a finite input to the distribution (3.280).

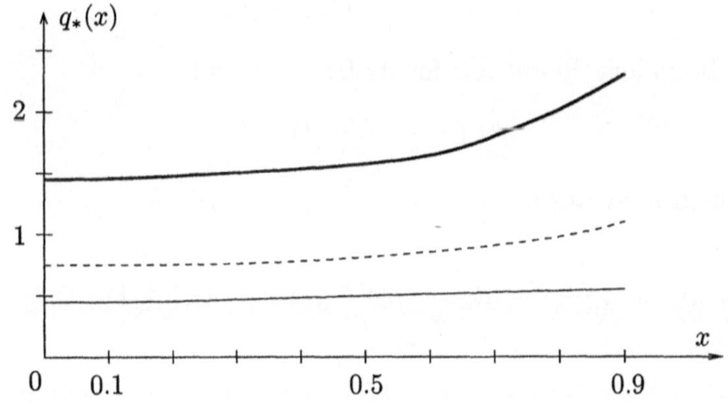

Figure 3.12: The function $q_*(x)$ for $n = 1/2$, $n = 1$ (dashed) and $n = 2$ (bold)

3.5.2. INTERACTION BETWEEN A CYLINDER AND A SHELL

We now consider the axisymmetric problem of an elastic circular cylindrical shell of finite length $2l$ fitted with the interference δ onto a circular infinite elastic cylinder of radius R. We assume that the external surface of the cylindrical shell is loaded by the pressure $p(z)$. We will also suppose that the contact between the shell and cylinder takes place along all the whole length of the shell ($|z| \leq l$), there are no friction forces in the contact region, and outside this region the cylinder is unloaded (Fig. 3.13a,b).

Let R_1 be the internal radius of the shell, $\delta = R - R_1 > 0$ and suppose δ has the order of an elastic displacement in the linear theory; h is the thickness of the shell; G, ν and G_1, ν_1 are the elastic constants of the cylinder and shell, respectively. Let $y(z)$ be the normal displacement (deflection) of the shell, and $q(z)$ the contact pressure between the surfaces of the shell and cylinder.

If we introduce the following dimensionless quantities and parameters:

$$z^* = z/l, \quad \lambda = R/l, \quad \mu = R_1/l, \quad \delta^* = \delta/l = \lambda - \mu, \tag{3.282}$$

$$y^*(z^*) = y(z)/l, \quad \kappa = h/l, \quad p^*(z^*) = \frac{6(1-\nu_1)\,p(z)}{G_1\kappa^3}, \tag{3.283}$$

$$q^*(z^*) = \frac{6(1-\nu_1)\,q(z)}{G_1\kappa^3}, \quad \rho = \frac{6(1-\nu_1)G}{(1-\nu)G_1\kappa^3}, \quad d^2 = \frac{3(1-\nu_1^2)}{\mu^2\kappa^2}, \tag{3.284}$$

then the problem can be reduced to the simultaneous solution of the differential equation for the shell deflection in Timoshenko (1940, formula (228))

$$y'''' + 4d^2 y = p(z) - q(z) \tag{3.285}$$

under the boundary conditions for its free end-walls

$$y''(\pm 1) = y'''(\pm 1) = 0, \tag{3.286}$$

and the integral equation

$$\int_{-1}^{1} q(\zeta)\, K\left(\frac{\zeta - z}{\lambda}\right) d\zeta = \pi\rho[\delta + y(z)] \qquad (|z| \le 1), \tag{3.287}$$

where $K(t)$ is given by formula $(3.17)_1$, where we omit the subscript. In (3.285)–(3.287) and later we omit the asterisks for the dimensionless quantities.

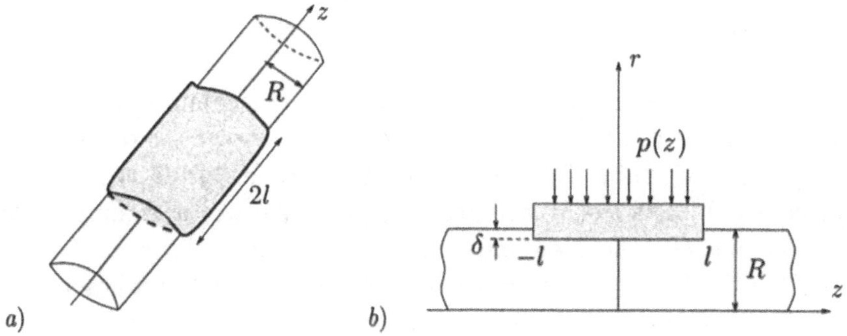

Figure 3.13: A cylindrical shell on a cylinder *a*) in 3-D, *b*) a one-half of axial section

An approximate solution of equation (3.285) can be presented in the form

$$y(z) = \sum_{n=0}^{N} A_n Q_n(z), \tag{3.288}$$

where $Q_n(z)$ is a special system of orthogonal polynomials, satisfying relations

$$\int_{-1}^{1} [Q_i''''(\zeta) + 4d^2 Q_i(\zeta)]Q_k(\zeta)d\zeta = \begin{cases} 1 & \text{for } i = k \\ 0 & \text{for } i \neq k \end{cases},$$

$$Q_k''(\pm 1) = Q_k'''(\pm 1) = 0.$$

To construct the $Q_n(z)$, we use Gram–Schmidt orthogonalization process, for polynomials, (see subsection 21.7-1 in Korn, G.A. and Korn, T.M. (1968)), taking $Q_0(z) = (2\sqrt{2}d)^{-1}$ as the original polynomial. The coefficients A_n in (3.288) have the form

$$A_n = \int_{-1}^{1} [p(\zeta) - q(\zeta)]Q_n(\zeta)d\zeta. \tag{3.289}$$

According to (3.288), we expand the solution of equation (3.287) in a series

$$q(z) = q_\delta(z) + \sum_{k=0}^{N} A_k q_k(z), \tag{3.290}$$

where $q_\delta(z)$ and $q_k(z)$ are solutions of equation (3.287) for the right-hand sides $\pi\rho\delta$ and $\pi\rho Q_k(z)$, respectively.

For simplicity, we confine ourselves to the case when the function $p(z)$ is even; then the functions $g(z)$ and $q(z)$ will be even too. For this case we take the functions $q_k(z)$ in (3.290), which correspond to the even polynomials $Q_k(z)$.

Below we give an algorithm for an approximate solution of (3.287) with an even right-hand side $\pi\rho g(z)$. We seek the solution in the form

$$q(z) = \rho\omega(z)(1 - z^2)^{-1/2},$$

where we take $\omega(z) \equiv \omega^\circ(\varphi)$, where $z = \cos\varphi$, in the form of a Lagrange interpolational polynomial

$$\omega^\circ(\varphi) = \frac{1}{i+1}\sum_{n=1}^{i+1} \omega^\circ(\theta_n)\left[[1 + 2\sum_{m=1}^{i} \cos(2m\theta_n)\cos(2m\varphi)\right]$$

with Chebyshev nodes

$$\theta_n = \pi(2n-1)[4(i+1)]^{-1} \qquad (n = 1, 2, ..., i+1).$$

Applying the collocation method by Chebyshev nodes (Erdogan (1969), Erdogan, Gupta and Cook (1973)) to the integral equation (3.287), we obtain the following linear algebraic system in the quantities $\omega^\circ(\theta_n)$:

$$\sum_{n=1}^{i+1} \omega^\circ(\theta_n)\{\ln(2\lambda) + \Phi_i(\theta_k,\theta_n) + 2^{-1}[F(t_{kn}^+) + F(t_{kn}^-)](i+1)g^\circ(\theta_k) \quad (3.291)$$

$$(t_{kn}^\pm = \lambda^{-1}(\cos\theta_k \pm \cos\theta_n); \qquad k = 1, 2, ..., i+1),$$

where

$$F(t) = \ln|t| + K(t), \quad (3.292)$$

$$g^\circ(\varphi) \equiv g(z), \qquad \Phi_0(\varphi,\psi) = 0,$$

$$\Phi_i(\varphi,\psi) = \sum_{m=1}^{i} m^{-1}\cos(2m\varphi)\cos(2m\psi).$$

We now focus attention on a scheme for calculating the function $F(t)$ $(3.292)_2$, or $K(t)$ $(3.17)_1$, at the points $t = t_{kn}^\pm$, taking $\nu = 0.3$ in calculations. For $|t| \geq 3$, we use the following expansion of the kernel $K(t)$[9]

$$K(t) = \sum_{n=1}^{\infty} b_n \exp(-|t|\gamma_n) \quad (3.293)$$

$$= 2\sum_{n=1}^{\infty} \exp(-|t|y_n)[A_n\cos(tx_n) - B_n\sin(|t|x_n)],$$

$$b_n = A_n + iB_n = \pi i\{[\xi_n/L(\xi_n)]'\}^{-1}, \quad \xi_n = i\gamma_n = x_n + iy_n. \quad (3.294)$$

Here, ξ_n are the poles of the meromorphic function $L(s)/s$ lying in the upper half-plane of the complex variable $s = u + iv$. We recall that $L(u)$ coincides with $L_1(u)$ in the form (3.18). The values of the constants x_n, y_n, A_n and B_n $(n = 1, 2, ..., 7)$ are given in Table 3.6.

Table 3.6: The constants from (3.293), (3.294)

n	x_n	y_n	A_n	B_n
1	1.3621971	2.7221755	0.2137338	0.0536358
2	1.6376249	6.0600720	0.0929762	0.0150262
3	1.8282558	9.2668351	0.0604835	0.00821272
4	1.9672412	12.442529	0.0449938	0.00532456
5	2.0762838	15.605441	0.0358676	0.00377888
6	2.1659332	18.761738	0.0298361	0.00284168
7	2.2420230	21.914138	0.0255482	0.00222556

[9]This expansion is obtained with help of representation (∗) in footnote 5 in subsection 3.3.2 for the meromorphic function $L(s)/s$ and by passing in $(3.17)_1$ to the integration in the complex s-plane.

For $|t| \le 3$, the interval of integration in expression $(3.17)_1$ for $K(t)$ is divided into two intervals $(0, A)$ and (A, ∞). On the interval $(0, A)$ the integral is calculated using the Gauss method with an automatic accuracy control. In the second integral we write

$$L_1(u) = 1 + \sum_{m=1}^{\infty} c_m u^{-m}, \tag{3.295}$$

(see (3.78)) so that

$$\int_A^{\infty} \frac{L(u)}{u} \cos(ut)\, du = G_1(t) + \sum_{m=1}^{\infty} c_m G_{m+1}(t),$$

$$G_1(t) = -\operatorname{Ci}(At), \qquad G_2(t) = \frac{\cos(At)}{A} + t \operatorname{Si}(At),$$

$$G_n(t) = \frac{\cos(At)}{(n-1)A^{n-1}} - \frac{t}{(n-1)(n-2)} \left[\frac{\sin(At)}{A^{n-2}} + t G_{n-2} \right] \quad (n = 3, 4, ...),$$

where

$$\operatorname{Si}(t) = \int_0^t \frac{\sin x}{x} dx, \qquad \operatorname{Ci}(t) = -\int_t^{\infty} \frac{\cos x}{x} dx$$

are the integral sine and cosine, respectively (Gradshteyn and Ryzhik (1994, formulae 8.230)). The values of the coefficients c_m are given in Table 3.7.

The summation limit in (3.295) and the constant A can be chosen according to the specified accuracy of $K(t)$ for a fixed value of t. Near $|t| = 3$ we checked the agreement between the two procedures ($|t| \ge 3$, $|t| \le 3$) up to four or five significant digits.

Table 3.7: The coefficients c_m

m	c_m	m	c_m	m	c_m
1	0.4	4	−2.9097125	7	−30.0101066
2	−0.965	5	−3.13526	8	−133.2626437
3	−2.336	6	−7.10831806	9	−610.0766252

After determination of the even functions $q_\delta(z)$ and $q_k(z)$ from equation (3.287), we substitute series (3.290) into relation (3.289) and obtain a linear algebraic system for A_k in the form

$$A_n + \rho \sum_{k=0}^{N} A_k \int_0^{\pi} \omega_k^{\circ}(\theta) Q_n(\cos\theta) d\theta \tag{3.296}$$

$$= \int_0^{\pi} [\rho\omega_\delta^\circ(\theta) + p(\cos\theta)\sin\theta] Q_n(\cos\theta)d\theta, \quad n = 0, 1, ..., N.$$

Solving system (3.296), we get the approximate expressions for the functions $y(z)$ and $q(z)$ by formulae (3.288) and (3.290).

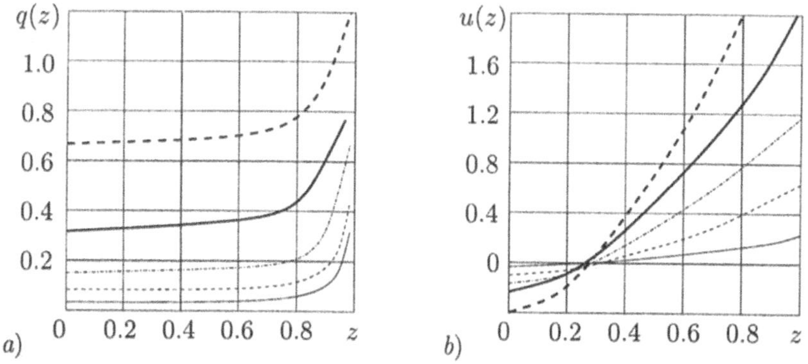

Figure 3.14: a) The contact pressure $q(z)$ and b) the deflection $u(z)$, for $\rho = 1/4$, 1/2 (dashed), 1 (dotted-dashed), 2 (bold) and 4 (bold dashed)

Analysis of results shows that, for $\lambda \geq 1/2$ and $\rho \leq 4$, the method converges well when the numbers of equations in systems (3.291) and (3.296) increase, however, the convergence worsens when the parameter λ is small and the parameter ρ is large. The convergence deteriorates first near the point $z = 0$. For $\rho \leq 1/4$, the results correspond well with those in section 3.2, for the interaction between the cylinder and a rigid sleeve.

Calculations have been made for the stress-free shell ($p(z) \equiv 0$) and for $\nu_1 = 0.3$, $\kappa = 0.1$, $\delta = 0.002$, $\lambda = 4$ and different ρ. The plots of the contact pressure $q(z)$ and deflection $u(z) = y(z) \times 10^4$ of the shell are shown in Fig. 3.14a,b. For $\nu_1 = \nu = 0.3$ and $G_1 = G$, we have from (3.282)–(3.284) that $\rho \sim (l/h)^3$. We see from Fig. 3.14a that the thicker or shorter is the shell, the smaller is the contact pressure. As we see from Fig. 3.14b, there is a part inside the contact zone, where we have a negative deflection, but the quantity $\delta + y(z) = 0.002 + u(z) \times 10^{-4}$ is nevertheless positive, so that the contact weakens but does not break. The thiner or longer the shell, the greater is the weakening of the contact inside the contact region.

In conclusion we note that we can change the order of the square root singularity of solution (3.290), due to the deformability of the shell. Following Aleksandrov, Grishin and Kovalenko (1985), one could paste together solution (3.290) and boundary layers like (3.274), near the points $z = \pm 1$, in the form

$$\varphi_1(z) = P_1(1+z)^{\kappa-1} + S_1, \quad \varphi_2(z) = P_2(1-z)^{\kappa-1} + S_2,$$

$$P_n, S_n = \text{const}, \quad n = 1, 2.$$

Suppose the functions $\varphi_n(z)$, $n = 1, 2$, determine the contact pressure in the small zones $(-1, -1+\varepsilon_1)$ and $(1-\varepsilon_2, 1)$, respectively. The quantity κ is the solution of equation (3.275). Then the values of P_n, S_n and ε_n, $n = 1, 2$, can be determined from a system of the equations

$$\int_{-1}^{-1+\varepsilon_1} [\varphi_1(1+z) - q(z)]dz + \int_{1-\varepsilon_2}^{1} [\varphi_2(1-z) - q(z)]dz = 0,$$

$$\int_{-1}^{-1+\varepsilon_1} z[\varphi_1(1+z) - q(z)]dz + \int_{1-\varepsilon_2}^{1} z[\varphi_2(1-z) - q(z)]dz = 0,$$

$$\varphi_1(\varepsilon_1) = q(-1+\varepsilon_1), \quad \varphi_2(\varepsilon_2) = q(1-\varepsilon_2),$$

$$\varphi_1'(\varepsilon_1) = q'(-1+\varepsilon_1), \quad \varphi_2'(\varepsilon_2) = q'(1-\varepsilon_2).$$

So we present more exactly the character of behaviour of the contact pressure near the endpoints of the contact region.

3.6 Contact Problems for a Circular Plate

3.6.1. FORMULATION AND REDUCTION TO AN INFINITE SYSTEM

We consider the axisymmetric contact problem for the impression of a punch of radius a into a planar surface of a circular plate, which is a cylinder of radius R and height h. The cylinder lies without friction on the rigid base $z = 0$, and its lateral surface $r = R$ is subjected to sliding support, e.g., is placed in a rigid box with smooth walls, Fig. 3.15a,b. There are no friction forces in the contact domain.

In cylindrical coordinates r, φ, z the boundary conditions of this problem have the form

$$\tau_{rz} = 0 \quad (z = h, 0 \le r \le R), \quad u_z = -\delta(r) \quad (z = h, 0 \le r \le a), \qquad (3.297)$$

$$\sigma_z = 0 \quad (z = h, a < r \le R), \qquad (3.298)$$

$$u_z = \tau_{rz} = 0 \qquad (z = 0,\, 0 \le r \le R), \tag{3.299}$$

$$u_r = \tau_{rz} = 0 \qquad (0 \le z \le h,\, r = R), \tag{3.300}$$

where $\delta(r)$ is a function describing the displacement of the punch and its shape.

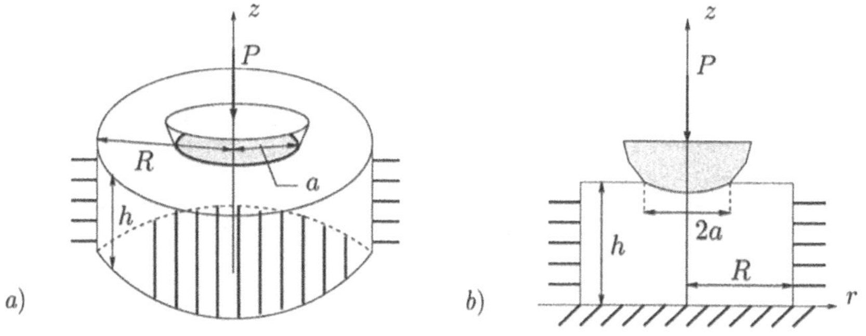

Figure 3.15: A punch on a circular plate; *a*) a 3-D view; *b*) an axial section

Using formulae (3.37), (3.42) and (3.43), we can reduce the problem to the investigation of the dual series-equations

$$\sum_{k=1}^{\infty} a_k K(u_k)\, \mathcal{J}_0(u_k x) = \delta_*(x) \qquad (0 \le x \le \lambda^{-1}), \tag{3.301}$$

$$\sum_{k=1}^{\infty} a_k\, \mathcal{J}_0(u_k x) = 0 \qquad (\lambda^{-1} < x \le b\lambda^{-1}), \tag{3.302}$$

where

$$x = rh^{-1}, \quad \lambda = ha^{-1}, \quad b = Ra^{-1}, \quad \delta_*(x) = \delta(r)h^{-1},$$

$$K(u) = \frac{\cosh(2u) - 1}{u[2u + \sinh(2u)]} = A\frac{P_1(u^2)}{P_2(u^2)} \tag{3.303}$$

$$= A \prod_{n=1}^{\infty} \left(1 + \frac{u^2}{\delta_n^2}\right)\left(1 + \frac{u^2}{\gamma_n^2}\right)^{-1}, \quad A = K(0).$$

Here, $a_k = Q_k\theta^{-1}$ are the unknown coefficients; $K(u)$ is the even meromorphic function, which can be presented in the form (3.303), where $\pm i\delta_n$ and

$\pm i\gamma_n$ are the denumerable sets of simple zeros and poles, respectively; u_k are the zeros of equation (3.29)

$$\mathcal{J}_1(u_k b\lambda^{-1}) = 0. \tag{3.304}$$

The unknown dimensionless normal contact pressure $\varphi(x) = q(r)\theta^{-1}$ is expressed by relation

$$\varphi(x) = \sum_{k=1}^{\infty} a_k\, \mathcal{J}_0(u_k x) \quad (0 \le x \le b\lambda^{-1}). \tag{3.305}$$

Note that the problem in which the plate is rigidly fixed along its base $z = 0$, can also be reduced to equations (3.301)–(3.305) with another function $K(u)$.

We use the method of reducing dual series-equations to infinite algebraic systems suggested in Aleksandrov (1975). Suppose $\delta_*(x) = \mathcal{J}_0(i\varepsilon x)$. The Bessel function $\mathcal{J}_0(ux)$ satisfies equation 8.403.2 in Gradshteyn and Ryzhik (1994)

$$(\mathcal{L} + u^2)\mathcal{J}_0(ux) = 0, \qquad \mathcal{L} = \frac{d^2}{dx^2} + \frac{1}{x}\frac{d}{dx}.$$

Making use of the fact that performing the operation $\mathcal{L}$ on the function $\mathcal{J}_0(ux)$ gives $-u^2\mathcal{J}_0(ux)$ and taking into account (3.303) and (3.305), we rewrite equation (3.301) as follows:

$$A\, P_1(-\mathcal{L})\varphi(x) = P_2(-\mathcal{L})\mathcal{J}_0(i\varepsilon x) = P_2(-\varepsilon^2)\mathcal{J}_0(i\varepsilon x), \tag{3.306}$$

where $x \in (0, \lambda^{-1})$; $P_1(\mathcal{L})$ and $P_2(\mathcal{L})$ denote differential operators in x of infinite order. The general solution of the differential equation (3.306), which is bounded at $x = 0$, has the form $(x \in (0, \lambda^{-1}))$

$$\varphi(x) = K^{-1}(i\varepsilon)\, \mathcal{J}_0(i\varepsilon x) + \sum_{n=1}^{\infty} x_n\, \mathcal{J}_0(i\delta_n x)\, \mathcal{I}_0^{-1}(\delta_n \lambda^{-1}), \tag{3.307}$$

where x_n are arbitrary constants, $\mathcal{I}_0(x) = \mathcal{J}_0(ix)$ is the modified Bessel function. Equation (3.302) says that $\varphi(x) = 0$ for $\lambda^{-1} < x \le b\lambda^{-1}$.

Consider the integrals 5.54 in Gradhsteyn and Ryzhik (1994):

$$\int x Z_p(\alpha x) Y_p(\beta x)\, dx = \frac{\beta x Z_p(\alpha x) Y_{p-1}(\beta x) + \alpha x Z_{p-1}(\alpha x) Y_p(\beta x)}{\alpha^2 - \beta^2}, \tag{3.308}$$

$$\int x[Z_p(\alpha x)]^2\,dx = \frac{x^2}{2}\left\{[Z_p(\alpha x)]^2 - Z_{p-1}(\alpha x)Z_{p+1}(\alpha x)\right\}, \qquad (3.309)$$

where $Z_p(x)$ and $Y_p(x)$ are any Bessel functions of order p. Since $\mathcal{J}_{-1}(x) = -\mathcal{J}_1(x)$, we see from (3.304), (3.308) and (3.309) that the functions $\mathcal{J}_0(u_k x)$, $k = 1, 2 \ldots$, are orthogonal on the interval $0 \le x \le b\lambda^{-1}$ with weight x (see (3.30)). Using the orthogonality conditions for $\mathcal{J}_0(u_k x)$ and formulae (3.305), (3.307), we can express a_k via x_k. Substituting the expression of a_k via x_k into equation (3.301), in which the meromorphic function $K(u)$ (3.303) is represented by the sum of principal values[10] (Lawrentjew, Schabat (1967))

$$K(u) = A - \frac{2}{\pi}\sum_{m=1}^{\infty}\frac{g_m\, u^2}{\gamma_m(u^2+\gamma_m^2)}, \quad \gamma_m = \pi i\{[K^{-1}(i\gamma_m)]'\}^{-1}, \qquad (3.310)$$

we reduce the dual series-equations (3.301), (3.302) to the following infinite system of linear algebraic equations

$$(B_1 + B_2)\,X(\varepsilon) = D_1(\varepsilon) + D_2(\varepsilon), \qquad (3.311)$$

$$B_1 = \left\{\frac{\gamma_m\,\mathcal{K}_1(\gamma_m\lambda^{-1})\,\mathcal{I}_0(\delta_n\lambda^{-1}) + \delta_n\,\mathcal{I}_1(\delta_n\lambda^{-1})\,\mathcal{K}_0(\gamma_m\lambda^{-1})}{(\delta_n^2-\gamma_m^2)\,\mathcal{K}_0(\gamma_m\lambda^{-1})\mathrm{I}_0(\delta_n\lambda^{-1})}\right\},$$

$$B_2 = -\left\{\frac{\gamma_m\,\mathcal{I}_1(\gamma_m\lambda^{-1})\,\mathcal{I}_0(\delta_n\lambda^{-1}) - \delta_n\,\mathcal{I}_1(\delta_n\lambda^{-1})\,\mathcal{I}_0(\gamma_m\lambda^{-1})}{(\delta_n^2-\gamma_m^2)\,\mathcal{K}_0(\gamma_m\lambda^{-1})\,\mathcal{I}_\mu(\delta_n\lambda^{-1})} \times \frac{\mathcal{K}_1(\gamma_m b\lambda^{-1})}{\mathcal{I}_1(\gamma_m b\lambda^{-1})}\right\},$$

$$D_1(\varepsilon) = \left\{\frac{\gamma_m\,\mathcal{K}_1(\gamma_m\lambda^{-1})\,\mathcal{I}_0(\varepsilon\lambda^{-1}) + \varepsilon\,\mathcal{I}_1(\varepsilon\lambda^{-1})\,\mathcal{K}_0(\gamma_m\lambda^{-1})}{(\gamma_m^2-\varepsilon^2)\,\mathcal{K}_0(\gamma_m\lambda^{-1})K(i\varepsilon)}\right\},$$

$$D_2(\varepsilon) = -\left\{\frac{\gamma_m\,\mathcal{I}_1(\gamma_m\lambda^{-1})\,\mathcal{I}_0(\varepsilon\lambda^{-1}) - \varepsilon\,\mathcal{I}_1(\varepsilon\lambda^{-1})\,\mathcal{I}_0(\gamma_m\lambda^{-1})}{(\gamma_m^2-\varepsilon^2)\,\mathcal{K}_0(\gamma_m\lambda^{-1})K(i\varepsilon)} \times \frac{\mathcal{K}_1(\gamma_m b\lambda^{-1})}{\mathcal{I}_1(\gamma_m b\lambda^{-1})}\right\}.$$

Here, $B_l = \{b_{mn}^l\}$ are the matrices, $D_l(\varepsilon) = \{d_m^l(\varepsilon)\}$, $X(\varepsilon) = \{x_n(\varepsilon)\}$ are vector columns of infinite order and $\mathcal{K}_l(x)$ is the other modified Bessel function ($l = 1, 2$). Note that in the general case the function $\delta_*(x)$ can be expanded into a series in the functions $\mathcal{J}_0(i\varepsilon_k x)$ and therefore we can obtain a number of such systems.

---

[10] In this section the prime means a derivative.

One can prove that $|\gamma_m|, |\delta_m| \sim m$ for $m \to \infty$ (cf. (3.381)), and this means that the matrix of system (3.311) is singular. We take into account the asymptotic representations for $\mathcal{K}_0(x)$ and $\mathcal{I}_0(x)$ for $x \to \infty$ (Gradshteyn and Ryzhik (1994)) and regularize system (3.311) by separating and exactly inverting the infinite matrix $A_0 = \{a_{mn}\}$ with the elements

$$a_{mn} = (\delta_n - \gamma_m)^{-1}. \tag{3.312}$$

To obtain the inverse matrix $A_0^{-1} = \{\tau_{nm}\}$, one should reduce auxiliary dual equations (7.1) in Aleksandrov (1975), having an exact solution, to a similar infinite system with the matrix A_0. As a result, we get for the elements of A_0^{-1} the formula

$$\tau_{nm} = [(K_-^{-1}(i\gamma_m))' \, K'_+(-i\delta_n)(\gamma_m - \delta_n)]^{-1}, \tag{3.313}$$

$$K_+(u) = K_-(-u) = \sqrt{A} \prod_{n=1}^{\infty} \left(1 + \frac{u}{i\delta_n}\right)\left(1 + \frac{u}{i\gamma_n}\right)^{-1}. \tag{3.314}$$

Thus we reduce system (3.311) to the infinite system of the second kind

$$X(\varepsilon) = A_0^{-1}(D_1(\varepsilon) + D_2(\varepsilon)) - A_0^{-1}(B_1 + B_2 - A_0)X(\varepsilon), \tag{3.315}$$

whose solution can be obtained by reduction to a finite system, and, for certain values of the parameters, by the method of successive approximations.

If in system (3.311) we pass to the limit as $R \to \infty$, then the matrix B_2 and vector column $D_2(\varepsilon)$ vanish, and we obtain the infinite system $B_1 X(\varepsilon) = D_1(\varepsilon)$ corresponding to the contact problem for an elastic layer.

If we exactly factorize the function $K(u)$, then we meet some difficulties in calculating the elements (3.313) of the matrix A_0^{-1}, namely the derivative of the infinite product (3.314). Therefore, we will approximately factorize the function $K(u)$, approximating it on the real axis by the function

$$K_0(u) = \frac{\tanh(Bu)}{u} \prod_{n=1}^{N} \frac{u^2 + p_n^2}{u^2 + q_n^2}, \qquad \frac{A}{B} = \prod_{n=1}^{N} \frac{p_n^2}{q_n^2}. \tag{3.316}$$

For the function $K(u)$ $(3.303)_1$ for $A = B = 0.5$, $N = 2$, $p_1 = 1.174$, $p_2 = 2.208$, $q_1 = 1.496$, and $q_2 = 1.732$, the relative error of approximation (3.315) does not exceed 2%.

We then obtain for the elements of the matrix A_0^{-1} the following simple expressions:

$$\tau_{nm} \approx [\Gamma_m \Delta_n (\delta_n - \gamma_m)]^{-1}, \tag{3.317}$$

$$\Delta_n = \frac{R_+(-ip_n)}{p_n - q_n} \prod_{k=1,k\neq n}^{N} \frac{p_k - p_n}{q_k - p_n} \qquad (n \leq N),$$

$$\Delta_n = \frac{2B\sqrt{B}(2n-2N-2)!!}{\pi(2n-2N-1)!!} \prod_{k=1}^{N} \frac{p_k - \alpha_n}{q_k - \alpha_n} \qquad (n > N),$$

$$\Gamma_m = \frac{1}{R_+(-iq_m)(p_m - q_m)} \prod_{k=1,k\neq m}^{N} \frac{q_k - q_m}{p_k - q_m} \qquad (m \leq N),$$

$$\Gamma_m = \sqrt{B}\frac{(2m-2N-2)!!}{(2m-2N-3)!!} \prod_{k=1}^{N} \frac{q_k - \beta_m}{p_k - \beta_m} \qquad (m > N),$$

$$R_+(-iu) = \sqrt{\frac{B}{\pi}}\Gamma\left(\frac{1}{2} - \frac{Bu}{\pi}\right)\Gamma^{-1}\left(1 - \frac{Bu}{\pi}\right).$$

Here, $\Gamma(x)$ is the gamma-function, $i\delta_n$ and $i\gamma_m$ are the zeros and poles of the function $K_0(u)$, respectively, moreover,

$$\delta_n = \begin{cases} p_n & \text{for } n \leq N \\ \alpha_n = \pi B^{-1}(n-N) & \text{for } n > N \end{cases}, \tag{3.318}$$

$$\gamma_m = \begin{cases} q_m & \text{for } m \leq N \\ \beta_m = \pi B^{-1}(m-N-1/2) & \text{for } m > N \end{cases}, \tag{3.319}$$

where α_n and β_m are, respectively, the zeros and poles of the meromorphic function $\tanh(Bu)$ in the complex u-plane.

We replace the infinite system (3.315)–(3.319) by the finite system

$$x_n(\varepsilon) = \sum_{m=1}^{M} \tau_{nm}(d_m^1(\varepsilon) + d_m^2(\varepsilon)) + \sum_{k=1}^{M}\left(\sum_{m=1}^{M} \tau_{nm}(b_{mn}^1 + b_{mn}^2)\right) x_k(\varepsilon), \tag{3.320}$$

where $n = 1, ..., M$, and the value of M can be chosen in dependence on a specified accuracy. Calculations show that the convergence of the method is improves as the parameter $(R-a)/h$ grows.

In Table 3.8, for the planar punch when $\delta(r) = \delta = \text{const}$, and one should pass to the limit as $\varepsilon \to 0$, we present the following dimensionless values of the contact pressure $q(r) = -\sigma_z(r,h)$, $0 \leq r \leq a$, and the force P acting onto the punch:

$$q_*(r/a) = q(r)a(\theta\delta)^{-1}, \quad N_0 = P(\theta\delta a)^{-1}, \tag{3.321}$$

$$q_*(r/a) = \frac{1}{A\lambda}\left[1 + \sum_{n=1}^{\infty} x_n(0)\frac{\mathcal{I}_0(\delta_n r h^{-1})}{\mathcal{I}_0(\delta_n \lambda^{-1})}\right], \tag{3.322}$$

$$N_0 = \frac{2\pi}{A}\left[\frac{1}{2\lambda} + \sum_{n=1}^{\infty} \frac{x_n(0)}{\delta_n}\frac{\mathcal{I}_1(\delta_n \lambda^{-1})}{\mathcal{I}_0(\delta_n \lambda^{-1})}\right], \tag{3.323}$$

calculated for various values of the dimensionless parameters λ and b. The coefficients $x_n(0)$ here satisfy system (3.311).

Table 3.8: The pressure and force for a planar punch

λ	0.3	0.3	0.3	0.5	0.5	0.5
b	1.1	1.2	1.5	1.1	1.3	1.5
$q_*(0)$	6.64	6.62	6.59	3.98	3.90	3.86
$q_*(0.5)$	6.62	6.56	6.51	3.98	3.88	3.83
$q_*(0.95)$	8.06	8.26	8.09	5.23	5.97	5.93
N_0	23.1	23.2	22.9	14.3	14.9	14.8
λ	0.5	1.0	1.0	1.0	1.0	1.0
b	2.0	1.3	1.5	1.7	2.0	3.0
$q_*(0)$	3.84	1.96	1.92	1.89	1.86	1.84
$q_*(0.5)$	3.80	2.01	1.98	1.95	1.93	1.90
$q_*(0.95)$	5.84	3.74	4.05	4.11	4.08	4.02
N_0	14.5	8.44	8.74	8.75	8.66	8.54

Analysis of the values of the dimensionless force N_0 from Table 3.8, when the value of λ is fixed, allows us to assert that, as the parameter b increases from 1 up to a certain value, the resistance of the cylinder to the embedding of the punch grows, but for further increase of b this resistance decreases and tends to a certain constant value. This fact may be used in practice for choosing the maximal stiffness of the elastic finite circular cylinder placed without interference in a rigid box with smooth walls. In other words, for a certain ratio between radius R of the cylinder and radius a of the contact zone, one must apply the greatest force to embed a planar punch for a specified value, getting the greatest strength intensity factor for the contact pressure (the coefficient at the square root singularity for the contact pressure; this singularity is expanded in the series (3.322), which diverges for $r = a$; see values of $q_*(0.95)$ in the table).

3.6.2. THE METHOD OF A SUPERPOSITION

We now solve problem (3.297)–(3.300) by the method of superposition. In this method, we seek the solution of the original problem in the form

$$u_r(r,z) = u_r^1(r,z) - u_r^2(r,z), \tag{3.324}$$

$$u_z(r,z) = u_z^1(r,z) - u_z^2(r,z), \tag{3.325}$$

where the upper index 1 corresponds to the solution of the problem for an elastic layer when

$$\tau_{rz}(r,h) = 0, \qquad \sigma_z(r,h) = \begin{cases} -q(r) & \text{for } 0 \le r \le a \\ 0 & \text{for } a < r \le R \end{cases},$$

$$\tau_{rz}(r,0) = u_z(r,0) = 0.$$

This solution is given by formulae (3.47)–(3.57). The index 2 in (3.324), (3.325) corresponds to the axisymmetric solution for a layer, which has the form (3.62)–(3.66) when

$$\sigma_z = \tau_{rz} = 0 \quad (z = h), \qquad u_z = \tau_{rz} = 0 \quad (z = 0).$$

To be exact,

$$u_r^2(r,z) = \sum_{k=1}^{\infty} D_k u_r^{(k)},$$

$$u_z^2(r,z) = \sum_{k=1}^{\infty} D_k u_z^{(k)},$$

$$\tau_{rz}^2(r,z) = \sum_{k=1}^{\infty} D_k \tau_{rz}^{(k)}.$$

The unknown coefficients D_k and the function $q(r)$ of the contact stresses are determined from the conditions

$$u_r(R,z) = 0, \qquad \tau_{rz}(R,z) = \tau_{rz}^1(R,z) - \tau_{rz}^2(R,z) = 0 \tag{3.326}$$

and from the contact condition

$$u_z(r,h) = -\delta(r) \qquad (0 \le r \le a). \tag{3.327}$$

Using formulae (3.67), (3.68) of generalized orthogonality of the homogeneous solutions for the layer, we find from (3.326) that

$$D_k = [GW_{kk}\, \mathcal{J}_1(\alpha_k R)]^{-1} \int_0^a q(x)\Omega_k(x)x\, dx, \tag{3.328}$$

where

$$\Omega_k(x) = \int_0^{\infty} M_k(u)\, \mathcal{J}_1(uR)\, \mathcal{J}_0(ux)u\, du, \tag{3.329}$$

$$M_k(u) = 2\int_0^h [L_1(u,z)L_{4k}^0(z) - L_3(u,z)L_{2k}^0(z)]\,dz.$$

To satisfy condition (3.327), we obtain the integral equation

$$K\{q\} = \int_0^a \rho q(\rho)K(r,\rho)\,d\rho = \theta\left[\delta(r) + \sum_{k=1}^{\infty} D_k L_{2k}^0(h)\,\mathcal{J}_0(\alpha_k r)\right] \qquad (3.330)$$

$$(0 \le r \le a),$$

whose kernel is expressed by formula (3.58). It is convenient to represent the solution

$$q(r) = \theta[q_0(r) + \sum_{k=1}^{\infty} D_k L_{2k}^0(h) q_k(r)], \qquad (3.331)$$

where $q_k(r)$ are the solutions of the integral equations $(0 \le r \le a)$

$$K\{q_k\} = \begin{cases} \delta(r) & \text{for } k=0 \\ \mathcal{J}_0(\alpha_k r) & \text{for } k \ge 1 \end{cases}. \qquad (3.332)$$

If we substitute (3.331) into (3.328), we obtain an infinite system of linear algebraic equations of the form

$$x_k = g_k + \sum_{n=1}^{\infty} a_{kn} x_n \qquad (k = 1, 2, ...), \qquad (3.333)$$

$$x_k = D_k L_{2k}^0(h)\mathcal{I}_1(\gamma_k c), \qquad \gamma_k = -i\alpha_k h, \qquad c = R/h, \qquad (3.334)$$

$$g_k = -2i\gamma_k\, W_{kk}^{-1}[L_{2k}^0(h)]^2\,\mathcal{K}_1(\gamma_k c)T_{0k}h^{-1}, \qquad (3.335)$$

$$a_{kn} = -2i\gamma_k\, W_{kk}^{-1}[L_{2k}^0(h)]^2\,\mathcal{K}_1(\gamma_k c)\,\mathcal{I}_1^{-1}(\gamma_n c)T_{nk}h^{-1}, \qquad (3.336)$$

$$T_{nk} = \int_0^a q_n(x)\,\mathcal{I}_0(\gamma_k x/h)x\,dx. \qquad (3.337)$$

In deriving relations (3.333)–(3.336), we have used the value of integral (3.329)

$$\Omega_k(x) = 2\gamma_k h^{-1} L_{2k}^0(h)\,\mathcal{I}_0(\gamma_k x/h)\,\mathcal{K}_1(\gamma_k c).$$

Taking into account that, as $n \to \infty$, the asymptotics of the numbers γ_n is determined by the right-hand side of formula (3.61) multiplied by $-i$, we can prove that, as $k, n \to \infty$,

$$|g_k| \sim k^{-1}\exp[-\pi k(R-a)/h],$$

$$|a_{kn}| \sim k^{-1} \exp[-\pi(k+n)(R-a)/h].$$

Hence, system (3.333) belongs to the class of normal Poincaré–Koch systems (Kantorovich and Krylov (1958, chapter 1, § 2, subsection 6)) and its solution can be obtained by the reduction to a finite system for any values of $(R-a)/h > 0$.

Thus, the contact problem is reduced to investigation of the infinite system (3.333) and integral equations (3.332).

We find the solutions of equations (3.332) by using the asymptotic method for *large* λ, $\lambda = h/a$, and we then substitute the expressions into formula (3.337) for T_{nk} and determine the coefficients in system (3.333). As a result (Chebakov (1980)), we determine the contact stress by formula (3.331); for the case $\delta(r) = \delta = \text{const}$ we have

$$q_k(\rho a) = \frac{2\delta_k}{\pi a}\left[\sum_{j=0}^{M} H_j^k(\lambda) S_j(\rho) + G_k(\rho)\right] \quad (0 \le \rho \le 1, \ \ M \to \infty), \tag{3.338}$$

$$\delta_0 = \delta, \quad \delta_k = 1 \ \ (k \ge 1); \quad H_j^k(\lambda) = \sum_{s=j}^{M} \lambda^{-(2s+1)} [\beta_{sj}^k + \lambda^{-1} \alpha_{s+1,j}^k], \tag{3.339}$$

$$G_0(\rho) = S_0(\rho), \quad G_k(\rho) = -\frac{d}{d\rho}\rho \int_\rho^1 \frac{\cos(a\alpha_k t)\, dt}{t\sqrt{t^2 - \rho^2}} \quad (k \ge 1), \tag{3.340}$$

$$S_j(\rho) = \frac{1}{\sqrt{1-\rho^2}} \sum_{k=0}^{j-1} \frac{(j-1)!\,(2j\rho^2 - 2j + 2k + 1)}{k!\,(j-k-1)!\,(2k+1)} (1-\rho^2)^k \rho^{2(j-k-1)}, \tag{3.341}$$

where the coefficients β_{sj}^k and $\alpha_{s+1,j}^k$ are determined from the simple reccurence relations

$$\alpha_{sj}^k = \frac{2}{\pi} \sum_{m=j}^{s-1} b_m z_{mj} \sum_{p=0}^{s-m-1} \frac{\beta_{s-m-1,p}^k}{2p + 2m - 2j + 1} \quad \begin{pmatrix} s \ge 1 \\ 0 \le j \le s-1 \end{pmatrix}, \tag{3.342}$$

$$\beta_{sj}^k = \frac{2}{\pi}\left[b_s z_{sj} F_{s-j}^k + \sum_{m=j}^{s-1} b_m z_{mj} \sum_{p=0}^{s-m-1} \frac{\alpha_{s-m,p}^k}{2p + 2m - 2j + 1}\right] \tag{3.343}$$

$$\begin{pmatrix} s \ge 0 \\ 0 \le j \le s \end{pmatrix},$$

$$\beta_{00}^k = 2\pi^{-1} b_0 z_{00} F_0^k, \quad z_{mj} = (2m)!\,[(2j)!\,(2m-2j)!]^{-1}, \tag{3.344}$$

$$b_k = \frac{(-1)^k}{(2k)!}\int\limits_0^\infty [1 - uK(u)]u^{2k}\,du, \quad F_k^0 = \frac{1}{2k+1}, \tag{3.345}$$

$$F_k^m = \frac{(2k)!}{(\gamma_m/\lambda)^{2k+1}}\left[\sum_{n=0}^{k}\frac{(\gamma_m/\lambda)^{2n}}{(2n)!}\sinh\frac{\gamma_m}{\lambda} \right. \tag{3.346}$$

$$\left. - \sum_{n=1}^{k}\frac{(\gamma_m/\lambda)^{2n-1}}{(2n-1)!}\cosh\frac{\gamma_m}{\lambda}\right] \quad (m \geq 1).$$

The connection between the force P, acting on the punch, and the displacement δ of the punch is determined by the formulae

$$P = 4a\delta\theta\left[P_0 + \sum_{k=1}^{\infty} x_k\,\mathcal{I}_1^{-1}(\gamma_k c)P_k\right],$$

$$P_k = R_k + \sum_{j=0}^{M} H_j^k(\lambda)(2j+1)^{-1},$$

$$R_0 = 1, \quad R_k = \lambda\gamma_k^{-1}\sinh(\gamma_k/\lambda) \quad (k \geq 1),$$

Here, x_k is the solution of system (3.333) in which one should place

$$g_k = \tan^2(\gamma_k)\,\mathcal{K}_1(\gamma_k c)T_{0k}, \quad a_{kn} = \tan^2(\gamma_k)\,\mathcal{K}_1(\gamma_k c)\,\mathcal{I}_1^{-1}(\gamma_k c)T_{nk}, \tag{3.347}$$

$$T_{nk} = \frac{2}{\pi\lambda}\left[\sum_{j=0}^{M} H_j^n(\lambda)F_j^k + t_{nk}\right] \quad (M \to \infty), \tag{3.348}$$

$$t_{0k} = \lambda\gamma_k^{-1}\,\sinh(\gamma_k/\lambda), \tag{3.349}$$

$$t_{nk} = \frac{\lambda\sinh[(\gamma_n+\gamma_k)/\lambda]}{2(\gamma_n+\gamma_k)} + \frac{\lambda\sinh[(\gamma_n-\gamma_k)/\lambda]}{2(\gamma_n-\gamma_k)}. \tag{3.350}$$

Some results for the dimensionless quantities (3.321)–(3.323) for various values of the parameters λ and $b = R/a$ are presented in Table 3.9. The infinite system (3.333), (3.347)–(3.350) has been replaced by a finite one. We note that the larger the parameter $(R-a)/h$, the fewer equations are needed in the finite system. Fig. 3.16 shows the plot of the dependence between N_0 and b for the fixed value $\lambda = 2$.

Table 3.9: The pressure and force

λ	2.0	2.0	2.0	4.0	4.0	4.0
b	1.5	2.0	3.0	1.5	2.0	3.0
$q_*(0.2)$	0.971	1.02	1.04	0.460	0.612	0.750
$q_*(0.95)$	2.51	3.00	3.00	1.35	2.00	2.44
N_0	5.43	6.10	6.08	2.95	4.07	4.84

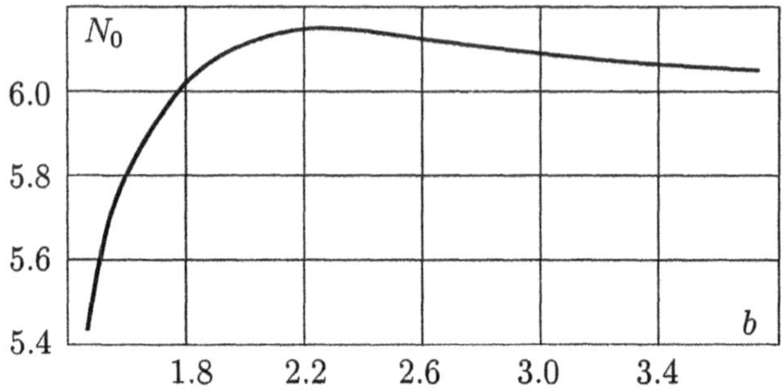

Figure 3.16: The dependence between the force N_0 and $b = R/a$

Unlike (3.322), in (3.338), (3.341) we have the explicit square root singularity for the contact pressure. The analysis of the quantity $q_*(0.95)$ in Tables 3.8 and 3.9 says that the contact pressure and its strength intensity factor at the point $r = a$ grow considerably as $a \to R$ and $\lambda = ha^{-1} \to 0$. For instance, for $b = Ra^{-1} = 1.5$ and $h = 0.3a$ the strength intensity factor is approximately four times greater than for $h = 4a$. At the same time, for small fixed λ the strength intensity factor and the force N_0 have a local maximum for a certain value of b. It is known that the strength intensity factor can characterize the local strength of the elastic body on the boundary of the contact zone. We now see how this depends on the geometry of the plate (parameters R and h) and of the punch (parameter a).

3.6.3. THE PLATE WITH A NONCYLINDRICAL LATERAL SURFACE

Using cylindrical coordinates (r, φ, z), we consider a body of revolution bounded by the surfaces $z = 0$, $z = h$ and $r = R(z)$; $R(z)$ is a smooth function describing the lateral surface of the body. This body may be called a plate with a noncylindrical lateral surface. Suppose the plate lies on the smooth rigid base $z = 0$, its lateral surface is free of stresses, and the upper plane side $z = h$ of the plate is in frictionless contact with the circular planar punch, the vertical settlement δ of which is given in the contact domain $r \leq a$ (Fig. 3.17a,b).

The problem can be reduced to the following axisymmetric boundary-value problem in cylindrical coordinates:

$$\tau_{rz} = 0, \qquad u_z = -\delta \qquad (z = h,\, r \leq a), \tag{3.351}$$

$$\tau_{rz} = 0, \qquad \sigma_z = 0 \qquad (z = h,\ a < r < R(h)), \tag{3.352}$$

$$\tau_{rz} = 0, \qquad u_z = 0 \qquad (z = 0), \tag{3.353}$$

and also

$$\sigma_r - R'(z)\tau_{rz} = 0 \qquad (0 < z < h,\ r = R(z)), \tag{3.354}$$

$$\sigma_z R'(z) - \tau_{rz} = 0 \qquad (0 < z < h,\ r = R(z)). \tag{3.355}$$

As before, we seek the solution of the boundary-value broblem (3.351)–(3.355) for Navier's equations in the form (3.324), (3.325). Boundary conditions (3.351)–(3.353), excepting condition $(3.351)_2$, will then be satisfied. Satisfying the latter condition, we get equations (3.330)–(3.332) for $\delta(r) \equiv \delta$. We will satisfy the other boundary conditions of (3.354), (3.355) in two ways: the boundary collocation method and the method of least squares.

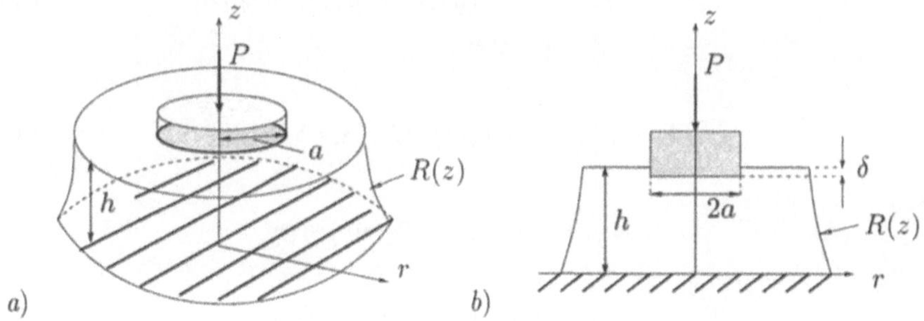

Figure 3.17: A punch on a circular plate with noncylindrical lateral surface $R(z)$; a) a 3-D view; b) an axial section

To construct the solutions of the integral equations (3.332), we use the asymptotic method for *large* λ where $\lambda = h/a$. The formulae (3.338)–(3.346) allow us to find the solution of the integral equations in the domain of convergence of the method of *large* λ with practically any chosen degree of accuracy.

Thus, the contact stress under the punch is determined by formula (3.331) in which the constants D_k are required. We find them by using the boundary collocation method, satisfying the boundary conditions (3.354), (3.355) at N points on the curve $r = R(z)$ $(0 < z < h)$.

First, we present the components of the stress tensor of the original problem, in accordance with (3.324), (3.325), in the form

$$\tau_{rz} = \tau_{rz}^1 - \tau_{rz}^2, \qquad \sigma_r = \sigma_r^1 - \sigma_r^2, \qquad \sigma_z = \sigma_z^1 - \sigma_z^2, \tag{3.356}$$

where the first terms on the right-hand sides, taking account of (3.49)–(3.57) and (3.331), can be presented as follows

$$\tau_{rz}^1 = \theta\left[T_1^0(r,z) + \sum_{k=1}^{2N} D_k T_1^k(r,z)\right], \tag{3.357}$$

$$\sigma_r^1 = \theta\left[T_2^0(r,z) + \sum_{k=1}^{2N} D_k T_2^k(r,z)\right], \tag{3.358}$$

$$\sigma_z^1 = \theta\left[T_3^0(r,z) + \sum_{k=1}^{2N} D_k T_3^k(r,z)\right]. \tag{3.359}$$

Here we have introduced the notation

$$T_1^k = L_{2k}^0(h)\int_0^\infty Q_k(y)L_3(z,y)\,\mathcal{J}_1(yr)y\,dy, \tag{3.360}$$

$$T_2^k = L_{2k}^0(h)\int_0^\infty Q_k(y)[L_4(z,y)\,\mathcal{J}_0(yr) \tag{3.361}$$

$$+(yr)^{-1}\,L_5(z,y)\,\mathcal{J}_1(yr)]y\,dy,$$

$$T_3^k = L_{2k}^0(h)\int_0^\infty Q_k(y)L_4(z,y)\,\mathcal{J}_0(yr)y\,dy, \tag{3.362}$$

and also

$$Q_k(y) = \int_0^a q_k(r)\,\mathcal{J}_0(yr)r\,dr. \tag{3.363}$$

We have retained $2N$ terms in the sums (3.357)–(3.359), since the boundary conditions on the lateral surface give $2N$ equations. Similarly, we obtain the following expressions for the second terms on the right-hand sides in (3.356) (G is the shear modulus):

$$\tau_{rz}^2 = G\sum_{k=1}^{2N} D_k E_1^k(r,z), \qquad \sigma_r^2 = G\sum_{k=1}^{2N} D_k E_2^k(r,z), \tag{3.364}$$

$$\sigma_z^2 = G\sum_{k=1}^{2N} D_k E_3^k(r,z), \tag{3.365}$$

$$E_1^k(r,z) = L_{3k}^0(z)\,\mathcal{J}_1(\alpha_k r), \qquad E_3^k(r,z) = L_{6k}^0(z)\,\mathcal{J}_0(\alpha_k r), \tag{3.366}$$

$$E_2^k(r,z) = L_{4k}^0(z)\,\mathcal{J}_0(\alpha_k r) + L_{5k}^0(z)\,\mathcal{J}_1(\alpha_k r)/(r\alpha_k). \tag{3.367}$$

We denote the collocation points by (r_k, z_k) $(k = 1, ..., N)$ where we have $r_k = R(z_k)$ $(0 \le z_k \le h)$. Satisfying the boundary conditions (3.354), (3.355) at the indicated collocation points, we obtain a system of $2N$ linear algebraic equations for the $2N$ required constants D_k in the form

$$\sum_{k=1}^{2N} D_k a_{km}^1 = b_m^1, \quad \sum_{k=1}^{2N} D_k a_{km}^2 = b_m^2 \quad (m = 1, ..., N), \tag{3.368}$$

$$\begin{aligned}
a_{km}^1 &= (1-\nu)E_2^k(r_m, z_m) - T_2^k(r_m, z_m) \\
&-R'(z_m)[(1-\nu)E_1^k(r_m, z_m) - T_1^k(r_m, z_m)], \\
b_m^1 &= T_2^0(r_m, z_m) - R'(z_m)T_1^0(r_m, z_m), \\
a_{km}^2 &= (1-\nu)E_1^k(r_m, z_m) - T_1^k(r_m, z_m) \\
&-R'(z_m)[(1-\nu)E_3^k(r_m, z_m) - T_3^k(r_m, z_m)], \\
b_m^2 &= T_1^0(r_m, z_m) - R'(z_m)T_3^0(r_m, z_m).
\end{aligned}$$

The formulae for calculation of the matrix and the right-hand side of the system (3.368) contain only single improper integrals of the type (3.360) in which the internal integral (3.363) can be explicitly calculated on account of (3.338).

To control the numerical results, we use the method of least squares as the second way to approximately satisfy of the boundary conditions on the lateral surface. The constants D_k are found by minimizing the functional

$$J = \int_\Gamma [(\sigma_r - R'\tau_{rz})^2 + (R'\sigma_z - \tau_{rz})^2]\,ds, \tag{3.369}$$

where Γ is an element of the lateral surface; the stress tensor components under the integral sign are determined by formulae (3.356)–(3.367). From the minimization of J (3.369) we get $\partial J/\partial D_m = 0$ $(m = 1, ..., 2N)$ and we obtain the following system of linear algebraic equations for D_k:

$$\sum_{k=1}^{2N} D_k a_{km} = d_m \qquad (m = 1, ..., 2N), \tag{3.370}$$

where

$$a_{kn} = \int_0^h [r_k(z)r_n(z) + p_k(z)p_n(z)]\,dz, \tag{3.371}$$

$$d_n = -\int_0^h [r_0(z)r_n(z) + p_0(z)p_n(z)]\,dz, \tag{3.372}$$

$$r_k(z) = T_2^k - (1-\nu)E_2^k - R'(z)[T_1^k - (1-\nu)E_1^k] \quad (k \geq 1), \tag{3.373}$$

$$p_k(z) = -T_1^k + (1-\nu)E_1^k + R'(z)[T_3^k - (1-\nu)E_3^k] \quad (k \geq 1), \tag{3.374}$$

$$r_0(z) = T_2^0 - R'(z)T_1^0, \quad p_0(z) = R'(z)T_3^0 - T_1^0. \tag{3.375}$$

On the right-hand sides of equalities (3.373)–(3.375) one should set $r = R(z)$. The elements of the matrix and of the right-hand side of system (3.370), in contrast to those of system (3.368), are calculated as double integrals. Taking into account the difficulties arising in calculation of the double integrals (3.371), (3.372), we replace the integrals in z by finite sums, choosing the points z_k on the segment $[0, h]$:

$$a_{kn} = \sum_{m=1}^{M} [r_k(z_m)r_n(z_m) + p_k(z_m)p_n(z_m)],$$

$$d_n = -\sum_{m=1}^{M} [r_0(z_m)r_n(z_m) + p_0(z_m)p_n(z_m)].$$

This considerably simplifies the calculations, without sacrificing accuracy.

As an example, Tsvetkov and Chebakov (1989) took a body of revolution whose axial section is an isosceles trapezoid. In calculations they set $2N$ =6–10, M=6–24; they checked the agreement between the results obtained by collocation and by the method of least squares. Calculations show that the convergence improves as R increases, and the collocation method gives plausible results for six equations. The collocation method provides stable results for a polynomial lateral surfaces and for a fairly small number of collocation points. The domain of applicability of the method of least squares is wider. Both approaches provide a quick solution.

We note that for the efficient numerical realization of formulae (3.47)–(3.51) one should separate the slowly decreasing terms from the integrals in these formulae and then integrate them analytically.

Dovnorovich (see reference in Gladwell (1980)) and Tsvetkov and Chebakov (1990) suggested satisfying the boundary conditions on the lateral surface by using the best approximation in the Chebyshev sense, and modified Remez methods. This method demonstrates a high efficiency for arbitrary values of the parameter λ.

3.7 Contact Problems for a Finite Cylinder

3.7.1. HOMOGENEOUS SOLUTIONS

Lur'e (1964, p. 397) considered the homogeneous solutions for an infinite cylinder, i.e., the solutions of the equilibrium equations for a solid cylinder which is unloaded on its surface (for zeroth boundary conditions). These solutions are based on the complex roots of the characteristic equation (3.15)

$$D(\beta) \equiv u^2[\mathcal{I}_0^2(\beta) - \mathcal{I}_1^2(\beta)] - 2(1-\nu)\mathcal{I}_1^2(\beta) = 0. \tag{3.376}$$

The expressions for the displacements, corresponding to the homogeneous solution for the root β_k of equation (3.376) in the first quadrant, have the form[11]

$$u_r^{(k)}(r,z) = L_k\, u(r,\beta_k)\exp(i\beta_k z), \tag{3.377}$$
$$u_z^{(k)}(r,z) = L_k\, w(r,\beta_k)\exp(i\beta_k z), \tag{3.378}$$

where

$$u(r,\beta_k) = \frac{1}{\mathcal{I}_1(\beta_k)}\left[-r\mathcal{I}_0(\beta_k r) + \left(\frac{2(1-\nu)}{\beta_k} + \frac{\mathcal{I}_0(\beta_k)}{\mathcal{I}_1(\beta_k)}\right)\mathcal{I}_1(\beta_k r)\right], \tag{3.379}$$

$$w(r,\beta_k) = -\frac{i}{\mathcal{I}_1(\beta_k)}\left[r\mathcal{I}_1(\beta_k r) + \left(\frac{2(1-\nu)}{\beta_k} - \frac{\mathcal{I}_0(\beta_k)}{\mathcal{I}_1(\beta_k)}\right)\mathcal{I}_0(\beta_k r)\right], \tag{3.380}$$

and $L_k = M_k + iN_k$ is a complex constant. The expressions for stresses can be obtained with help of Hooke's law.

As we can see from (3.377)–(3.380), there are two independent real homogeneous solutions corresponding to each such root β_k. The first few zeros $\beta_k = \gamma_k + i\gamma_k$ are given in Table 3.10 for $\nu = 0.3$.

Table 3.10: First roots for (3.376)

k	β_k	k	β_k
1	$1.362 + i\,2.722$	4	$1.966 + i\,12.44$
2	$1.637 + i\,6.060$	5	$2.074 + i\,15.61$
3	$1.827 + i\,9.266$	6	$2.162 + i\,18.76$

For the large zeros β_k, $k = 7, 8, \ldots$, one should use the asymptotics (7.4.21) in Lur'e (1964):

$$\beta_k \approx k\pi i + \frac{1}{2}\ln(4k\pi) - i\left[\frac{\ln(4k\pi)}{4k\pi} - \frac{1}{2k\pi}\left(\frac{1}{4} - 2(1-\nu)\right)\right]. \tag{3.381}$$

[11] The subscript (k) here does not mean a differentiation.

We will use these homogeneous solutions to solve boundary-value problems for a finite cylinder.

3.7.2. CONTACT PROBLEMS

We consider the axisymmetric frictionless contact between a finite elastic cylinder of length $2l$ and radius R, and a rigid cylindrical sleeve fitted with the interference δ for $|z| \leq a$. We suppose that the cylinder end-walls are either $\mathcal{A}$) rigidly fixed or $\mathcal{B}$) free of stresses. It is necessary to determine the normal contact stresses $\sigma_r = -q(z)$ $(r = R,\ |z| \leq a)$ as well as the connection between the interference and the integral of the function $q(z)$ over the contact area. The problems can be reduced to boundary-value problems for the axisymmetric Navier's equations under the following boundary conditions on the lateral surface of the cylidner:

$$u_r = -\delta \quad (|z| \leq a), \qquad \sigma_r = 0 \quad (a < |z| < l); \qquad \tau_{rz} = 0, \tag{3.382}$$

and under the following boundary conditions on the end-walls of the cylinder:

$$\mathcal{A})\ u_r = u_z = 0; \qquad \text{or } \mathcal{B})\ \tau_{rz} = \sigma_z = 0. \tag{3.383}$$

Besides, the stresses are bounded at $r = 0$.

We will seek the solution of the boundary-value problems (3.382), (3.383) in the form $u_r = u^1 - u^2$, $u_z = w^1 - w^2$ where u^1, w^1 are the solutions of the Navier's equations for an infinite cylinder under the boundary conditions

$$\sigma_r = -q(r) \quad (|z| \leq a), \qquad \sigma_r = 0 \quad (|z| > a); \qquad \tau_{rz} = 0, \tag{3.384}$$

and u^2, w^2 are the superpositions of the axisymmetric homogeneous solutions for the infinite cylinder whose surface is subjected to the conditions $\sigma_r = \tau_{rz} = 0$. Similarly, we present the stresses as the differences $\tau_{rz} = \tau^1 - \tau^2$, $\sigma_z = \sigma^1 - \sigma^2$.

Using formulae (3.2)–(3.15), we write the solution of inhomogeneous problem (3.384) in the form

$$u^1 = \frac{1}{2\pi} \int_{-\infty}^{\infty} [c_1 \mathcal{I}_1(\alpha r) - c_2 \alpha r \mathcal{I}_0(\alpha r)] \exp(-i\alpha z)\, d\alpha, \tag{3.385}$$

$$w^1 = \frac{i}{2\pi} \int_{-\infty}^{\infty} \{c_2 \alpha r \mathcal{I}_1(\alpha r) - [a_1 - 4(1-\nu)c_2]\mathcal{I}_0(\alpha r)\} \exp(-i\alpha z)\, d\alpha, \tag{3.386}$$

$$\sigma^1 = \frac{G}{\pi} \int_{-\infty}^{\infty} \{[2(2-\nu)c_2 - c_1]\mathcal{I}_0(\alpha r) + c_2 \alpha r \mathcal{I}_1(\alpha r)\}\alpha \exp(-i\alpha z)\, d\alpha, \quad (3.387)$$

$$\tau^1 = \frac{iG}{\pi} \int_{-\infty}^{\infty} \{c_2 \alpha r \mathcal{I}_0(\alpha r) - [c_1 - 2(1-\nu)c_2]\mathcal{I}_1(\alpha r)\}\alpha \exp(-i\alpha z)\, d\alpha, \quad (3.388)$$

$$c_1 = -\frac{Q(\alpha)}{2G\alpha^3 RD(\alpha R)} \left[\mathcal{I}_0(\alpha R) + \frac{2(1-\nu)}{\alpha R}\mathcal{I}_1(\alpha R)\right],$$

$$c_2 = -\frac{Q(\alpha)}{2G\alpha^3 RD(\alpha R)} \frac{\mathcal{I}_1(\alpha R)}{\alpha R}, \quad Q(\alpha) = \int_{-a}^{a} q(\xi)\exp(i\alpha\xi)\, d\xi.$$

Satisfying the first boundary condition of (3.382), we obtain an integral equation for the function $q(z)$ with the integral operator as in formulae (3.16)–(3.18). Introducing the notation $x = za^{-1}$, $\theta = G(1-\nu)^{-1}$, $\lambda = Ra^{-1}$, $\mu = la^{-1}$, $\varphi(x) = q(z)a(\delta\theta)^{-1}$ etc., we can write this equation as follows

$$\int_{-1}^{1} \varphi(\xi)\, K\left(\frac{\xi - x}{\lambda}\right) d\xi = \pi - \pi F(x) \qquad (|x| \leq 1); \quad (3.389)$$

$$K(t) = \int_0^{\infty} \frac{L(u)}{u} \cos(ut)\, du, \quad (3.390)$$

$$F(x) = \sum_{n=1}^{\infty} \left(\cosh\frac{\delta_n \mu}{\lambda}\right)^{-1} \left\{A_n\left[a_n \cosh\frac{\delta_n x}{\lambda} \cos\frac{\gamma_n x}{\lambda}\right.\right.$$

$$\left.\left. + b_n \sinh\frac{\delta_n x}{\lambda} \sin\frac{\gamma_n x}{\lambda}\right] - B_n\left[b_n \cosh\frac{\delta_n x}{\lambda}\cos\frac{\gamma_n x}{\lambda} - a_n \sinh\frac{\delta_n x}{\lambda}\sin\frac{\gamma_n x}{\lambda}\right]\right\},$$

$$a_n + ib_n = \frac{2(1-\nu)}{\gamma_n^2 + \delta_n^2}[\gamma_n - i\delta_n].$$

The function $L(u)$ coincides with the function $L_1(u)$ from (3.18).

It is convenient to seek the solution of the integral equation (3.389) in the form

$$\varphi(x) = \varphi_0(x) - \sum_{n=1}^{\infty}[A_n \varphi_n^A(x) - B_n \varphi_n^B(x)], \quad (3.391)$$

where $\varphi_0(x)$ is the solution of equation (3.389) when $F(x) \equiv 0$.

We will approximately satisfy the other boundary conditions of (3.383) using the method of least squares in order to determine the values of A_n, B_n $(n = 1, 2, \dots)$.

To find the approximate solution of equation (3.389), one can apply the asymptotic methods for *large* and *small* λ from section 3.2. Here we use the method for *small* λ when the function-symbol $L(u)u^{-1}$ in (3.390) is specially approximated by the formula (Aleksandrov (1986))

$$K^*(u) \approx \frac{\sqrt{u^2 + B^2}}{u^2 + C^2} \exp\left(\frac{D|u|}{u^2 + E^2}\right), \quad \frac{B}{C^2} = K^*(0) = A, \tag{3.392}$$

instead of approximation (3.85) used in section 3.2. On the whole interval $0 \le u < \infty$, the error of this approximation does not exceed 5% for $\nu = 0.3$, $B = 1.0962$, $C = 1.6882$, $D = 0.4$ and $E = 3.0193$. If the right-hand side of equation (3.389) is equal to $\pi \cosh \frac{\delta x}{\lambda} \cos \frac{\gamma x}{\lambda}$, then we obtain the asymptotic solution of equation (3.389) with (3.392), for small λ, in the form

$$\varphi(x) = \frac{1}{4\lambda}\left[\tilde{w}\left(\frac{1+x}{\lambda}\right) + \tilde{w}\left(\frac{1-x}{\lambda}\right)\right], \tag{3.393}$$

$$\tilde{w}(s) = \exp(\eta/\lambda)w(s,\eta) + \exp(\overline{\eta}/\lambda)w(s,\overline{\eta}), \quad \eta = \delta + i\gamma, \quad \overline{\eta} = \delta - i\gamma, \tag{3.394}$$

where the function $w(s,\eta) \equiv w(s)$ is found from the Wiener–Hopf integral equation

$$\int_0^\infty w(s)k(t-s)\,ds = \pi \exp(-\eta t) \quad (0 \le t < \infty). \tag{3.395}$$

The solution of (3.395) is

$$w(s) = \frac{C+\eta}{\sqrt{B+\eta}} \exp[g(\eta)][w_*(s) + I(s)], \quad g(\eta) = \frac{-D\eta}{\pi(\eta^2 - E^2)} \ln\frac{\eta}{E}, \tag{3.396}$$

$$w_*(s) = \frac{\exp(-Bs)}{\sqrt{\pi s}} + \frac{C-\eta}{\sqrt{B-\eta}} \exp(-\eta s)\Phi(\sqrt{(B-\eta)s}), \tag{3.397}$$

$$I(s) = \frac{D}{2\pi}\int_0^s w_*(s-\tau)[\exp(E\tau)\,\mathrm{Ei}\,(-E\tau) + \exp(-E\tau)\,\mathrm{Ei}\,(E\tau)]\,d\tau.$$

Here $\Phi(x)$ is the probability integral (2.101), $\mathrm{Ei}\,(\pm x)$ is the integral exponential function 8.211.1, 8.213.2 in Gradshteyn and Ryzhik (1994):

$$\mathrm{Ei}\,(x) = -\int_{-x}^\infty \frac{\exp(-t)}{t}\,dt \quad \text{for } x < 0,$$

$$\mathrm{Ei}\,(x) = \mathrm{C} + \ln x + \sum_{n=1}^{\infty} \frac{x^n}{nn!} \qquad \text{for } x > 0,$$

where C is Euler's constant.

If we set $\eta = 0$ in (3.396), (3.397), then we will have the solution of equation (3.389) with (3.392) for $F(x) \equiv 0$.

Similarly, when the right-hand side of equation (3.389) is equal to $\pi \sinh \frac{\delta x}{\lambda} \sin \frac{\gamma x}{\lambda}$, we obtain the asymptotic solution of equation (3.389) with (3.392), for small λ, in the form (3.393) where it should be taken

$$\tilde{w}(s) = -i[\exp(\eta/\lambda)w(s,\eta) - \exp(\overline{\eta}/\lambda)w(s,\overline{\eta})]. \tag{3.398}$$

According to the method of least squares, we find the required constants A_n, B_n $(n = 1, 2, ...)$ by minimizing functional $(\rho = rR^{-1})$

$$\mathcal{A}) \qquad \mathcal{F} = \frac{1}{a^2} \int_0^1 \{[u^1(\rho,\mu) - u^2(\rho,\mu)]^2 + [w^1(\rho,\mu) - w^2(\rho,\mu)]^2\}\, d\rho,$$

or

$$\mathcal{B}) \qquad \mathcal{F} = \frac{1}{4G^2} \int_0^1 \{[\sigma^1(\rho,\mu) - \sigma^2(\rho,\mu)]^2 + [\tau^1(\rho,\mu) - \tau^2(\rho,\mu)]^2\}\, d\rho,$$

where the integration takes place along the generator of the end-wall surface of the cylinder.

In expressions (3.385)–(3.388) for u^1, w^1, σ^1, τ^1, the integral in α can be calculated using the theory of residues and Jordan's lemma, but the integral in ξ can be found analytically on the basis of representations (3.391), (3.392), (3.394), (3.396)–(3.398) by using the well-known properties of a Laplace transform. To be exact, using the convolution theorem for a Laplace transform, we find that

$$\int_0^{2/\lambda} w(t,\eta) \exp[-a(2/\lambda - t)]\, dt$$

$$= \frac{C+\eta}{\sqrt{B+\eta}} \exp[g(\eta)] \left[w^*\left(\frac{2}{\lambda}, \eta, a\right) + J\left(\frac{2}{\lambda}\right)\right],$$

$$w^*(t,\eta,a) = \frac{C-a}{\eta - a} \frac{\exp(-at)}{\sqrt{B-a}} \Phi(\sqrt{(B-a)t})$$

$$-\frac{C-\eta}{\eta-a}\frac{\exp(-\eta t)}{\sqrt{B-\eta}}\Phi(\sqrt{(B-\eta)t}) \qquad (\eta \neq a),$$

$$w^*(t,a,a) = \frac{C-a}{B-a}\sqrt{\frac{t}{\pi}}\exp(-Bt) + \frac{\exp(-at)}{\sqrt{B-a}}\Phi(\sqrt{(B-a)t})$$

$$\times\left[1 - \frac{C-a}{2(B-a)} + (C-a)t\right],$$

$$J(t) = \frac{D}{2\pi}\int_0^t w^*(t-\tau,\eta,a)[\exp(E\tau)\,\mathrm{Ei}\,(-E\tau) + \exp(-E\tau)\,\mathrm{Ei}\,(E\tau)]\,d\tau.$$

Table 3.11: Contact pressure and its integral value

			problem $\mathcal{A}$				problem $\mathcal{B}$			
λ	μ	K	φ_1	φ_2	φ_3	S	φ_1	φ_2	φ_3	S
1	1.1	6	4.06	2.67	2.68	6.27	2.29	2.40	2.40	4.73
1	1.2	6	3.25	2.62	2.61	5.66	2.60	2.50	2.49	5.06
1	∞	6	2.84	2.52	2.51	5.25	2.84	2.52	2.51	5.25
1	1.1	9	4.26	2.81	2.78	6.19	2.15	2.39	2.40	4.68
1	1.2	9	3.42	2.63	2.60	5.56	2.69	2.53	2.52	5.10
1	1.3	9	3.19	2.57	2.55	5.38	2.92	2.56	2.54	5.23
1	∞	9	3.06	2.55	2.53	5.29	3.06	2.55	2.53	5.29
1.5	1.1	9	2.92	1.75	1.73	3.99	1.44	1.61	1.62	3.15
1.5	1.2	9	2.68	1.80	1.77	3.93	1.59	1.65	1.65	3.27
1.5	1.3	9	2.48	1.76	1.74	3.78	1.75	1.69	1.69	3.39
1.5	∞	9	2.22	1.73	1.71	3.63	2.22	1.73	1.71	3.63

We retain $2N$ homogeneous solutions in formulae (3.389) and (3.391), i.e., we will sum from $n = 1$ to $n = N$. To determine the constants A_n, B_n ($n = 1, 2, ..., N$) by minimizing the functional $\mathcal{F}$, we obtain the system of $2N$ linear algebraic equations $\partial\mathcal{F}/\partial A_n = 0$, $\partial\mathcal{F}/\partial B_n = 0$. To calculate the integrals with respect to the variable $\rho \in (0;1)$ in this system, we use Chebyshev's equal weights integration formula 25.4.28 in Abramowitz and Stegun (1972) with number of nodes $K = 6$ or $K = 9$. The nearness of the results obtained for different values of K and N attests to the stability of the computation. For the results given here we set $K = N$, $\nu = 0.3$, so that we have the exact satisfaction of the boundary conditions at the chosen nodes. Table 3.11 shows the values of $\varphi_k = \varphi(x_k)$ at the k-nodes of the corresponding quadrature formula and also the values of the integral characteristic $S = \int_{-1}^{1} \varphi(x)dx$, for various λ, μ, K. The value $\mu = \infty$ corresponds to the infinite cylinder. One can see from the table that, when

$\lambda = 1$, $\mu = \infty$, the values of S, as K increases, tend to the value $S = 5.48$ determined on the basis of the asymptotic solution for the infinite cylinder. For a fixed value of λ and $\mu \to 1$, the quantity S increases if the cylinder end-walls are fixed; it decreases if the end-walls are free of stress.

In Fig. 3.18 we give the plots of the contact pressure $\varphi(x)$ for some values of $\lambda = R/a$ and $\mu = l/a$, for the both types of boundary conditions (3.383). One can see from this figure that for problem $\mathcal{B}$, when λ is fixed and μ is close to one, there are segments near the endpoints $x = \pm 1$ where the contact pressure falls (i.e., the contact is weakened). To avoid the weakening of the contact, the cylinder length (i.e., μ) should be increased, when the values of the length of the contact region and the cylinder radius are fixed (i.e., λ is fixed). When the values of the lengths of the contact region and of the cylinder are fixed (i.e., μ is fixed), one should diminish the cylinder radius (i.e., the value of λ); this considerably weakens the influence of the end-walls on the distribution of the contact pressure.

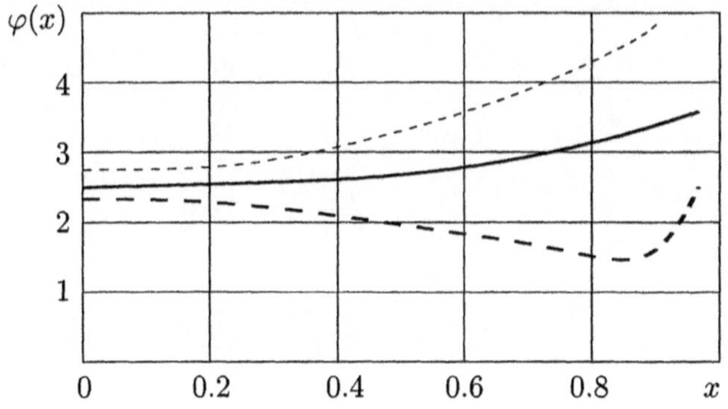

Figure 3.18: The contact pressure $\varphi(x)$ for $\lambda = 1$, $\mu = \infty$ (bold) and $\mu = 1.1$ (dashed for problem $\mathcal{A}$ and bold dashed for problem $\mathcal{B}$)

It is worth mentioning that the integral characteristic S equals the total circular load caused by a given interference between the cylinder and sleeve. In practice, this interference is usually a result of heating and cooling the sleeve. Sometimes it may lead to a collapse of the sleeve due to too great contact load S. Analyzing the result obtained for the two types of the boundary conditions on the end-walls of the finite cylinder, we see that these conditions determine the behaviour of S for a fairly wide sleeve, whose length $2a$ is comparable with cylinder length $2l$. For $l/a = \mu = 1.1$, the stress-free end-walls give value of S less almost by a quarter than the fixed end-walls (see Table 3.11). Therefore, for a contact of a finite cylinder with a wide sleeve, the question of the end-wall boundary conditions is crucial, whereas for a fairly narrow sleeve we can simply use the solution for an infinite cylinder, neglecting the end-wall boundary conditions.

Chapter 4

CONTACT PROBLEMS FOR A WEDGE

There are considerable differences between the first, stress, and the second, displacement, boundary-value problems for a three-dimensional wedge. Ufliand (1965) found the exact solution of the second problem by using the Kontorovoch–Lebedev integral transform on the real axis (see Ditkin and Prudnikov (1965)):

$$f_*(\tau) = \int_0^\infty \frac{f(r)}{r} \mathcal{K}_{i\tau}(r)\, dr,$$

$$f(r) = \frac{2}{\pi^2} \int_0^\infty f_*(\tau) \sinh(\pi\tau)\, \mathcal{K}_{i\tau}(r)\, d\tau.$$

Here, $\mathcal{K}_{i\tau}(r)$ is the modified Bessel function, whose integral representation can be found in Abramowitz and Stegun (1972, formula 9.6.24)

$$\mathcal{K}_{i\tau}(r) = \int_0^\infty \exp(-r \cosh s) \cos(\tau s)\, ds. \tag{4.1}$$

An effective method for solving the first problem, which consists in reducing it to a generalized[1] Hilbert boundary-value problem, was proposed by Ulitko (1979). This method uses the Kontorovich–Lebedev integral trans-

[1]In 1942 I.N. Vekua generalized the Hilbert problem by introducing integral terms with Fredholm kernels in the boundary condition.

form in the complex plane

$$f_*(\mu) = \int_0^\infty \frac{f(r)}{r} \mathcal{K}_\mu(r)\, dr,$$

$$f(r) = \frac{1}{\pi i} \int_{c-i\infty}^{c+i\infty} f_*(\mu)\mu \mathcal{I}_\mu(r)\, d\mu, \tag{4.2}$$

based on the relation 9.6.2 in Abramowitz and Stegun (1972)

$$\sin(\pi\mu)\, \mathcal{K}_\mu(z) = (\pi/2)[\mathcal{I}_{-\mu}(z) - \mathcal{I}_\mu(z)]. \tag{4.3}$$

Here, $\mathcal{I}_\mu(z)$ is the other modified Bessel function. Lubyagin, Pozharskii and Chebakov (1991) developed this method and derived exact formulae, which contain series in powers of $(1 - 2\nu)$ for the displacements and stresses in a three-dimensional wedge, one face of which is subject to the action of normal and tangential concentrated forces, and the other has various boundary conditions. The feasibility of such a representation was predicted by Ufliand (1972). As particular cases, these formulae give the known solutions for the Boussinesq and Cerruti problems for a half-space (Love (1927), Timoshenko and Goodier (1951), Gladwell (1980)). Earlier similar problems were solved in closed form for $\nu = 1/2$ (Ufliand (1972), Efimov, A.B. and Efimov, D.G. (1987), Aleksandrov and Pozharskii (1989)), when the equations for the determination of three harmonic functions in the Papkovich–Neuber representation become uncoupled. For $\nu \neq 1/2$ there are numerous approximate solutions (e.g., Hetenyi (1970), Keer, Lee and Mura (1983), Hanson and Keer (1990)), which are generally constructed only for the special case in which the wedge is a quarter-space. Gerber (1968) was probably the first who examined a 3-D contact problem for a quarter-space. Keer, Lee and Mura (1984) investigated a contact problem for a quarter-space for the case of a rectangular contact area using the finite element method. The Green functions obtained in Lubyagin, Pozharskii and Chebakov (1991) were applied in Lubyagin, Pozharskii and Chebakov (1992), Aleksandrov and Pozharskii (1992, 1994, 1999a), Pozharskii (1994, 1995) to a wedge with an arbitrary aperture angle and for different areas of contact.

In this chapter we investigate three-dimensional contact problems for a wedge for the cases of a strip, wedge-shaped, elliptic or arbitrary contact area.

4.1 Equilibrium under Given Normal Loads

We will investigate problems for a three-dimensional wedge with an aperture angle 2α using cylindrical coordinates r, φ, z, where the z-axis is directed along the edge of the wedge so that the system of coordinates is a right-handed system. As is known (Timoshenko and Goodier (1951)), the solution of Navier's equations of equilibrium (1.3) in coordinates r, φ, z can be expressed by means of three arbitrary harmonic functions $\Phi_n = \Phi_n(r, \varphi, z)$, $n = 0, 1, 2$, in the form

$$u_r = \frac{\partial \Phi_0}{\partial r} + \frac{1}{4(1-\nu)}\frac{\partial}{\partial r}(r\omega_1) - \omega_1, \quad \omega_1 = \sin\varphi\,\Phi_1 - \cos\varphi\,\Phi_2, \tag{4.4}$$

$$u_\varphi = \frac{1}{r}\frac{\partial \Phi_0}{\partial \varphi} + \frac{1}{4(1-\nu)}\frac{\partial \omega_1}{\partial \varphi} - \omega_2, \quad \omega_2 = \cos\varphi\,\Phi_1 + \sin\varphi\,\Phi_2, \tag{4.5}$$

$$u_z = \frac{\partial \Phi_0}{\partial z} + \frac{r}{4(1-\nu)}\frac{\partial \omega_1}{\partial z}. \tag{4.6}$$

With the help of Hooke's law from formulae (4.4)–(4.6) one obtains the relations for the stresses:

$$\frac{\sigma_\varphi}{G} = \frac{2}{r}\frac{\partial \Phi_0}{\partial r} + \frac{2}{r^2}\frac{\partial^2 \Phi_0}{\partial \varphi^2} + \frac{1-2\nu}{2(1-\nu)r}\left(\frac{\partial \omega_2}{\partial \varphi} + \frac{\partial}{\partial r}(r\omega_1)\right) \tag{4.7}$$

$$-\frac{2}{r}\left(\omega_1 + \frac{\partial \omega_2}{\partial \varphi}\right) - \frac{1}{2(1-\nu)r}\frac{\partial}{\partial \varphi}\left(\omega_2 - \frac{\partial \omega_1}{\partial \varphi}\right),$$

$$\frac{\tau_{r\varphi}}{G} = \frac{\partial}{\partial r}\left(\frac{2}{r}\frac{\partial \Phi_0}{\partial \varphi} + \frac{1}{2(1-\nu)}\frac{\partial \omega_1}{\partial \varphi} - \omega_2\right) + \frac{1}{r}\left(\omega_2 - \frac{\partial \omega_1}{\partial \varphi}\right), \tag{4.8}$$

$$\frac{\tau_{\varphi z}}{G} = \frac{\partial}{\partial z}\left(\frac{2}{r}\frac{\partial \Phi_0}{\partial \varphi} + \frac{1}{2(1-\nu)}\frac{\partial \omega_1}{\partial \varphi} - \omega_2\right). \tag{4.9}$$

We will assume for simplicity that boundary-value problems for the wedge are even with respect to the z coordinate and we will search for the harmonic functions Φ_n in the form of the Fourier and Kontorovich–Lebedev integrals in the complex plane:

$$\Phi_n(r, \varphi, z) = \frac{1}{\pi i}\int_0^\infty \cos(\lambda z)\,d\lambda \int_{c-i\infty}^{c+i\infty} [a_n(\mu)\cos(\mu\varphi) + b_n(\mu)\sin(\mu\varphi)]\,\mathcal{I}_\mu(\lambda r)\,d\mu, \tag{4.10}$$

$$|\operatorname{Re}\mu| < c, \quad a_n(-\mu) = -a_n(\mu), \quad b_n(-\mu) = b_n(\mu), \; n = 0, 1, 2.$$

In the following we will use for the modified Bessel function $\mathcal{I}_\mu(r)$ relations 9.6.26 in Abramowitz and Stegun (1972):

$$\frac{\partial}{\partial r}\mathcal{I}_\mu(\lambda r) = \frac{\lambda}{2}[\mathcal{I}_{\mu-1}(\lambda r) + \mathcal{I}_{\mu+1}(\lambda r)], \tag{4.11}$$

$$\frac{\mu}{r}\mathcal{I}_\mu(\lambda r) = \frac{\lambda}{2}[\mathcal{I}_{\mu-1}(\lambda r) - \mathcal{I}_{\mu+1}(\lambda r)], \tag{4.12}$$

and for the two-dimensional Dirac function $\delta_*(x)$ its expansion as Fourier and Kontorovich–Lebedev integrals:

$$r\delta_*(r-x)[\delta_*(z-y) + \delta_*(z+y)] \tag{4.13}$$

$$= \frac{2}{\pi^2 i}\int_0^\infty \cos(\lambda y)\cos(\lambda z)\,d\lambda \int_{c-i\infty}^{c+i\infty} \mu\,\mathcal{K}_\mu(\lambda x)\,\mathcal{I}_\mu(\lambda r)\,d\mu.$$

One-half, $z \le 0$, of the wedge is shown in Fig 4.1.

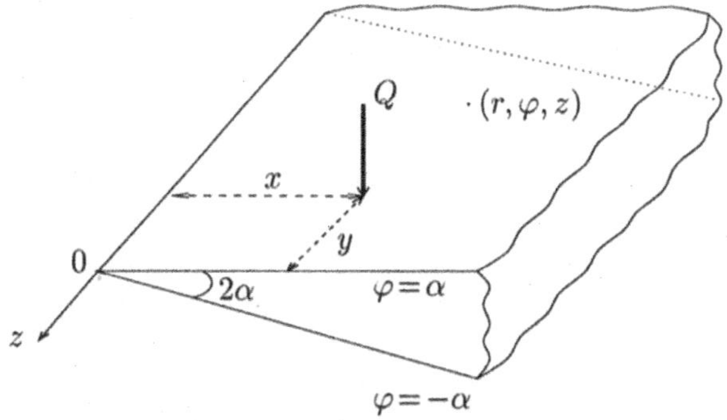

Figure 4.1: A one-half of the elastic wedge under concentrated load Q

4.1.1. ONE FACE OF THE WEDGE IS STRESS-FREE

Consider problem $\mathcal{A}$, where we suppose that one face of an elastic three-dimensional wedge is stress-free and the other face is subject to the action of normal concentrated forces Q at points $r = x$, $z = \pm y$. The boundary conditions of this problem have the form

$$\varphi = -\alpha: \quad \sigma_\varphi = \tau_{r\varphi} = \tau_{\varphi z} = 0, \tag{4.14}$$

$$\varphi = \alpha: \quad \sigma_\varphi = -Q\delta_*(r-x)[\delta_*(z-y)+\delta_*(z+y)], \quad \tau_{r\varphi} = \tau_{\varphi z} = 0. \tag{4.15}$$

Besides, the stresses vanish at infinity.

To satisfy the boundary conditions, we use formulae (4.4)–(4.13) and shift the integration contour with respect to μ in representation (4.10) into the regularity strip of the functions $a_n(\zeta,\beta)$, $b_n(\zeta,\beta)$. For example, in accordance with (4.11), (4.12), we have

$$\int_{c-i\infty}^{c+i\infty} \frac{\partial}{\partial r} a(\mu)\, \mathcal{I}_\mu(\lambda r)\, d\mu = \frac{\lambda}{2} \int_{c-i\infty}^{c+i\infty} [a(\mu+1) + a(\mu-1)]\, \mathcal{I}_\mu(\lambda r)\, d\mu,$$

$$\int_{c-i\infty}^{c+i\infty} \frac{\mu}{r} a(\mu)\, \mathcal{I}_\mu(\lambda r)\, d\mu = \frac{\lambda}{2} \int_{c-i\infty}^{c+i\infty} [a(\mu+1) - a(\mu-1)]\, \mathcal{I}_\mu(\lambda r)\, d\mu.$$

As a result, we obtain a system of six equations for six unknown functions $A_n(\mu)$, $B_n(\mu)$, $n = 0, 1, 2$, which are connected with the functions $a_n(\mu)$, $b_n(\mu)$ by the formulae

$$A_0(\mu) = a_0(\mu)\sin(\mu\alpha), \quad B_0(\mu) = b_0(\mu)\cos(\mu\alpha), \tag{4.16}$$

$$A_1(\mu) = [b_1(\mu) - a_2(\mu)]\sin[(\mu-1)\alpha], \quad B_1(\mu) = [b_1(\mu) + a_2(\mu)]\sin[(\mu+1)\alpha], \tag{4.17}$$

$$A_2(\mu) = [b_2(\mu) + a_1(\mu)]\cos[(\mu-1)\alpha], \quad B_2(\mu) = [b_2(\mu) - a_1(\mu)]\cos[(\mu+1)\alpha]. \tag{4.18}$$

Thus, this system has the form

$$\lambda[A_0(\mu+1) - A_0(\mu-1)] = -\frac{1-2\nu}{2(1-\nu)} A_1(\mu), \quad A_1(\mu) = B_1(\mu), \tag{4.19}$$

$$\lambda[B_0(\mu+1) - B_0(\mu-1)] = \frac{1-2\nu}{2(1-\nu)} A_2(\mu), \quad A_2(\mu) = -B_2(\mu), \tag{4.20}$$

$$\begin{aligned} &\cot(\mu\alpha)\left\{\frac{\lambda}{8(1-\nu)}[(\mu+2)A_1(\mu+1) + (\mu-2)A_1(\mu-1)] - \lambda^2 A_0(\mu)\right\} \\ &+ \cot[(\mu+2)\alpha]\left\{\frac{\lambda}{2}\left(1 - \frac{\mu+2}{4(1-\nu)}\right) A_1(\mu+1) + \frac{\lambda^2}{2} A_0(\mu+2)\right\} \\ &- \cot[(\mu-2)\alpha]\left\{\frac{\lambda}{2}\left(1 + \frac{\mu-2}{4(1-\nu)}\right) A_1(\mu-1) - \frac{\lambda^2}{2} A_0(\mu-2)\right\} = \Pi(\mu), \end{aligned} \tag{4.21}$$

$$\tan(\mu\alpha)\left\{\frac{\lambda}{8(1-\nu)}[(\mu+2)A_2(\mu+1)+(\mu-2)A_2(\mu-1)]+\lambda^2 B_0(\mu)\right\}$$

$$+\tan[(\mu+2)\alpha]\left\{\frac{\lambda}{2}\left(1-\frac{\mu+2}{4(1-\nu)}\right)A_2(\mu+1)-\frac{\lambda^2}{2}B_0(\mu+2)\right\} \tag{4.22}$$

$$-\tan[(\mu-2)\alpha]\left\{\frac{\lambda}{2}\left(1+\frac{\mu-2}{4(1-\nu)}\right)A_2(\mu-1)+\frac{\lambda^2}{2}B_0(\mu-2)\right\}=-\Pi(\mu),$$

where

$$\Pi(\mu)=Q\cos(\lambda y)\,\mu\,\mathcal{K}_\mu(\lambda x)(\pi G x)^{-1}.$$

We use the first four equations, (4.19), (4.20), to express $A_n(\mu)$, $B_n(\mu)$, $n=1,2$, in terms of $A_0(\mu)$, $B_0(\mu)$, and substitute equations (4.21), (4.22) to obtain equations for $A_0(\mu)$ and $B_0(\mu)$:

$$\alpha_1(\mu)A_0(\mu-2)+\gamma_1(\mu)A_0(\mu)+\beta_1(\mu)A_0(\mu+2)=2\lambda^{-2}\,\Pi(\mu), \tag{4.23}$$

$$\alpha_2(\mu)B_0(\mu-2)+\gamma_2(\mu)B_0(\mu)+\beta_2(\mu)B_0(\mu+2)=-2\lambda^{-2}\,\Pi(\mu), \tag{4.24}$$

where

$$\alpha_1(\mu)=\frac{(\mu-2)\cot(\mu\alpha)-\mu\cot[(\mu-2)\alpha]}{2(1-2\nu)},\qquad \beta_1(\mu)=\alpha_1(\mu+2),$$

$$\gamma_1(\mu)=-\alpha_1(\mu)-\beta_1(\mu)-\frac{4\cot(\mu\alpha)\sin^2(2\alpha)}{\cos(2\mu\alpha)-\cos(4\alpha)},$$

$$\alpha_2(\mu)=\frac{-(\mu-2)\tan(\mu\alpha)+\mu\tan[(\mu-2)\alpha]}{2(1-2\nu)},\qquad \beta_2(\mu)=\alpha_2(\mu+2),$$

$$\gamma_2(\mu)=-\alpha_2(\mu)-\beta_2(\mu)-\frac{4\tan(\mu\alpha)\sin^2(2\alpha)}{\cos(2\mu\alpha)+\cos(4\alpha)}.$$

We introduce new functions $F_n(\mu)$, $n=1,2$, by the formulae

$$F_1(\mu)=\alpha_1(\mu+1)[A_0(\mu-1)-A_0(\mu+1)], \tag{4.25}$$

$$F_2(\mu)=\alpha_2(\mu+1)[B_0(\mu-1)-B_0(\mu+1)] \tag{4.26}$$

and write the functional equations (4.23), (4.24) in the form

$$F_1(\mu-1)-F_1(\mu+1)=\frac{4\cot(\mu\alpha)\sin^2(2\alpha)}{\cos(2\mu\alpha)-\cos(4\alpha)}A_0(\mu)+\frac{2}{\lambda^2}\Pi(\mu), \tag{4.27}$$

$$F_2(\mu-1)-F_2(\mu+1)=\frac{4\tan(\mu\alpha)\sin^2(2\alpha)}{\cos(2\mu\alpha)+\cos(4\alpha)}B_0(\mu)-\frac{2}{\lambda^2}\Pi(\mu). \tag{4.28}$$

Taking the evenness of the functions $A_0(\mu), B_0(\mu), F_{1,2}(\mu)$ and the oddness of the function $\Pi(\mu)$ into account, we can represent equations (4.27), (4.28) on the imaginary $\mu = i\tau$ axis in the form

$$\operatorname{Im} F_1(1+i\tau) = \frac{2\coth(\alpha\tau)\sin^2(2\alpha)}{\cosh(2\alpha\tau)-\cos(4\alpha)} \operatorname{Re} A_0(i\tau) - \frac{1}{\lambda^2} \operatorname{Im}\Pi(i\tau), \quad (4.29)$$

$$\operatorname{Im} F_2(1+i\tau) = \frac{-2\tanh(\alpha\tau)\sin^2(2\alpha)}{\cosh(2\alpha\tau)+\cos(4\alpha)} \operatorname{Re} B_0(i\tau) + \frac{1}{\lambda^2} \operatorname{Im}\Pi(i\tau). \quad (4.30)$$

Using the Schwarz type formula (Lawrentjew and Schabat (1967))

$$\operatorname{Re} F_{1,2}(i\tau) = \int_0^\infty \frac{\sinh(\pi t) \operatorname{Im} F_{1,2}(1+it)}{\cosh(\pi t)+\cosh(\pi\tau)}\, dt, \quad (4.31)$$

we rewrite equations (4.29), (4.30) as follows:

$$\operatorname{Re} F_1(i\tau) = -\int_0^\infty \frac{\sinh(\pi t) \operatorname{Im}\Pi(it)}{\lambda^2\,[\cosh(\pi t)+\cosh(\pi\tau)]}\, dt \quad (4.32)$$

$$+2\sin^2(2\alpha)\int_0^\infty \frac{\sinh(\pi t)\coth(\alpha t) \operatorname{Re} A_0(it)\, dt}{[\cosh(\pi t)+\cosh(\pi\tau)][\cosh(2\alpha t)-\cos(4\alpha)]},$$

$$\operatorname{Re} F_2(i\tau) = \int_0^\infty \frac{\sinh(\pi t) \operatorname{Im}\Pi(it)}{\lambda^2\,[\cosh(\pi t)+\cosh(\pi\tau)]}\, dt \quad (4.33)$$

$$-2\sin^2(2\alpha)\int_0^\infty \frac{\sinh(\pi t)\,\tanh(\alpha t) \operatorname{Re} B_0(it)\, dt}{[\cosh(\pi t)+\cosh(\pi\tau)][\cosh(2\alpha t)+\cos(4\alpha)]}.$$

If we put $\mu = i\tau$ ($\tau = \operatorname{Im}\mu$) in formulae (4.25), (4.26), we have

$$\operatorname{Im} A_0(1+i\tau) = \frac{1-2\nu}{2}\,\frac{\cosh(2\alpha\tau)-\cos(2\alpha)}{\sinh(2\alpha\tau)+\tau\sin(2\alpha)} \operatorname{Re} F_1(i\tau), \quad (4.34)$$

$$\operatorname{Im} B_0(1+i\tau) = \frac{1-2\nu}{2}\,\frac{\cosh(2\alpha\tau)+\cos(2\alpha)}{\sinh(2\alpha\tau)-\tau\sin(2\alpha)} \operatorname{Re} F_2(i\tau). \quad (4.35)$$

Now, with the help of formulae (4.32)–(4.35) we obtain Fredholm integral equations of the second kind in new auxiliary functions $\Phi_n(t)$, $n = 1, 2$, which are defined by formulae

$$\frac{1-2\nu}{2\lambda}\,\frac{\cos(\lambda y)}{\pi G}\,\Phi_1(t) = \cosh\frac{\pi t}{2}\,\frac{\sinh(2\alpha t)+t\sin(2\alpha)}{\cosh(2\alpha t)-\cos(2\alpha)} \operatorname{Im} A_0(1+it), \quad (4.36)$$

$$\frac{1-2\nu}{2\lambda}\frac{\cos(\lambda y)}{\pi G}\Phi_2(t) = -\cosh\frac{\pi t}{2}\frac{\sinh(2\alpha t) - t\sin(2\alpha)}{\cosh(2\alpha t)+\cos(2\alpha)}\operatorname{Im} B_0(1+it). \tag{4.37}$$

To do this we at first express the functions $\operatorname{Re} A_0(it)$ and $\operatorname{Re} B_0(it)$ on the right-hand sides in formulae (4.32), (4.33) through the functions $\operatorname{Im} A_0(1+it)$ and $\operatorname{Im} B_0(1+it)$, respectively, using the Schwarz type formulae, which are similar to the formula (4.31). Then we replace the functions on the right-hand sides in formulae (4.34), (4.35) by the right-hand sides of formulae (4.32), (4.33). Taking formulae (4.36), (4.37) into account, we obtain integral equations in the form ($0 \le t < \infty$, $\beta x > 0$, $n = 1, 2$)

$$\Phi_n(t) - (1-2\nu)\int_0^\infty L_n(t,y)\,\Phi_n(y)\,dy = -Q\cosh\frac{\pi t}{2}\,\mathcal{K}_{it}(\beta x), \tag{4.38}$$

where

$$L_n(t,y) = 2\cosh\frac{\pi t}{2}\,\sinh\frac{\pi y}{2}\,W_n(y,\alpha) \tag{4.39}$$

$$\times\int_0^\infty \frac{\sinh(\pi u)\,g_n(u,\alpha)\,du}{[\cosh(\pi u)+\cosh(\pi t)][\cosh(\pi u)+\cosh(\pi y)]},$$

$$W_1(y,\alpha) = \frac{\cosh(2\alpha y)-\cos(2\alpha)}{\sinh(2\alpha y)+y\sin(2\alpha)},\quad W_2(y,\alpha) = -\frac{\cosh(2\alpha y)+\cos(2\alpha)}{\sinh(2\alpha y)-y\sin(2\alpha)}, \tag{4.40}$$

$$g_1(u,\alpha) = \frac{\coth(\alpha u)\,\sin^2(2\alpha)}{\cosh(2\alpha u)-\cos(4\alpha)},\quad g_2(u,\alpha) = \frac{\tanh(\alpha u)\,\sin^2(2\alpha)}{\cosh(2\alpha u)+\cos(4\alpha)}. \tag{4.41}$$

In the deriving the right-hand side of (4.38), we have used the integral

$$\int_0^\infty \frac{u\,\sinh(\pi u)\,\mathcal{K}_{iu}(x)}{\cosh(\pi u)+\cosh(\pi t)}\,du = x\,\mathcal{K}_{it}(x).$$

One can easily show that all the desired functions of the form (4.16)–(4.18) can be expressed in terms of the solutions of the integral equations (4.38).

4.1.2. OTHER TYPES OF BOUNDARY CONDITIONS

Now we consider the cases when one wedge face rests without friction on a rigid base (problem $\mathcal{B}$) or is rigidly clamped (problem $\mathcal{C}$). For problem $\mathcal{B}$ the boundary conditions take the form

$$\varphi = 0: \quad u_\varphi = \tau_{r\varphi} = \tau_{\varphi z} = 0. \tag{4.42}$$

$$\varphi = 2\alpha: \quad \sigma_\varphi = -Q\delta_*(r-x)[\delta_*(z-y)+\delta_*(z+y)], \quad \tau_{r\varphi} = \tau_{\varphi z} = 0.$$

Repeating the previous formulae, we see that in this case we must put $A_2(\mu) \equiv B_2(\mu) \equiv B_0(\mu) \equiv 0$ in formulae (4.16)–(4.18). For the function $A_0(\mu)$ we obtain a Fredholm integral equation of the second kind in an auxiliary function $\Phi_3(t)$, which has the form (4.38), for $n = 3$ and with doubled right-hand side, and (4.39) where $W_3(y, \alpha) = W_1(y, 2\alpha)$, $g_3(u, \alpha) = g_1(u, 2\alpha)$. The relation between the functions $\Phi_3(t)$ and $A_0(\mu)$ is the same as that between the functions $\Phi_1(t)$ and $A_0(\mu)$ (see (4.36)).

Now, consider problem C for the wedge, one side of which is rigidly fixed with the boundary conditions:

$$\varphi = 0: \quad u_r = u_\varphi = u_z = 0, \tag{4.43}$$

$$\varphi = 2\alpha: \quad \sigma_\varphi = -Q\delta_*(r-x)[\delta_*(z-y)+\delta_*(z+y)], \quad \tau_{r\varphi} = \tau_{\varphi z} = 0. \tag{4.44}$$

It is convenient to reformulate boundary conditions (4.43) for the functions $(\partial u_r/\partial z - \partial u_z/\partial r)$, u_φ, u_z. Using formulae (4.4)–(4.6), (4.10) and the condition $\partial u_r/\partial z - \partial u_z/\partial r = 0$, we get $a_2(\mu) \equiv 0$, and with the help of the condition $u_z = 0$ we conclude that $a_0(\mu) \equiv 0$. Using the other boundary conditions (4.43), (4.44), we derive a system of four equations for $a_1(\mu)$, $b_n(\mu)$, $n = 0, 1, 2$, in the form

$$a_1(\mu) = \frac{4(1-\nu)}{4\nu-3}\left[\frac{\mu\, b_2(\mu)}{4(1-\nu)} - \frac{\lambda}{2}(b_0(\mu+1) - b_0(\mu-1))\right], \tag{4.45}$$

$$\begin{gathered} b_1(\mu)\sin[2(\mu+1)\alpha] - (b_2(\mu)+a_1(\mu))\cos[2(\mu-1)\alpha] \\ = b_1(\mu)\sin[2(\mu-1)\alpha] + (b_2(\mu)-a_1(\mu))\cos[2(\mu+1)\alpha], \end{gathered} \tag{4.46}$$

$$\begin{gathered} \left(\frac{\mu+1}{2(1-\nu)} - 1\right) b_1(\mu)\sin[2(\mu+1)\alpha] \\ - \left(\frac{\mu-1}{2(1-\nu)} + 1\right) b_1(\mu)\sin[2(\mu-1)\alpha] \\ + \left[2\lambda b_0(\mu+1) - \left(\frac{\mu+1}{2(1-\nu)} - 1\right)(b_2(\mu)-a_1(\mu))\right]\cos[2(\mu+1)\alpha] \\ - \left[2\lambda b_0(\mu-1) + \left(\frac{\mu-1}{2(1-\nu)} + 1\right)(b_2(\mu)+a_1(\mu))\right]\cos[2(\mu-1)\alpha] = 0, \end{gathered} \tag{4.47}$$

$$\begin{gathered} \frac{\cos(2\mu\alpha)}{4(1-\nu)}[2(b_1(\mu+1) - b_1(\mu-1)) + \mu(b_1(\mu+1) + b_1(\mu-1))] \\ + \cos[2(\mu+2)\alpha]\left(1 - \frac{\mu+2}{4(1-\nu)}\right) b_1(\mu+1) \end{gathered} \tag{4.48}$$

$$-\cos[2(\mu-2)\alpha]\left(1+\frac{\mu-2}{4(1-\nu)}\right)b_1(\mu-1)$$

$$-\sin(2\mu\alpha)\left[\frac{1}{2(1-\nu)}\left(b_2(\mu+1)+b_2(\mu-1)+a_1(\mu+1)-a_1(\mu-1)\right)\right.$$

$$\left.+\frac{\mu}{4(1-\nu)}\left(b_2(\mu+1)-b_2(\mu-1)+a_1(\mu+1)+a_1(\mu-1)\right)+2\lambda\, b_0(\mu)\right]$$

$$+\sin[2(\mu+2)\alpha]\left[\left(1-\frac{\mu+2}{4(1-\nu)}\right)\left(b_2(\mu+1)-a_1(\mu+1)\right)+\lambda\, b_0(\mu+2)\right]$$

$$+\sin[2(\mu-2)\alpha]\left[\left(1+\frac{\mu-2}{4(1-\nu)}\right)\left(b_2(\mu-1)-a_1(\mu-1)\right)+\lambda\, b_0(\mu-2)\right]$$

$$=4\Pi(\mu)/\lambda.$$

We use the first three equations, (4.45)–(4.47), to express the functions $a_1(\mu), b_{1,2}(\mu)$ in terms of the function $b_0(\mu)$, and then we substitute in the fourth equation (4.48). After that, we obtain an equation which can be written in terms of a new function $B(\mu) = b_0(\mu)\cos(2\mu\alpha)$ in the form

$$\alpha(\mu)B(\mu-2)+\gamma(\mu)B(\mu)+\beta(\mu)B(\mu+2)=\Pi(\mu)/\lambda^2, \tag{4.49}$$

where ($\kappa = 3-4\nu$)

$$\alpha(\mu)=\left[\frac{(\mu-2\nu)(\mu-1)}{2(1-\nu)\kappa}\,\frac{\sin(2\alpha)}{\cos[2(\mu-1)\alpha]}-\frac{\cos[2(\mu-1)\alpha]}{\sin(2\alpha)}\right]$$

$$\times\frac{\kappa(1-\nu)\sin(2\alpha)\,\cos[2(\mu-1)\alpha]}{(1-2\nu)(\kappa\sin[4(\mu-1)\alpha]-(\mu-1)\sin(4\alpha))}$$

$$\times\left[\frac{1-2\nu}{\kappa}\,\frac{\cos(2\alpha)}{\cos[2(\mu-1)\alpha]\,\cos[2(\mu-2)\alpha]}-1\right]+\frac{1-\nu}{2\kappa}\tan[2(\mu-1)\alpha]$$

$$+\frac{(\mu-2)\sin(2\alpha)}{4\kappa\cos[2(\mu-1)\alpha]\,\cos[2(\mu-2)\alpha]}+\frac{1-2\nu}{4\kappa}\tan[2(\mu-2)\alpha],$$

$$\beta(\mu)=\alpha(\mu+2),\quad \gamma(\mu)=\gamma_*(\mu)-\alpha(\mu)-\beta(\mu),$$

$$\gamma_*(\mu)=\frac{(\mu+2)\,\sin^2(2\alpha)\,\sin^2[2(\mu+1)\alpha]}{(\kappa\sin[4(\mu+1)\alpha]-(\mu+1)\sin(4\alpha))\,\cos(2\mu\alpha)\,\cos[2(\mu+2)\alpha]}$$

$$-\frac{(\mu-2)\,\sin^2(2\alpha)\,\sin^2[2(\mu-1)\alpha]}{(\kappa\sin[4(\mu-1)\alpha]-(\mu-1)\sin(4\alpha))\,\cos(2\mu\alpha)\,\cos[2(\mu-2)\alpha]}$$

$$+2(1-\nu)\left\{\frac{\sin(2\alpha)\,\sin[2(\mu-1)\alpha]}{(\kappa\sin[4(\mu-1)\alpha]-(\mu-1)\sin(4\alpha))\,\cos[2(\mu-2)\alpha]}\right.$$

$$-\frac{\sin(2\alpha)\sin[2(\mu+1)\alpha]}{(\kappa\sin[4(\mu+1)\alpha]-(\mu+1)\sin(4\alpha))\cos[2(\mu+2)\alpha]}\Bigg\}$$

$$+\frac{\tan(2\mu\alpha)\sin^2(2\alpha)}{\cos(4\mu\alpha)+\cos(8\alpha)}.$$

Introducing a new analytical function

$$F(\mu)=\alpha(\mu+1)(B(\mu-1)-B(\mu+1)), \tag{4.50}$$

we rewrite equation (4.49) as follows

$$F(\mu-1)-F(\mu+1)=\lambda^{-2}\Pi(\mu)-\gamma_*(\mu)B(\mu). \tag{4.51}$$

Due to the evenness of the function $F(\mu)$ on the imaginary axis in the plane of the complex argument $\mu=\sigma+i\tau$, i.e. when $\mu=i\tau$, from (4.51) we obtain the relation

$$\operatorname{Im}F(1+i\tau)=-\frac{1}{2\lambda^2}\operatorname{Im}\Pi(i\tau)+\frac{1}{2}\operatorname{Im}\gamma_*(i\tau)\operatorname{Re}B(i\tau). \tag{4.52}$$

In a similar way, from formula (4.50) when $\mu=i\tau$ we have

$$\operatorname{Re}F(i\tau)=2\operatorname{Im}\alpha(1+i\tau)\operatorname{Im}B(1+i\tau). \tag{4.53}$$

We use relation (4.52) and Schwarz type formula (4.31) to obtain

$$\operatorname{Re}F(i\tau)=-\frac{1}{2\lambda^2}\int_0^\infty\frac{\sinh(\pi t)\operatorname{Im}\Pi(it)}{\cosh(\pi t)+\cosh(\pi\tau)}\,dt \tag{4.54}$$

$$+\frac{1}{2}\int_0^\infty\frac{\sinh(\pi t)\operatorname{Im}\gamma_*(it)\operatorname{Re}B(it)}{\cosh(\pi t)+\cosh(\pi\tau)}\,dt.$$

There is also a Schwarz type formula for the function $B(\mu)$:

$$\operatorname{Re}B(i\tau)=\int_0^\infty\frac{\sinh(\pi t)\operatorname{Im}B(1+it)}{\cosh(\pi t)+\cosh(\pi\tau)}\,dt. \tag{4.55}$$

Now, we substitute relation (4.55) for the function $\operatorname{Re}B(i\tau)$ into the right-hand side of equality (4.54). Eliminating the function $\operatorname{Re}F(i\tau)$ from formulae (4.53) and (4.54), we obtain a Fredholm integral equation of the second kind for the function $\Phi_4(t)$, which is related with the function $B(\mu)$ by the formula

$$\frac{\cos(\lambda y)}{\lambda\pi G}\,\Phi_4(t) = 4\cosh\frac{\pi t}{2}\,\mathrm{Im}\,\alpha(1+it)\,\mathrm{Im}\,B(1+it).$$

This integral equation has the form (4.38), (4.39) with $n = 4$ and

$$W_4(y,\alpha) = \frac{2\kappa\,\sinh(4\alpha y) - 2y\sin(4\alpha)}{2\kappa\,\cosh(4\alpha y) + 2y^2 - 2y^2\cos(4\alpha) + \kappa^2 + 1}, \tag{4.56}$$

$$\begin{aligned} g_4(u,\alpha) = -g_2(u,2\alpha) + \{\sin^2(2\alpha)\,(f_0(u)[2f_1(u) - uf_2(u)] \\ -f_3(u)[2f_2(u) + uf_1(u)]) - 2(1-\nu)\sin(2\alpha)(f_0(u)[\sin(6\alpha) \\ -\sin(2\alpha)\cosh(4\alpha u)] - f_3(u)\cos(2\alpha)\sinh(4\alpha u))\}/f_4(u), \end{aligned} \tag{4.57}$$

$$f_0(u) = \kappa\sinh(4\alpha u)\cos(4\alpha) - u\sin(4\alpha), \tag{4.58}$$

$$f_1(u) = \cos(4\alpha) + \sin^2(4\alpha) - \cosh(4\alpha u), \tag{4.59}$$

$$f_2(u) = \sin(4\alpha)\tanh(2\alpha u)[1 + \cos(4\alpha)], \tag{4.60}$$

$$f_3(u) = [\kappa\cosh(4\alpha u) - 1]\sin(4\alpha), \tag{4.61}$$

$$f_4(u) = [f_0^2(u) + f_3^2(u)][\sinh^2(2\alpha u) + \cos^2(4\alpha)]. \tag{4.62}$$

All the stresses and displacements in a 3-D wedge, one face of which is rigidly fixed, can be expressed in terms of the function $\Phi_4(t)$.

4.1.3. ANALYSIS OF THE FREDHOLM EQUATIONS

Let us analyze the integral equations. The kernel (4.39) of the integral equation (4.38), $n = 1,2,3,4$, depends on the boundary conditions on one wedge face, and the right-hand side of this equation depends on the load acting on the other face. For $\nu = 1/2$ or $2\alpha = \pi$ (half-space) the integral equation (4.38) reduces to an explicit formula for the function $\Phi_n(t)$, $n = 1,2,3,4$, but in these cases the problem can be solved by simple methods. Let us investigate the applicability of the successive approximation method (Cochran (1972)) for solving equation (4.38) in the space $C_M(0,\infty)$ of the functions continuous and bounded on the semiaxis. The right-hand side of (4.38) belongs to $C_M(0,\infty)$ due to the asymptotic estimate (Lebedev (1965))

$$K_{it}(x) \approx \left(\frac{2\pi}{t}\right)^{1/2}\exp(-\pi t/2)\sin\left(\frac{\pi}{4} + t\ln t - t - t\ln\frac{x}{2}\right), \quad t\to\infty. \tag{4.63}$$

Farther on, we will use the uniform convergence of the Neumann functional series in powers of $(1-2\nu)$, which represent the solutions of equations (4.38), $n = 1, 2, 3, 4$, e.g. to justify term-by-term integration of these series.

We recall that $C_M(0,\infty)$ is a Banach space and the integral operator in (4.38) is continuous linear operator acting from $C_M(0,\infty)$ into $C_M(0,\infty)$. The norm $\|U\|$ of this operator is found by formula

$$\|U\| = \sup_{0 \le u < \infty} (1-2\nu) \int_0^\infty |L_n(u,y)|\, dy,$$

where we write the modulus inside the integral because the function $L_n(u,y)$, $n = 1, 2, 3, 4$, retains one sign when $0 \le u, y < \infty$.

Proposition 4.1. *For all $\alpha \in [0,\pi], u > 0, y > 0$ we have*

$$W_3(y,\alpha) g_3(u,\alpha) \le (1+y^2)/(2uy). \tag{4.64}$$

The inequality (4.64) follows from the inequalities $g_3(u,\alpha) \le (4u\alpha)^{-1}$, $W_3(y,\alpha) \le 2\alpha(1+y^2)/y$; these, in turn, arise from the inequalities

$$\alpha t \le \sinh(\alpha t), \quad 2\sin^2(2\alpha)\cosh(\alpha t) \le \cosh(2\alpha t) - \cos(4\alpha),$$

$$\frac{\alpha^3}{3!} + \sin\alpha \ge 0, \quad \frac{\alpha^2}{2} + \frac{\alpha \sin\alpha}{2} \ge 1 - \cos\alpha.$$

From this proposition and the estimate

$$\sup_{0 \le u < \infty} \cosh\frac{\pi u}{2} \int_0^\infty\int_0^\infty \frac{\sinh(\pi y/2)\sinh(\pi t)\,(1+y^2)\,dy\,dt}{ty[\cosh(\pi t)+\cosh(\pi u)][\cosh(\pi t)+\cosh(\pi y)]} \le 1.118, \tag{4.65}$$

it follows that, for problem $\mathcal{B}$, and for all α, the corresponding integral operator on the left-hand side of equation (4.38) is a contractive operator in the space $C_M(0,\infty)$, at least when $\nu > 0.053$.

Proposition 4.2. *For any $u \ge 0$ we have*

$$\frac{\cosh(\pi u/2)}{\cosh(\pi t)+\cosh(\pi u)} \le \begin{cases} 1/[\cosh(\pi t)+1] & \text{for } 0 \le t \le \ln(3+2\sqrt{2})/\pi \\ 1/[4\sinh(\pi t/2)] & \text{for } t \ge \ln(3+2\sqrt{2})/\pi \end{cases} \tag{4.66}$$

Inequality (4.66), which is not difficult to prove, allows us to obtain quite acceptable estimates for the norm of the integral operator in (4.38). Calculations show that usually these estimates give sufficiently wide intervals of the values of Poisson's ratio ν ($\nu_* < \nu \le 1/2$), in which the successive approximation process can be used for solving equation (4.38). The norm estimates of the integral operators

$$\mathcal{A}_n[u, f(y)] = \int_0^\infty L_n(u, y)\, f(y)\, dy, \qquad n = 1, 2, \tag{4.67}$$

for problem $\mathcal{A}$ and of the integral operator $(1-2\nu)\mathcal{A}_4[u, f(y)]$ for problem $\mathcal{C}$, where $\mathcal{A}_4[u, f(y)]$ has the form (4.67), are given in Table 4.1 for the angles $2\alpha = \pi k/4$, $k = 1, 2, ..., 7$, denoted as I_+ $(n = 1)$, I_- $(n = 2)$ and I_0 $(\nu = 0.3)$, respectively.

Table 4.1: Estimates of norms

k	1	2	3	4	5	6	7
I_+	0.669	0.516	0.309	0	0.135	0.103	0.0522
I_-	1.68	1.44	0.0524	0	0.0224	0.0909	0.0166
I_0	0.0417	0.0356	0.0103	0	0.000229	0.00656	0.00171

For a fixed angle α in problems $\mathcal{A}$ and $\mathcal{C}$ the corresponding Neumann series converge, as a rule, for all values of Poisson's ratio of practical importance. However, for problem $\mathcal{A}$ there exist values of α and ν, for which the method of successive approximation is not applicable for solving equation (4.38) with $n = 2$. For example, when $2\alpha = \pi/2$ (quarter-space) using the integral 3.264.2 in Gradshteyn and Ryzhik (1994)

$$4\int_0^\infty \frac{\cosh(\pi t/2)\, dt}{[\cosh(\pi t) + \cosh(\pi u)][\cosh(\pi t) + \cosh(\pi y)]}$$

$$= \left[\cosh\frac{\pi u}{2}\cosh\frac{\pi y}{2}\left(\cosh\frac{\pi u}{2} + \cosh\frac{\pi y}{2}\right)\right]^{-1},$$

we find that operators (4.67) can be represented in the form $(n = 1, 2)$

$$\mathcal{A}_n[u, f(y)] = \int_0^\infty \frac{\pm f(y)\sinh(\pi y/2)\, dy}{[\sinh(\pi y/2) \pm y][\cosh(\pi u/2) + \cosh(\pi y/2)]}.$$

Here and below, the upper sign corresponds to $n = 1$ while the lower sign corresponds to $n = 2$. The norms of these operators in the space $C_M(0, \infty)$ are attained at $u = 0$. Since

$$\int_0^\infty \frac{\sinh(\pi y/2)\, dy}{[\sinh(\pi y/2) \pm y][1 + \cosh(\pi y/2)]} \approx \begin{cases} 0.440 & \text{for } n = 1 \\ 1.30 & \text{for } n = 2 \end{cases},$$

then for the quarter-space, one face of which is stress-free, the corresponding Neumann series converge under the condition $(1-2\nu)1.30 < 1$ or $\nu > 0.116$.

It is interesting that for problem $\mathcal{A}$, when $\alpha \to 0$, the method of succesive approximations is applicable for $\nu > 0.092$. To prove this, it is necessary to put α zero in the integral operator (4.67) for $n = 2$ and to use the estimate

$$\sup_{0\le u<\infty} \cosh\frac{\pi u}{2} \int_0^\infty\int_0^\infty \frac{t\,\sinh(\pi t)}{y(1+y^2)[\cosh(\pi t)+\cosh(\pi u)]}$$

$$\times \frac{\sinh(\pi y/2)\,dt\,dy}{[\cosh(\pi t)+\cosh(\pi y)]} < 1.225.$$

Representing the solutions of the integral equations (4.38) by the Neumann series, one can obtain explicit analytical expressions for the stresses and displacements in the 3-D wedge. Let us write the expression for the normal displacement on the face of the wedge, where the normal concentrated forces Q act at the points $r = x$, $z = \pm y$:

$$u_\varphi = -\frac{4Q}{\pi^3\,\theta} \int_0^\infty\int_0^\infty \sinh\frac{\pi u}{2}\, W(u,\alpha,\beta x)\mathcal{K}_{iu}(\beta r)\cos(\beta y)\cos(\beta z)\,d\beta\,du, \tag{4.68}$$

where

$$W(u,\alpha,\beta x) = W_1(u,\alpha)\,\mathcal{B}_1\left[u, \cosh\frac{\pi y}{2}\,\mathcal{K}_{iy}(\beta x)\right] \tag{4.69}$$

$$-W_2(u,\alpha)\,\mathcal{B}_2\left[u, \cosh\frac{\pi y}{2}\,\mathcal{K}_{iy}(\beta x)\right]$$

for problem $\mathcal{A}$ and

$$W(u,\alpha,\beta x) = 2W_n(u,\alpha)\,\mathcal{B}_n\left[u, \cosh\frac{\pi y}{2}\,\mathcal{K}_{iy}(\beta x)\right] \tag{4.70}$$

for problems $\mathcal{B}$ ($n = 3$) and $\mathcal{C}$ ($n = 4$). In formulae (4.69) and (4.70) we have introduced the notation

$$\mathcal{B}_n = \sum_{m=0}^{\infty} (1-2\nu)^m\,\mathcal{A}_n^m, \quad n = 1,2,3,4, \quad 0 \le \nu_* \le \nu \le 1/2.$$

Here the integral operators $\mathcal{A}_n$ are defined by formula (4.67) and the value of ν_* is defined by the norm of the corresponding integral operator.

In deriving (4.68) we have used the fact that, for the function $G(\mu)$, which is analytic in the band $|\operatorname{Re}\mu| \leq c$, and even, we have the integral (see formula (4.3))

$$\int_{c-i\infty}^{c+i\infty} G(\mu)\,\mathcal{I}_\mu(\lambda r)\,d\mu = \frac{2}{\pi}\int_0^\infty \sinh(\pi\tau)\,G(i\tau)\,\mathcal{K}_{i\tau}(\lambda r)\,d\tau. \tag{4.71}$$

For problem $\mathcal{A}$ with $2\alpha = \pi$, using the integrals 2.16.14.4 (for $\nu = 0$) and 2.16.52.6 in Prudnikov, Brychkov and Marichev (1992, Vol. 2)

$$\frac{4}{\pi^2}\int_0^\infty\int_0^\infty \cosh(\pi u)\,\mathcal{K}_{iu}(\beta x)\,\mathcal{K}_{iu}(\beta r)\,\cos(\beta z)\,d\beta\,du = \frac{1}{\sqrt{(r-x)^2+z^2}} \tag{4.72}$$

and taking the evenness of the problem with respect to z into account, we see that formulae (4.68) and (4.69) give the known Green function (Love (1927), Gladwell (1980))

$$u_\varphi = -\frac{Q}{2\pi\theta}\left(\frac{1}{R_+} + \frac{1}{R_-}\right), \quad R_\pm^2 = (r-x)^2 + (z \pm y)^2.$$

For $\nu = 1/2$ formulae (4.68)–(4.70) give the Green functions for an incompressible wedge in Aleksandrov and Pozharskii (1989). If we multiply the right-hand side of (4.68) by $\cos(\varepsilon z)$, integrate it with respect to z from 0 to ∞ and then put ε zero, we arrive at the corresponding formulae for the two-dimensional problems for a wedge in Tranter (1948) and Ufliand (1965). Note that the functions

$$W(u,\alpha) = \frac{W_1(u,\alpha) - W_2(u,\alpha)}{2} = \frac{\sinh(4\alpha u) + u\sin(4\alpha)}{\cosh(4\alpha u) - u^2[1-\cos(4\alpha)] - 1}, \tag{4.73}$$

$W_n(u,\alpha)$, $n = 3, 4$, coincide with the known functions for the two-dimensional problems for a wedge. That is why, in particular, the behaviour of the normal stress σ_φ when $r \to 0$ in our case is similar to that in the two-dimensional case.

4.1.4. EXAMPLE: PROBLEM $\mathcal{A}$ FOR A QUARTER-SPACE

In this subsection we give an example showing the effectiveness of the formulas obtained above. Sheveleva (1984) has suggested an approximation of the Green function for a quarter-space (problem $\mathcal{A}$), and, omitting some terms, presupposed that this function differs no more than by 10% from the exact Green function. In the following, a normal concentrated force Q should act on one face of an elastic quarter-space ($\alpha = \pi/4$) at the point $r = x \neq 0$, $z = 0$, whereas the other face should be stress-free; instead of the normal displacement $u_\varphi(r)$ for $\varphi = \pi/4$ and $z = 0$, we consider the dimensionless function $f(y) = -2\pi u_\varphi(r)x\theta/Q$ with $y = r/x$. Using the values of integrals 6.794.1 and 6.671.14 in Gradshteyn and Rhyzik (1994)

$$\frac{4}{\pi^2}\int_0^\infty\int_0^\infty \mathcal{K}_{iu}(\beta x)\,\mathcal{K}_{iu}(\beta r)\,\cos(\beta z)\,d\beta\,du = \frac{1}{\sqrt{(r+x)^2+z^2}},$$

together with (4.68) and (4.69), we get

$$f(y) = \frac{1}{|y-1|} + \frac{1}{|y+1|} + f_*(y), \tag{4.74}$$

where

$$f_*(y) = \frac{4}{\pi^2}\int_0^\infty\int_0^\infty \mathcal{K}_{iu}(ty)\left\{\frac{2u^2(\cosh(\pi u)+1)}{\cosh(\pi u)-1-2u^2}\,\mathcal{K}_{iu}(t)\right. \tag{4.75}$$

$$\left.+\frac{\sinh(\pi u)}{2}\left[\frac{\Phi_+(u,t)}{\sinh(\pi u/2)+u} + \frac{\Phi_-(u,t)}{\sinh(\pi u/2)-u}\right]\right\} du\,dt,$$

$$\Phi_\pm(u,t) = (1-2\nu)\int_0^\infty L_\pm(u,y)\left[\Phi_\pm(y,t) + \cosh\left(\frac{\pi y}{2}\right)\mathrm{K}_{iy}(t)\right] dy, \tag{4.76}$$

$$0 \le u < \infty,$$

$$L_\pm(u,y) = \pm\sinh\left(\frac{\pi y}{2}\right)\left[\left(\sinh\left(\frac{\pi y}{2}\right) \pm y\right)\left(\cosh\left(\frac{\pi u}{2}\right) + \cosh\left(\frac{\pi y}{2}\right)\right)\right]^{-1}.$$

The first two terms on the right-hand side of formula (4.74) correspond to the approximative Green function in Sheveleva (1984). In Table 4.2 we give the quantities

$$a_1 = f_*(y)/f(y), \quad a_2 = [f_*(y) + (y+1)^{-1}]/f(y),$$

for various $y = 0.1k$, $k = 0, ..., 49$. At the edge of the wedge we have $y = 0$ and $\mathcal{K}_{iu}(0) = \pi\delta_*(u)$ (Dirac impulse function). In view of the value of integral 2.16.2.1 in Prudnikov, Brychkov and Marichev (1992, Vol. 2)

$$\int_0^\infty \mathcal{K}_{iu}(t)\,dt = \frac{\pi}{2\cosh(\pi u/2)},$$

we use instead of (4.75) and (4.76) the special limit formulae

$$f_*(0) = \frac{8}{\pi^2 - 4} + \pi \left[\frac{\Phi_+(0)}{\pi + 2} + \frac{\Phi_-(0)}{\pi - 2} \right],$$

$$\Phi_\pm(0) = (1 - 2\nu) \int_0^\infty L_\pm(0, y)[\Phi_\pm(y) + 1]\, dy,$$

$$\Phi_\pm(u) = (1 - 2\nu) \int_0^\infty L_\pm(u, y)[\Phi_\pm(y) + 1]\, dy, \quad 0 \le u < \infty.$$

Table 4.2: A comparison of Green functions.

k	a_1	a_2	k	a_1	a_2
0	0.1916	0.5968	25	0.1826	0.4278
1	0.2077	0.5642	26	0.1841	0.4351
2	0.2098	0.5259	27	0.1860	0.4422
3	0.2017	0.4811	28	0.1879	0.4489
4	0.1860	0.4302	29	0.1913	0.4562
5	0.1647	0.3735	30	0.1932	0.4621
6	0.1390	0.3112	31	0.1946	0.4674
7	0.1098	0.2433	32	0.1967	0.4728
8	0.07636	0.1687	33	0.1954	0.4758
9	0.03983	0.08784	34	0.1963	0.4800
10	—	—	35	0.1971	0.4839
11	0.03631	0.08011	36	0.1988	0.4881
12	0.06408	0.1421	37	0.2004	0.4922
13	0.08840	0.1936	38	0.2007	0.4952
14	0.1044	0.2323	39	0.1991	0.4969
15	0.1176	0.2646	40	0.1982	0.4989
16	0.1309	0.2938	41	0.2002	0.5026
17	0.1411	0.3179	42	0.2001	0.5048
18	0.1496	0.3385	43	0.2002	0.5071
19	0.1568	0.3565	44	0.2008	0.5096
20	0.1624	0.3718	45	0.2008	0.5116
21	0.1686	0.3863	46	0.2001	0.5131
22	0.1722	0.3980	47	0.2005	0.5152
23	0.1749	0.4081	48	0.2005	0.5170
24	0.1797	0.4190	49	0.2025	0.5199

The calculations show that the fraction $f_*(y)/f(y)$ lies inside the interval $0.18 \ldots 0.21$ for $\nu = 0.3$, $0 \le y \le 0.4$ and $2.5 \le y \le 10$, i.e. the additional term $f_*(y)$ gives a contribution up to 21% to the exact Green function, which corrects Sheveleva's (1984) presupposition of 10% accuracy of her approximate Green function. It is also interesting that, for the same values of ν and y, the first term on the right-hand side of (4.74), which corresponds to the Green function for a half-space, contributes approximately 50% to the value of $f(y)$ (see the values of a_2 in Table 4.2).

Remark. *Here and below, when numerical methods are applied for solving the integral equations* (4.38), *we do not use the formulae* (4.69) *and* (4.70) *but we use the collocation method and Gaussian quadrature. For this, we recommend changing the unknown function in* (4.38) *by the formula*

$$\Phi_n(t) = \Phi_n^*(t) - Q\cosh\frac{\pi t}{2}\,\mathcal{K}_{it}(\beta x), \quad n = 1, 2, \tag{4.77}$$

providing an exponential decay of the right-hand side of the Fredholm integral equation (see (4.63)*).*

4.1.5. INVERSION OF INTEGRAL OPERATORS

We investigate now the question of invertibility of the integral operators (4.69), (4.70). The inverse operators are useful, e.g. in crack problems for a 3-D wedge (Pozharskii (1993)). It is obvious that in the cases of problems $\mathcal{B}$ and $\mathcal{C}$ we have

$$\mathcal{B}_n^{-1} = I - (1-2\nu)\mathcal{A}_n \quad (n = 3, 4), \tag{4.78}$$

where I is the identity operator. In the case of problem $\mathcal{A}$ the question of invertibility was solved by Pozharskii (1994) in the following theorem.

Theorem 4.1. *The operator of the form (see* (4.69), (4.73)*)*

$$\frac{W_1(u,\alpha)}{2W(u,\alpha)}\,\mathcal{B}_1[u, f(y)] - \frac{W_2(u,\alpha)}{2W(u,\alpha)}\,\mathcal{B}_2[u, f(y)] : \; C_M(0,\infty) \to C_M(0,\infty)$$

has an inverse, which is equal to $I - (1-2\nu)\mathcal{A}_0$, *where the operator* $\mathcal{A}_0$ *is defined by the formulae*

$$\mathcal{A}_0[u, f(y)] = \int_0^\infty L(u,y) f(y)\,dy + \frac{W_1(u,\alpha) + W_2(u,\alpha)}{4W(u,\alpha)}\,\mathcal{B}[u, f(y)]$$

$$+ \cosh\frac{\pi u}{2} \int_0^\infty\!\!\int_0^\infty \frac{\sinh(\pi t)\, g(t,\alpha)\, \sinh(\pi y/2)}{[\cosh(\pi t) + \cosh(\pi u)][\cosh(\pi t) + \cosh(\pi y)]}$$

$$\times \langle (W_1(y,\alpha) + W_2(y,\alpha)) f(y) + (1-2\nu) W_3(y,\alpha)\,\mathcal{B}[y, f(s)] \rangle\, dt\, dy,$$

$$L(u,y) = 2\cosh\frac{\pi u}{2}\,\sinh\frac{\pi y}{2}\,W(y,\alpha)$$

$$\times\int_0^\infty \frac{\sinh(\pi t)\,g_3(t,\alpha)\,dt}{[\cosh(\pi t)+\cosh(\pi u)][\cosh(\pi t)+\cosh(\pi y)]},$$

$$\mathcal{B}[u,f(y)] = \sum_{m=0}^{\infty}(1-2\nu)^m\mathcal{A}_3^m[u,\mathcal{C}[t,f(y)]], \tag{4.79}$$

$$\mathcal{C}[t,f(y)] = 4\cosh\frac{\pi t}{2}\int_0^\infty\int_0^\infty \frac{\sinh(\pi s)\,g(s,\alpha)\,\sinh(\pi y/2)W(y,\alpha)}{(\cosh(\pi s)+\cosh(\pi t))(\cosh(\pi s)+\cosh(\pi y))}$$

$$\times f(y)\,ds\,dy + \int_0^\infty L_3(t,y)\,[W_1(y,\alpha)+W_2(y,\alpha)]\,f(y)/W_3(y,\alpha)\,dy,$$

$$g(y,\alpha) = (g_1(y,\alpha)+g_2(y,\alpha))/2.$$

This representation holds for arbitrary values of α, at least when $\nu > 0.053$.

To prove this theorem, we consider an auxiliary boundary-value problem for the wedge:

$$\varphi = -\alpha: \quad \sigma_\varphi = \tau_{r\varphi} = \tau_{\varphi z} = 0, \tag{4.80}$$

$$\varphi = \alpha: \quad u_\varphi = \delta_*(r-x)[\delta_*(z-y)+\delta_*(z+y)],\ \ \tau_{r\varphi} = \tau_{\varphi z} = 0. \tag{4.81}$$

Solving this problem, we use the methods developed above. As a result, we change from boundary conditions (4.80), (4.81) to a generalized Hilbert boundary-value problem, in the sense of I.N.Vekua, in which functional equations are reduced to one Fredholm integral equation of the second kind in an auxiliary function $\Phi_5(t)$. The stresses and displacements in the wedge can then be expressed in terms of this function. The left-hand side of this equation is the same as in (4.38), $n = 5$, and the right-hand side has the form

$$X(t) = 2(1-2\nu)\beta x\,\cosh\frac{\pi t}{2}\int_0^\infty \frac{\sinh(\pi y)\,\mathcal{K}_{iy}(\beta x)}{\cosh(\pi y)+\cosh(\pi t)}\,g(y,\alpha)\,dy$$

$$-(1-2\nu)\int_0^\infty L_5(t,y)y\,\cosh\frac{\pi y}{2}\,\mathcal{K}_{iy}(\beta x)\,\frac{W_1(y,\alpha)+W_2(y,\alpha)}{W_1(y,\alpha)\,W_2(y,\alpha)}\,dy.$$

The kernel of this integral equation has the form

$$L_5(t, y) = -\cosh\frac{\pi t}{2}\sinh\frac{\pi y}{2}\frac{W_1(y,\alpha)W_2(y,\alpha)}{W(y,\alpha)}$$

$$\times\int_0^\infty \frac{\sinh(\pi u)\,(g_1(u,\alpha) - g_2(u,\alpha))\,du}{(\cosh(\pi u) + \cosh(\pi t))(\cosh(\pi u) + \cosh(\pi y))}.$$

Using the identities

$$-W_1(y,\alpha)W_2(y,\alpha)/W(y,\alpha) = W_1(y, 2\alpha),$$

$$(g_1(u,\alpha) - g_2(u,\alpha))/2 = g_1(u, 2\alpha),$$

we get $L_5(t, y) \equiv L_3(t, y)$, where the function $L_3(t, y)$ serves as a kernel of equation (4.38) for problem $\mathcal{B}$. This is due to the fact that in both problems, $\mathcal{B}$ and (4.80), (4.81), the same components of the stress tensor and displacement vector are given. Consequently, for solving the Fredholm integral equation for the function $\Phi_5(t)$ one can use again the method of successive approximations, at least when $\nu > 0.053$ (see (4.65)). Now, to obtain the formulae of the theorem, we express the stress $\sigma_\varphi(r, \alpha, z)$ through the function $\Phi_5(t)$ and use the fact that this expression is inverse to the expression of u_φ (4.68) and (4.69) in problem $\mathcal{A}$.

4.1.6. PAPKOVICH–NEUBER FUNCTIONS

Of course, to determine a complete stress-strain state of a 3-D wedge, we need explicit expresions for all Papkovich–Neuber functions. Using the former technique and a superposition of even and odd solutions, with respect to z, we obtain such complete solutions.

For problem $\mathcal{A}^*$ with the boundary conditions

$$\varphi = -\alpha: \quad \sigma_\varphi = \tau_{r\varphi} = \tau_{\varphi z} = 0;$$

$$\varphi = \alpha: \quad \sigma_\varphi = -Q\delta_*(r - x)\delta_*(z - y), \quad \tau_{r\varphi} = \tau_{\varphi z} = 0,$$

the solution is determined by formulae (4.4)–(4.6), where the Papkovich–Neuber functions have the form ($n = 0, 1, 2$ and $m = 1, 2$)

$$\Phi_n(r, \varphi, z) = -\frac{Q}{G\pi^3}\int_0^\infty\int_0^\infty \sinh(\pi\tau)[A_n(\tau, \beta)\cosh(\varphi\tau)$$

$$+B_n(\tau,\beta)\sinh(\varphi\tau)]\,\mathcal{K}_{i\tau}(\beta r)\cos[\beta(z-y)]\,d\tau\,d\beta,$$

$$A_0(\tau,\beta) = -\frac{1-2\nu}{\beta\sinh(\alpha\tau)}\int_0^\infty W_1(u,\alpha)\Psi_1(u,\beta)\frac{\sinh(\pi u/2)\,du}{\cosh(\pi u)+\cosh(\pi\tau)},$$

$$B_0(\tau,\beta) = \frac{1-2\nu}{\beta\cosh(\alpha\tau)}\int_0^\infty W_2(u,\alpha)\Psi_2(u,\beta)\frac{\sinh(\pi u/2)\,du}{\cosh(\pi u)+\cosh(\pi\tau)},$$

$$A_1(\tau,\beta) = R_1(\tau,\alpha)C_2(\tau,\beta), \quad B_1(\tau,\beta) = R_2(\tau,\alpha)C_1(\tau,\beta),$$

$$A_2(\tau,\beta) = S_2(\tau,\alpha)C_1(\tau,\beta), \quad B_2(\tau,\beta) = S_1(\tau,\alpha)C_2(\tau,\beta),$$

$$C_m(\tau,\beta) = (-1)^m 2(1-\nu)\frac{W_m(\tau,\alpha)}{\cosh(\pi\tau/2)}\Psi_m(\tau,\beta),$$

$$\Psi_m(\tau,\beta) = \Psi_m^*(\tau,\beta) + F(\tau,\beta), \quad F(\tau,\beta) = \cosh\frac{\pi\tau}{2}\,\mathcal{K}_{i\tau}(\beta x),$$

$$R_1(\tau,\alpha) = \frac{2\cosh(\alpha\tau)\,\cos\alpha}{\cosh(2\alpha\tau)+\cos(2\alpha)}, \quad R_2(\tau,\alpha) = \frac{2\sinh(\alpha\tau)\,\cos\alpha}{\cosh(2\alpha\tau)-\cos(2\alpha)},$$

$$S_1(\tau,\alpha) = \frac{2\sinh(\alpha\tau)\,\sin\alpha}{\cosh(2\alpha\tau)+\cos(2\alpha)}, \quad S_2(\tau,\alpha) = \frac{2\cosh(\alpha\tau)\,\sin\alpha}{\cosh(2\alpha\tau)-\cos(2\alpha)}.$$

The functions $W_m(u,\alpha)$, $m=1,2$, are determined by formulae (4.40). The functions $\Psi_m^*(\tau,\beta)$, $m=1,2$, satisfy the Fredholm integral equations of the second kind $(0\le\tau<\infty)$

$$\Psi_m^*(\tau,\beta) = (1-2\nu)\int_0^\infty L_m(\tau,u)[\Psi_m^*(u,\beta)+F_m(u,\beta)]\,du, \tag{4.82}$$

where the kernels $L_m(\tau,u)$, $m=1,2$, are determined by formulae (4.39)–(4.41). Note that equations (4.82) were obtained from (4.38) by using formula (4.77).

For problem $\mathcal{B}^*$ with the boundary conditions

$$\varphi=0: \quad u_\varphi = \tau_{r\varphi} = \tau_{\varphi z} = 0;$$

$$\varphi=2\alpha: \quad \sigma_\varphi = -Q\delta_*(r-x)\delta_*(z-y), \quad \tau_{r\varphi} = \tau_{\varphi z} = 0,$$

the Papkovich–Neuber functions have the form $(n=0,1,2)$

$$\Phi_n(r,\varphi,z) = -\frac{Q}{G\pi^3}\int_0^\infty\int_0^\infty \sinh(\pi\tau)[A_n(\tau,\beta)\cosh(\varphi\tau)$$

$$+B_n(\tau,\beta)\sinh(\varphi\tau)]\,\mathcal{K}_{i\tau}(\beta r)\cos[\beta(z-y)]\,d\tau\,d\beta,$$

$$A_1(\tau,\beta)=0,\quad B_0(\tau,\beta)=0,\quad B_2(\tau,\beta)=0,$$

$$A_0(\tau,\beta)=-\frac{2(1-2\nu)}{\beta\sinh(2\alpha\tau)}\int_0^\infty W_1(u,\alpha)\Psi_1(u,\beta)\frac{\sinh(\pi u/2)\,du}{\cosh(\pi u)+\cosh(\pi\tau)},$$

$$B_1(\tau,\beta)=R_2(\tau,2\alpha)C_3(\tau,\beta),\quad A_2(\tau,\beta)=S_2(\tau,2\alpha)C_3(\tau,\beta),$$

$$C_3(\tau,\beta)=-4(1-\nu)\frac{W_3(\tau,\alpha)}{\cosh(\pi\tau/2)}\Psi_3(\tau,\beta),\quad \Psi_3(\tau,\beta)=\Psi_3^*(\tau,\beta)+F(\tau,\beta),$$

$$W_3(u,\alpha)=W_1(u,2\alpha)=\frac{\cosh(4\alpha u)-\cos(4\alpha)}{\sinh(4\alpha u)+u\sin(4\alpha)}. \tag{4.83}$$

The functions $\Psi_3^*(\tau,\beta)$ satisfy the Fredholm integral equation (4.82) with $m=3$, whose kernel is determined by formulae (4.39) with $n=3$ and $g_3(u,\alpha)=g_1(u,2\alpha)$.

For problem C^* with the boundary conditions

$$\varphi=0:\quad u_r=u_\varphi=u_z=0;$$

$$\varphi=2\alpha:\quad \sigma_\varphi=-Q\delta_*(r-x)\delta_*(z-y),\quad \tau_{r\varphi}=\tau_{\varphi z}=0,$$

we obtain the following Papkovich–Neuber functions ($n=0,1,2$):

$$\Phi_n(r,\varphi,z)=-\frac{Q}{G\pi^3}\int_0^\infty\int_0^\infty \sinh(\pi\tau)[A_n(\tau,\beta)\cosh(\varphi\tau)$$

$$+B_n(\tau,\beta)\sinh(\varphi\tau)]\,\mathcal{K}_{i\tau}(\beta r)\cos[\beta(z-y)]\,d\tau\,d\beta,$$

$$A_0(\tau,\beta)=0,\qquad A_2(\tau,\beta)=0,$$

$$B_0(\tau,\beta)=-\frac{2(1-2\nu)}{\beta\cosh(2\alpha\tau)}\int_0^\infty W_4(u,\alpha)\Psi_4(u,\beta)\frac{\sinh(\pi u/2)\,du}{\cosh(\pi u)+\cosh(\pi\tau)},$$

$$A_1(\tau,\beta)=\tau S_4(\tau,\alpha)C_1^*(\tau,\beta)-T_1(\tau,\alpha)C_2^*(\tau,\beta),$$

$$B_1(\tau,\beta)=-[\kappa R_3(\tau,\alpha)+\tau S_3(\tau,\alpha)]C_1^*(\tau,\beta)+T_2(\tau,\alpha)C_2^*(\tau,\beta),$$

$$B_2(\tau,\beta)=-\kappa S_4(\tau,\alpha)C_1^*(\tau,\beta)+T_3(\tau,\alpha)C_2^*(\tau,\beta),\quad \kappa=3-4\nu,$$

$$C_1^*(\tau,\beta)=4(1-\nu)\frac{W_4(\tau,\alpha)}{\cosh(\pi\tau/2)}\Psi_4(\tau,\beta),\quad \Psi_4(\tau,\beta)=\Psi_4^*(\tau,\beta)+F(\tau,\beta),$$

$$C_2^*(\tau,\beta)=(1-2\nu)\Big[R_1(\tau,2\alpha)C_1^*(\tau,\beta)$$

$$+S_1(\tau, 2\alpha) \int_0^\infty C_1^*(u, \beta) \frac{\sinh(\pi u)\, du}{\cosh(\pi u) - \cosh(\pi\tau)}\Bigg],$$

$$R_3(\tau, \alpha) = \frac{2\cos(2\alpha)\,\cosh(2\alpha\tau)}{h(\tau, \alpha)}, \quad h(\tau, \alpha) = \kappa \sinh(4\alpha\tau) - \tau \sin(4\alpha),$$

$$S_3(\tau, \alpha) = \frac{2\sin(2\alpha)\,\sinh(2\alpha\tau)}{h(\tau, \alpha)}, \quad S_4(\tau, \alpha) = \frac{2\sin(2\alpha)\,\cosh(2\alpha\tau)}{h(\tau, \alpha)},$$

$$T_1(\tau, \alpha) = \frac{\sinh(4\alpha\tau)}{h(\tau, \alpha)}, \quad T_2(\tau, \alpha) = \frac{\cosh(4\alpha\tau) + \cos(4\alpha)}{h(\tau, \alpha)},$$

$$T_3(\tau, \alpha) = \frac{\sin(4\alpha)}{h(\tau, \alpha)}.$$

The function $W_4(u, \alpha)$ is determined by formula (4.56). The function $\Psi_4^*(\tau, \beta)$ satisfies the Fredholm integral equation (4.82) with $m = 4$, whose kernel is determined by formulae (4.39) with $n = 4$ and (4.56)–(4.62).

The solutions of problems $\mathcal{A}^*$, $\mathcal{B}^*$ and $\mathcal{C}^*$ play the same role for wedges as the Boussinesq solution does for a half-space, which is a particular case of the solution of problem $\mathcal{A}^*$. These solutions can be applied for various (crack, contact etc.) problems for elastic wedges with any aperture angle.

Example. From the boundary conditions $\tau_{r\varphi} = \tau_{\varphi z} = 0$ on that face of the wedge, where the concentrated force acts, we get $u_\varphi = -\omega_2/2$, on the same face. Here ω_2 is determined by formula $(4.5)_2$. For $2\alpha = \pi$, a half-space, we derive the following formulae for the normal displacement on the face:

$$u_\varphi(r, \pi/2, z) = -\frac{\Phi_2(r, \pi/2, z)}{2}$$

$$= -\frac{2(1-\nu)P}{G\pi^3} \int_0^\infty \int_0^\infty \sinh(\pi\tau) \coth(\pi\tau)\, \mathcal{K}_{i\tau}(\beta x)\, \mathcal{K}_{i\tau}(\beta r) \cos[\beta(z - y)]\, d\tau\, d\beta \tag{4.84}$$

$$= -\frac{(1-\nu)P}{2\pi G\sqrt{(r - x)^2 + (z - y)^2}}$$

for problem $\mathcal{A}^*$ ($W(\tau, \pi/2) = \coth(\pi\tau)$; see integral (4.72)), this is in agreement with the Boussinesq solution;

$$u_\varphi(r, \pi, z) = \frac{\Phi_1(r, \pi, z)}{2}$$

$$= -\frac{2(1-\nu)P}{G\pi^3} \int_0^\infty \int_0^\infty \sinh(\pi\tau) W_m\left(\tau, \frac{\pi}{2}\right) \mathcal{K}_{i\tau}(\beta x)\, \mathcal{K}_{i\tau}(\beta r) \cos[\beta(z-y)]\, d\tau\, d\beta$$

for problems $\mathcal{B}^*$ ($m = 3$) and $\mathcal{C}^*$ ($m = 4$).

4.2 Strip Punch on a Wedge Face

Consider the interaction between a strip punch and an elastic three-dimensional wedge. The contact area Ω is defined by the inequalities $a \le r \le b$, $|z| < \infty$, and the base of the punch is defined by the function $f(r,z)$, which is periodic with respect to z with a period $2l$ (for $z \le 0$ the contact area is shown hatched in Fig. 4.2). The punch presses into the wedge face $\varphi = \alpha$ by forces P acting at a distance H from the edge of the wedge and related to the length of the period. We neglect friction between the punch and the wedge; the face $\varphi = \alpha$ is stress-free outside the contact area. It is assumed that one of the three boundary conditions considered above (stress-free, slipping support and rigid support; problems $\mathcal{A}$, $\mathcal{B}$ and $\mathcal{C}$, respectively) holds on the face $\varphi = -\alpha$ of the wedge[2]. It is required to determine the distribution of the normal contact stresses under the punch $\sigma_\varphi(r,\alpha,z) = -q(r,z), (r,z) \in \Omega$. Then with the help of the two integral conditions of equilibrium of the punch, we can find the relations between the quantites P, H and δ, the settlement or indentation of the punch, and also the angle γ of rotation of the punch about the z-axis.

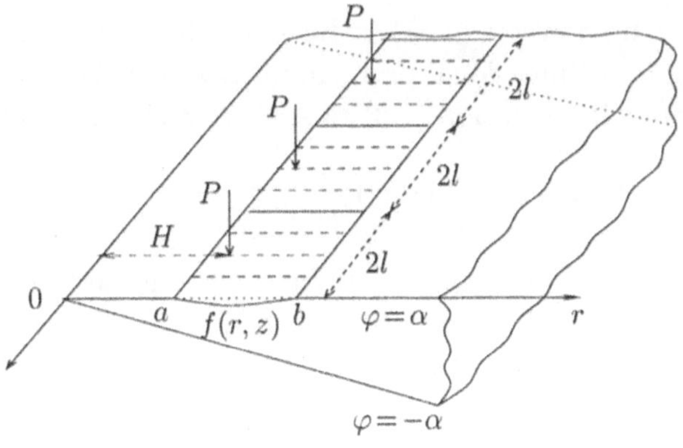

Figure 4.2: A periodic strip contact zone, with period $2l$, applied to the punch forces P and the punch base function $f(r,z)$

For simplicity, we will assume that the punch base function $f(r,z)$ is even with respect to z and can be represented as a Fourier series. It then suffices to solve the problem for the case $f(r,z) = f(r)\cos(cz)$ $(c = \pi m/l)$

[2]For problems $\mathcal{B}$ and $\mathcal{C}$ we have considered the wedge $0 \le \varphi \le 2\alpha$, but here we need only an expression for u_φ on the upper face of the wedge.

and to construct a superposition of the solutions obtained for different values of $m \geq 1$ as well as of the solution of the contact problem of the planar deformation of a wedge ($m = 0$) (Aleksandrov (1967b)). Knowing formulae (4.68)–(4.70), we can write the integral equation for the function $q(r)$ ($q(r,z) = q(r)\cos(cz)$) as follows:

$$\frac{1}{\pi^2}\int_a^b q(x)\,dx\int_0^\infty \sinh(\pi u)\,W(u,\alpha,cx)\,\mathcal{K}_{iu}(cr)\,du = \theta f(r) \quad (a \leq r \leq b). \tag{4.85}$$

In the case $2\alpha = \pi$ for problem $\mathcal{A}$, we have

$$W(u,\pi/2,cx) = 2\coth(\pi u)\,\mathcal{K}_{iu}(cx),$$

and the kernel of the integral equation (4.85) reduces to the known kernel for a half-space due to the integral 6.794.1 in Gradshteyn and Ryzhik (1994)

$$\frac{2}{\pi}\int_0^\infty \mathcal{K}_{iu}(cx)\,\mathcal{K}_{iu}(cr)\cosh(\pi u)\,du = \mathcal{K}_0(c|r-x|).$$

When $c \to 0$ equation (4.85) reduces to the integral equation of the corresponding two-dimensional problem for a wedge. For $\nu = 1/2$ equation (4.85) coincides with equation (1.8) in Aleksandrov and Pozharskii (1989).

4.2.1. REDUCTION TO INFINITE SYSTEMS

In the expression for the function $W(u,\alpha,cx)$ we have the meromorphic function $W(u,\alpha)$ (4.73) for problem $\mathcal{A}$ or $W_n(u,\alpha)$ for problems $\mathcal{B}$ ($n = 3$, (4.83)) and $\mathcal{C}$ ($n = 4$, (4.56)). These meromorphic functions can be expanded into infinite products like (3.303) and series like (3.310). Using the expansions into series and the complex integration formula (4.71), we expand the kernel of the integral equation (4.85) as series of residues of the corresponding function at its poles $i\gamma_m$ ($\mathrm{Im}\,(i\gamma_m) \geq 0$; cf. integral 6.794.10 in Gradshteyn and Ryzhik (1994)):

$$\frac{1}{\pi^2}\int_0^\infty \sinh(\pi u)\,W(u,\alpha,cx)\,\mathcal{K}_{iu}(cr)\,du \tag{4.86}$$

$$= \sum_{m=1}^\infty s_m \begin{Bmatrix} \mathcal{K}_{\gamma_m}(cr)\mathcal{I}_{\gamma_m}(cx) & \text{for } x < r \\ \mathcal{I}_{\gamma_m}(cr)\mathcal{K}_{\gamma_m}(cx) & \text{for } x > r \end{Bmatrix}$$

$$+\sum_{m=1}^{\infty} s_m \mathcal{F}\left[i\gamma_m, \cosh\frac{\pi y}{2}\, \mathcal{K}_{iy}(cx)\right] \mathcal{I}_{\gamma_m}(cr) \Big/ \cosh\frac{i\pi\gamma_m}{2},$$

where for problem $\mathcal{A}$ ($n = 1, 2,\ \gamma_1 = 0$)

$$\mathcal{F}[0, f(y)] = \varepsilon_1(\mathcal{B}_1[0, f(y)] - I) + \varepsilon_2(\mathcal{B}_2[0, f(y)] - I),\quad \varepsilon_n = \frac{1 \mp \cos(2\alpha)}{2s_1[2\alpha \pm \sin(2\alpha)]},$$

$$\mathcal{F}[i\gamma_m, f(y)] = \mathcal{B}_n[i\gamma_m, f(y)] - I,\quad \text{if}\quad \sinh(2\alpha i\gamma_m) \pm i\gamma_m \sin(2\alpha) = 0,$$

$$m = 2, 3, \ldots,$$

and for problems $\mathcal{B}$ ($n = 3$) and $\mathcal{C}$ ($n = 4$)

$$\mathcal{F}[i\gamma_m, f(y)] = \mathcal{B}_n[i\gamma_m, f(y)] - I,\quad m = 1, 2, \ldots .$$

For problem $\mathcal{B}$ with $2\alpha = \pi$ and for problem $\mathcal{C}$ the functions $W_n(u, \alpha)$ are of $\tanh u$ type, and we have

$$s_m^{-1} = \frac{1}{2}\frac{d}{du}[W_n^{-1}(u, \alpha)]_{u = i\gamma_m},\quad n = 3, 4,\quad m = 1, 2, \ldots .$$

For problem $\mathcal{A}$ with any value of α and for problem $\mathcal{B}$ with $2\alpha \neq \pi$ the corresponding functions are of $\coth u$ type, and we have, respectively,

$$s_m^{-1} = t_m \frac{d}{du}[W^{-1}(u, \alpha)]_{u = i\gamma_m},\qquad s_m^{-1} = t_m \frac{d}{du}[W_3^{-1}(u, \alpha)]_{u = i\gamma_m},$$

$$t_1 = 1;\quad t_m = 1/2,\ m = 2, 3, \ldots .$$

Suppose that the function $f(r)$ is represented in the form of a Kontorovich–Lebedev integral in the complex plane (4.2), then it is sufficient to consider the case when

$$f(r) = \mathcal{I}_\eta(cr)/\mathcal{I}_\eta(cb)\quad (a \leq r \leq b). \tag{4.87}$$

Using Theorem 4.1 from subsection 4.1.5 and formula (4.78), one can prove that for the case (4.87) the solution of equation (4.85) has the following structure (see also Babeshko and Berkovich (1972)):

$$\frac{xq(x)}{\theta} = \sum_{m=0}^{\infty} x_m \mathcal{I}_{\delta_m}(cx)\mathcal{I}_{\delta_m}^{-1}(cb) + \sum_{m=1}^{\infty} y_m \mathcal{K}_{\delta_m}(cx)\mathcal{K}_{\delta_m}^{-1}(ca), \tag{4.88}$$

where

$$x_0 = i\eta W^{-1}(i\eta, \alpha),\quad x_0 = i\eta W_n^{-1}(i\eta, \alpha),\quad \delta_0 = \eta. \tag{10}$$

for problems $\mathcal{A}$, $\mathcal{B}$ ($n = 3$) and $\mathcal{C}$ ($n = 4$), respectively. Here $i\delta_m$ ($m = 1, 2, \ldots$) are zeros of the function $W(u, \alpha)$ or $W_n(u, \alpha)$ ($\operatorname{Im}(i\delta_m) > 0$) in the plane of the complex variable u.

Substituting the representation (4.88) into equation (4.85) with kernel (4.86), using the integral 5.55 in Gradshteyn and Ryzhik (1994)

$$\int f_\mu(cx) g_\nu(cx) \frac{dx}{x} = \frac{cx}{\nu^2 - \mu^2} G\left(f_\mu(cx), g_\nu(cx)\right),$$

$$G\left(f_\mu(y), g_\nu(y)\right) = f_\mu(y) \frac{d}{dy} g_\nu(y) - g_\nu(y) \frac{d}{dy} f_\mu(y),$$

where $f_\mu(cx)$ and $g_\nu(cx)$ are any functions among $\mathcal{I}_\mu(cx)$, $\mathcal{I}_\nu(cx)$, $\mathcal{K}_\mu(cx)$ and $\mathcal{K}_\nu(cx)$, using identities

$$\sum_{n=1}^{\infty} \frac{s_n}{\delta_m^2 - \gamma_n^2} = 0, \quad m = 1, 2, \ldots$$

and taking a linear independence of the system of functions $\mathcal{I}_{\delta_m}(cx)$ and $\mathcal{K}_{\delta_m}(cx)$, $m = 1, 2, \ldots$, into account, we get two infinite systems of linear algebraic equations of the second kind [3] for the unknown columns $X = \{x_m\}$, $Y = \{y_m\}$, $m = 1, 2, \ldots$, in the form

$$X = A^{-1}(A - A_{11})X - A^{-1}A_{12}Y - A^{-1}A_{13}X \tag{4.89}$$

$$-A^{-1}A_{14}Y + A^{-1}B_1 + A^{-1}B_3,$$

$$Y = A^{-1}(A - A_{22})Y - A^{-1}A_{21}X + A^{-1}B_2, \tag{4.90}$$

where elements of the matrices $A = (A_{nm})$, $A^{-1} = (A_{nm}^{-1})$, $A_{kl} = (A_{klnm})$ ($k = 1, 2$; $l = 1, 2, 3, 4$; $n, m = 1, 2, \ldots$) and of the columns $B_j = (B_{jn})$ ($j = 1, 2, 3$; $n = 1, 2, \ldots$), when the symbol of the kernel of equation (4.85) is of the $\tanh u$ type (similarly, when it is of the $\coth u$ type), are calculated by formulae

$$A_{nm} = (\delta_m - \gamma_n)^{-1}, \quad A_{nm}^{-1} = (K_+'(-i\delta_n)(K_-^{-1}(i\gamma_m))'(\gamma_m - \delta_n))^{-1},$$

$$K_+(u) = F \prod_{n=1}^{\infty} \left(1 + \frac{u}{i\delta_n}\right)\left(1 + \frac{u}{i\gamma_n}\right)^{-1}; \quad K_-(u)F = K_+(-u),$$

$$F = \lim_{u \to 0} W_3(u, \pi/2)/u; \quad F = \lim_{u \to 0} W_4(u, \alpha)/u,$$

[3]The technique is similar to that which was used in subsection 3.6.1, cf. formulae (3.312)–(3.315).

$$A_{11nm} = (\delta_m^2 - \gamma_n^2)^{-1}\left(\mathcal{I}'_{\delta_m}(\lambda_2)\mathcal{I}^{-1}_{\delta_m}(\lambda_2) - \mathcal{K}'_{\gamma_n}(\lambda_2)\mathcal{K}^{-1}_{\gamma_n}(\lambda_2)\right),$$

$$A_{12nm} = (\delta_m^2 - \gamma_n^2)^{-1}\left(\mathcal{K}_{\delta_m}(\lambda_2)\mathcal{K}'_{\gamma_n}(\lambda_2)\mathcal{K}^{-1}_{\delta_m}(\lambda_1)\mathcal{K}^{-1}_{\gamma_n}(\lambda_2)\right.$$
$$\left.-\mathcal{K}'_{\delta_m}(\lambda_2)\mathcal{K}^{-1}_{\delta_m}(\lambda_1)\right),$$

$$A_{21nm} = (\delta_m^2 - \gamma_n^2)^{-1}\left(\mathcal{I}'_{\delta_m}(\lambda_1)\mathcal{I}^{-1}_{\delta_m}(\lambda_2) - \mathcal{I}_{\delta_m}(\lambda_1)\mathcal{I}'_{\gamma_n}(\lambda_1)\mathcal{I}^{-1}_{\delta_m}(\lambda_2)\mathcal{I}^{-1}_{\gamma_n}(\lambda_1)\right),$$

$$A_{22nm} = (\delta_m^2 - \gamma_n^2)^{-1}\left(\mathcal{I}'_{\gamma_n}(\lambda_1)\mathcal{I}^{-1}_{\gamma_n}(\lambda_1) - \mathcal{K}'_{\delta_m}(\lambda_1)\mathcal{K}^{-1}_{\delta_m}(\lambda_1)\right),$$

$$A_{13nm} = \left[\mathcal{I}_{\delta_m}(\lambda_2)\mathcal{K}_{\gamma_n}(\lambda_2)\cosh\frac{i\pi\gamma_n}{2}\right]^{-1}\mathcal{F}\left\{i\gamma_n, \frac{\cosh \pi y/2}{y^2 + \delta_m^2}\right.$$
$$\left.\times[G(\mathcal{I}_{\delta_m}(\lambda_1), \mathcal{K}_{iy}(\lambda_1)) - G(\mathcal{I}_{\delta_m}(\lambda_2), \mathcal{K}_{iy}(\lambda_2))]\right\},$$

$$A_{14nm} = \left[\mathcal{K}_{\delta_m}(\lambda_1)\mathcal{I}_{\gamma_n}(\lambda_1)\cosh\frac{i\pi\gamma_n}{2}\right]^{-1}\mathcal{F}\left\{i\gamma_n, \frac{\cosh \pi y/2}{y^2 + \delta_m^2}\right.$$
$$\left.\times[G(\mathcal{K}_{\delta_m}(\lambda_1), \mathcal{K}_{iy}(\lambda_1)) - G(\mathcal{K}_{\delta_m}(\lambda_2), \mathcal{K}_{iy}(\lambda_2))]\right\},$$

$$B_{1n} = \frac{-x_0}{\gamma_n^2 - \delta_0^2}G(\mathcal{I}_{\delta_0}(\lambda_2), \mathcal{K}_{\gamma_n}(\lambda_2))\mathcal{I}^{-1}_{\delta_0}(\lambda_2)\mathcal{K}^{-1}_{\gamma_n}(\lambda_2),$$

$$B_{2n} = \frac{-x_0}{\gamma_n^2 - \delta_0^2}G(\mathcal{I}_{\delta_0}(\lambda_1), \mathcal{I}_{\gamma_n}(\lambda_1))\mathcal{I}^{-1}_{\delta_0}(\lambda_2)\mathcal{I}^{-1}_{\gamma_n}(\lambda_1),$$

$$B_{3n} = -x_0\left[\mathcal{I}_{\delta_0}(\lambda_2)\mathcal{K}_{\gamma_n}(\lambda_2)\cosh\frac{i\pi\gamma_n}{2}\right]^{-1}\mathcal{F}\left\{i\gamma_n, \frac{\cosh \pi y/2}{y^2 + \delta_0^2}\right.$$
$$\left.\times[G(\mathcal{I}_{\delta_0}(\lambda_1), \mathcal{K}_{iy}(\lambda_1)) - G(\mathcal{I}_{\delta_0}(\lambda_2), \mathcal{K}_{iy}(\lambda_2))]\right\}, \quad \lambda_1 = ca, \quad \lambda_2 = cb.$$

For the analysis of the matrices in system (4.89), (4.90) there are asymptotic formulae ($|p| \to \infty$):

$$\mathcal{I}_p(z) = O[(z^2 + p^2)^{-1/4}\exp(\xi(z,p))],$$

$$\mathcal{K}_p(z) = O[(z^2 + p^2)^{-1/4}\exp(-\xi(z,p))],$$

$$\xi(z,p) = \sqrt{z^2 + p^2} - p\,\mathrm{Arcsinh}\left(\frac{p}{z}\right),$$

$$|\arg p - k\pi| \le \frac{\pi}{2} - \delta \quad (k = 1, 2), \quad 0 < \delta \ll 1,$$

$$\frac{\mathcal{I}'_p(z)}{\mathcal{I}_p(\zeta)} \sim (z^2 + p^2)^{1/4}(\zeta^2 + p^2)^{1/4}\left(\frac{z}{\zeta}\right)^p,$$

$$\frac{\mathcal{K}_p'(z)}{\mathcal{K}_p(\zeta)} \sim -(z^2+p^2)^{1/4}(\zeta^2+p^2)^{1/4}\left(\frac{z}{\zeta}\right)^{-p},$$

$$\frac{\mathcal{I}_p(z)}{\mathcal{I}_p(\zeta)} \sim (z^2+p^2)^{-1/4}(\zeta^2+p^2)^{1/4}\left(\frac{z}{\zeta}\right)^{p},$$

$$\frac{\mathcal{K}_p(z)}{\mathcal{K}_p(\zeta)} \sim (z^2+p^2)^{-1/4}(\zeta^2+p^2)^{1/4}\left(\frac{z}{\zeta}\right)^{-p}.$$

In the general case, one can show that the infinite system (4.89), (4.90) is quasiregular (see definition in footnote 3 in subsection 2.4.1). For small values of α, as in Babeshko and Berkovich (1972), the system can be solved by the method of successive approximations. The zero approximation to the solution of the system (4.89), (4.90) has the form

$$X^{(0)} = A^{-1}B_1 + A^{-1}B_3, \quad Y^{(0)} = A^{-1}B_2,$$

and it can be transformed to form (3.1) from Babeshko and Berkovich (1972), where it is necessary to add the term $A^{-1}B_3$ on the right-hand side of the first formula of (3.1).

4.2.2. METHOD OF DUAL INTEGRAL EQUATIONS

We will investigate separately the case when $a = 0$, that is, when the punch presses from the edge of the elastic wedge. Here for problems $\mathcal{B}$ and $\mathcal{C}$ it is necessary that $f(0) = 0$; suppose that this condition holds also for problem $\mathcal{A}$. We introduce the functions

$$Q(u) = \sinh\frac{\pi u}{2}\left[W_1(u,\alpha)\,\mathcal{B}_1\left\{u, \cosh\frac{\pi y}{2}\int_0^b q(x)\mathcal{K}_{iy}(cx)\,dx\right\}\right. \tag{4.91}$$

$$\left. -W_2(u,\alpha)\,\mathcal{B}_2\left\{u, \cosh\frac{\pi y}{2}\int_0^b q(x)\mathcal{K}_{iy}(cx)\,dx\right\}\right],$$

$$Q(u) = 2\sinh\frac{\pi u}{2}\,W_n(u,\alpha)\,\mathcal{B}_n\left\{u, \cosh\frac{\pi y}{2}\int_0^b q(x)\mathcal{K}_{iy}(cx)\,dx\right\}, \quad n = 3, 4, \tag{4.92}$$

for problems $\mathcal{A}$, $\mathcal{B}$ $(n = 3)$ and $\mathcal{C}$ $(n = 4)$, respectively,

Theorem 4.1 from subsection 4.1.5 allow us to express the function $q(x)$ from formula (4.91), (4.92) in the form

$$xq(x) = \frac{2}{\pi^2} \int_0^\infty Q(u) \frac{u}{W^n(u)} K_{iu}(cx)\, du \tag{4.93}$$

$$- \frac{2}{\pi^2}(1-2\nu) \int_0^\infty \frac{u \sinh(\pi u)}{\cosh(\pi u/2)} \mathcal{A}_m \left\{ u, \frac{Q(y)}{\sinh(\pi y)} \frac{\cosh(\pi y/2)}{W^n(y)} \right\} K_{iu}(cx)\, du,$$

where $n=1$, $m=0$ for problem $\mathcal{A}$, $n=2$, $m=3$ for problem $\mathcal{B}$, $n=3$, $m=4$ for problem $\mathcal{C}$; we use the notation $W^1(u) = W(u,\alpha)$ and $W^n(u) = W_{n+1}(u,\alpha)$, $n=2,3$.

The integral equation (4.85), with $a=0$, is equivalent to the following dual integral equations:

$$\int_0^\infty Q(u)\, K_{iu}(cr)\, du = \frac{\pi^2}{2} \theta f(r), \quad 0 \le r \le b, \tag{4.94}$$

$$\int_0^\infty Q(u) \frac{u}{W^n(u)} K_{iu}(cr)\, du - (1-2\nu) \int_0^\infty \frac{u \sinh(\pi u)}{\cosh(\pi u/2)} \tag{4.95}$$

$$\times \mathcal{A}_m \left\{ u, \frac{Q(y)}{\sinh(\pi y)} \frac{\cosh(\pi y/2)}{W^n(y)} \right\} K_{iu}(cr)\, du = 0, \quad b < r < \infty.$$

This pair can be reduced to a Fredholm integral equation of the second kind with a symmetric kernel. We seek the function $Q(u)$ in the form

$$Q(u) = N(u) + M(u), \tag{4.96}$$

$$N(u) = \theta u \sinh(\pi u) \int_0^b \frac{f(r)}{r} K_{iu}(cr)\, dr, \tag{4.97}$$

$$M(u) = \left(\frac{2}{\pi}\right)^{3/2} \sinh(\pi u) \int_b^\infty \varphi(t)\, \mathrm{Im}\, K_{1/2+iu}(ct)\, dt. \tag{4.98}$$

On introducing representations (4.96)–(4.98) into (4.94), (4.95), we satisfy equation (4.94) identically. Equation (4.95) is transformed with help of the

Abel inversion formulae to a Fredholm integral equation of the second kind for the function $\varphi(t)$:

$$\varphi(t) + \lambda \int_b^\infty F(t,s)\,\varphi(s)\,ds = p(t), \quad b \le t < \infty, \tag{4.99}$$

$$F(t,s) = \int_0^\infty \Big[\Big(\frac{\sinh(\pi u)}{W^n(u)} - \cosh(\pi u)\Big) \operatorname{Im} \mathcal{K}_{1/2+iu}(cs)$$

$$-2(1-2\nu)\sinh\frac{\pi u}{2}\,\mathcal{A}_m\Big\{u, \frac{\cosh(\pi y/2)}{W^n(y)} \operatorname{Im}\mathcal{K}_{1/2+iy}(cs)\Big\}\Big] \operatorname{Im}\mathcal{K}_{1/2+iu}(ct)\,du,$$

$$p(t) = \sqrt{\frac{2}{\pi}}\,c \int_0^\infty [(1-2\nu)\sinh\frac{\pi u}{2}\,\mathcal{A}_m\Big\{u, \frac{N(y)}{\sinh(\pi y/2)\,W^n(y)}\Big\}$$

$$-N(u)/W^n(u)] \operatorname{Im}\mathcal{K}_{1/2+iu}(ct)\,du, \quad \lambda = 4c/\pi^2.$$

Lemma 4.1. *The kernel of integral equation* (4.99) *is symmetric, that is,* $F(t,s) = F(s,t)$.

This lemma is clear for problems $\mathcal{B}$ and $\mathcal{C}$ but for problem $\mathcal{A}$ the validity of the lemma is established by an analysis of each term of the Neumann series from formula (4.79), permutation of the integrals and a change of the variables of integration.

Remark. *Symmetry of the kernel of integral equation* (4.85) *and of the kernels of other integral equations of contact problems in this chapter can be checked by similar way.*

Thus, a solution of the integral equation of the contact problem for a strip punch, which reaches the edge of the wedge, reduces for problem $\mathcal{A}$ to the sequential solution of two Fredholm integral equations of the second kind (the solution of one of them is represented by the Neumann series in (4.79)), and for problems $\mathcal{B}$ and $\mathcal{C}$ to one equation of the form (4.99).

Using formulae (4.93), (4.96)–(4.98) and integrating by parts, one can show that

$$\lim_{r\to b-0} \sqrt{c(b-r)}\,q(r)/\theta = -\frac{2}{\pi^2}\varphi(b)/\theta = \varepsilon. \tag{4.100}$$

The contact pressure $q(r)$, of course, has a square root singularity at the point $r = b$, and a factor at this singularity is expressed in terms of the function $\varphi(t)$ at the very same point $t = b$. Below for problem $\mathcal{A}$ with $2\alpha = \pi/2$, $\nu = 1/2$ and $f(r) = r(1 - r/b)$ we give the values of the dimensionless parameter ε (4.100) as a function of the dimensioless parameter cb:

Table 4.3: Factor (4.100)

cb	0.01	0.1	0.5	1	2	3
ε	0.624	−0.00136	−0.229	−0.187	0.145	0.598

We recall that the contact pressure is determined by the formula $q(r,z) = q(r)\cos(cz)$, in the strip contact zone, i.e., the function $q(r)$ corresponds to the c-harmonic of the Fourier decomposition. To find a real contact pressure for an arbitrary periodic strip punch, we should take a superposition of all the harmonics $c = \pi m/l$, $m = 0, 1, ...$, where $2l$ is the period of the punch base. Similarly, we should take a superposition of the corresponding factors ε in order to determine the real strength intensity factor which characterizes the contact pressure near $r = b$. We can see from Table 4.3 that some values of c give negative factors ε, for a fixed value of b. This means that there is no contact near the point $r = b$ if the punch base is exactly the c-harmonic $f(r,z) = f(r)\cos(cz)$.

Note that for problem $\mathcal{B}$ with $2\alpha = \pi$ and problem $\mathcal{C}$ with $2\alpha = \pi$ and $\nu = 1/2$ we have $F(t,s) \equiv 0$. This means that the contact problem has an exact solution in these cases.

4.3 Wedge-Shaped Punch on a Wedge Face

We will no longer assume symmetry with respect to the z coordinate. Using the solutions from subsection 4.1.6, we reduce the contact problem for a punch whose planform is a region Ω, and one of the three boundary conditions (problems $\mathcal{A}$ $(n = 1)$, $\mathcal{B}$ $(n = 2)$ or $\mathcal{C}$ $(n = 3)$) holds on the other face of the wedge, to the following integral equation $((r,z) \in \Omega)$:

$$\frac{2}{\pi^3\theta}\int_\Omega q(x,y)\,d\Omega \int_0^\infty\int_0^\infty \sinh(\pi u)\,W^n(u)\,\mathcal{K}_{iu}(tr)\Big[\mathcal{K}_{iu}(tx) \tag{4.101}$$

$$+\frac{1}{\cosh(\pi u/2)}\mathcal{D}_n\Big\{u, \cosh\frac{\pi s}{2}\,\mathcal{K}_{is}(tx)\Big\}\Big]\cos[t(z-y)]\,du\,dt = f(r,z),$$

$$\mathcal{D}_1 = \frac{W_1(u,\alpha)}{2W(u,\alpha)}(\mathcal{B}_1 - I) - \frac{W_2(u,\alpha)}{2W(u,\alpha)}(\mathcal{B}_2 - I); \quad \mathcal{D}_n = \mathcal{B}_{n+1} - I, \ n = 2,3. \tag{4.102}$$

Here $q(r,z)$ is the contact stress under the punch, $f(r,z)$ is the function defining the shape of punch base and the degree to which it penetrates into the wedge (Fig. 4.3a).

4.3.1. ASYMPTOTIC METHOD

Let Ω be an infinite wedge of angle 2β $(0 < \beta \le \pi/2)$, described in polar coordinates ρ, ψ $(r = \rho\cos\psi,\ z = \rho\sin\psi)$ by inequalities $0 \le \rho < \infty$, $|\psi| \le \beta$. The half of the contact area is shown hatched in Fig. 4.3*b* for $z \le 0$. Although the contact domain is symmetric with respect to z, the contact problem itself is not necessarily symmetric with respect to z because of possible asymmetry of the base of the punch.

We will use Mellin transforms defined by the formulae (Sneddon (1972))

$$F(s) = \int_0^\infty \rho^{s-1} f(\rho)\,d\rho, \quad f(\rho) = \frac{1}{2\pi i}\int_{c-i\infty}^{c+i\infty} F(s)\rho^{-s}\,ds.$$

More information on Mellin transforms is given in Gladwell (1980, subsection 9.2).

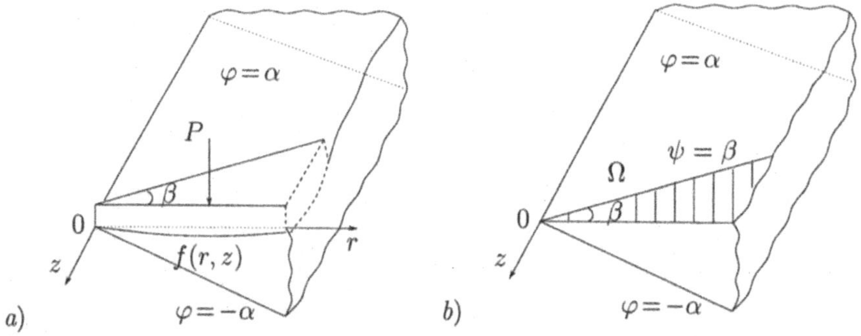

Figure 4.3: The one-half, $z \le 0$, of the wedge: *a*) in contact with a wedge-shaped punch; *b*) the wedge-shaped contact zone Ω

To eliminate solutions of equation (4.101) with infinite energy, we will confine our attention to the case in which both functions $q_*(\rho,\psi) = q(r,z)/\theta$ and $f_*(\rho,\psi) = f(r,z)$ have Mellin transforms with variable ρ and

$$\int_{-\beta}^{\beta} d\psi \int_0^\infty q_*(\rho,\psi)\rho\,d\rho < \infty, \quad \int_{-\beta}^{\beta} d\psi \int_0^\infty f_*(\rho,\psi)\rho\,d\rho < \infty. \tag{4.103}$$

To use the asymptotic method, we separate out the principal part $1/R =$

$1/\sqrt{(r-x)^2+(z-y)^2}$ in the kernel of equation (4.101), using the identity

$$W^n(u) = W^n(u) - \coth(\pi u) + \coth(\pi u) \tag{4.104}$$

and integral (4.84). We write now equation (4.101) in terms of ρ and ψ and take Mellin transforms with respect to ρ of both sides. For the principal part $1/R$ of the kernel the Mellin transform has the form (Erdelyi (1954, Vol. 1, formula (17) on p. 310))

$$K_s^0(t) = \frac{1}{2}\int_0^\infty \frac{\rho^{s-1/2}\,d\rho}{\sqrt{1+2\rho\cos(\pi-t)+t^2}} \tag{4.105}$$

$$= \frac{\pi}{2\cos(\pi s)}\mathcal{P}_{s-1/2}(-\cos t), \quad |\operatorname{Re} s| < \frac{1}{2},$$

where $\mathcal{P}_s(x) = \mathcal{P}_s^0(x)$ is the Legendre function 8.704 in Gradshteyn and Ryzhik (1994). For the other part of the kernel we use formula 2.16.14.3 in Prudnikov, Brychkov and Marichev (1992, Vol. 2) in order to get a Mellin transform for $\cos(t\rho)\mathcal{K}_{iu}(t\rho)$, with respect to ρ. As a result, we obtain

$$\int_{-1}^{1} q_s(\xi) K_s\left(\frac{\xi}{\lambda}, \frac{x}{\lambda}\right) d\xi = \pi f_s(x), \quad |x| \le 1, \tag{4.106}$$

where

$$K_s = K_s(t,p) = \frac{\pi}{2\cos(\pi s)}\mathcal{P}_{s-1/2}(-\cos(t-p)) + K_s^*(t,p), \tag{4.107}$$

$$K_s^*(t,p) = \frac{1}{2}\int_0^\infty \sinh(\pi u)\,[W^n(u) - \coth(\pi u)]$$

$$\times [R_+(-s,u,t)R_+(s,u,p) + R_-(-s,u,t)R_-(s,u,p)]\,du$$

$$+ \int_0^\infty \sinh\frac{\pi u}{2}\,W^n(u)\Big[R_+(s,u,p)\,\mathcal{D}_n\Big\{u, \cosh\frac{\pi y}{2}\,R_+(-s,y,t)\Big\}$$

$$+R_-(s,u,p)\,\mathcal{D}_n\Big\{u, \cosh\frac{\pi y}{2}\,R_-(-s,y,t)\Big\}\Big]\,du \quad (|\operatorname{Re} s| < 1/2),$$

$$R_+(s,u,t) = \frac{1}{2}\Gamma\left(\frac{1}{2}+s+iu\right)\operatorname{cosec}[\pi(1/2+s-iu)/2] \tag{4.108}$$

$$\times[\mathcal{P}^{-iu}_{s-1/2}(\sin t) + \mathcal{P}^{-iu}_{s-1/2}(-\sin t)],$$

$$R_-(s,u,t) = \frac{1}{2}\Gamma\left(\frac{1}{2} + s + iu\right)\sec\left[\pi(1/2 + s - iu)/2\right]$$

$$\times[\mathcal{P}^{-iu}_{s-1/2}(\sin t) - \mathcal{P}^{-iu}_{s-1/2}(-\sin t)],$$

$$x = \psi/\beta, \quad \lambda = 1/\beta, \quad q_s(x) = q^*_s(\psi), \quad f_s(x) = f^*_s(\psi)/\beta,$$

$$\frac{1}{2\pi i}\int\limits_{\Gamma} q^*_s(\psi)\rho^{-s-3/2}ds = q_*(\rho,\psi), \quad \frac{1}{2\pi i}\int\limits_{\Gamma} f^*_s(\psi)\rho^{-s-1/2}ds = f_*(\rho,\psi).$$

Here Γ is a straight line parallel to the imaginary axis in the complex s-plane, and $\Gamma(s)$ is the gamma function. Note that the comparison of the two different Mellin transforms above gives the integral for a half-space ($n = 1$, $2\alpha = \pi$)

$$\int\limits_0^\infty \cosh(\pi u)\left[R_+(-s,u,t)R_+(s,u,p) + R_-(-s,u,t)R_-(s,u,p)\right]du$$

$$= \frac{\pi}{\cos(\pi s)}\mathcal{P}_{s-1/2}(-\cos(t-p)), \quad |\operatorname{Re} s| < 1/2.$$

Kernel (4.107) of integral equation (4.106) admits the following asymptotic expansion:

$$K_s\left(\frac{\xi}{\lambda},\frac{x}{\lambda}\right) = \sum_{m=0}^{\infty}\frac{1}{\lambda^{2m}}\left\{\left(a_m(s) + b_m(s)\ln\left|\frac{\xi - x}{\lambda}\right|\right)(\xi - x)^{2m}\right. \qquad (4.109)$$

$$\left. + f^s_{2m}(\xi,x)\right\},$$

where $f^s_{2m}(\xi,x)$ are homogeneous polynomials of degree $2m$.

Formula (4.109) is obtained by expanding the integral term $K^*_s(t,p)$ in (4.107) in Maclaurin series, and the first term $K^0_s(t-p)$ in series (Gradshteyn and Ryzhik (1994, formula 8.704), Erdelyi (1955, formula (2) on p. 74))

$$K^0_s(t) = \frac{\pi}{2\cos(\pi s)}\mathcal{F}\left(\frac{1}{2} + s, \frac{1}{2} - s, 1; \cos^2\frac{t}{2}\right)$$

$$= \frac{\cos(\pi s)}{2\pi}\sum_{m=0}^{\infty}\frac{\Gamma(m+1/2+s)\Gamma(m+1/2-s)}{(m!)^2}\left(h_m - 2\ln\sin\frac{t}{2}\right)\sin^2\frac{mt}{2},$$

$$h_m = 2\psi(m+1) - \psi(m+1/2+s) - \psi(m+1/2-s),$$

where $\mathcal{F}(a,b,c;x)$ is the hypergeometric function and $\psi(s) = \Gamma'(s)/\Gamma(s)$ is the ψ-function. It can be shown that series (4.109) converges uniformly in $|\xi|,|x| \le 1$ for $\lambda > \max(1/(4\alpha), 2/\pi)$, i.e., for small α we must have $\alpha > \beta/4$.

The first few terms of series (4.109) are

$$a_0(s) = -\mathrm{C} - \frac{1}{2}\,\psi(1/2+s) - \frac{1}{2}\,\psi(1/2-s) + \ln 2,$$

$$a_1(s) = \frac{1}{8}\,(1/4-s^2)[2\psi(2) - \psi(3/2+s) - \psi(3/2-s) + 2\ln 2] + 373/9000,$$

$$b_0(s) = -1, \quad b_1(s) = -\frac{1}{4}\,(1/4-s^2),$$

$$f_0^s(\xi,x) = d_0(s) = \int\limits_0^\infty F(u,s)\,du \quad (|\operatorname{Re} s| < 1/2), \tag{4.110}$$

$$F(u,s) = \frac{1}{4\pi^2}\,\Gamma\left(\frac{1}{4}+\frac{s}{2}+i\frac{u}{2}\right)\Gamma\left(\frac{1}{4}+\frac{s}{2}-i\frac{u}{2}\right) \tag{4.111}$$

$$\times\left[\sinh(\pi u)\,[W^n(u) - \coth(\pi u)]\,\Gamma\left(\frac{1}{4}-\frac{s}{2}+i\frac{u}{2}\right)\Gamma\left(\frac{1}{4}-\frac{s}{2}-i\frac{u}{2}\right)\right.$$

$$\left.+2\sinh\frac{\pi u}{2}\,W^n(u)\,\mathcal{D}_n\left\{u, \cosh\frac{\pi y}{2}\,\Gamma\left(\frac{1}{4}-\frac{s}{2}+i\frac{y}{2}\right)\Gamma\left(\frac{1}{4}-\frac{s}{2}-i\frac{y}{2}\right)\right\}\right],$$

$$f_2^s(\xi,x) = d_2^1(s)(\xi^2+x^2) + d_2^2(s)\xi x,$$

$$d_2^1(s) = \int\limits_0^\infty \left(\frac{1}{4} - s^2 + u^2\right) F(u,s)\,du \quad (|\operatorname{Re} s| < 1/2),$$

$$d_2^2(s) = -\int\limits_0^\infty \frac{\Gamma(1/2+s+iu)\,(1/2+s+iu)}{\cos[\pi(1/2+s-iu)/2]}\,\mathcal{P}_{1/2+s}^{-iu}(0)\left[\frac{1}{2}\sinh(\pi u)\,[W^n(u)\right.$$

$$-\coth(\pi u)]\,\frac{\Gamma(1/2-s+iu)\,(1/2-s+iu)}{\cos[\pi(1/2-s-iu)/2]}\,\mathcal{P}_{1/2-s}^{-iu}(0) + \sinh\frac{\pi u}{2}\,W^n(u)$$

$$\left.\times\mathcal{D}_n\left\{u, \cosh\frac{\pi y}{2}\,\frac{\Gamma(1/2-s+iy)\,(1/2-s+iy)}{\cos[\pi(1/2-s-iy)/2]}\,\mathcal{P}_{1/2-s}^{-iy}(0)\right\}\right]du,$$

where C is Euler's constant.

An asymptotic solution of the integral equation (4.106) with kernel (4.109) for small β can be obtained by the method of *large* λ (Aleksandrov and Babeshko (1972)). Henceforth, to fix ideas and ensure a solution with finite elastic energy (see conditions (4.103)), we will confine ourselves to the case

$$f_*(\rho,\psi) = f\rho^{\mu}\exp(-\gamma\rho) \quad (\mu \geq \delta - 1,\ \delta > 0,\ \gamma > 0). \tag{4.112}$$

If $\mu = 0$, $\gamma \to 0$ the punch degenerates into a flat punch. Using integral 3.381.4 in Gradshteyn and Ryzhik (1994), we deduce from (4.112) that

$$f_s^*(\psi) = f\gamma^{(-s+1/2+\mu)}\,\Gamma(s+1/2+\mu), \qquad \operatorname{Re} s > -1/2-\mu.$$

Suppose that β is so small that one can ignore terms of order λ^{-2} and higher. Then the solution of the problem can be expressed in the form

$$q_*(\rho,\psi) = \frac{f\gamma^{-\mu+1}}{2\pi i\sqrt{\beta^2-\psi^2}} \int\limits_{\Gamma} \frac{(\rho\gamma)^{-s-3/2}\,\Gamma(s+1/2+\mu)}{g(s)}\,ds, \tag{4.113}$$

where

$$g(s) = \ln(4\lambda) - \mathrm{C} - \frac{1}{2}\,\psi(1/2+s) - \frac{1}{2}\,\psi(1/2-s) + d_0(s). \tag{4.114}$$

Using the theory of residues, one can obtain from (4.113) an approximate solution of the problem for small β, provided that the zeros of the function $g(s)$ (4.114) are known. The position of the straight line Γ is chosen so as to ensure that condition $(4.103)_1$ is satisfied and the integral

$$q_s^*(\psi) = \int\limits_0^{\infty} q_*(\rho,\psi)\rho^{s+1/2}\,d\rho \tag{4.115}$$

converges (Fig. 4.4).

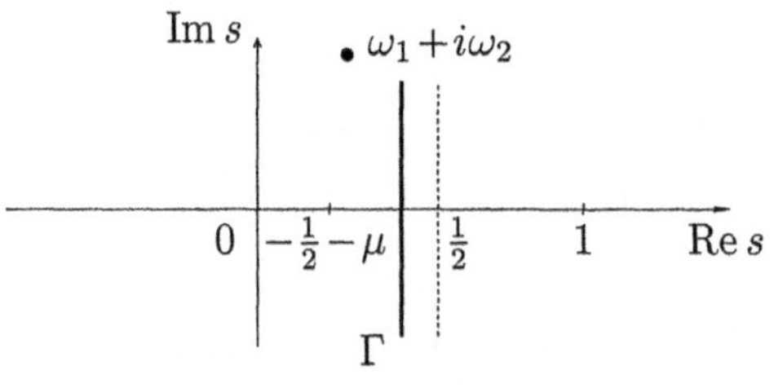

Figure 4.4: The contour Γ in (4.113)

Let us study the zeros of the function $g(s)$ defined by (4.114) in the strip $|\operatorname{Re} s| \leq 3/2$, taking into account that, as follows from the symmetry of the kernel of integral equation (4.106) $K_s(t,p) = K_s(p,t)$, it takes real values on the real and imaginary axes, and $g(-s) = g(s)$. The zeros of the function $g_0(s) = g(s) - d_0(s)$, correpsonding to an elastic half-space, in the strip have been investigated in Aleksandrov and Babeshko (1972) as the following theorem:

Theorem 4.2. *For $\lambda > 1$ the function*

$$g_0(s) = \kappa_* - \frac{1}{2}\psi(1/2+s) - \frac{1}{2}\psi(1/2-s), \quad \kappa_* = \ln(4\lambda) - \mathrm{C},$$

has only four single zeros in the strip $|\operatorname{Re} s| \leq 3/2$:

$$s_m = \pm[\eta_0(\beta) + 1/2], \ m = 1, 2, \quad \eta_0(\beta) = O(1/(2\kappa_*)) > 0,$$

$$s_m = \pm i\theta_0(\beta), \ m = 3, 4, \quad \theta_0(\beta) = O(1/\beta).$$

The function $d_0(s)$ defined by (4.110) in $|\operatorname{Re} s| < 1/2$ must be considered in the region $|\operatorname{Re} s| \geq 1/2$ as an analytic continuation of (4.110). The function $g(s)$ has a simple pole on the real axis at $s = 1/2$, with residue $-V_n$ $(n = 1, 2, 3)$, where

$$V_n = \frac{1}{2} + A_n(1 + \mathcal{B}_n\{0,\, 1\}) - \frac{1}{\pi}, \quad A_n = \lim_{u\to 0} uW^n(u). \tag{4.116}$$

Obviously, $V_3 = 1/2 - 1/\pi \approx 0.182$. Table 4.4 lists the values of V_n, $n = 1, 2$, as functions of $\alpha = \pi k/8$ for $\nu = 0.3$:

Table 4.4: Residues (4.116)

k	1	2	3	4	5	6	7
V_1	8.27	0.969	0.583	0.500	0.493	0.406	0.350
V_2	0.649	0.818	0.462	0.182	0.296	0.394	0.283

In addition $V_1 \to +\infty$ and $V_2 \to 1/2 - 1/\pi$ as $\alpha \to 0$.

It follows from these calculations that, as a rule, $g(1/2 \pm 0) = \mp\infty$ on the real axis. It can be shown that $g(3/2 - 0) = +\infty$, $g(-i\infty) = -\infty$. For sufficiently large values of λ, $g(0) > 0$. Therefore, for a fixed angle α and $\lambda > \lambda_*(\alpha)$, $g(s)$ will have in the strip $|\operatorname{Re} s| \leq 3/2$ two simple zeros on the imaginary axis and two on the real axis: $s_m = \pm i\theta_*$, $m = 1, 2$, $\theta_* = \theta_*(\alpha, \lambda) = O(\lambda)$ $(\lambda \to \infty)$, $s_m = \pm(1/2 + \eta)$, $m = 3, 4$, $\eta = \eta(\alpha, \lambda) \in (0; 1)$, and moreover $\lambda_*(\alpha) = O(\exp[1/(8\alpha)])$ as $\alpha \to 0$. Below we will use the following Rouché's theorem (Korn, G.A. and Korn, T.M. (1968, subsection 7.6-1)).

Theorem 4.3. *If $f_1(s)$ and $f_2(s)$ are analytic throughout a simply connected bounded open region D and its boundary contour C, and if $|f_2(s)| < |f_1(s)| \neq 0$ on C, then $f_1(s)$ and $f_1(s) + f_2(s)$ have the same number of zeros in the region D.*

Using Theorem 4.2 on the zeros of $g_0(s)$ and Theorem 4.3, we conclude that for fixed α, as $\lambda \to \infty$ the zeros of $g(s)$ on the imaginary axis are the only ones in the strip $|\operatorname{Re} s| < 1/2$. For example, if $\lambda = 5$, $\nu = 0.3$, $\alpha = \pi k/8$ $(k = 1, 2, ..., 7)$ then for all types of boundary conditions $(n = 1, 2, 3)$ $s_m \approx \pm i11.2$, $m = 1, 2$, because of the exponential decrease of the function $|\Gamma(x + iy)|$, in expressions (4.110), (4.111) for $d_0(s)$ as $|y| \to \infty$, $x, y \in \Re$ ($\Re$ is the set of the real numbers). For small 2α, comparable with $\beta = 0.2$ (and the same λ, ν), the values of θ_* are as follows:

Table 4.5: Zeros θ_*

2α	0.1	0.2	0.4
$n = 1$	19.3	13.0	11.3
$n = 2$	0.0960	10.2	11.2
$n = 3$	—	8.38	11.2

We recall that the values of $n = 1, 2, 3$ in Table 4.5 correspond to problems $\mathcal{A}$, $\mathcal{B}$ and $\mathcal{C}$, respectively.

When λ is fixed and $\alpha \to 1/(4\lambda) + 0$, the equation $g(s) = 0$ may have additional complex zeros in the strip $|\operatorname{Re} s| < 1/2$. Thus, for an elastic wedge with one rigidly fastened side $(n = 3)$, at $2\alpha = 0.166$, $\lambda = 5$, $\nu = 0.3$, the function $g(s)$ has eight pure imaginary zeros in the interval $(i2.1, i5.1)$ and three real zeros in $(0, 1/2)$. At $2\alpha = 2\beta = 0.4$ ($\lambda = 5$, the elastic wedge angle equals the wedge angle of the punch), $\nu = 0.3$, $n = 3$, there are two additional real roots in the interval $(0.46, 0.5)$.

Let us suppose that the contour Γ in formula (4.113) is contained in the strip $0 < \operatorname{Re} s < 1/2$ and intersects the real axis to the right of the zero of $g(s)$ lying in this strip, say $s = \omega_1 + i\omega_2$, with the largest real part ω_1 (if there are any such zeros), and also to the right of the point $-(1/2 + \mu)$ if it lies in the interval $(0, 1/2)$ (Fig. 4.4). One can then use the theory of residues to find the leading terms of the asymptotic expansion of $q_*(\rho, \psi)$ as $\rho \to 0$. Let us assume from the start that in the strip $|\operatorname{Re} s| < 1/2$ the function $g(s)$ has zeros only on the imaginary axis $(\lambda > \lambda_*(\alpha))$. Then, if $\delta - 1 \leq \mu < -1/2$, the principal singularity of $q_*(\rho, \psi)$ will be $\rho^{\mu-1}$, second to which come oscillatory singularities $\rho^{-3/2} \cos(\theta_* \ln[\rho\gamma])$ and $\rho^{-3/2} \sin(\theta_* \ln[\rho\gamma])$. If $\mu \geq -1/2$, the oscillatory singularities prevail. Thus, in the neighbourhood of the apex of a wedge-shaped punch pressed into the face of an elastic wedge, the contact conditions may be violated. For an elastic wedge with one stress-free side ($n = 1$), the frequency of these oscillations will increase as $\alpha \to 1/(4\lambda) + 0$.

Now suppose that in the strip $0 < \operatorname{Re} s < 1/2$ the equation $g(s) = 0$ has an additional complex root $s = \omega_1 + i\omega_2$, $\omega_2 \neq 0$ (α and β are of the same order of magnitude; $g(s) \neq 0$ for $\operatorname{Re} s = 1/2$). In that case, if $\delta - 1 \leq \mu < -(\omega_1 + 1/2)$, the principal singularity of the contact stress function will be $\rho^{\mu-1}$, second to which will be oscillatory singularities which are stronger than the former and of the order $\rho^{-(\omega_1+i\omega_2)-3/2}$. If $\mu \geq -(\omega_1 + 1/2)$, these oscillatory singularities will prevail.

Using Rouché's theorem, one can show that this qualitative picture remains unchanged if, while using the *large* λ method, one does not ignore terms of order λ^2 and higher. Note that the functions $a_1(s)$, $d_2^1(s)$, $d_2^2(s)$, unlike $a_0(s)$ and $d_0(s)$, do not have poles at $s = \pm 1/2$.

Similar arguments show that $q_*(\rho, \psi) \sim O(\rho^{-2-\eta})$ as $\rho \to \infty$, where $\operatorname{Re} s = 1/2 + \eta$ is the least real part of any root of the equation $g(s) = 0$ in the strip $1/2 < \operatorname{Re} s < 3/2$.

Noting the behaviour of the function $q_*(\rho, \psi)$ as $\rho \to 0$ and $\rho \to \infty$, one sees that the corresponding integral $(4.103)_1$ converges. Integral (4.115) is clearly also convergent, provided that the straight line Γ intersects the real axis slightly to the left of $s = 1/2$.

4.3.2. BUBNOV–GALERKIN METHOD

Babeshko, Glushkov and Zinchenko (1989) used the Bubnov–Galerkin method for the problem of a wedge-shaped punch on a half-space, and obtained results which are in agreement with those in Keer and Parihar (1978) and Parihar and Keer (1979). This method was also used in Glushkov and Glushkova (1992) for a wedge-shaped crack. First of all, we note that the exponent of singularity of the contact pressure function $q_*(\rho, \psi)$ when $\rho \to 0$ is related to the points of the spectrum of the integral operator on the left-hand side of equation (4.106). The poles of $q_s(\xi)$ coincide with those values of s for which there may exist nontrivial solutions of the corresponding homogeneous equation, i.e. points of the spectrum of the integral operator in (4.106). Here such values of s_k do not depend on ξ.

For determination of s_k we carry out a discretization of (4.106) using the Bubnov–Galerkin method: the solution in sought in the form of an expansion in a system of basic functions $v_m(\xi)$:

$$q_s(\xi) = \sum_{m=0}^{\infty} t_m(s) v_m(\xi)$$

while to determine $t_m = t_m(s)$ we use a second basis $\{u_l\}_{l=0}^{\infty}$. As a result, we arrive at the following infinite system:

$$\sum_{m=0}^{\infty} a_{lm} t_m = f_l, \quad l = 0, 1, ..., \tag{4.117}$$

$$a_{lm} = (K_s v_m, w_l)|_{\mathcal{L}_2} = \frac{1}{\pi} \int_{-1}^{1} \int_{-1}^{1} K_s(\beta\xi, \beta x)\, v_m(\xi) w_l(x)\, d\xi\, dx,$$

$$f_l = \int_{-1}^{1} f_s(x) w_l(x)\, dx.$$

Here $\{w_k\}_{k=0}^{\infty}$ is a system of projectors on the basis $\{u_l\}_{l=0}^{\infty}$, i.e. $(u_l, w_k)|_{\mathcal{L}_2} = \delta_{kl}$, where δ_{kl} is the Kronecker delta.

The function $q_s(\xi)$ at $\xi = \pm 1$ has a weak singularity of the form $(1 - \xi^2)^{-1/2}$. To regularize the integral equation (4.106), it is necessary to take singularity into account in the coordinate functions. Hence, it is natural to select as the basis the system of functions

$$v_m(\xi) = \frac{\mathrm{T}_m(\xi)}{\sqrt{1-\xi^2}}; \quad u_l(x) = c_l \mathrm{T}_l(x), \quad c_0 = 1, c_l = 2\,(l \geq 1),$$

where $\mathrm{T}_m(x)$ are Chebyshev polynomials of the first kind. In the second system we do not introduce the singularity, provided that the right-hand side is a smooth function. In view of the condition of orthogonality of the Chebyshev polynomials 8.949.9 in Gradshteyn and Ryzhik (1994)

$$\int_{-1}^{1} \frac{\mathrm{T}_m(x)\,\mathrm{T}_l(x)}{\sqrt{1-x^2}}\, dx = \begin{cases} 0 & \text{for } m \neq l \\ \pi/2 & \text{for } m = l \neq 0 \\ \pi & \text{for } m = l = 0 \end{cases},$$

we obtain

$$w_k(x) = \frac{\mathrm{T}_k(x)}{\pi\sqrt{1-x^2}}.$$

We derive another form for the Mellin transform of the principal part of the kernel in (4.106). This principal part can be presented by formula (1.15)

$$\int_{0}^{\infty} \mathcal{J}_0(tR)\, dt = \frac{1}{R}. \tag{4.118}$$

We expand the Bessel function $\mathcal{J}_0(tR)$ in a product of Bessel functions, in polar coordinates, by the formula 8.531.1 in Gradshteyn and Ryzhik (1994)

$$\mathcal{J}_0(tR) = \mathcal{J}_0(t\rho)\mathcal{J}_0(t\rho_*) + 2\sum_{k=1}^{\infty} \mathcal{J}_k(t\rho)\mathcal{J}_0(t\rho_*)\cos(k\phi), \tag{4.119}$$

$$R = \sqrt{\rho^2 + \rho_*^2 - 2\rho\rho_* \cos\phi}, \tag{4.120}$$

and then use the Mellin transform of the Bessel function (Erdelyi (1954, Vol. 1, formula (1) on p. 326))

$$\int_0^\infty \mathcal{J}_k(t\rho)\rho^{s-1}\, d\rho = \frac{2^{s-1}\,\Gamma(k/2+s/2)}{t^s\,\Gamma(k/2-s/2+1)}, \qquad -k < \operatorname{Re} s < \frac{3}{2}. \tag{4.121}$$

If we compare this with the Mellin transform (4.105), we see that the Legendre function can be expanded in the form

$$\frac{1}{2\cos(\pi s)}\mathcal{P}_{s-1/2}(-\cos(t-p)) = \frac{1}{2\pi}\sum_{k=-\infty}^{\infty} G_k(s)\exp[ik(t-p)], \tag{4.122}$$

$$G_k(s) = \frac{\Gamma(s/2+|k|/2+1/4)\,\Gamma(-s/2+|k|/2+1/4)}{2\Gamma(s/2+|k|/2+3/4)\,\Gamma(-s/2+|k|/2+3/4)}.$$

Using formulae 7.355, 8.940 in Gradshteyn and Ryzhik (1994), we get

$$\int_{-1}^{1} \exp(\pm iax)\frac{\mathrm{T}_m(x)}{\sqrt{1-x^2}}\,dx = \int_0^\pi \exp(\pm ia\cos\vartheta)\cos(m\vartheta)\,d\vartheta \tag{4.123}$$

$$= \pi\exp\left(\frac{\pm im\pi}{2}\right)\mathcal{J}_m(a).$$

Formulae (4.122) and (4.123) give the following expression for the elements of the matrix of system (4.117) $a_{lm} = a_{lm}(s)$ $(l, m = 0, 1, ...)$:

$$a_{lm}(s) = {\sum_{n=0}^{\infty}}' G_n(s)\,\mathcal{J}_m(\beta n)\mathcal{J}_l(\beta n)\cos\left(\frac{l-m}{2}\pi\right) \tag{4.124}$$

$$+\frac{1}{\pi^2}\int_{-1}^{1}\int_{-1}^{1} K_s^*(\beta x, \beta y)\,\frac{\mathrm{T}_m(x)\,\mathrm{T}_l(y)}{\sqrt{1-x^2}\,\sqrt{1-y^2}}\,dx\,dy.$$

Here the prime denotes that the first term $(n = 0)$ of the series in (4.124) has the coefficient $1/2$. If the determinant of the infinite-dimensional matrix

$D(s)$ with elements (4.124) is zero for $s = s_k$, then $q_*(\rho, \psi) \sim \rho^{-\gamma}$, $\gamma = 3/2 + s_k$ as $\rho \to 0$. For $\beta > 0.1\pi$ it suffices to truncate the matrix with elements (4.124) up to dimension 4–5 to guarantee three correct significant digits for zeros of its determinant on the real axis. Note that the functions $R_\pm(s, u, t)$ (4.108) take real values for $s \in \Re$; for calculation of values of Legendre functions, which are contained in expressions for $R_\pm(s, u, t)$, one can use their representation through the hypergeometric function 8.704 in Gradshteyn and Ryzhik (1994)

$$\mathcal{P}^{-iu}_{s-1/2}(x) = \frac{1}{\Gamma(1+iu)} \left(\frac{1+x}{1-x}\right)^{-iu/2} \mathcal{F}\left(\frac{1}{2} - s, \frac{1}{2} + s, 1 + iu; \frac{1-x}{2}\right). \tag{4.125}$$

In Tables 4.6–4.8 for problems $\mathcal{A}$, $\mathcal{B}$ and $\mathcal{C}$, respectively, we give the values of the largest singularity exponent γ, corresponding to real zeros of $D(s)$, $s \in (-3/2; -1/2)$, for various α and β (in degrees) and $\nu = 0.3$.

Table 4.6: Singularity exponent for problem $\mathcal{A}$

2α	45	90	105	120	135	180
$2\beta = 45$	0.999	0.985	0.972	0.868	0.862	0.788
$2\beta = 90$	0.999	0.987	0.966	0.785	0.740	0.703
$2\beta = 135$	0.999	0.990	0.761	0.662	0.632	0.611

Table 4.7: Singularity exponent for problem $\mathcal{B}$

2α	45	90	135	180	225	270	315
$2\beta = 45$	0.448	0.528	0.447	0.390	0.411	0.422	0.412
$2\beta = 90$	0.224	0.295	0.224	0.180	0.194	0.204	0.195
$2\beta = 135$	0.158	0.080	0.246	0.416	0.353	0.310	0.350

Table 4.8: Singularity exponent for problem $\mathcal{C}$

2α	45	90	135	180	225	270	315
$2\beta = 45$	0.626	0.685	0.704	0.711	0.714	0.719	0.721
$2\beta = 90$	0.555	0.602	0.622	0.629	0.632	0.637	0.640
$2\beta = 135$	0.520	0.542	0.555	0.559	0.561	0.565	0.567

For problem $\mathcal{A}$ with $2\alpha = 225°$, $270°$, $315°$ and for the same values of 2β as in Table 4.6, the values of γ differ less than 6% from the corresponding values given in the last column of this table; the last column is taken from Babeshko, Glushkov and Zinchenko (1989) (a wedge-shaped punch on a half-space). For wedges with the angles $2\alpha > \pi$ the norms of the integrals operators in (4.101) are very small (see Table 4.1), and in $W^n(u) - \coth(\pi u) \sim \exp(-2\alpha u)$ as $u \to \infty$, so that in representation (4.104) the term $\coth(\pi u)$, corresponding to a half-space, plays the main role. That is why the solution of the contact problem for a wedge with big angle is close to that for a half-space.

Using the theorem on implicit functions we conclude that the roots $s_k(\alpha, \beta)$ of the equation $D(s) = 0$ depend analytically on α and β. Therefore, these roots describe continuous curves in the s-plane as α and β change continuously.

Calculations show that for problem $\mathcal{A}$ with $2\beta = \pi$ and $2\alpha \approx 100°$, two additional zeros of $D(s)$ ($\gamma_1 = 0.98$, $\gamma_2 = 0.96$) appear on the interval $s \in (-3/2; -1/2)$ near the point $s = -1/2$. If we fix α and decrease 2β, these zeros coalesce into a double root, which give the singularity of order $\rho^{-\gamma}(C_1 + C_2 \ln \rho)$, and then the zeros get off the real axis and become complex conjugate (Fig. 4.5). This brings oscillations in the contact pressure as $\rho \to 0$, and separation of the apex of the punch from the surface of the wedge.

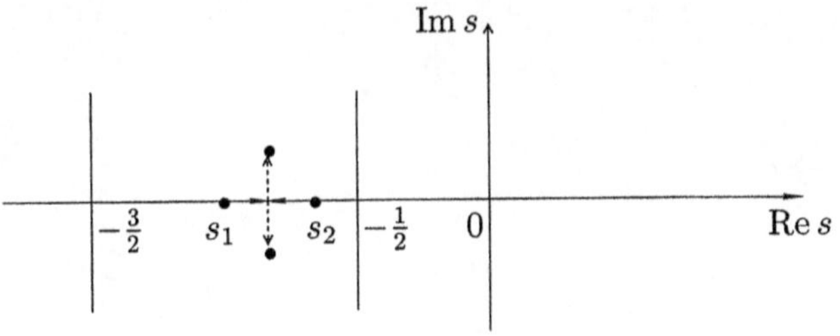

Figure 4.5: Two zeros coalesce and then become complex conjugate

In problem $\mathcal{C}$ for fairly acute angles α one can see zeros of $D(s)$ as $s \in (0; 1/2)$. For example, when $2\alpha = 0.1\pi$, $2\beta = 45°$ we have $s_1 = 0.45$, $s_2 = 0.47$. On combining and going into the complex plane, these roots bring the principal singularity of order $\gamma = \omega_1 + i\omega_2 + 3/2$, $\omega_1 \in (0; 1/2)$, and were tested with help of the asymptotic method.

Note that problem $\mathcal{B}$ is equivalent to the problem of a symmetric embedding of two identical punches into the different faces of the elastic wedge with double aperture angle. Moreover, this holds for arbitrary, not only for wedge-shaped, identical punches (Fig. 4.6a,b). In particular, a comparison of the last column of Table 4.6 (one punch on a half-space) with the column $2\alpha = 90°$ of Table 4.7 (two punches on a half-space) enables us to estimate a degree of correlation with two wedge-shaped punches on a half-space. It is worth mentioning that there is Rvachev's exact solution for two wedge-shaped plane punches on a half-space (problem $\mathcal{B}$ with $2\alpha = \pi/2^4$, Rvachev and Protsenko (1977)).

This solution has no singularities as $\rho \to 0$ because the plane shape of the punch ($f(r, z)$ = const in (4.101)) violate condition $(4.103)_2$, integral $(4.101)_1$ and the elastic energy integral

$$\int_{-\beta}^{\beta} d\psi \int_{0}^{\infty} q_*(\rho, \psi) f_*(\rho, \psi) \rho \, d\rho$$

diverge for Rvachev's solution. In other words, this solution gives infinite elastic energy, whereas Kondrat'ev (1967) in his famous paper showed that the uniqueness of the solution for a wedge-shaped domain is provided only by the finiteness of the energy integral.

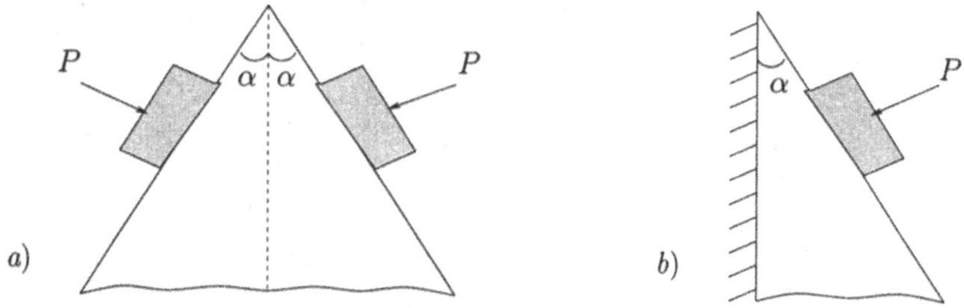

Figure 4.6: The symmetric problem for two identical punches (section *a*) is equivalent to the problem for one punch, with the sliding support on the other face of the wedge (problem $\mathcal{B}$), the angle of which is twice smaller (section *b*)

In conclusion we underline that the knowledge of the stress singularity exponent is useful and important for investigation of contact problems with polygonal punches. For instance, using the finite or boundary element method for such complicated problems, one must exactly fix a functional space in which the solution is sought. Moreover, if we know the behaviour of the contact stress near the angular point, we can correctly choose basis functions to solve the complicated problems numerically.

[4]It can be shown that for contact problem $\mathcal{B}$ with $2\alpha = \pi/2$ the kernel of the integral equation (4.127) is $1/R + 1/R_+$, $R_+ = \sqrt{(r+x)^2 + (z-y)^2}$. The same kernel appears in the integral equation for the contact of two punches with a half-space, if the punches are symmetric.

Galin (1961) was the first who considered the contact problem for a wedge-shaped punch. In his investigation the effect of the punch on a half-space is accompanied by the effect of some loading outside the contact region, and the contact pressure has a ρ^{-1} singularity at the wedge apex.

The principal conclusion of this section is that there is an oscillation of the contact pressure at the punch apex with a fairly acute angle. This oscillation is found analytically (theorem 4.2), and its physical meaning is that the acute tip of the rigid punch will separate from the elastic base.

More comments on stress singularities at a sharp corner can be found in Gladwell (1980, chapter 9).

4.4 Elliptic Punch on a Wedge Face

Let a rigid punch whose surface is an elliptic paraboloid

$$f(r,z) = (r-a)^2/(2R_1) + z^2/(2R_2) \tag{4.126}$$

be pressed into the side $\varphi = \alpha$ of a wedge by a force P applied along the $z = 0$ axis at a distance H from the edge. Due to the action of the force P, the punch settles by an amount δ and rotates through an angle γ about the straight line $r = a$, $\varphi = \alpha$. Suppose that $R_1 < R_2$, the unknown contact zone is an ellipse Ω: $(r-a)^2/c^2 + z^2/b^2 \leq 1$ (in Fig. 4.7 only half of the contact ellipse is shown), $a > c$, $b \geq c$. It is required to determine, for specified values of P, R_1, R_2, a, the distribution $\sigma_\varphi(r,\alpha,z) = -q(r,z)$, $(r,z) \in \Omega$, of the normal contact stress under the punch and also the quantities $\lambda = a/b$, b, c, δ, γ and H. We again consider problems $\mathcal{A}$, $\mathcal{B}$ and $\mathcal{C}$, corresponding, respectively, to conditions (4.14), (4.42) and (4.43) in the $\varphi = -\alpha$ plane.

The dimensionless parameter λ, which has been introduced here, characterizes the relative remoteness of the contact region from the edge of the wedge.

By using identity (4.104) and integral (4.84), we separate out the singular part $1/R$ of the kernel of the integral equation. We obtain the integral equation in the form

$$\int_\Omega q(x,y)\left[\frac{1}{R} + T(x,y,r,z)\right] d\Omega = 2\pi\theta g(r,z), \quad (r,z) \in \Omega, \tag{4.127}$$

$$g(r,z) = \delta + \gamma(r-a) - f(r,z), \quad R = \sqrt{(r-x)^2 + (z-y)^2},$$

$$T(x,y,r,z) = \frac{4}{\pi^2}\int_0^\infty\int_0^\infty \left[\sinh\frac{\pi u}{2}\, W(u,\alpha,\beta x) - \cosh(\pi u)\, \mathcal{K}_{iu}(\beta x)\right] \tag{4.128}$$

$$\times \mathcal{K}_{iu}(\beta r) \cos[\beta(z-y)]\, d\beta\, du.$$

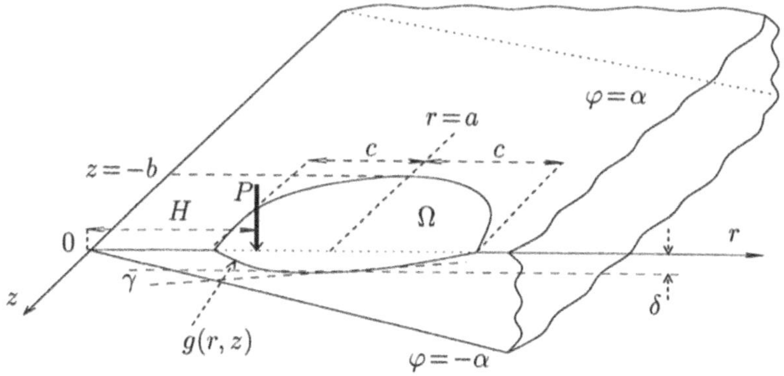

Figure 4.7: A half of the elliptic contact area Ω; the force P, settlment δ, rotation angle γ; the function $g(r, z)$ from (4.127)

Next, we use the following dimensionless quantities and notation (we shall omit the primes):

$$r' = \frac{r-a}{b}, \quad x' = \frac{x-a}{b}, \quad z' = \frac{z}{b}, \quad y' = \frac{y}{b}, \quad \delta' = \frac{\delta}{b}, \quad c' = \frac{c}{b}, \tag{4.129}$$

$$R' = \frac{R}{b}, \quad H' = \frac{H}{b}, \quad P' = \frac{P}{2\theta b^2}, \quad q'(x', y') = \frac{q(x,y)}{2\theta}, \quad A = \frac{b}{2R_1}, \tag{4.130}$$

$$B = \frac{b}{2R_2}, \quad T'(x', y', r', z') = bT(x, y, r, z), \quad \Omega' : \frac{r'^2}{c'^2} + z'^2 \le 1. \tag{4.131}$$

Lemma 4.2. *The function $T(x, y, r, z)$ and all its derivatives are continuous when (x, y), $(r, z) \in \Omega$. When $\lambda > 1 + c$ $(1 \le 2\alpha < 2\pi)$, $\lambda > (2\alpha)^{-1} + c$ $(c/2 \le \alpha \le 1)$, $\lambda > \sqrt{1 + c^2(1 + 4\alpha^2)}(2\alpha)^{-1}$ $(0 < 2\alpha \le c/2)$, the function $T(x, y, r, z)$ $[(x, y), (r, z) \in \Omega]$ can be represented by the absolutely convergent series*

$$T(x, y, r, z) = \sum_{n=1}^{\infty} \frac{f_n(x, y, r, z)}{\lambda^n}, \tag{4.132}$$

where $f_n(x, y, r, z)$ are certain polynomials.

Expansion (4.132) is obtained by term-by-term integration of the functional series (4.69), (4.70) for $W(u,\alpha,\beta x)$ in formula (4.128) using representations (4.1), (4.125) and the integrals 2.16.36.2,3 in Prudnikov, Brychkov and Marichev (1992, Vol. 2)

$$\frac{4}{\pi^2}\int_0^\infty \mathcal{K}_{iu}(\beta x)\,\mathcal{K}_{iu}(\beta r)\cos[\beta(z-y)]\,d\beta = \frac{1}{\sqrt{xr}\cosh(\pi u)}\mathcal{P}_{iu-1/2}\left(1+\frac{R^2}{2rx}\right),$$

$$\mathcal{P}_{iu-1/2}(1+\omega^2) = \mathcal{F}\left(\frac{1}{2}-iu,\frac{1}{2}+iu,1;-\frac{\omega^2}{2}\right), \tag{4.133}$$

$$x\int_0^\infty \mathcal{K}_{ig}(\beta x)\,\mathcal{K}_{iu}(\beta x)\,\cos[\beta(z-y)]\,d\beta = \sum_{n=0}^\infty(-1)^n\left(\frac{z-y}{x}\right)^{2n} \tag{4.134}$$

$$\times\int_0^\infty\int_0^\infty \frac{\cos(us)\,\cos(gt)\,ds\,dt}{(\cosh s+\cosh t)^{2n+1}} = \sum_{n=0}^\infty\left(\frac{z-y}{x}\right)^{2n}$$

$$\times\frac{(-1)^n\,\pi^2\,p(n)}{2^{2n+1}\,[(2n)!]^2\,[\cosh(\pi u)+\cosh(\pi g)]}\quad (|(z-y)/x|<2),\quad p(0)=0,$$

$$p(n) = \prod_{k=0}^{n-1}[(1+2k)^2+(u+g)^2][(1+2k)^2+(u-g)^2],\ \ n=1,2,...,$$

$$\int_0^\infty \mathcal{K}_{ig}[\beta(x+\lambda)]\,\mathcal{K}_{iu}[\beta(r+\lambda)]\,\cos[\beta(z-y)]\,d\beta$$

$$=\sum_{n=0}^\infty(-1)^n\frac{(z-y)^{2n}}{\lambda^{2n+1}}\int_0^\infty\int_0^\infty\frac{\cos(us)\,\cos(gt)\,ds\,dt}{[(1+x/\lambda)\cosh t+(1+r/\lambda)\cosh s]^{2n+1}}.$$

We must also expand the hypergeometric function in a series in (4.133). Terms in which r or x occur in the denominator are expanded in Taylor series in powers of r/λ or x/λ. Since all the series converge absolutely for the λ indicated in the lemma, they can be regrouped in the form of (4.132). Doing so, it is easy to determine the explicit form of the functions $f_n(x,y,r,z)$, $n=1,2,\ldots$.

Now, to construct the solution of the integral equation (4.127) we use the asymptotic method for *large* λ, which is effective when the punch is sufficiently far from the edge of the wedge. By expanding the solution of the integral equation (4.127) in the form

$$q(x,y) = \sum_{n=0}^{\infty} \frac{q_n(x,y)}{\lambda^n}, \tag{4.135}$$

taking formulae (4.126), (4.129)–(4.132) into account and equating terms of similar powers of λ, we obtain an infinite system of integral equations for $q_n(x,y)$ $(n = 0, 1, ...)$:

$$\int_\Omega \frac{q_0(x,y)}{R}\, d\Omega = \pi(\delta + \gamma r - Ar^2 - Bz^2), \tag{4.136}$$

$$\int_\Omega \frac{q_1(x,y)}{R}\, d\Omega = -(a_0 + \kappa_1) \int_\Omega q_0(x,y)\, d\Omega, \tag{4.137}$$

$$\int_\Omega \frac{q_2(x,y)}{R}\, d\Omega = \int_\Omega \left\{ \left[x\left(\frac{a_0}{2} + \kappa_2^{01}\right) + r\left(\frac{a_0}{2} + \kappa_2^{10}\right)\right] q_0(x,y) \right. \tag{4.138}$$

$$\left. -(a_0 + \kappa_1) q_1(x,y) \right\} d\Omega,$$

$$\int_\Omega \frac{q_3(x,y)}{R}\, d\Omega = -\int_\Omega \left\{ \left[x^2\left(\frac{3}{8}a_0 + \kappa_3^{02}\right) + r^2\left(\frac{3}{8}a_0 + \kappa_3^{20}\right) \right. \right. \tag{4.139}$$

$$\left. + rx\left(\frac{a_0}{4} + 2\kappa_3^{11}\right) + \frac{a_1}{2}R^2 - (z-y)^2\kappa_3 \right] q_0(x,y)$$

$$\left. - \left[x\left(\frac{a_0}{2} + \kappa_2^{01}\right) + r\left(\frac{a_0}{2} + \kappa_2^{10}\right)\right] q_1(x,y) + (a_0 + \kappa_1) q_2(x,y) \right\} d\Omega$$

etc., where $(r,z) \in \Omega$,

$$a_0 = \int_0^\infty (\tanh(\pi u)\, W^m(u) - 1)\, du, \tag{4.140}$$

$$a_1 = -\int_0^\infty (\tanh(\pi u)\, W^m(u) - 1) \left(\frac{1}{4} + u^2\right) \frac{du}{2}, \quad m = 1, 2, 3. \tag{4.141}$$

The constants a_0, a_1 correspond to the term $W^m(u) - \coth(\pi u)$ in identity (4.104). The constants κ_i^{jk} are connected with the other terms in the kernel of the integral equation, and can be found by means of the formulae (operators $\mathcal{D}_m$ are given in formulae (4.102), $m = 1, 2, 3$)

$$\kappa_i^{jk} = \frac{8}{\pi^2} \int_0^\infty \sinh \frac{\pi u}{2} W^m(u)\, \mathcal{D}_m \left\{ u, \cosh \frac{\pi y}{2} J_i^{jk}(u, y) \right\} du, \quad \kappa_i^{00} = \kappa_i,$$

$$J_i^{jk}(u, y) = \int_0^\infty \int_0^\infty \frac{\cos(us)\cos(yt)}{(\cosh s + \cosh t)^i} \cosh^j s \cosh^k t \, ds \, dt, \quad j + k < i. \tag{4.142}$$

With help of formula (4.134) it is possible to find the integrals $J_i^{jk}(u, y)$ in the form

$$J_1^{00}(u, y) = \frac{\pi^2}{2[\cosh(\pi u) + \cosh(\pi y)]}, \quad J_2^{01}(u, y) = \frac{\pi^2(1 + u^2 - y^2)}{4([\cosh(\pi u) + \cosh(\pi y)]},$$

$$J_2^{10}(u, y) = J_2^{01}(y, u), \quad J_3^{00}(u, y) = \frac{\pi^2(1 + (u + y)^2)(1 + (u - y)^2)}{32[\cosh(\pi u) + \cosh(\pi y)]},$$

$$J_3^{02}(u, y) = \frac{\pi^2}{64[\cosh(\pi u) + \cosh(\pi y)]}[(1 + (u + y)^2)(1 + (u - y)^2)$$
$$+2((u + y)^2 - 3)((u - y)^2 - 3) + 32(u^2 - y^2)],$$

$$J_3^{11}(u, y) = \frac{\pi^2}{64[\cosh(\pi u) + \cosh(\pi y)]}[(3 - (u + y)^2)(1 + (u - y)^2)$$
$$+(3 - (u - y)^2)(1 + (u + y)^2)], \quad J_3^{20}(u, y) = J_3^{02}(y, u).$$

It is easily to check analytically that the term $(u^2-y^2)/[\cosh(\pi u)+\cosh(\pi y)]$ in the formulae for $J_2^{01}(u, y)$, $J_2^{10}(u, y)$, $J_3^{02}(u, y)$, $J_3^{20}(u, y)$ gives zero contribution to the corresponding constants κ_i^{jk}. Therefore,

$$\kappa_2^{01} = \kappa_2^{10} = \kappa_1/2, \quad \kappa_3^{02} = \kappa_3^{20}. \tag{4.143}$$

Thus, only four among the seven constants κ_i^{jk} in equations (4.136)–(4.139) are independent. Formulae (4.143) are directly connected to the symmetry of the kernel of the integral equation (4.127): $T(x, y, r, z) = T(r, y, x, z)$.

We can see that equations (4.136)–(4.139) are similar to equation (1.29) for the contact problem for a half-space. We know from subsection 1.5.1 that this equation has an exact solution for an elliptic domain Ω and a polynomial right-hand side. Due to formulae (1.51)–(1.54) both these conditions are satisfied for equations (4.136)–(4.139).

The solution in form (4.135) for the wedge will depend on the constants a_i and κ_i^{jk}. To calculate the constants κ_i^{jk} it is reasonable, instead of summing the Neumann series for the operators $\mathcal{D}_m$, to solve the corresponding Fredholm integral equations of the second kind using the collocation method and Gaussian quadrature formula. For example, for problem $\mathcal{A}$

$$\kappa_i^{jk} = \frac{4}{\pi^2} \int_0^\infty \sinh \frac{\pi u}{2} \left[W_1(u,\alpha)\Phi_1(u) - W_2(u,\alpha)\Phi_2(u)\right] du. \tag{4.144}$$

The integral in formula (4.144) can be calculated by Gaussian quadrature formula where the values $\Phi_l(u)$, $l = 1, 2$, at the nodes u_n of the quadrature formula are determined from a Fredholm integral equation of the second kind:

$$\Phi_l(u) = (1-2\nu)\int_0^\infty L_l(u,y)\left[\Phi_l(y) + \cosh\frac{\pi y}{2} J_i^{jk}(u_n, y)\right] dy, \quad 0 \le u < \infty.$$

The values of the constants a_0, a_1, κ_1, κ_3, κ_3^{02} and κ_3^{11}, calculated for $\alpha = \pi k/12$, $k = 1, 2, \ldots, 11$ and $\nu = 0.3$ are given in Tables 4.9–4.11 for problems $\mathcal{A}$, $\mathcal{B}$ and $\mathcal{C}$, respectively. The symbol E-n in these tables means multiplication by 10^{-n}.

Table 4.9: Factors (4.140)–(4.142) for problem $\mathcal{A}$

k	a_0	a_1	κ_1	κ_3	κ_3^{02}	κ_3^{11}
1	78.7	−46.1	−16.8	−4.60	−5.34	−4.57
2	7.68	−3.19	−1.94	−0.509	−0.654	−0.518
3	1.38	−0.450	−0.455	−0.105	−0.171	−0.116
4	0.235	−6.36E-2	4.91E-2	1.50E-2	1.65E-2	1.26E-2
5	1.93E-2	−4.52E-3	4.56E-2	1.31E-2	1.54E-2	1.18E-2
6	0.00	0.00	0.00	0.00	0.00	0.00
7	−9.33E-3	1.78E-3	2.47E-2	6.76E-3	8.69E-3	6.31E-3
8	−4.93E-2	8.83E-3	1.91E-2	5.41E-3	6.60E-3	4.91E-3
9	−0.103	1.77E-2	4.42E-3	1.10E-3	1.72E-3	1.08E-3
10	−0.141	2.38E-2	3.90E-3	1.69E-3	5.33E-4	1.26E-3
11	−0.157	2.62E-2	2.68E-3	1.24E-3	2.52E-4	9.02E-4

In the successive solution of equations (4.136)–(4.139), it can be shown that their right-hahd sides are always polynomials in r and z and their solution can therefore be found in closed form using the formulae from subsection 1.5.1. According to formula (1.52), each function $q_n(x,y)$ ($n = 0, 1, \ldots$) has a square root singularity on the boundary $\partial\Omega$ of the contact ellipse. As we know from subsection 1.5.2, for the elliptic paraboloid (4.126) the contact pressure must be zero on the boundary. It is necessary to set $q(x, y) = 0$, $(x, y) \in \partial\Omega$, by virtue of the smoothness of the shape of the base of the punch. The last condition gives additional relations between the parameters of the contact.

Table 4.10: Factors (4.140)–(4.142) for problem $\mathcal{B}$

k	a_0	a_1	κ_1	κ_3	κ_3^{02}	κ_3^{11}
1	−1.38	4.39	0.152	0.173	−9.35E-2	7.96E-2
2	0.129	0.121	0.158	6.82E-2	2.81E-2	4.80E-2
3	0.500	−0.125	0.00	0.00	0.00	0.00
4	0.222	−5.45E-2	4.82E-2	1.63E-2	1.32E-2	1.34E-2
5	−0.190	2.94E-2	1.71E-2	7.36E-3	2.58E-3	5.42E-3
6	−0.318	5.31E-2	0.00	0.00	0.00	0.00
7	−0.255	4.22E-2	7.90E-3	3.42E-3	1.12E-3	2.53E-3
8	−0.157	2.64E-2	7.24E-3	2.84E-3	1.45E-3	2.19E-3
9	−0.115	1.99E-2	0.00	0.00	0.00	0.00
10	−0.154	2.55E-2	4.70E-3	1.81E-3	9.84E-4	1.41E-3
11	−0.222	3.51E-2	3.43E-3	1.40E-3	5.99E-4	1.07E-3

Table 4.11: Factors (4.140)–(4.142) for problem $\mathcal{C}$

k	a_0	a_1	κ_1	κ_3	κ_3^{02}	κ_3^{11}
1	−2.49	8.29	1.71E-2	7.84E-2	−8.07E-2	2.93E-2
2	−1.08	0.894	2.46E-2	3.46E-2	−2.48E-2	1.60E-2
3	−0.611	0.220	1.74E-2	1.18E-2	−2.75E-3	7.13E-3
4	−0.419	9.07E-2	−3.63E-3	−2.61E-3	7.57E-4	−1.54E-3
5	−0.350	6.17E-2	−3.61E-3	−1.92E-3	−6.02E-5	−1.30E-3
6	−0.327	5.50E-2	0.00	0.00	0.00	0.00
7	−0.313	5.16E-2	−9.58E-4	−4.40E-4	−1.00E-4	−3.19E-4
8	−0.296	4.76E-2	1.95E-3	8.92E-4	2.10E-4	6.47E-4
9	−0.277	4.39E-2	3.44E-3	1.50E-3	4.63E-4	1.11E-3
10	−0.264	4.14E-2	9.83E-5	3.20E-5	2.63E-5	2.80E-5
11	−0.256	4.01E-2	−7.72E-4	−3.31E-4	−1.13E-4	−2.47E-4

By invoking the equilibrium conditions

$$\int_{\Omega} q(x, y)\, d\Omega = P, \quad \int_{\Omega} q(x, y) x\, d\Omega = P(H - \lambda)$$

and introducing notation by means of the formulae

$$S_{00} = \mathcal{K}, \quad S_{01} = \frac{\mathcal{E} - c^2\mathcal{K}}{c^2(1 - c^2)}, \quad S_{10} = \frac{\mathcal{K} - \mathcal{E}}{1 - c^2},$$

$$S_{11} = \frac{(1+c^2)\mathcal{E} - 2c^2\mathcal{K}}{3c^2(1-c^2)^2}, \quad S_{02} = \frac{2(1-2c^2)\mathcal{E} + c^2(3c^2-1)\mathcal{K}}{3c^4(1-c^2)^2},$$

$$S_{20} = \frac{(3-c^2)\mathcal{K} - 2(2-c^2)\mathcal{E}}{3(1-c^2)^2}, \quad f_0 = 3c^3(S_{11}^2 - S_{02}S_{20}),$$

$$f_1 = (2S_{20} - c^2 S_{11})/f_0, \quad f_2 = (2S_{11} - c^2 S_{02})/f_0,$$

$$f_3 = (2c^2 S_{02} - S_{11})/f_0, \quad f_4 = (2c^2 S_{11} - 2S_{20})/f_0,$$

$$f_5 = \frac{(a_0 + 2\kappa_2^{10})((a_0+\kappa_1)S_{00}^{-1}\lambda^{-3} - \lambda^{-2})}{1 + (a_1 - a_0/4 - 2\kappa_3^{11})/(3\lambda^3 S_{01})},$$

$$f_6 = \frac{(a_0/2 + \kappa_2^{10})f_5}{6S_{00}S_{11}} + \frac{(a_0+\kappa_1)^2}{S_{00}^2}, \quad \varepsilon = -\frac{3a_0}{8} - \frac{a_1}{2} - \kappa_3^{20},$$

$$f_7 = \frac{f_6}{\lambda^2} - \frac{a_0+\kappa_1}{\lambda S_{00}}\left(1 + \frac{f_6}{\lambda^2}\right) - \frac{cS_{10}}{2\lambda^3 S_{00}}((a_1 - 2\kappa_3)(c^2 f_3 - f_2) + 2\varepsilon(f_1 - c^2 f_4)),$$

$$f_8 = 1 + f_7 + \frac{\varepsilon c^2}{3\lambda^3 S_{00}} - \frac{a_1}{6\lambda^3 S_{00}}, \quad f_9 = c^2\left(\frac{f_7}{3} + \frac{\varepsilon c^2 - a_1/6}{5\lambda^3 S_{00}}\right),$$

$$f_{10} = \frac{f_7}{3} + \frac{\varepsilon c^2/3 - a_1/2}{5\lambda^3 S_{00}}, \quad f_{11} = \frac{3S_{00}}{\pi f_8}, \quad f_{14} = -\frac{\pi c f_{12}}{3\lambda^3 S_{00}},$$

$$f_{12} = (a_1 - 2\kappa_3)f_3 - 2\varepsilon f_4, \quad f_{13} = 2\varepsilon f_1 - (a_1 - 2\kappa_3)f_2,$$

$$f_{15} = -\frac{2\pi c}{3}\left[-f_4 + \frac{f_{12}c}{\lambda^3}\left(-\frac{f_1 S_{10}}{2S_{00}} + \frac{c^2 f_4 S_{01}}{2S_{00}} + \frac{f_1 - f_4}{3}\right)\right],$$

$$f_{16} = -\frac{2\pi c}{3}\left[f_3 + \frac{f_{12}c}{\lambda^3}\left(\frac{f_2 S_{10}}{2S_{00}} - \frac{c^2 f_3 S_{01}}{2S_{00}} - \frac{f_2 - f_3}{3}\right)\right],$$

$$f_{17} = \frac{2cS_{00}}{f_8}\left[f_4\left(f_{10} - \frac{c^2 S_{01}}{2S_{00}}f_8\right) - \frac{f_1}{c^2}\left(f_9 - \frac{c^2 S_{10}}{2S_{00}}f_8\right)\right],$$

$$f_{18} = \frac{2cS_{00}}{f_8}\left[\frac{f_2}{c^2}\left(f_9 - \frac{c^2 S_{10}}{2S_{00}}f_8\right) - f_3\left(f_{10} - \frac{c^2 S_{01}}{2S_{00}}f_8\right)\right],$$

$$f_{19} = f_1 + f_4 + \frac{c(f_{13} - f_{12})}{\lambda^3}\left[f_1\left(\frac{1}{3} - \frac{S_{10}}{2S_{00}}\right) - f_4\left(\frac{1}{3} - \frac{c^2 S_{01}}{2S_{00}}\right)\right],$$

$$f_{20} = f_2 + f_3 + \frac{c(f_{13} - f_{12})}{\lambda^3}\left[f_2\left(\frac{1}{3} - \frac{S_{10}}{2S_{00}}\right) - f_3\left(\frac{1}{3} - \frac{c^2 S_{01}}{2S_{00}}\right)\right],$$

where $\mathcal{K} = \mathcal{K}(\sqrt{1-c^2})$, $\mathcal{E} = \mathcal{E}(\sqrt{1-c^2})$ are the complete elliptic integrals of the first and second kind, respectively, we finally obtain that

$$q(x,y) = \frac{3P}{2\pi c}\sqrt{1 - \frac{x^2}{c^2} - y^2}, \quad H = \lambda, \tag{4.145}$$

$$\frac{B}{A} = \frac{2S_{00}f_{19}(1 - f_{11}f_{14}) - (f_{12} - f_{13})(f_{11}f_{15} + f_{17})/\lambda^3}{2S_{00}f_{20}(1 - f_{11}f_{14}) - (f_{12} - f_{13})(f_{11}f_{16} + f_{18})/\lambda^3}, \tag{4.146}$$

$$\frac{P}{B} = \frac{(f_{12}f_{14} + f_{15})A/B + f_{14}f_{18} + f_{16}}{1 - f_{11}f_{14}}, \tag{4.147}$$

$$\frac{\delta}{B} = \frac{(f_{11}f_{15} + f_{17})A/B + f_{11}f_{16} + f_{18}}{1 - f_{11}f_{14}}, \tag{4.148}$$

$$\frac{\gamma}{B} = f_5\left[\frac{\delta}{2S_{00}B} + c\left(\frac{A}{B}f_1 - f_2\right)\left(\frac{1}{3} - \frac{S_{10}}{2S_{00}}\right) + c\left(f_3 - \frac{A}{B}f_4\right)\left(\frac{1}{3} - \frac{c^2 S_{01}}{2S_{00}}\right)\right]. \tag{4.149}$$

Formulae (4.145)–(4.149) determine the solution of the problem with an accuracy up to $O(\lambda^{-4})$. Equations (4.146) and (4.147) serve for the determination of λ and c. But in practical analysis, instead of solving this system of two equations, it is more convenient to assume temporarily that the value of λ is known and the value of a is not yet given. Then equation (4.146) serves for the determination of c or the eccentricity $e = \sqrt{1-c^2}$ of the contact ellipse. It is then possible to find b from equation (4.147) and this means also the values of a and c (dimensional see (4.129)). The extent of the embedding and the skewness of the punch are determined from equations (4.148) and (4.149).

As $\lambda \to \infty$, solution (4.145)–(4.149) reduces to Lur'e's well-known solution for an elliptic paraboloid pressed into an elastic half-space (formulae (1.64)–(1.69)).

Let us carry out a numerical analysis of solution (4.145)–(4.149), taking problem $\mathcal{A}$ with $2\alpha = \pi/2$ (quarter-space) and $\nu = 0.3$. The plots of relation (4.146), which relates the ratio, c of the semiaxis of the contact ellipse, to the ratio of the radii of curvature of the punch $R_1/R_2 = B/A$ are shown in Fig. 4.8*a* for $\lambda = 2$ and $\lambda = \infty$. The calculations show that, when $0.1 \leq c \leq 0.9$, the difference between the corresponding values of R_1/R_2 for $\lambda = 6$ and $\lambda = \infty$ does not exceed 0.1%. The quantity δ/B from equation (4.148) when $\lambda \to \infty$ tends extremely slowly to its limiting value and, for example, when $\lambda = 6$, $c = 0.1$, differs from it by 2.2%. Hence, the closeness of the contact region to the edge of the wedge has a far greater effect on the extent of embedding of the punch than on the eccentricity of the contact ellipse.

The quantity γ/B is negative when $\lambda = 2, 4, 6$; $0.1 \le c \le 0.9$ and $\gamma/B \to 0$ when $\lambda \to \infty$. It follows from (4.129), (4.130), (4.146)–(4.148) that the quantity δ is proportional to $P^{2/3}$ (here, δ and P are dimensional, the same proportion for a half-space is given, e.g. in Johnson (1985, appendix A3)), and the quantity γ is proportional to $P^{1/3}$. Plots of the dependence of the quantities

$$\alpha_1 = 2\delta R_2^{1/3}(\theta/P)^{2/3}, \quad \alpha_2 = -2\gamma R_2^{2/3}(\theta/P)^{1/3} \tag{4.150}$$

(δ and P are dimensional here) on c are shown for different λ in Fig. 4.8*b* by the solid and dashed lines, respectively. It is seen that the embedding of the punch becomes larger as the shape of the contact region approaches a circular shape (c increases).

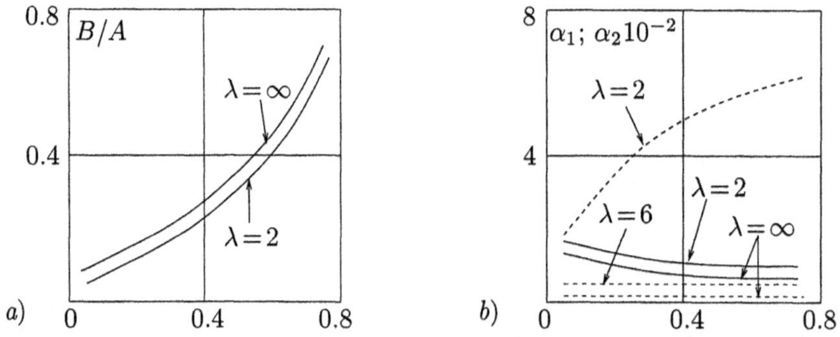

Figure 4.8: *a*) Dependence between $R_1/R_2 = B/A$ and c; *b*) dependence of quantities (4.150) on c

Formulae (4.145) show that the force is applied at the centre of the contact ellipse and that the distribution of the contact stresses is symmetrical about the axes of symmetry of this ellipse. It turns out that the terms which take account of the asymmetry of the function $q(x, y)$ and of the point of application of the force have an order of magnitude of λ^{-4}. In order to take account of them, that is to solve integral equation (4.127) with an accuracy of $O(\lambda^{-5})$ it is possible to specify only one of the quantities R_1 or R_2 and, here, R_1/R_2 will be determined during the course of the construction of a bounded solution of type (4.145)–(4.149).

In conclusion, we note that the method of *large* λ only enables one, within the framework of a reasonable formulation of the problem and subject to the condition $q(x,y) = 0$, $(x,y) \in \partial\Omega$, to determine the contact ellipse and other quantities to a limited degree of accuracy. For a punch with planar base, when the solution of (4.127) is sought in a class of functions having a square root singularity on $\partial\Omega$, all the required quantities can be found to any degree of accuracy (in this case, the contact ellipse is assumed to be known).

4.5 Contact Problem for a Wedge with Unknown Contact Area

When the punch (an elliptic paraboloid) approaches the edge of the wedge, it is evident that the contact area loses its elliptic form and becomes asymmetric. In this case we shall use the Galanov–Newton method (section 2.9), which enables the normal contact pressures and the unknown contact area to be determined simultaneously.

We will use the dimensional notation from the previous section, in particular formula (4.126), but we will change the formulation of the problem given there. Now, for specified values of δ, γ, a, R_1, R_2, it is required to determine the contact area Ω, the distribution of the normal contact stresses $q(r,z)$, $(r,z) \in \Omega$, and also the quantities P and H.

We will assume that the area Ω is completely contained within a rectangle S, two sides of which are parallel to the edge of the wedge, with centre at the r axis and semiaxes b and c $(b \geq c)$. The integral equation and inequality, to which the solution of this problem can be reduced, have the form

$$\theta^{-1} \int_S K(M,N) q(N)\, d\Omega_N = g(M), \quad q(M) \geq 0, \quad M \in \Omega, \tag{4.151}$$

$$\theta^{-1} \int_S K(M,N) q(N)\, d\Omega_N > g(M), \quad q(M) = 0, \quad M \in (S \setminus \Omega), \tag{4.152}$$

$$M = (r,z), \quad N = (x,y), \quad g(r,z) = 2\pi(\delta + \gamma(r-a) - f(r,z)),$$

$$K(r,z,x,y) = 1/R + T(r,z,x,y), \tag{4.153}$$

where the function $T(r,z,x,y)$ is determined by expression (4.128).

To determine the quantities P and H we must add to system (4.151), (4.151) the following equilibrium conditions:

$$\int_{\Omega} q(M)\, d\Omega_M = P, \quad \int_{\Omega} rq(M)\, d\Omega_M = PH.$$

When calculating the values of the function $K(M, N)$ of the form (4.153) we smooth its singularity at $M = N$ using the formulae

$$1/R \to 1/R_*, \quad R_* = \sqrt{(r-x)^2 + (z-y)^2 + \delta_0}, \tag{4.154}$$

and on the edge (for problem $\mathcal{A}$, $n = 1, 2$)

$$K(0, z, x, y) \to A_0/R_0, \quad R_0 = \sqrt{x^2 + (z-y)^2 + \delta_0}, \tag{4.155}$$

$$A_0 = \pi \frac{4\alpha + \sin(4\alpha)}{2(4\alpha^2 - \sin^2(2\alpha))} + \frac{2}{\pi} \int_0^\infty \tanh\frac{\pi u}{2} \{W_1(u,\alpha)F_1(u) \tag{4.156}$$

$$-W_2(u,\alpha)F_2(u)\} \cos[u \ln((R_0 + |z-y|)/(x + \sqrt{\delta_0}))]\, du,$$

$$F_n(u) - (1-2\nu)\int_0^\infty L_n(u,y)F_n(y)\, dy = \frac{\pi}{2}(1-2\nu)L_n(u,0), \; 0 \le u < \infty,$$

$$L_n(u,0) = \pm\pi \frac{1 \mp \cos(2\alpha)}{2\alpha \pm \sin(2\alpha)} \cosh\frac{\pi u}{2} \int_0^\infty \frac{\tanh(\pi t/2)\, g_{1,2}(t,\alpha)}{\cosh(\pi t) + \cosh(\pi u)}\, dt.$$

There are similar formulae for problems $\mathcal{B}$ and $\mathcal{C}$. When deriving the formulae on the edge we have used the fact that $\mathcal{K}_{iu}(0) = \pi\delta_*(u)$ (see formula (4.1), δ_* is the Dirac function). It can be shown that the regularizing parameter δ_0 in formulae (4.154)–(4.156) must be related to the net spacings h_1 and h_2 (we assumed $\delta_0 = h_1h_2/16$ in the calculations).

If the rectangle S does not reach the edge of the wedge, we place its centre at the point $r = a$, $z = 0$ and introduce the following dimensionless quantities and notation

$$r - a = r'b, \quad x - a = x'b, \quad z = z'b, \quad y = y'b, \quad \delta = \delta'b, \quad H = H'b, \tag{4.157}$$

$$A = b/(2R_1), \quad B = b/(2R_2), \quad \lambda = a/b, \quad \varepsilon = c/b, \tag{4.158}$$

$$q(r,z)/\theta = 2\pi q'(r',z'), \quad P/\theta = 2\pi b^2 P', \quad S' \to S, \quad \Omega' \to \Omega. \tag{4.159}$$

Formulae (4.157)–(4.159) hold when $\lambda > \varepsilon$ (Fig. 4.9a) when the rectangle S is elongated along the z axis ($R_1 \le R_2$) or $\lambda > 1$ (Fig. 4.9b) when it is elongated along the r axis ($R_1 \geq R_2$). When $\lambda \leq \varepsilon$ (Fig. 4.9c, $R_1 \leq R_2$) or $\lambda \leq 1$ (Fig. 4.9d, $R_1 \geq R_2$) we will assume that one side of the rectangle is situated on the edge of the wedge, and we will take as its centre the point $r = c$, $z = 0$ when $R_1 \leq R_2$ or the point $r = b$, $z = 0$ when $R_1 \geq R_2$. In these cases we will also use notation (4.157)–(4.159), replacing the first two equations (4.157) by $r - c = r'b$, $x - c = x'b$ when $R_1 \leq R_2$ or $r - b = r'b$, $x - b = x'b$ when $R_1 \geq R_2$. We will henceforth omit the primes. The dimensionless parameter λ represents, as before, the relative remoteness of the punch from the edge of the wedge.

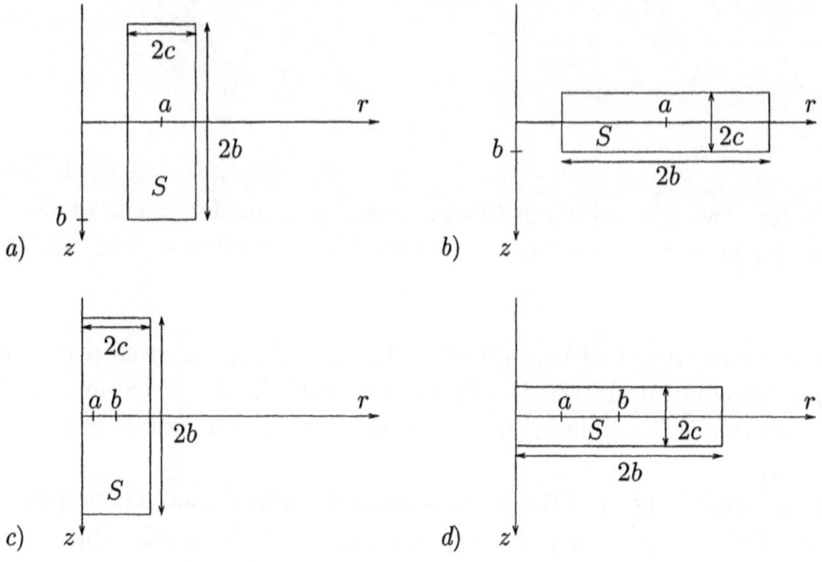

Figure 4.9: Different locations of the rectanfgle S; ($r = a, z = 0$) is the apex of the elliptic paraboloid; a) $\lambda > \varepsilon$, or $a > c$, and $R_1 \leq R_2$; b) $\lambda > 1$, or $a > b$, and $R_1 \geq R_2$; c) $\lambda \leq \varepsilon$, or $a \leq c$, and $R_1 \leq R_2$; d) $\lambda \leq 1$, or $a \leq b$, and $R_1 \geq R_2$; the z-axis coincides with the edge

To test the computer program (see appendix) we used the asymptotic solution obtained in the previous section for problem $\mathcal{A}$ with $2\alpha = \pi/2$, $\nu = 0.3$, $\lambda = 2$, $\varepsilon = 0.5$, $\delta = 1$, $\gamma = -0.0450$, $A = 2.19$ and $B = 1$, which is found by the formula

$$q(r, z) = 0.326\sqrt{1 - (r/0.422)^2 - (z/0.704)^2}.$$

We carried out numerical analysis for the case when one wedge face is stress-free (problem $\mathcal{A}$), which has an interesting application for calculation of Novikov's tooth gear (Korotkin (1997)). Two characteristic cases of the appearance of edge effects in Novikov's gear transmissions are shown in Fig. 4.10:

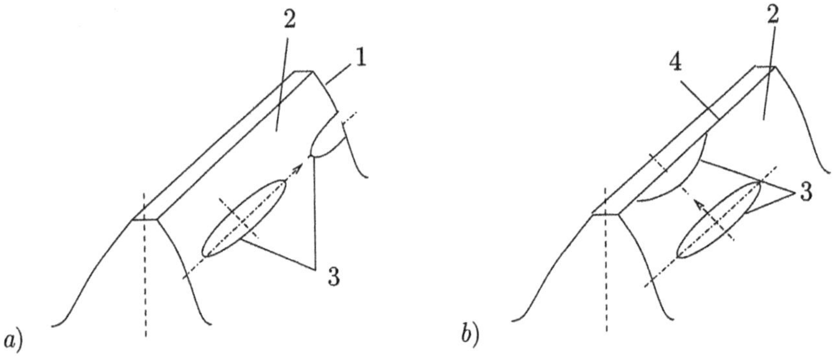

Figure 4.10: Two edge effects in Novikov gear; *a*) a contact area is prolate in direction perpendicular to the edge; *b*) a contact area is prolate along the edge

– at the end edge 1 of tooth 2 when the elliptic (or near-elliptic) contact zone, moving along the length of the tooth in the direction of the arrow, approaches the edge in the direction of the major axis of the ellipse (Fig. 4.10*a*);

– at the crown edge 4 of the tooth when the contact zone, moving in the direction of the arrow along the working height of the tooth, approaches the edge in the direction of the minor axis of the ellipse (Fig. 4.10*b*).

In calculating the contact strength of Novikov's transmissions it is necessary to know the stress-strain state at the contact along the rib (the edges of the teeth), under the condition that the angle of surface intersection at the rib is considerably different from $\pi/2$. For instance, there are angles $110°$, $135°$ at the edge because of convexity of the teeth, so that the wedge can be chosen as a simplified and suitable model for the transmission.

Values of the indenting force $10^3 P$ as a function of the settlement $10^3 \delta$ are given in Table 4.12 for different values of 2α (deg) and two orientations of the punch with respect to the edge. Here $\nu = 0.3$, $\lambda = 0$, $\varepsilon = 0.15$, $\gamma = 0$ and $A = 0.005$ and $B = 0.1$ (up to values of $2\alpha = 180°$ inclusive; the punch is elongated along the r-axis) or $A = 0.1$ and $B = 0.005$ (below the row corresponding to $2\alpha = 180°$; the punch is alongated along the edge); for $2\alpha = 180°$ we give the exact values using the solution from subsection 1.5.2.

Table 4.12: Values of the force $P(\delta)$

$10^3\delta$ / 2α	4	4.5	5	5.5	6	6.5
65	0.116	0.141	0.167	0.194	0.222	0.250
90	0.210	0.255	0.302	0.351	0.400	0.453
110	0.253	0.306	0.361	0.419	0.478	0.541
135	0.311	0.374	0.441	0.510	0.581	0.657
180	0.601	0.717	0.840	0.969	1.10	1.24
135	0.440	0.527	0.616	0.713	0.812	0.913
110	0.314	0.376	0.441	0.509	0.581	0.654
90	0.215	0.257	0.302	0.348	0.396	0.447
65	0.0813	0.0967	0.113	0.131	0.150	0.168

The smaller the angle of the wedge, the smaller is the force P needed to press the punch by a fixed settlement, regardless of the orientation of the punch. This is due to the boundary condition of the other face of the wedge, which is stress-free but will have non-zero displacements. In view of the regularities of (4.154)–(4.156), the value $\lambda = 0$ corresponds to the case when the point at which the punch and the wedge initially touch almost reaches the edge. An analysis of these results, and also of the corresponding calculations, carried out for $\varepsilon = 0.1$ and $\varepsilon = 0.25$ shows that when $2\alpha \approx 90°$ the value $P = P(\delta)$ as $\lambda \to 0$ is independent of which axes of coordinates (r or z) the elliptic paraboloid is elongated along. In other words, a prolate along the edge of a quarter-space punch is pressed near the edge by the same force as a punch which is prolate in the direction perpendicular to the edge (Fig. 4.11*a*,*b*).

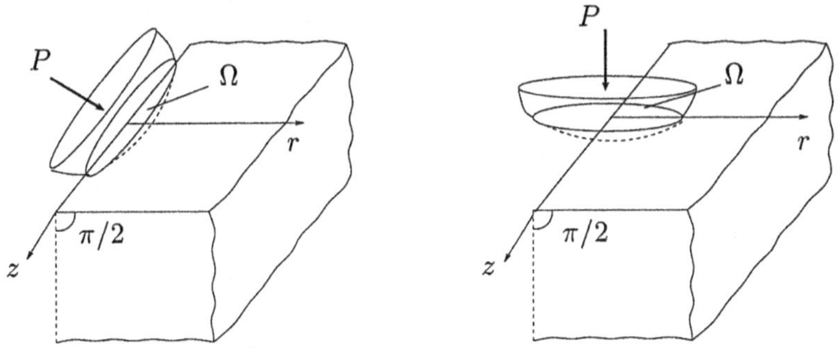

a) *b*)

Figure 4.11: For a quarter-space, the force for a prolate along the edge punch (*a*) is the same as for a punch which is prolate in the direction perpendicular to the edge (*b*)

The other important conclusion is that for a fixed values of R_1/R_2, ε and α, the dimensional settlement δ of the punch near the edge, $\lambda \to 0$, is proportional to the dimensional force $P^{2/3}$, as for a regular surface in the Hertz problem (see, e.g. Jonhson (1985, appendix A3)).

The upper half of the contact zones Ω are shown hatched in Fig. 4.12*a* for $2\alpha = 65°$ and in Fig. 4.12*b* for $2\alpha = 135°$, inside the upper half of the contact rectangles S. The left side, $r = -\varepsilon$ (see $(4.157)_1$), of the rectangles coincides with the edge of the wedge. Here $\nu = 0.3$, $\varepsilon = 0.15$, $\delta = 0.005$, $\gamma = 0$, $A = 0.1$ and $B = 0.005$; for $\lambda = 0$ the boundary of the region Ω is shown by the continuous curve, while for $\lambda = \varepsilon$ it is shown by dashed curve. It can be seen that for $2\alpha = 65°$ the area of the region Ω is considerably less than for $2\alpha = 135°$ (this also occurs in the case when $A < B$). For fairly acute angles 2α and $\lambda \to 0$ breakdown of the contact is observed in the neighbourhood of the point where the punch and the wedge initially touch (as though the edge withdraws), particularly when the punch is elongated along the edge (Fig. 4.12*a*).

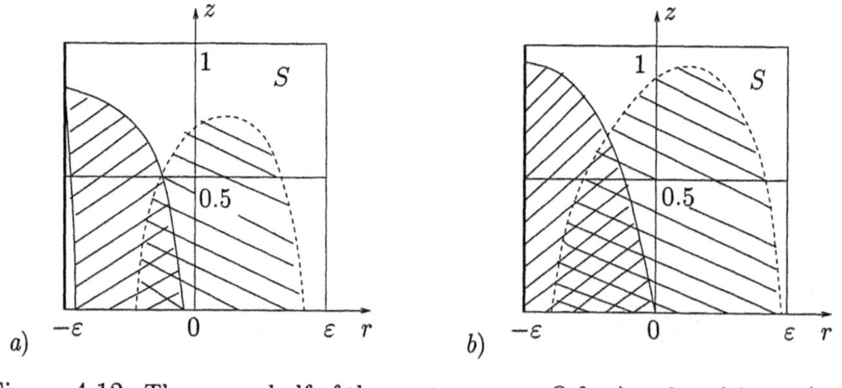

Figure 4.12: The upper half of the contact zones Ω for $\lambda = 0$ and $\lambda = \varepsilon$ (dashed), in the rectangle S; the bold side of the rectangle coincides with the edge; *a*) $2\alpha = 65°$; *b*) $2\alpha = 135°$

After solving the contact problem, knowing the function $q(r, z)$ and the contact area Ω, we can determine the effective dimensionless stress $\sigma'_e = \sigma_e/(2\pi\theta)$, which plays an important role in applications. The Huber–Mises plasticity criterion is based on this stress. As an example (in the frame of the concept of surface strength) we will determine σ'_e at the point where the punch and the wedge initially touch $r = a_0$, $\varphi = \alpha$, $z = 0$ ($a_0 = \lambda - \varepsilon$ when $\lambda \leq \varepsilon$ and $R_1 \leq R_2$; $a_0 = \lambda - 1$ when $\lambda \leq 1$ and

$R_1 \geq R_2$; while in other cases $a_0 = 0$) from the following formulae (we omit the primes)

$$\sigma_e = \frac{1}{\sqrt{2}}[(\sigma_1 - \sigma_2)^2 + (\sigma_2 - \sigma_3)^2 + (\sigma_1 - \sigma_3)^2]^{1/2}, \tag{4.160}$$

where

$$\sigma_1 = \frac{\nu}{1-\nu}\left(\frac{\partial u_r}{\partial r} + \frac{\partial u_z}{\partial z}\right) + \frac{\partial u_r}{\partial r} - \frac{\nu}{1-\nu} q_0, \quad \sigma_2 = -q_0, \tag{4.161}$$

$$\sigma_3 = \frac{\nu}{1-\nu}\left(\frac{\partial u_r}{\partial r} + \frac{\partial u_z}{\partial z}\right) + \frac{\partial u_z}{\partial z} - \frac{\nu}{1-\nu} q_0, \quad q_0 = q(a_0, 0), \tag{4.162}$$

$$\frac{\partial u_r}{\partial r} = \frac{\partial^2 \Phi_0}{\partial r^2} + \frac{\lambda}{4(1-\nu)}\chi_1 - \frac{1-2\nu}{2(1-\nu)}\chi_2, \tag{4.163}$$

$$\frac{\partial u_z}{\partial z} = \frac{\partial^2 \Phi_0}{\partial z^2} + \frac{\lambda}{4(1-\nu)}\chi_3, \tag{4.164}$$

$$\frac{\partial^2 \Phi_0}{\partial r^2} = -\frac{2(1-2\nu)}{\pi^2}\frac{d^2}{dr^2}\left\{\int_0^\infty\int_0^\infty\int_0^\infty E_1(p,t,\tau)\,\frac{\mathcal{K}_{it}(rp)}{p}\,dp dt d\tau\right\}_{r=\lambda}, \tag{4.165}$$

$$\frac{\partial^2 \Phi_0}{\partial z^2} = \frac{2(1-2\nu)}{\pi^2}\int_0^\infty\int_0^\infty\int_0^\infty E_1(p,t,\tau)\,p\,\mathcal{K}_{it}(\lambda p)\,dp dt d\tau, \tag{4.166}$$

$$\chi_1 = \sin\alpha\,\frac{\partial^2 \Phi_1}{\partial r^2} - \cos\alpha\,\frac{\partial^2 \Phi_2}{\partial r^2} \tag{4.167}$$

$$= -\frac{4(1-\nu)}{\pi^2}\frac{d^2}{dr^2}\left\{\int_0^\infty\int_0^\infty E_2(p,t)\mathcal{K}_{it}(rp)\,dp dt\right\}_{r=\lambda},$$

$$\chi_2 = \sin\alpha\,\frac{\partial \Phi_1}{\partial r} - \cos\alpha\,\frac{\partial \Phi_2}{\partial r} \tag{4.168}$$

$$= -\frac{4(1-\nu)}{\pi^2}\frac{d}{dr}\left\{\int_0^\infty\int_0^\infty E_2(p,t)\,\mathcal{K}_{it}(rp)\,dp dt\right\}_{r=\lambda},$$

$$\chi_3 = \sin\alpha\,\frac{\partial^2 \Phi_1}{\partial z^2} - \cos\alpha\,\frac{\partial^2 \Phi_2}{\partial z^2} = \frac{4(1-\nu)}{\pi^2}\int_0^\infty\int_0^\infty E_2(p,t)\,p^2\,\mathcal{K}_{it}(\lambda p)\,dp dt, \tag{4.169}$$

$$E_1(p,t,\tau) = \frac{\sinh(\pi t)\sinh(\pi\tau)}{\cosh(\pi t) + \cosh(\pi\tau)}[W_1(\tau,\alpha)\coth(\alpha t)E_+(\tau,p) \tag{4.170}$$

$$-W_2(\tau,\alpha)\tanh(\alpha t)E_-(\tau,p)],$$

$$E_2(p,t) = -2\sin(2\alpha)\sinh(\pi t)\left[\frac{E_+(t,p)}{\sinh(2\alpha t)+t\sin(2\alpha)}\right. \tag{4.171}$$

$$\left.-\frac{E_-(t,p)}{\sinh(2\alpha t)-t\sin(2\alpha)}\right], \quad E_\pm(t,p) = \frac{F_\pm(t,p)}{\cosh(\pi t/2)} + S(t,p), \tag{4.172}$$

$$S(t,p) = -\frac{1}{2\pi}\int_\Omega q(r,z)\,\mathcal{K}_{it}[p(r+b_0)]\cos(pz)\,drdz, \tag{4.173}$$

$$F_\pm(t,p) = (1-2\nu)\int_0^\infty L_m(t,y)\left[\Phi_\pm(y,p)+\cosh\frac{\pi y}{2}S(y,p)\right]dy, \tag{4.174}$$

$$0 \le t < \infty, \quad m = 1,2.$$

Here σ_n $(n = 1,2,3)$ are the principal stresses, $\Phi_n = \Phi_n(r,\varphi,z)$ $(n = 0,1,2)$ are the functions which occur in the Papkovich–Neuber representation (4.4)–(4.6), and $b_0 = \lambda - a_0$. Differentiation with respect to r, which occurs in formulae (4.165), (4.167) and (4.168) can be carried out inside the integrals when $\lambda > 0$ using the well-known relations for cylindrical functions (Gradshteyn and Ryzhik (1994)).

If we put $2\alpha = \pi$ in (4.160)–(4.174) and assume that the function $q(r,z)$ is defined in the elliptic region Ω by the relation

$$q(r,z) = q_0\sqrt{1-(r-a_0)^2/a_*^2 - z^2/b_*^2}, \tag{4.175}$$

we obtain the following formula (Belyaev (1957)),

$$\sigma_e = (1-2\nu)q_0\sqrt{1-\beta+\beta^2}/(1+\beta), \tag{4.176}$$

where $\beta = b_*/a_*$ if $a_* \ge b_*$ and $\beta = a_*/b_*$ if $a_* \le b_*$.

When deriving (4.176) we have used the values of the integrals

$$\frac{2}{\pi^2}\int_0^\infty\int_0^\infty\int_0^\infty \frac{\sinh(\pi t)\sinh(\pi\tau)}{\cosh(\pi t)+\cosh(\pi\tau)}\left[\coth\frac{\pi\tau}{2}\coth\frac{\pi t}{2}+\tanh\frac{\pi\tau}{2}\tanh\frac{\pi t}{2}\right]$$

$$\times p\,\mathcal{K}_{it}(rp)\,\mathcal{K}_{i\tau}(a_0p)\cos(pz)\,dpdtd\tau = \frac{d}{dz}\left[\frac{z}{(r-a_0)^2+z^2}\right],$$

$$\frac{1}{2\pi}\int_\Omega q(r,z)\,\frac{d}{dz}\left[\frac{z}{(r-a_0)^2+z^2}\right]d\Omega = \frac{a_*q_0}{a_*+b_*}. \tag{4.177}$$

In integral (4.177) the function $q(r,z)$ is defined by (4.175), and Ω is the corresponding ellipse with centre at the point $r = a_0$, $z = 0$.

Table 4.13: The effective stress

$10^3\delta$ / 2α	4	4.5	5	5.5	6	6.5
65	0.501	0.525	0.549	0.573	0.595	0.610
90	0.832	0.894	0.943	0.996	1.05	1.10
110	0.952	1.02	1.08	1.15	1.21	1.28
135	1.03	1.08	1.15	1.21	1.26	1.33
180	1.17	1.24	1.31	1.37	1.43	1.49

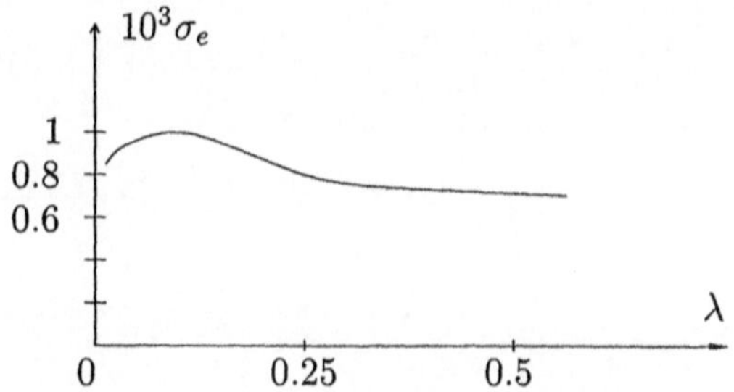

Figure 4.13: The effective stress σ_e versus λ

Table 4.13 shows values of the effective stress $10^3\sigma_e$ at the point $r = a_0$, $\varphi = \alpha$, $z = 0$ as a function of $10^3\delta$ for different 2α (deg) when the punch is elongated along the edge. Here $\nu = 0.3$, $\varepsilon = \lambda = 0.15$, $\gamma = 0$, $A = 0.1$ and $B = 0.005$. When $2\alpha = 180°$ the calculations were carried out using (4.176) in accordance with the exact solution of the problem. At some distance from the edge, the greater the settlement δ, or the force P, the larger is the stress σ_e. At the same time, calculations show that, in the region of the edge, the dependence of σ_e on δ and also on λ may be non-monotonic. When the elliptic paraboloid approaches the edge along its semi-major axis, the values of σ_e are usually larger than when approaching along the semi-minor axis. It can be seen from Table 4.13 that for acute aperture angles of the wedge the effective stresses are less dangerous than for obtuse aperture angles. This fact may be explained by more compliance of an elastic material of the wedge in the region of its edge (the value of the settlment of the punch increases when α diminishes, see Table 4.12). When $\lambda \to 0$ σ_e tends to a certain finite value ($2\alpha < \pi$), which equals zero for $q_0 = 0$, i.e. when contact is broken in the region of the point where the punch and the wedge initially touch (for fairly acute angles 2α).

Calculations using formulae (4.160)–(4.174) show that, when $\pi/2 < 2\alpha < \pi$, maximum of the surface effective stresses at the point where the punch and the wedge initially touch as a function of parameter λ takes place near the edge or at the edge of the wedge ($\lambda \to 0$), where breaking-off of tooth gears is observed. A plot of typical dependence of σ_e on λ, at the point where the punch and the wedge initially touch, is given in Fig. 4.13. Here $2\alpha = 110°$, $\nu = 0.3$, $\varepsilon = 0.15$, $\gamma = 0$, $10^3 A = 4.8$; $10^3 B = 48$ and $10^3 P = 0.374$, the punch approaches the edge along its semi-major axis. Note that these values of parameters correspond well to an actual Novikov transmission, shown in Fig. 4.10*a*.

Chapter 5

CONTACT PROBLEMS FOR A CONE

Lur'e (1964, section 2.10) obtained the 3-D solution for the elastic equilibrium of a cone loaded by a force applied at its apex. The solution of the general 3-D problem for of a circular cone involves the integral expansion of vector functions in terms of the complete system of vector harmonics on the cone surface with help of a Mellin integral transformation and a Fourier series (Ulitko (1979)). The solution of the static thermoelasticity problem was obtained for a transversally-isotropic cone in Podil'chuk (1992).

The axisymmetric contact problem for an infinite elastic cone was studied in Nuller (1970) using the method of piecewise-homogeneous solutions.

In this chapter we derive and solve an integral equation for the nonaxisymmetric contact problem for an infinite cone. For an annular sleeve we use asymptotic methods (Pozharskii (1997)). We separately consider, as an important special case, the contact problem for an annular punch imbedding a half-space, for which the exact solution is known (Antipov (1989)). We derive simple asymptotic formulae which are more convenient than this tedious exact solution. For a wedge-shaped punch acting on an elastic cone we emphasize, as in section 4.3, the asymptotics of the contact stresses at the apex of the punch. These asymptotics are constructed analytically for small angles of the punch, and for other angles we use a numerical method (Pozharskii (1998), Pozharskii and Chebakov (1998)). The exact solution of the perturbed integral equation for a wedge-shaped punch impressing a half-space (a special case of a cone) was obtained in Vasil'yev (1995). However, it is difficult to separate the singularities of the contact stresses in this solution. We also investigate the contact problem for a cone with unknown contact area (Aleksandrov and Pozharskii (1998)).

In the last section of this chapter we consider contact problems for a spherical layer and for a sector of a spherical layer (truncated cone). The torsional contact problem for a spherical layer was treated in Aleksandrov and Karpenko (1980).

5.1 Equilibrium under Given Normal Loads

Consider an infinite cone of angle 2α $(0 < \alpha < \pi)$ in the spherical coordinate system r, η, φ. Let the concentrated normal force of intensity Q be applied at the point $r = x$, $\varphi = \psi$ of the cone surface $\eta = \alpha$. The axial section of the cone, which contains the force vector, is shown in Fig. 5.1.

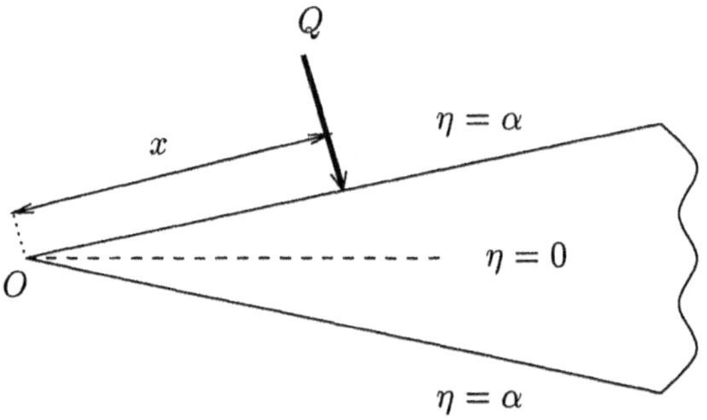

Figure 5.1: An axial section of the cone

The boundary conditions can be presented in the form

$$\eta = \alpha: \quad \tau_{\eta r} = \tau_{\eta\varphi} = 0, \quad \sigma_\eta = -Q\delta_*(r - x)\delta_*(\varphi - \psi), \tag{5.1}$$

where $\delta_*(x)$ is the Dirac delta function. The stresses must vanish at infinity.

The solution of problem (5.1) is based on the representation of the harmonic function $\Phi(r, \eta, \varphi)$ in the form of a Mellin integral over r and a Fourier series in φ $(0 < r < \infty,\ 0 \le \varphi < 2\pi)$

$$\Phi(r, \eta, \varphi) = \frac{1}{(2\pi)^{3/2}\, i} \sum_{k=-\infty}^{\infty} \int_{c-i\infty}^{c+i\infty} \omega_k(\eta, s) r^{1-s} \exp(ik\varphi)\, ds \quad (\sigma_1 < c < \sigma_2),$$

$$\omega_k(\eta, s) = \frac{1}{\sqrt{2\pi}} \int_0^{2\pi} d\varphi \int_0^{\infty} \Phi(r, \eta, \varphi) r^{s-2} \exp(-ik\varphi)\, dr.$$

The functions $\omega_k(\eta, s)$, $k \in \mathcal{Z}$ ($\mathcal{Z}$ is the set of integer numbers) are holomorphic in the strip $\sigma_1 < \mathrm{Re}\, s < \sigma_2$ of the complex variable s; the values σ_1 and σ_2 are determined from the convergence conditions for certain integrals.

Using the Papkovich–Neuber representation in spherical coordinates, we reduce problem (5.1) to the determination of the auxiliary functions $a_k(s)$, $b_k(s)$ and $c_k(s)$, $k \in \mathcal{Z}$, from the following system of three linear algebraic equations

$$s\left\{\frac{s^2-2(1-\nu)}{1-2\nu}a_k(s)\frac{d\bar{\mathcal{P}}_s^k}{d\eta}+sb_k(s)\frac{d\bar{\mathcal{P}}_{s-2}^k}{d\eta}-\frac{s+1}{2}\frac{k^2c_k(s)}{\sin\eta}\bar{\mathcal{P}}_{s-1}^k\right\}_{\eta=\alpha}=0, \tag{5.2}$$

$$\left\{\left[s\frac{s+3-2\nu(2s+1)}{1-2\nu}\bar{\mathcal{P}}_s^k-\frac{s-4(1-\nu)}{1-2\nu}\frac{d^2\bar{\mathcal{P}}_s^k}{d\eta^2}\right]a_k(s)\right. \tag{5.3}$$

$$\left.+\left[(s-1)\bar{\mathcal{P}}_{s-2}^k-\frac{d^2\bar{\mathcal{P}}_{s-2}^k}{d\eta^2}\right]b_k(s)+k^2c_k(s)\frac{d}{d\eta}\left(\frac{\bar{\mathcal{P}}_{s-1}^k}{\sin\eta}\right)\right\}_{\eta=\alpha}=1,$$

$$-k^2\left\{\frac{s-4(1-\nu)}{1-2\nu}a_k(s)\frac{d}{d\eta}\left(\frac{\bar{\mathcal{P}}_s^k}{\sin\eta}\right)+b_k(s)\frac{d}{d\eta}\left(\frac{\bar{\mathcal{P}}_{s-2}^k}{\sin\eta}\right)\right. \tag{5.4}$$

$$\left.-c_k(s)\left[\frac{d^2\bar{\mathcal{P}}_{s-1}^k}{d\eta^2}+\frac{s(s-1)}{2}\bar{\mathcal{P}}_{s-1}^k\right]\right\}_{\eta=\alpha}=0,$$

where the bar denotes the associated Legendre functions $\mathcal{P}_s^k$ which are nonsingular on the cone axis $\eta = 0$ (Gradshteyn and Ryzhik (1994, formulae 8.751.1 and 8.752.2)) and have the normalizing factor $[(s+1/2)\Gamma(s-k+1)/\Gamma(s+k+1)]^{1/2}$:

$$\bar{\mathcal{P}}_s^k=\left[(s+1/2)\frac{\Gamma(s-k+1)}{\Gamma(s+k+1)}\right]^{1/2}\mathcal{P}_s^k, \tag{5.5}$$

$$\mathcal{P}_s^k(x)=(-1)^k\frac{\Gamma(s+k+1)(1-x^2)^{k/2}}{2^k\Gamma(s-k+1)\,k!} \tag{5.6}$$

$$\times\mathcal{F}\left(k-s,k+s+1,k+1;\frac{1-x}{2}\right),$$

$$\mathcal{P}_s^{-k}(x)=(-1)^k\frac{\Gamma(s-k+1)}{\Gamma(s+k+1)}\mathcal{P}_s^k(x),\quad k\le 0, \tag{5.7}$$

where $\Gamma(s)$ is the gamma function. The argument $\cos\eta$ of the Legendre functions is omitted in formulae (5.2)–(5.5). Due to formula (5.7), the functions $\bar{\mathcal{P}}_s^k$ and $\bar{\mathcal{P}}_s^{-k}$ are related by

$$\bar{\mathcal{P}}_s^{-k}=\bar{\mathcal{P}}_s^{-k}(\cos\eta)=(-1)^k\bar{\mathcal{P}}_s^k(\cos\eta).$$

The displacement vector can be expressed via the functions $a_k(s)$, $b_k(s)$ and $c_k(s)$. We give only the normal displacement u_η, which is expressed at the cone surface as

$$u_\eta(r,\alpha,\varphi) = -\frac{Q}{2G}\frac{1}{4\pi^2 i}\int_{c-i\infty}^{c+i\infty}\left\{\sum_{k=-\infty}^{\infty} K_k(s,\alpha)\exp(ik[\varphi-\psi])\right\}\left(\frac{x}{r}\right)^s ds, \tag{5.8}$$

$$K_k(s,\alpha) = \left\{-\frac{s-4(1-\nu)}{1-2\nu}a_k(s)\frac{d\bar{\mathcal{P}}_s^k}{d\eta} - b_k(s)\frac{d\bar{\mathcal{P}}_{s-2}^k}{d\eta} + \frac{k^2 c_k(s)}{\sin\eta}\bar{\mathcal{P}}_{s-1}^k\right\}_{\eta=\alpha}. \tag{5.9}$$

Expression (5.8) will be used later when solving the contact problem. We pass from $\bar{\mathcal{P}}_s^k(\cos\eta)$ to $\mathcal{P}_s^k(\cos\eta)$ in formula (5.9), using formula (5.5) and the well-known relations 8.733 and 8.735 in Gradshteyn and Ryzhik (1994)

$$\frac{d\mathcal{P}_s^k}{d\eta} = -k\cot\eta\,\mathcal{P}_s^k - (s-k+1)(s+k)\mathcal{P}_s^{k-1},$$

$$\frac{d^2\mathcal{P}_s^k}{d\eta^2} = [k^2 + k(k+1)\cot^2\eta - s(s+1)]\mathcal{P}_s^k$$

$$+(s-k+1)(s+k)\cot\eta\,\mathcal{P}_s^{k-1},$$

$$\mathcal{P}_s^{k-1} = \frac{s+k-1}{s-k+1}\cos\eta\,\mathcal{P}_{s-1}^{k-1} + \frac{\sin\eta}{s-k+1}\mathcal{P}_{s-1}^k,$$

$$\mathcal{P}_{s-2}^{k-1} = \frac{s-k}{s+k-2}\cos\eta\,\mathcal{P}_{s-1}^{k-1} - \frac{\sin\eta}{s+k-2}\mathcal{P}_{s-1}^k,$$

$$\mathcal{P}_s^k = -(s+k-1)\sin\eta\,\mathcal{P}_{s-1}^{k-1} + \cos\eta\,\mathcal{P}_{s-1}^k,$$

$$\mathcal{P}_{s-2}^k = (s-k)\sin\eta\,\mathcal{P}_{s-1}^{k-1} + \cos\eta\,\mathcal{P}_{s-1}^k,$$

where the argument is omitted.

With the help of these relations, we pass in formulae (5.2)–(5.4) and (5.9) to the functions $\mathcal{P} = \mathcal{P}_{s-1}^{k-1}(\cos\alpha)$ and $\mathcal{P}^1 = \mathcal{P}_{s-1}^k(\cos\alpha)$. By solving system (5.2)–(5.4), we finally obtain

$$K_k(s,\alpha) = \frac{f_1(s)(\mathcal{P})^3 + f_2(s)(\mathcal{P})^2\mathcal{P}^1 + f_3(s)\mathcal{P}(\mathcal{P}^1)^2 + f_4(s)(\mathcal{P}^1)^3}{g_1(s)(\mathcal{P})^3 + g_2(s)(\mathcal{P})^2\mathcal{P}^1 + g_3(s)\mathcal{P}(\mathcal{P}^1)^2 + g_4(s)(\mathcal{P}^1)^3}, \tag{5.10}$$

$$f_1(s) = -2(1-\nu)\cos^2\alpha\cot\alpha\,(s-k)^2(s+k-1)^2,$$

$$f_2(s) = (s-k)(s+k-1)\Big[(1-\nu)\cos^2\alpha\, s(s-1)$$

$$+\frac{1}{2}k^2\cot^2\alpha - 2(1-\nu)\cos^2\alpha(1+3k\cot^2\alpha)\Big],$$

$$f_3(s) = s(s-1)\left[\frac{3}{2}+k\cot^2\alpha\right](1-\nu)\sin 2\alpha + \frac{1}{2}k^2\cot\alpha(1+2k\cot^2\alpha)$$

$$-2(1-\nu)\left[\left(\frac{3}{2}k^2-k\right)\sin(2\alpha)+\left(2k-\frac{5}{2}k^2\right)\cot\alpha+2k^2\cot^3\alpha\right],$$

$$f_4(s) = \frac{1}{2}k^2\cot^2\alpha + 2(1-\nu)\Big[-\frac{1}{2}s(s-1)\sin^2\alpha$$

$$+\frac{1}{2}k^2(1+\cot^2\alpha)+k\left(\frac{1}{2}+\cos^2\alpha\right)\Big],$$

$$g_1(s) = -\cot^2\alpha\,(s-k)^2(s+k-1)^2[s(s-1)\sin^2\alpha+2(1-\nu)(k^2-1)],$$

$$g_2(s) = (s-k)(s+k-1)\Big\{\frac{1}{4}s^2(s-1)^2\sin(2\alpha)$$

$$+s(s-1)\left[\left(\frac{3}{2}k-\frac{1}{2}\right)\sin(2\alpha)-\cot\alpha(1-\nu+k(k+3))\right]$$

$$-\frac{1}{2}k^2(1-k^2)\cot\alpha(1+\cot^2\alpha)+6(1-\nu)k(1-k^2)\cot^3\alpha\Big\},$$

$$g_3(s) = \frac{1}{2}s^2(s-1)^2[1+(2k-3)\cos^2\alpha]+s(s-1)k[(3k-2)\cos^2\alpha$$

$$-k-(k+1)\cot^2\alpha(2k+2(1-\nu))]+k^2(k^2-1)\Big[\frac{1}{2}$$

$$+\cot^2\alpha\left(k+\frac{1}{2}+2(1-\nu)\right)+\cot^4\alpha(k-4(1-\nu))\Big],$$

$$g_4(s) = \frac{s^2(s-1)^2}{4}\sin(2\alpha)-\frac{s(s-1)}{2}(k^2+k(1+k)\cot^2\alpha)\sin(2\alpha)$$

$$-k^2(1+k)\cot\alpha\left[\frac{1-k}{2}(1+\cot^2\alpha)+2(1-\nu)\cot^2\alpha\right].$$

One can show that the denominator of the function $K_k(s,\alpha)$ in formula (5.10) has no zeros on the line $s = 1/2 + is'$ in the plane of the complex variable s. In this connection, we will assume $c = 1/2$ in the integral in (5.8) and pass to the integration over s' (Fig. 5.2). We will omit the prime. As a result, formulae (5.8) and (5.10) acquire the real form

$$u_\eta(r,\alpha,\varphi) = -\frac{Q}{2\pi^2\theta}\sqrt{\frac{x}{r}}\int_0^\infty\Big\{K_0(s,\alpha) \tag{5.11}$$

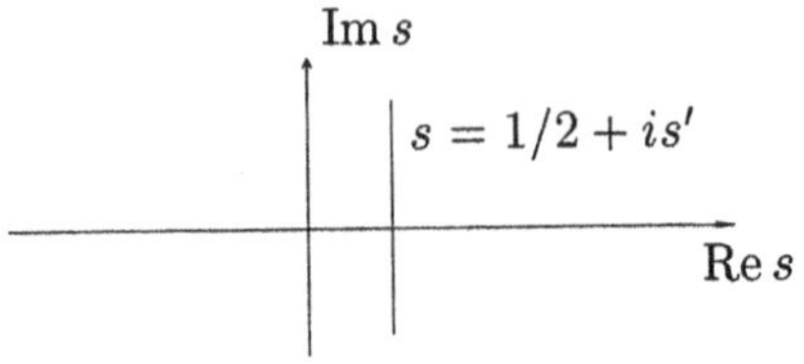

Figure 5.2: The integration contour in formula (5.8)

$$+2\sum_{k=1}^{\infty} K_k(s,\alpha)\cos(k[\varphi-\psi])\Big\}\cos\left(s\ln\frac{x}{r}\right)ds,$$

$$K_k(s,\alpha)=\frac{f_1(s)(\mathcal{P})^3+f_2(s)(\mathcal{P})^2\mathcal{P}^1+f_3(s)\mathcal{P}(\mathcal{P}^1)^2+f_4(s)(\mathcal{P}^1)^3}{g_1(s)(\mathcal{P})^3+g_2(s)(\mathcal{P})^2\mathcal{P}^1+g_3(s)\mathcal{P}(\mathcal{P}^1)^2+g_4(s)(\mathcal{P}^1)^3}, \tag{5.12}$$

$$\mathcal{P}=\mathcal{P}^{k-1}_{is-1/2}(\cos\alpha),\quad \mathcal{P}^1=\mathcal{P}^{k}_{is-1/2}(\cos\alpha), \tag{5.13}$$

$$f_1(s)=\cos^2\alpha\,\cot\alpha\,(s^2+(k-1/2)^2)^2, \tag{5.14}$$

$$f_2(s)=(s^2+(k-1/2)^2)\left[-\frac{1}{2}\cos^2\alpha\left(s^2+\frac{1}{4}\right)\right. \tag{5.15}$$

$$\left.-\cos^2\alpha(1+3k\cot^2\alpha)+\frac{k^2\cot^2\alpha}{4(1-\nu)}\right],$$

$$f_3(s)=\left(s^2+\frac{1}{4}\right)\left(\frac{3}{4}+\frac{1}{2}k\cot^2\alpha\right)\sin(2\alpha)-\frac{k^2\cot\alpha}{4(1-\nu)} \tag{5.16}$$

$$\times(1+2k\cot^2\alpha)+\left(\frac{3}{2}k^2-k\right)\sin(2\alpha)+\left(2k-\frac{5}{2}k^2\right)\cot^2\alpha+2k^2\cot^3\alpha,$$

$$f_4(s)=-\left[\frac{s^2+1/4}{2}\sin^2\alpha+\frac{k^2}{2}(1+\cot^2\alpha)+k\left(\frac{1}{2}+\cos^2\alpha\right)+\frac{k^2\cot^2\alpha}{4(1-\nu)}\right], \tag{5.17}$$

$$g_1(s)=\cot^2\alpha\left(s^2+\left(k-\frac{1}{2}\right)^2\right)^2\left[2(1-\nu)(k^2-1)-\left(s^2+\frac{1}{4}\right)\sin^2\alpha\right], \tag{5.18}$$

$$g_2(s)=\left(s^2+\left(k-\frac{1}{2}\right)^2\right)\left\{\frac{1}{4}\left(s^2+\frac{1}{4}\right)^2\sin(2\alpha)\right. \tag{5.19}$$

$$-\left(s^2+\frac{1}{4}\right)\left[\left(\frac{3}{2}k-\frac{1}{2}\right)\sin(2\alpha)-\cot\alpha\,(1-\nu+k(k+3))\right]$$

$$\left.-\frac{1}{2}k^2(1-k^2)\cot\alpha(1+\cot^2\alpha)+6(1-\nu)k(1-k^2)\cot^3\alpha\right\},$$

$$g_3(s) = -\frac{1}{2}\left(s^2+\frac{1}{4}\right)^2[1+(2k-3)\cos^2\alpha] \tag{5.20}$$
$$+\left(s^2+\frac{1}{4}\right)k[(3k-2)\cos^2\alpha - k - (k+1)\cot^2\alpha(2k+2(1-\nu))]$$
$$-k^2(k^2-1)\left[\frac{1}{2}+\cot^2\alpha\left(k+\frac{1}{2}+2(1-\nu)\right)+\cot^4\alpha\,(k-4(1-\nu))\right],$$
$$g_4(s) = -\left\{\frac{(s^2+1/4)^2}{4}\sin(2\alpha)+\frac{s^2+1/4}{2}(k^2+k(1+k)\cot^2\alpha) \right. \tag{5.21}$$
$$\left. \times\sin(2\alpha) - k^2(1+k)\cot\alpha\left[\frac{1-k}{2}(1+\cot^2\alpha)+2(1-\nu)\cot^2\alpha\right]\right\}.$$

In the derivation of formulae (5.11), (5.12), we took into account that $K_{-k}(s,\alpha) = K_k(s,\alpha)$, $k \in \mathcal{Z}$, and the cone functions $\mathcal{P}^{k-1}_{is-1/2}(\cos\alpha)$ and $\mathcal{P}^{k}_{is-1/2}(\cos\alpha)$ are real and even with respect to s.

For $\alpha = \pi/2$ the cone becomes a half-space, and the spherical coordinates r and φ on the surface of the half-space become polar coordinates. If we put $\alpha = \pi/2$ in (5.12) and use formula 8.756.1 in Gradshteyn and Ryzhik (1994) for $\mathcal{P}^k_s(0)$, we obtain

$$K_k\left(s,\frac{\pi}{2}\right) = \frac{\Gamma(is/2+k/2+1/4)\Gamma(-is/2+k/2+1/4)}{2\Gamma(is/2+k/2+3/4)\Gamma(-is/2+k/2+3/4)}. \tag{5.22}$$

Using the relation

$$\int_0^\infty \mathcal{J}_k(sx)\mathcal{J}_k(sr)\,ds = \frac{1}{\pi\sqrt{rx}}\int_0^\infty K_k\left(s,\frac{\pi}{2}\right)\cos\left(s\ln\frac{x}{r}\right)ds, \tag{5.23}$$

and formulae (4.118)–(4.120), we can see that the normal displacement $u_\eta(r,\pi/2,\varphi)$ in the form (5.11) with (5.22) coincides with the well-known term $1/R$, in the polar coordinates (see (4.120)), for a half-space. Formula (5.23) can be obtained on the basis of the relation

$$\int_0^\infty \mathcal{J}_k(t)\mathcal{J}_k\left(t\frac{r}{x}\right)dt = \frac{1}{2\pi i}\int_{c-i\infty}^{c+i\infty} g(s)\left(\frac{r}{x}\right)^{-s}ds,$$

$$g(s) = \frac{\Gamma(s/2+k/2)\Gamma(1/2-s/2+k/2)}{2\Gamma(1-s/2+k/2)\Gamma(1/2+s/2+k/2)},$$

derived with the help of the direct and inverse Mellin transforms of the Bessel function (formulae (4.121) in subsection 4.3.2 and (13) on p. 308 in Erdelyi (1954, Vol. 1)).

For the axisymmetric case, when the displacement u_η does not depend on φ, we retain only the function $K_0(s, \alpha)$ in the curly brackets in (5.11). This function has the form

$$K_0(s,\alpha) = [(s^2+1/4)\sin^2\alpha\,(\mathcal{P})^2 - \sin\alpha\cos\alpha\,\mathcal{P}\mathcal{P}^1 + \cos^2\alpha\,(\mathcal{P}^1)^2] \tag{5.24}$$
$$/[(s^2+1/4)^2\sin\alpha\cos\alpha\,(\mathcal{P})^2 + (s^2+1/4)\sin^2\alpha\,\mathcal{P}\mathcal{P}^1$$
$$-\sin\alpha\cos\alpha(s^2+2(1-\nu)\sin^{-2}\alpha+1/4)(\mathcal{P}^1)^2],$$
$$\mathcal{P} = \mathcal{P}_{is-1/2}(\cos\alpha), \quad \mathcal{P}^1 = \mathcal{P}^1_{is-1/2}(\cos\alpha).$$

Similarly, all the components of the displacement vector and of the stress tensor for an elastic infinite cone can be expressed in terms of the functions $a_k(s)$, $b_k(s)$ and $c_k(s)$, satisfying the system (5.2)–(5.4).

5.2 Rigid or Deformable Sleeve on a Cone

5.2.1. INTEGRAL EQUATIONS

Let us proceed to the contact problem for the interaction of a rigid sleeve (band) with an elastic cone of angle 2α. We assume that no friction forces act between the sleeve and the cone, and the cone surface outside the sleeve is unloaded. The sleeve is pressed into the cone with the force Q parallel to the cone axis. The scheme of contact is shown in Fig. 5.3.

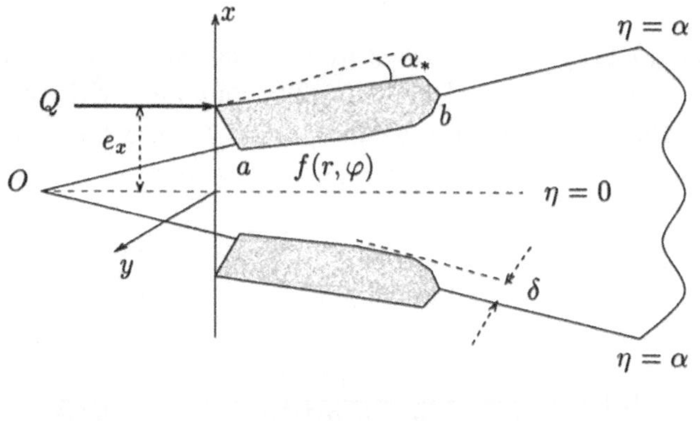

Figure 5.3: Contact between a rigid sleeve and a cone; the axial section, which contains the force Q

The contact region Ω is described by the relations $\eta = \alpha$, $a \le r \le b$, $0 \le \varphi < 2\pi$, (r, η, and φ are the spherical coordinates), and the shape of the sleeve base in the contact region is described by the function $f(r,\varphi)$. Acted on by the force Q, the sleeve undergoes a translational displacement δ perpendicular to the cone surface and rotates by the angles α_* and β_* around the y- and x-axes belonging to the plane perpendicular to the cone axis and passing through the point of application of the force Q. Thus, it is necessary to find the distribution of the normal contact stress $\sigma_\eta(r,\alpha,\varphi) = -q(r,\varphi)$, $(r,\varphi) \in \Omega$, under the sleeve. Then, with the help of the three integral conditions of the sleeve equilibrium, we can establish the relationship between the values Q and δ, e_x and α_*, and e_y and β_*, where e_x and e_y are the projections of the point of application of the applied force Q on the x- and y-axes. If we choose the x-axis in the plane which contains the cone axis and the force Q, as shown in Fig. 5.3, then $e_y = \beta_* = 0$. Showing the angle α_* in this figure, we suppose that the upper surface of the sleeve is parallel to the cone surface before the contact.

Let us expand the functions $q(r,\varphi)$ and $\delta(r,\varphi) = \delta + \alpha_* r \sin\alpha\cos\varphi + \beta_* r \sin\alpha \sin\varphi - f(r,\varphi)$ into Fourier series

$$q(r,\varphi) = q_0(r) + 2\sum_{n=1}^{\infty}[q_n^+(r)\cos(n\varphi) + q_n^-(r)\sin(n\varphi)], \tag{5.25}$$

$$\delta(r,\varphi) = \delta_0(r) + 2\sum_{n=1}^{\infty}[\delta_n^+(r)\cos(n\varphi) + \delta_n^-(r)\sin(n\varphi)].$$

Using the boundary condition $u_\eta(r,\alpha,\varphi) = -\delta(r,\varphi)$, $(r,\varphi) \in \Omega$, of the contact problem and formulae (5.12)–(5.21), we obtain the integral equations

$$\int_a^b q_n^{\pm}(x)\, k_n(r,x) x\, dx = \pi\theta\, \delta_n^{\pm}(r) \quad (a \le r \le b), \tag{5.26}$$

where

$$k_n(r,x) = \frac{1}{\sqrt{rx}} \int_0^\infty K_n(u,\alpha)\cos\left(u \ln\frac{x}{r}\right) du, \tag{5.27}$$

for the unknown functions occurring in series (5.25) ($n = 0, 1, ...$; for $n = 0$ we denote $q_0^{\pm}(r) \equiv q_0(r)$, $\delta_0^{\pm}(r) \equiv \delta_0(r)$).

Let us introduce the notation (the primes are omitted in what follows)

$$\xi = \lambda \ln\frac{r}{a} - 1, \quad x^* = \lambda\ln\frac{x}{a} - 1, \quad \lambda = 2\left(\ln\frac{b}{a}\right)^{-1}, \quad L_n(u) = uK_n(u,\alpha), \tag{5.28}$$

$$k_n^*\left(\ln\frac{x}{r}\right) = \sqrt{rx}k_n(r,x), \quad \varphi_n^{\pm}(\xi) = \frac{q_n^{\pm}(r)}{\theta}\left(\frac{r}{a}\right)^{3/2}, \quad f_n^{\pm}(\xi) = \frac{\lambda\sqrt{r}\delta_n^{\pm}(r)}{a^{3/2}}. \tag{5.29}$$

The dimensionless parameter λ characterizes the relative distance of the sleeve from the cone vertex: for fairly large λ the sleeve is sufficiently far from the cone vertex. Formulae (5.26) and (5.27) in the notation of (5.28), (5.29) can be written in the form

$$\int_{-1}^{1} \varphi_n^{\pm}(\xi)k_n\left(\frac{\xi - x}{\lambda}\right)\, d\xi = \pi f_n^{\pm}(x) \quad (|x| \le 1), \tag{5.30}$$

where

$$k_n(t) = \int_0^{\infty} \frac{L_n(u)}{u}\cos(ut)\, du. \tag{5.31}$$

Let us apply the asymptotic methods for *large* and *small* λ to solve the integral equations (5.30) with (5.31) and confine the calculations to the cases where either $n = 0$ (this corresponds to axial symmetry, if the load Q is distributed along the circle) or $n = 1$. In these cases, $L_0(u) = uK_0(u, \alpha)$, where the function $K_0(u, \alpha)$ has the form (5.24); the function $L_1(u)$ is

$$L_1(u) = u\,\frac{f_1(u)(\mathcal{P})^3 + f_2(u)(\mathcal{P})^2\mathcal{P}^1 + f_3(u)\mathcal{P}(\mathcal{P}^1)^2 + f_4(u)(\mathcal{P}^1)^3}{g_1(u)(\mathcal{P})^3 + g_2(u)(\mathcal{P})^2\mathcal{P}^1 + g_3(u)\mathcal{P}(\mathcal{P}^1)^2 + g_4(u)(\mathcal{P}^1)^3},$$

$$\mathcal{P} = \mathcal{P}_{iu-1/2}(\cos\alpha), \quad \mathcal{P}^1 = \mathcal{P}^1_{iu-1/2}(\cos\alpha)$$

$$f_1(u) = (u^2 + 1/4)^2\cos^2\alpha\cot\alpha,$$

$$f_2(u) = \left(u^2 + \frac{1}{4}\right)\left[-\frac{u^2 + 1/4}{2}\cos^2\alpha + \cot^2\alpha\left(\frac{1}{4(1-\nu)} - 1 - 2\cos^2\alpha\right)\right],$$

$$f_3(u) = \left(u^2 + \frac{1}{4}\right)\left(\frac{1}{4}\sin(2\alpha) + \cot\alpha\right) + \frac{1}{2}\sin(2\alpha)$$

$$+\cot\alpha\left[2\cot^2\alpha - \frac{1}{2} - \frac{1 + 2\cot^2\alpha}{4(1-\nu)}\right],$$

$$f_4(u) = -\frac{u^2 + 1/4}{2}\sin^2\alpha - 1 - \cos^2\alpha - \cot^2\alpha\,\frac{3 - 2\nu}{4(1-\nu)},$$

$$g_1(u) = -(u^2 + 1/4)^3\cos^2\alpha,$$

$$g_2(u) = \frac{1}{4}\left(u^2 + \frac{1}{4}\right)^3\sin(2\alpha) - \left(u^2 + \frac{1}{4}\right)^2[\sin(2\alpha) - (5-\nu)\cot\alpha],$$

$$g_3(u) = -\frac{1}{2}\left(u^2 + \frac{1}{4}\right)^2 \sin^2\alpha - \left(u^2 + \frac{1}{4}\right)[\sin^2\alpha + 4(2-\nu)\cot^2\alpha],$$

$$g_4(u) = -\frac{1}{4}\left(u^2 + \frac{1}{4}\right)^2 \sin(2\alpha) - \frac{1}{2}\left(u^2 + \frac{1}{4}\right)\sin(2\alpha)\,(1 + 2\cot^2\alpha)$$
$$+4(1-\nu)\cot^3\alpha.$$

5.2.2. SOLUTION FOR LARGE λ

For the functions $L_n(u)$ $(n = 1, 2, ...)$, the asymptotic relations

$$L_n(u) = Au + O(u^3), \quad u \to 0, \quad A = \text{const}, \tag{5.32}$$

$$L_n(u) = 1 + c_1 u^{-1} + c_2 u^{-2} + O(u^{-3}), \quad u \to +\infty, \quad c_{1,2} = \text{const}, \tag{5.33}$$

are valid. Taking formulae (5.32), (5.33) into account, we can represent the kernel (5.31) of the integral equation (5.30) in the form

$$k_n(t) = \ln|t|F_1(t) + |t|F_2(t) + F_3(t), \tag{5.34}$$

$$F_l(t) = \sum_{m=0}^{\infty} a_{lm} t^{2m}, \quad a_{10} = -1 \quad (l = 1, 2, 3).$$

Using the method of *large* λ, as $t \to 0$, we retain in the expansion (5.34) the terms up to $O(t^3)$, so that

$$k_n(t) = -\ln|t| + a_{20}|t| + a_{30} + a_{11}t^2\ln|t| + a_{31}t^2 + O(t^3), \tag{5.35}$$

$$a_{20} = -\frac{\pi}{2}c_1, \quad a_{30} = \int_0^\infty \frac{K_n(u) - 1 + \exp(-u)}{u}\,du, \quad a_{11} = \frac{c_2}{2}, \tag{5.36}$$

$$a_{31} = -\frac{3}{4}c_2 + \frac{1}{2}\int_0^\infty \left[\left(1 + \frac{c_1}{u} - K_n(u)\right)u^2 + c_2(1 - \exp(-u))\right]\frac{du}{u}. \tag{5.37}$$

When calculating the constants c_1 and c_2 in formulas (8) for $n = 0$ and $n = 1$, one should use the asymptotic expansions for the Legendre functions as $u \to +\infty$ (Zhurina and Karmazina (1966), Gradshteyn and Ryzhik (1994, formulae 8.72)):

$$\mathcal{P}_{iu-1/2}(\cos\alpha) = \frac{\exp(\alpha u)}{\sqrt{2\pi u \sin\alpha}}\left[1 + \frac{\cot\alpha}{8u} + \left(\frac{9}{128}\cot^2\alpha + \frac{1}{16}\right)\frac{1}{u^2}\right. \tag{5.38}$$

$$+\left(\frac{75}{1024}\cot^3\alpha+\frac{5}{64}\cot\alpha\right)\frac{1}{u^3}+O\left(\frac{1}{u^4}\right)\Bigg],$$

$$\mathcal{P}^1_{iu-1/2}(\cos\alpha)=\frac{\exp(\alpha u)}{\sqrt{2\pi u\sin\alpha}}\Bigg[u-\frac{3}{8}\cot\alpha-\frac{1}{16u\sin^2\alpha} \tag{5.39}$$

$$-\left(\frac{105}{1024}\cot^3\alpha+\frac{3}{32}\cot\alpha\right)\frac{1}{u^2}+O\left(\frac{1}{u^3}\right)\Bigg].$$

Taking formulae (5.38), (5.39) into account, we obtain

$$c_1=\cot\alpha\left[1-2\nu+\frac{7}{64}\cos^2\alpha\right], \tag{5.40}$$

$$c_2=\frac{1}{8}\sin^2\alpha+2(1-\nu)\cot\alpha\left[c_1-\frac{3}{4}\cot\alpha\right]-\frac{37}{128}\cot^2\alpha \tag{5.41}$$

$$-\frac{1}{8}c_1\cot\alpha\left[2-\frac{7}{8}\cos^2\alpha\right]$$

for $n=0$, and

$$c_1=\cot\alpha\left[1-2\nu+\frac{7}{64}\cos^2\alpha\right], \tag{5.42}$$

$$c_2=-\frac{3}{8}-\frac{227}{256}\cos^2\alpha-\cot^2\alpha\left[4+\frac{207}{256}+\frac{3}{4}\cos^2\alpha-\frac{33}{4}\nu\right] \tag{5.43}$$

$$+\frac{7}{128}c_1\sin(2\alpha)+c_1\cot\alpha\left(\frac{321}{64}-2\nu\right)$$

for $n=1$.

Tables 5.1 and 5.2 presents the values of the coefficients a_{lm} occuring in representation (5.35) for $\nu=0.3$ and various angles $\alpha=\pi k/8$.

Table 5.1: Constants (5.36), (5.37) for $n=0$

k	1	2	3	4	5	6	7
a_{20}	−1.871	−0.714	−0.271	0	0.271	0.714	1.871
a_{30}	−0.574	−0.981	1.981	2.079	1.127	−1.097	−1.254
a_{11}	−2.106	−0.364	−0.0199	0.0625	−0.0199	−0.364	−2.106
a_{31}	11.567	2.291	0.259	−0.109	−0.204	−1.037	−5.936

Table 5.2: Constants (5.36), (5.37) for $n=1$

k	1	2	3	4	5	6	7
a_{20}	−1.871	−0.714	−0.271	0	0.271	0.714	1.871
a_{30}	32.428	2.280	0.248	0.0794	−0.0247	−0.367	−1.091
a_{11}	−2.861	−0.747	−0.301	−0.188	−0.301	−0.747	−2.861
a_{31}	−2.392	1.670	0.647	0.286	−0.0383	−1.099	−5.777

According to the asymptotic methods for *large* λ, which was described in subsections 3.2.1–3.2.3, we obtain the solution of the integral equation (5.30) with kernel (5.35) in the form

$$\varphi_n^{\pm}(x) = \frac{1}{\sqrt{1-x^2}}\left\{\omega_{00}^{\pm}(x) + \frac{1}{\lambda}\omega_{10}^{\pm}(x) + \frac{1}{\lambda^2}\omega_{20}^{\pm}(x)\right. \tag{5.44}$$

$$\left. + \frac{\ln\lambda}{\lambda^2}\omega_{21}^{\pm}(x) + O(\lambda^{-3}\ln\lambda)\right\},$$

$$\omega_{00}^{\pm}(x) = \frac{1}{\pi}\left[S^{\pm} - \int_{-1}^{1}\frac{d}{d\xi}f_n^{\pm}(\xi)\frac{\sqrt{1-\xi^2}}{\xi - x}\,d\xi\right], \quad S^{\pm} = \int_{-1}^{1}\varphi_n^{\pm}(\xi)\,d\xi, \tag{5.45}$$

$$\omega_{10}^{\pm}(x) = -\frac{a_{20}}{\pi}\int_{-1}^{1}\frac{\sqrt{1-\xi^2}}{\xi - x}\,d\xi\int_{-1}^{1}\frac{\omega_{00}^{\pm}(t)\,\mathrm{sign}\,(t-\xi)}{\sqrt{1-t^2}}\,dt, \tag{5.46}$$

$$\omega_{20}^{\pm}(x) = -\frac{1}{\pi^2}\int_{-1}^{1}\frac{\sqrt{1-\xi^2}}{\xi - x}\,d\xi\int_{-1}^{1}[(t-\xi)(2a_{11}\ln|t-\xi| + 2a_{31} \tag{5.47}$$

$$+a_{11})\omega_{00}^{\pm}(t) + a_{20}\,\mathrm{sign}\,(t-\xi)\omega_{10}^{\pm}(t)]\,\frac{dt}{\sqrt{1-t^2}},$$

$$\omega_{21}^{\pm}(x) = \frac{2a_{11}}{\pi^2}\int_{-1}^{1}\frac{\sqrt{1-\xi^2}}{\xi - x}\,d\xi\int_{-1}^{1}\frac{(t-\xi)\omega_{00}^{\pm}(t)}{\sqrt{1-t^2}}\,dt. \tag{5.48}$$

In the simple special case $f_n^{\pm}(x) = f = \text{const}$, we find from formulae (5.44)–(5.48) that

$$\omega_{00}^{\pm}(x) = \frac{S}{\pi}, \quad \omega_{10}^{\pm}(x) = \frac{4.}{\pi^3}a_{20}S\,S_1(x), \quad \omega_{21}^{\pm}(x) = -\frac{S}{\pi}a_{11}(1-2x^2), \tag{5.49}$$

$$\omega_{20}^{\pm}(x) = \frac{S}{\pi}\left[(a_{11}(3/2 - \ln 2) + a_{31})(1-2x^2) + \frac{32}{\pi^4}a_{20}^2(S_2(x) - S_0)\right], \tag{5.50}$$

$$S_0 = \sum_{m=0}^{\infty}\frac{4m}{(4m^2-1)^3} \approx 0.1508, \tag{5.51}$$

$$S_1(x) = (1-2x^2) + 2\sqrt{1-x^2}\sum_{m=1}^{\infty}\frac{\sin[(2m+1)\arccos x]}{(2m+1)^2}, \tag{5.52}$$

$$S_2(x) = (1-x^2)\sum_{m=1}^{\infty}\frac{\nu_{2m}(x)}{(4m^2-1)^2}, \quad \nu_0(x)=0, \quad \nu_2(x) = 4+2x\ln\frac{1-x}{1+x}, \tag{5.53}$$

$$\nu_{2m+2}(x) = -2(1-2x^2)\nu_{2m}(x) + 16m/(4m^2-1) - \nu_{2m-2}(x), \quad m=1,2..., \tag{5.54}$$

$$S = \pi f[a_{30} + 0.8106a_{20}\lambda^{-1} + (a_{11}+a_{31}-0.03287a_{20}^2)\lambda^{-2} \tag{5.55}$$
$$+\ln(2\lambda)(1-a_{11}\lambda^{-2}) + O(\lambda^{-3}\ln\lambda)]^{-1},$$

where the function $S_2(x)$ for $|x| \le 1$ can be approximated with an error of 0.7% by the expression

$$S_2(x) \approx \left[0.4356 + 0.1321x^2 + 0.2494x\ln\frac{1-x}{1+x}\right](1-x^2).$$

5.2.3. SOLUTION FOR SMALL λ

If λ is small and $f_n^{\pm}(x) = f\exp(i\varepsilon x)$, then the zeroth term in the asymptotic representation of the solution of the integral equation (5) can be represented as (see subsection 3.2.4)

$$\varphi_n^{\pm}(x) = \varphi_+\left(\frac{1-x}{\lambda}\right) + \varphi_-\left(\frac{1+x}{\lambda}\right) - \frac{f\exp(i\varepsilon x)}{\lambda K_n(\lambda\varepsilon,\alpha)}, \tag{5.56}$$

where the functions $\varphi_{\pm}(t)$ can be determined from the solution of the integral equation

$$\int_0^{\infty}\varphi_{\pm}(\tau)k_n(t-\tau)\,d\tau = \frac{f}{\lambda}\exp(\pm i\varepsilon[1-\lambda t]) \tag{5.57}$$

on the semi-infinite interval $0 \le t < \infty$.

To obtain the solution of equation (5.57) by the Wiener–Hopf method, it is necessary to approximate the kernel $K_n(u,\alpha) = L_n(u)/u$ by a factorizable expression. Taking into account the behaviour of $L_n(u)$ as $u \to +\infty$ (see formulae (5.32), (5.33)), we use the approximation (Aleksandrov and Belokon' (1968))

$$K_n(u,\alpha) \approx \frac{\sqrt{u^2+B^2}(u^2+D^2)}{(u^2+C^2)(u^2+E^2)} + \frac{c_1u^2}{(u^2+C^2)(u^2+E^2)}, \tag{5.58}$$

where $BD^2/(C^2E^2) = A = K_n(0,\alpha)$ and the absolute value of the second term is much less that of the first term for any u. Tables 5.3 and 5.4

present the values of the constants occuring in approximation (5.58) and the relative error θ_* (in percent) of this approximation for $\nu = 0.3$ and various angles $\alpha = \pi k/8$. The smaller the value of the angle α, the worse is the accuracy of the approximation (5.58). To obtain a high-accuracy approximation to the function $K_1(u, \pi k/8)$ for $k = 1$, one should use a more complicated expression on the right-hand side of (5.58).

Table 5.3: Constants in approximation (5.58) for $n = 0$

k	B	C	D	E	c_1	θ_*
1	2.400	4.801	1.699	0.467	1.191	5
2	1.600	2.400	1.502	0.472	0.455	3
3	1.800	0.500	1.401	1.855	0.172	2
4	1.000	1.120	1.112	0.475	0.000	2
5	0.800	0.398	0.302	0.393	−0.172	2
6	1.000	0.500	0.500	0.875	−0.455	2
7	2.500	0.601	0.501	2.115	−1.191	2

Table 5.4: Constants in approximation (5.58) for $n = 1$

k	B	C	D	E	c_1	θ_*
2	1.551	0.510	2.770	3.165	0.455	3
3	0.900	0.091	0.100	1.010	0.172	2
4	1.000	1.500	1.662	1.159	0.000	1
5	0.990	0.210	0.201	0.928	−0.172	2
6	0.310	0.450	0.852	1.063	−0.455	2
7	2.830	0.551	0.630	2.387	−1.191	3

In the particular case $\varepsilon = 0$, the approximate solution of the Wiener–Hopf equation (5.57) with (5.58) can be represented in the form

$$\varphi_\pm(t) = \psi^0_\pm(t) + \psi^1_\pm(t), \tag{5.59}$$

$$\psi^0_\pm(t) = f\lambda^{-1}A^{-1/2}[\exp(-Bt)(\pi t)^{-1/2} + A^{-1/2}\Phi(\sqrt{Bt}) \tag{5.60}$$

$$-r_0 \exp(-Dt)\Phi(\sqrt{(B-D)t})],$$

$$\psi^1_\pm(t) = c_1 f\lambda^{-1}A^{-1/2}\{-Q_0(t) - r_1 J_0^+(D,t) + r_3 \exp(-Dt)\Phi(\sqrt{(B-D)t}) \tag{5.61}$$

$$+r_2 \int_0^t \exp(-D\tau)\Phi(\sqrt{(B-D)\tau})\mathcal{K}_0(B[t-\tau])\,d\tau - \frac{Dr_0}{2} J_1^-(-D,t)$$

$$-0.5r_0\, D(B-D)^{-1/2}[Q_1(t) + g_+(iD)\pi^{-1/2}t^{1/2}\exp(-Bt)]\},$$

$$r_0 = \frac{(C-D)(E-D)}{D\sqrt{B-D}}, \quad r_2 = \frac{r_0}{\pi}\left[\frac{B}{4(B-D)} - \frac{C+E-2D}{2} + \frac{1}{D}\right], \tag{5.62}$$

$$r_1 = \frac{(C+D)(E+D)}{4D\sqrt{B+D}}, \quad r_3 = r_0 g_+(iD)\left[\frac{2D-B}{4(B-D)} - \frac{C+E-2D}{2}\right], \tag{5.63}$$

$$Q_k(t) = \frac{1}{\pi^{3/2}}\int_0^t \tau^{k-1/2}\exp(-B\tau)\mathcal{K}_0(B[t-\tau])\,d\tau, \tag{5.64}$$

$$g_+(ix) = g_-(-ix) = \ln[(x+\sqrt{x^2-B^2})/B]/(\pi\sqrt{x^2-B^2}), \tag{5.65}$$

$$J_k^{\pm}(x,t) = \frac{1}{\pi}\int_0^t \exp(x\tau)\Phi(\sqrt{(B+x)\tau})\mathcal{K}_0(B[t-\tau])\tau^k\,d\tau \tag{5.66}$$

$$\mp g_{\pm}(ix)t^k\exp(xt)\Phi(\sqrt{(B+x)t}),$$

where $\Phi(x)$ is the probability integral (2.101) and $\mathcal{K}_0(t)$ is the modified Bessel function.

Calculations show that, in a certain range of λ depending on α, the solutions obtained with the help of the asymptotic methods for *large* and *small* λ converge to one another. Table 5.5 presents the values of $k_* = k_*^{\pm} = \lim\sqrt{1-x^2}\varphi_0^{\pm}(x)f^{-1}$ $(x \to 1-0)$ for $f_0^{\pm}(x) = f$, $\nu = 0.3$ and $\alpha = \pi k/8$ calculated by formulae (5.44), (5.49)–(5.55) for *large* λ, and (5.56), (5.59)–(5.66) for *small* λ.

Table 5.5: The contact stress intensity factor k_*

k	1	2	3	4	5	6	7
λ	5	1.55	1.2	1	2	6	10
large λ	0.304	0.382	0.359	0.382	0.327	0.285	0.405
small λ	0.308	0.374	0.322	0.394	0.387	0.395	0.507

We can see from this table that the value of λ, for which there is an agreement between the two solutions, is especially big for small and large cone angles ($k = 1$ and $k = 7$). Calculations show that, for a fixed cone angle 2α, the closer the sleeve to the cone vertex (the smaller is λ), the larger are the contact stress $\varphi_n^{\pm}(x)$, its intensity factor and integral characteristic $S^{\pm}$ $(5.45)_2$, which is equivalent to a total external normal load applied to the sleeve. This means that near the cone vertex there are more dangerous contact stresses. In practice, this may lead to collapse of such a construction near the conical point.

5.2.4. DEFORMABLE SLEEVE

Now, consider the axisymmetric interaction of a deformable sleeve with elastic characteristics G_1 and ν_1 with an elastic circular cone (G, ν) of angle 2α. The sleeve is set on the cone with an interference fit $\delta(r)$ (Fig. 5.4).

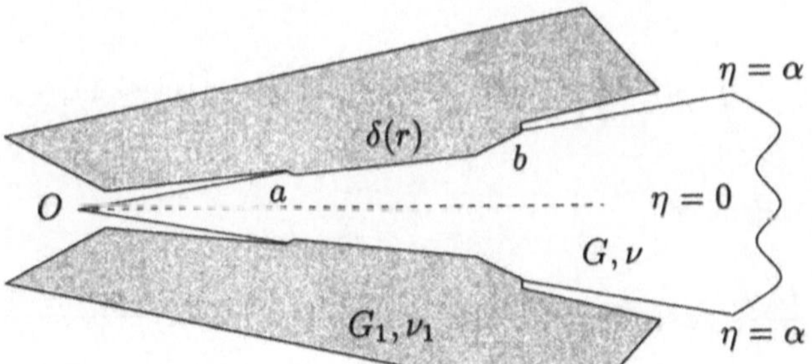

Figure 5.4: Contact between a deformable sleeve and a cone (an axial section)

The condition of contact can be written in the form

$$u_\eta(r,\alpha) - u_\eta^1(r,\alpha) = -\delta(r) \quad (a \le r \le b), \tag{5.67}$$

where $u_\eta(r,\alpha)$ and $u_\eta^1(r,\alpha)$ are the angular displacements of the cone and sleeve surfaces, respectively. We also suppose that the elastic sleeve is a sufficiently large body so that its dimensions considerably exceed the size of the contact region. In accordance with the Hertz theory of contact, the angular displacements of the sleeve surface due to the pressure $q(r)$ can be approximated with good accuracy by the angular displacements (with opposite sign) of the infinite cone surface of the angle $2\pi - 2\alpha$, due to the same pressure. Then, using condition (5.67), formulae (5.11) and (5.24), and taking the axial symmetry of the problem into account, we arrive at the following integral equation for unknown contact pressure $\sigma_\eta(r,\alpha) = -q(r)$:

$$\int_a^b q(x)\, k(r,x) x\, dx = \frac{\pi}{2}\theta\delta(r) \quad (a \le r \le b), \tag{5.68}$$

$$k(r,x) = \frac{1}{\sqrt{rx}} \int_0^\infty \frac{K_0(u,\alpha) + n_* K_0(u,\pi-\alpha)}{2} \cos\left(u \ln\frac{x}{r}\right) du, \quad n_* = \frac{\theta}{\theta_1}, \tag{5.69}$$

where $\theta_1 = G_1/(1-\nu_1)$; the function $K_0(u,\alpha)$ is defined by formula (5.24), and one should change ν to ν_1 in the corresponding expression for $K_0(u,\pi-\alpha)$.

For simplicity, we will assume that the cone and sleeve have the same elastic characteristics ($G_1 = G$, $\nu_1 = \nu$), i.e. $n_* = 1$. In dimensionless notation of the type (5.28), (5.29) equation (5.68) with (5.69) can be rewritten in the form

$$\int_{-1}^{1} \varphi(\xi)\, k\left(\frac{\xi - x}{\lambda}\right) d\xi = \pi f(x) \quad (|x| \le 1), \tag{5.70}$$

where

$$k(t) = \int_0^\infty \frac{L(u)}{u} \cos(ut)\, du, \quad L(u) = u\frac{K_0(u,\alpha) + K_0(u,\pi-\alpha)}{2}. \tag{5.71}$$

For calculations, it is more convenient to pass from the pair of functions $\mathcal{P}$ and $\mathcal{P}^1$, to the functions $\mathcal{P}$ and $\mathcal{P}^{-1}$ in expression (5.24). Due to relation 8.752.2 in Gradshteyn and Ryzhik (1994) we have $\mathcal{P}^1 = (u^2 + 1/4)\mathcal{P}^{-1}$, where $\mathcal{P}^{\pm 1} = \mathcal{P}^{\pm 1}_{iu-1/2}(\cos\alpha)$, and we obtain

$$L_0(u,\alpha) = \frac{(\mathcal{P})^2 - \cot\alpha\, \mathcal{P}\mathcal{P}^{-1} + \cot^2\alpha\, (u^2+1/4)(\mathcal{P}^{-1})^2}{(u^2+1/4)[\cot\alpha(\mathcal{P})^2 + \mathcal{P}\mathcal{P}^{-1} - \cot\alpha\,(u^2+\kappa_*)(\mathcal{P}^{-1})^2]},$$

$$\kappa_* = 2(1-\nu)\sin^{-2}\alpha + 1/4.$$

To calculate the pair $\mathcal{P}$ and $\mathcal{P}^{-1}$ we use formula (5.6) and the formula 8.704 in Gradshteyn and Ryzhik (1994)

$$\mathcal{P}_{iu-1/2}(\cos\alpha) = \mathcal{F}\left(\frac{1}{2} - iu, \frac{1}{2} + iu, 1; \sin^2\frac{\alpha}{2}\right),$$

$$\mathcal{P}^{-1}_{iu-1/2}(\cos\alpha) = \tan\frac{\alpha}{2}\, \mathcal{F}\left(\frac{1}{2} - iu, \frac{1}{2} + iu, 2; \sin^2\frac{\alpha}{2}\right),$$

containing the hypergeometric function $\mathcal{F}$.

One can show, using formulae (5.40)–(5.43) for $c_{1,2}$, that the function $L(u)$ of the form $(5.71)_2$ possesses properties (3.145), (3.146) for certain values of c_1 and c_2. Therefore, we can use the methods which were applied in subsection 3.3.1 for the interaction of an elastic cylinder with an elastic sleeve. Due to the structure of $L(u)$, it is sufficient to study equation (5.70) for $0 < \alpha < \pi/2$. Later we put $\nu = 0.3$. The solution of equation (5.70) for the special approximation of the function $L(u)$ by $\tanh(Au)$ can be found in closed form (see formulae (3.147)–(3.150)). For a conical sleeve with $f(x) \equiv f$ it has the form (3.149). However the relative error θ_* of this approximation for any $u \in [0, \infty)$ is not acceptable for sufficiently small angles α. For example, for $\alpha = 3\pi/8$ the error $\theta_* = 5\%$.

For large values of λ the solution of equation (5.70) can be found by formulae (3.153)–(3.158). The values of the constants a_0, a_1 and b_1 occurring in these formulae are presented in Table 5.6 for various α. Later, in calculations by formulae (3.156)–(3.158), we will neglect the terms of order $O(\lambda^{-4}\ln^2\lambda)$.

Table 5.6: Constants in (3.153)–(3.158)

α	a_0	a_1	b_1
$\pi/8$	−0.914	2.816	−2.106
$\pi/4$	−1.039	0.627	−0.364
$3\pi/8$	1.554	0.0275	−0.0199

One can obtain the solution of equation (5.70), using the asymptotic method for *small* λ with approximation (3.165). For the conical sleeve, this solution is determined by formulae (3.166)–(3.168). Here, we give the solution for the more accurate approximation

$$\frac{L(u)}{u} \approx \frac{\sqrt{u^2+B^2}}{u^2+C^2}\,\frac{u^2+D^2}{u^2+E^2} \qquad \left(\frac{BD^2}{C^2E^2} = A\right). \tag{5.72}$$

The values of the constants in the approximation (5.72) and its error θ_* (in percent) are given in Table 5.7 for various α. For the conical sleeve $(f(x) \equiv f)$ and for approximation (5.72), the approximate solution of the problem for small λ has the form

$$\varphi(x) = \frac{f}{\lambda}\left[\zeta^0\left(\frac{1-x}{\lambda}\right) + \zeta^0\left(\frac{1+x}{\lambda}\right) - \frac{1}{A}\right], \tag{5.73}$$

$$\zeta^0(t) = \frac{1}{A}\Phi(\sqrt{Bt}) + \frac{\exp(-Bt)}{\sqrt{\pi A t}} - \frac{(C-D)(E-D)}{D\sqrt{A(B-D)}}\exp(-Dt)\Phi\sqrt{(B-D)t}. \tag{5.74}$$

For $B < D$ in (5.74) one can substitute the probability integral $\Phi(x)$ by the function

$$-i\Phi(ix) = \frac{2}{\sqrt{\pi}}\int_0^x \exp(u^2)\,du.$$

Note that the solution of the problem for small λ can be written also in the multiplicative form (3.167) with the function $\zeta^0(t)$ (5.74).

Table 5.7: Constants in approximation (5.72)

α	A	B	C	D	E	θ_*
$\pi/8$	0.8820	0.6820	0.4710	1.871	3.493	3.5
$\pi/4$	2.0592	0.1746	0.2410	1.951	2.357	2
$3\pi/8$	3.5432	0.6985	0.4440	1	1	3

Calculations show that, in a certain range of λ depending on α, the solutions obtained with the help of the asymptotic methods for *large* and *small* λ converge to one another. The results of calculations for the case when $2\alpha = 3\pi/4$ ($E = D$) are shown in Table 5.8. One can see that the agreement of the asymptotics occurs here when $\lambda \approx 1$.

Note that we have obtained only global asymptotics for the conical sleeve which are erroneous near the boundary points of the contact region (see the remark made in the end of section 3.3).

Table 5.8: Calculations for the cone with $2\alpha = 3\pi/4$

λ	1/4	1/2	1	2	3	4
$\varphi(0)/f$ (3.149)	1.16	0.675	0.458	0.347	0.304	0.279
$\varphi(0)/f$ (3.157)	—	—	0.446	0.341	0.300	0.277
$\varphi(0)/f$ (5.73)	1.20	0.720	0.398	—	—	—
N_0/f (3.150)	3.14	2.01	1.42	1.08	0.952	0.876
N_0/f (3.158)	—	—	1.38	1.07	0.938	0.864
N_0/f (3.168)	3.13	2.00	1.46	—	—	—

Suppose the boundary point $r = b$ of the contact zone is fixed. Then the smaller the value of $r = a$, or the closer is the sleeve to the cone vertex, the smaller is the value of λ because $b/a = \exp(2/\lambda)$ due to formula $(5.28)_3$. Let the dimensional quantity b be 1 (m). Then $a = 0.607$ (m) for $\lambda = 4$, $a = 0.135$ (m) for $\lambda = 1$, and $a = 3.35 \cdot 10^{-4}$ (m) for $\lambda = 1/4$, so that the sleeve almost reaches the cone vertex for the last value of λ. We can see from Table 5.8 that the contact pressure $\varphi(0)$ at the centre of the contact zone is approximately four times greater for $\lambda = 1/4$ than for $\lambda = 4$, and the total external load N_0 is approximately 3.5 times greater. That is why we should avoid the neighbourhood of conical points for such constructions in practice because of extremely large contact loads.

5.3 Plane Sloping Annular Punch on a Half-Space

Let us study the important special case $\alpha = \pi/2$, where the cone is a half-space (Aleksandrov (1996); Fig. 5.5a,b), the function $K_n(u, \pi/2)$ has the form (5.22), and kernel (5.27) of the integral equation (5.26), using (5.23) and formulae 6.512.1, 9.134.3 in Gradshteyn and Ryzhik (1994), can be written in the form

$$k_n(r,x) = \pi \int_0^\infty \mathcal{J}_n(ur)\mathcal{J}_n(ux)\,du \tag{5.75}$$

$$= \frac{(2n-1)!!}{2^{2n}(2n)!!}\frac{e^{2n}\pi}{r+x}\mathcal{F}\left(n+\frac{1}{2}, n+\frac{1}{2}, 2n+1; e^2\right), \quad e = \frac{2\sqrt{rx}}{r+x}.$$

By using integral representation 9.111 in Gradshteyn and Ryzhik (1994) for the hypergeometric series $\mathcal{F}$, we can rewrite (5.75) as

$$k_n(r,x) = \frac{e^{2n}}{r+x}\int_0^1 t^{n-1/2}(1-t)^{n-1/2}(1-e^2t)^{-n-1/2}\,dt. \tag{5.76}$$

The integral in formula (5.76) for any n can be expressed as a linear combination of the complete elliptic integrals $\mathcal{K}(e)$ and $\mathcal{E}(e)$. For example, for $n=0$ and $n=1$ we have

$$k_0(r,x) = \frac{2}{r+x}\mathcal{K}(e), \quad k_1(r,x) = \frac{2}{(r+x)e^2}[(2-e^2)\mathcal{K}(e) - 2\mathcal{E}(e)], \tag{5.77}$$

respectively.

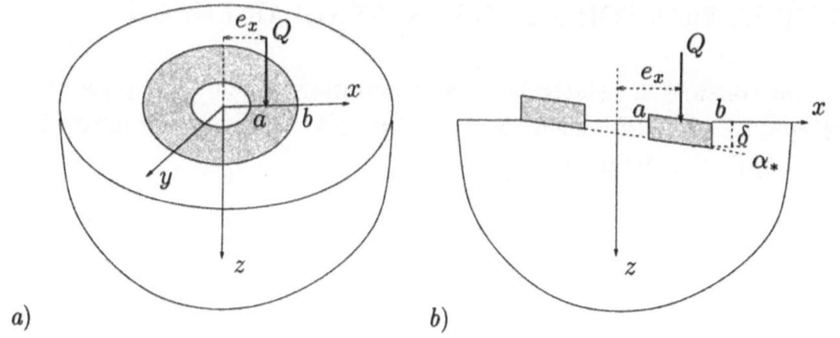

Figure 5.5: Plane sloping annular punch on a half-space: *a*) a 3-D view; *b*) the x-axial section

The application of the asymptotic methods for solving the integral equation of the contact problem for $\alpha = \pi/2$ becomes simpler for the following two reasons. In the first place, we have (cf. (5.33))

$$L_n(u) = uK_n(u,\pi/2) = 1 + c_2u^{-2} + c_4u^{-4} + O(u^{-6}), \quad c_n = \text{const}, \quad n = 2,4,$$

as $u \to \infty$. Therefore, if in the method of *large* λ we retain in the expansion of the kernel the same number of terms as we did earlier, then we obtain a more accurate solution. In the second place, the function $K_n(u, \pi/2)$ of the form (5.22) has a simple exact factorization, i.e. there is no need to approximate it by expression (5.58).

Thus, for the flat sloping annular punch we have only three equations (5.26); since $\delta_0(r) = \delta$, $\delta_1^+(r) = \alpha_* r$, $\delta_1^-(r) = \beta_* r$, all other values $\delta_n^\pm$ vanish, and this means that the corresponding functions $q_n^\pm(x)$ (for $n \geq 2$) vanish too. In the latter formulas δ is the settlement of the punch along the z-axis perpendicular to the half-space surface, α_* and β_* are the angles of rotation of the punch around the y- and x-axes. The equilibrium conditions have the form

$$Q = 2\pi \int_a^b q_0(r) r\, dr, \tag{5.78}$$

$$M_x = Qe_y = \pi \int_a^b q_1^-(r) r^2\, dr, \quad M_y = Qe_x = \pi \int_a^b q_1^+(r) r^2\, dr. \tag{5.79}$$

Note that if we choose the x-axis as shown in Fig. 5.5, then $e_y = \beta_* = M_x = 0$.

5.3.1. SOLUTION FOR A NARROW ANNULAR PUNCH

First, we consider a relatively narrow annular punch. We pass in the integral equations (5.26) with (5.27) to the following dimensionless notation (we will omit the primes)

$$\xi = \lambda \ln(r/a) - 1, \quad x' = \lambda \ln(x/a) - 1, \quad \lambda = 2[\ln(b/a)]^{-1}, \tag{5.80}$$

$$\varphi_0(\xi) = \frac{x^{3/2} q_0(x)}{\theta \delta a^{1/2}}, \quad \varphi_1(\xi) = \frac{x^{3/2} q_1^+(x)}{\theta \alpha_* a^{3/2}} = \frac{x^{3/2} q_1^-(x)}{\theta \beta_* a^{3/2}}, \tag{5.81}$$

$$M_0(t) = \operatorname{sech}(t/2) \mathcal{K}[\operatorname{sech}(t/2)], \quad t = (x - \xi)/\lambda, \tag{5.82}$$

$$M_1(t) = \{[2 - \operatorname{sech}^2(t/2)]\mathcal{K}[\operatorname{sech}(t/2)] - \mathcal{E}[\operatorname{sech}(t/2)]\}[\operatorname{sech}(t/2)]^{-1}, \tag{5.83}$$

where sech is the hyperbolic secant. Then we can rewrite (5.26) for $n = 0, 1$ as

$$\int_{-1}^{1} \varphi_0(\xi) M_0\left(\frac{x - \xi}{\lambda}\right) d\xi = \pi\lambda \exp\frac{1 + x}{2\lambda}, \tag{5.84}$$

$$\int_{-1}^{1} \varphi_1(\xi) M_1\left(\frac{x-\xi}{\lambda}\right) d\xi = \pi\lambda \exp\frac{3(1+x)}{2\lambda}. \tag{5.85}$$

Due to notation $(5.81)_2$, two equations of (5.26) for $q_1^+(x)$ and $q_1^-(x)$ reduce to one equation (5.85).

For a relatively narrow punch, when the ratio b/a is small, we have a large value of λ $(5.80)_3$ and small values of t $(5.82)_2$, so that the argument $\operatorname{sech}(t/2)$ of the complete elliptic integrals is nearly one. Due to expansions 8.113.3 and 8.114.3 in Gradshteyn and Ryzhik (1994), the complete elliptic integrals have a logarithmic singularity for such an argument. The kernels $M_0(t)$ and $M_1(t)$ of equations (5.84), (5.85) can be presented, using expressions (5.82), (5.83) and these expansions of the complete elliptic integrals, in the same form

$$\sum_{k=0}^{\infty} a_k t^{2k} + \ln|t| \sum_{k=0}^{\infty} b_k t^{2k}, \tag{5.86}$$

where the series converge absolutely for $|t| < \pi$. In representation (5.86) for $M_0(t)$ we have $a_0 = 2.079$, $a_1 = -0.1091$, $a_2 = 0.005352$, $b_0 = -1$, $b_1 = 0.06250$, and $b_2 = -0.003581$, and for $M_1(t)$ we have $a_0 = 0.07944$, $a_1 = 0.2857$, $a_2 = 0.004494$, $b_0 = -1$, $b_1 = -0.1875$, and $b_2 = 0.0009766$. Thus, we will now consider only one equation

$$\int_{-1}^{1} \varphi(\xi) \left[\sum_{k=0}^{\infty} a_k \left(\frac{x-\xi}{\lambda}\right)^{2k} + \ln\left|\frac{x-\xi}{\lambda}\right| \sum_{k=0}^{\infty} b_k \left(\frac{x-\xi}{\lambda}\right)^{2k}\right] d\xi \tag{5.87}$$

$$= \lambda \exp\left(\mu \frac{1+x}{\lambda}\right),$$

where $\varphi(\xi) = \varphi_0(\xi)$ and $\mu = 1/2$ for equation (5.84) and $\varphi(\xi) = \varphi_1(\xi)$ and $\mu = 3/2$ for equation (5.85).

For sufficiently large values of the dimensionless parameter λ $(\lambda > 2/\pi)$, the asypmtotic solution of the integral equation (9) can be constructed by the method of *large* λ, described in subsection 3.2.3. We finally obtain the relations

$$\varphi(x) = \varphi_0(\lambda, x) + \lambda^{-2}\varphi_1(\lambda, x) + \lambda^{-4}\varphi_2(\lambda, x) + O(\lambda^{-6}), \tag{5.88}$$

$$\varphi_0(\lambda, x) = \frac{1}{\sqrt{1-x^2}}\left[\frac{P}{\pi} - (c_0 x Q_0 + 2c_1 Q_1 + 3c_2 x Q_1\right.$$

$$\left. +4c_3 Q_2 + 5c_4 x Q_2 + c_5 Q_3)\right],$$

$$\varphi_1(\lambda,x)=\frac{1}{\sqrt{1-x^2}}\left\{\frac{2P}{\pi}A\left(\frac{3}{2}\right)Q_1-c_0[A(1)xQ_0+b_1xQ_1]\right.$$

$$\left.+c_1b_1\left(Q_1-\frac{2}{3}Q_2\right)-c_2\left[\frac{3}{4}A\left(\frac{5}{4}\right)xQ_0+\frac{1}{2}b_1xQ_2\right]+c_3b_1\left(\frac{3}{4}Q_1-\frac{2}{5}Q_3\right)\right\},$$

$$\varphi_2(\lambda,x)=\frac{1}{\sqrt{1-x^2}}\left\{\frac{4P}{\pi}\left[B\left(\frac{25}{12}\right)Q_2+\frac{3}{2}B\left(\frac{13}{12}\right)Q_1\right.\right.$$

$$\left.+\frac{1}{2}b_1A\left(\frac{3}{2}\right)\left(-\frac{1}{2}Q_1+\frac{1}{3}Q_2\right)\right]-c_0\left\langle\frac{3}{2}B\left(\frac{5}{6}\right)xQ_0+6B\left(\frac{19}{12}\right)xQ_1\right.$$

$$\left.+b_2xQ_2+\left[A\left(\frac{1}{2}\right)A\left(\frac{5}{4}\right)+\frac{1}{2}A\left(\frac{11}{8}\right)\right]xQ_0+b_1A(1)xQ_1+\frac{1}{6}b_1^2xQ_2\right\rangle$$

$$\left.+c_1\left[3B\left(\frac{4}{3}\right)Q_1+2b_2\left(Q_2-\frac{1}{5}Q_3\right)+b_1^2\left(-\frac{3}{8}Q_1+\frac{1}{3}Q_2-\frac{1}{15}Q_3\right)\right]\right\},$$

$$P=\pi\vartheta(\lambda)[\Theta(\lambda)]^{-1},$$

$$\vartheta(\lambda)=c_{-1}+\frac{1}{2}c_1+\frac{3}{8}c_3+\frac{5}{16}c_5-\frac{1}{4\lambda^2}(c_1+c_3)A\left(\frac{3}{2}\right)$$

$$+\frac{1}{12\lambda^4}c_1b_1A\left(\frac{3}{2}\right)-\frac{1}{\lambda^4}c_1B\left(\frac{4}{3}\right),$$

$$\Theta(\lambda)=\ln(2\lambda)+a_0+\frac{1}{\lambda^2}A(1)-\frac{1}{4\lambda^4}\left[A\left(\frac{3}{2}\right)\right]^2+\frac{9}{4\lambda^4}B\left(\frac{7}{6}\right),$$

$$c_{-1}=\lambda+\mu+\frac{\mu^2}{2\lambda}+\frac{\mu^3}{6\lambda^2}+\frac{\mu^4}{24\lambda^3}+\frac{\mu^5}{120\lambda^4}+\frac{\mu^6}{720\lambda^5},$$

$$c_0=\mu+\frac{\mu^2}{\lambda}+\frac{\mu^3}{2\lambda^2}+\frac{\mu^4}{6\lambda^3}+\frac{\mu^5}{240\lambda^4}+\frac{\mu^6}{120\lambda^5},$$

$$c_1=\frac{\mu^2}{2\lambda}+\frac{\mu^3}{2\lambda^2}+\frac{\mu^4}{4\lambda^3}+\frac{\mu^5}{12\lambda^4}+\frac{\mu^6}{48\lambda^5},$$

$$c_2=\frac{\mu^3}{6\lambda^2}+\frac{\mu^4}{6\lambda^3}+\frac{\mu^5}{12\lambda^4}+\frac{\mu^6}{36\lambda^5},\quad c_3=\frac{\mu^4}{24\lambda^3}+\frac{\mu^5}{24\lambda^4}+\frac{\mu^6}{48\lambda^5},$$

$$c_4=\frac{\mu^5}{120\lambda^4}+\frac{\mu^6}{120\lambda^5},\quad c_5=\frac{\mu^6}{720\lambda^5},$$

$$Q_0=-1,\quad Q_1=-x^2+\frac{1}{2},\quad Q_2=-x^4+\frac{1}{2}x^2+\frac{1}{8},$$

$$Q_3=-x^6+\frac{1}{2}x^4+\frac{1}{8}x^2+\frac{1}{16},$$

$$A(\gamma)=a_1+\gamma b_1-b_1\ln(2\lambda),\quad B(\gamma)=a_2+\gamma b_2-b_2\ln(2\lambda).$$

Using formulae (5.78) and (5.79), we find the indenting force

$$\frac{Q}{\theta b \delta} = \frac{2\pi}{\lambda} \exp\left(-\frac{2}{\lambda}\right) \left\{ P \left[1 + \frac{1}{2\lambda} + \frac{3}{16\lambda^2} + \frac{5}{96\lambda^3} + \frac{35}{3072\lambda^4} + \frac{21}{10240\lambda^5} \right.\right. \tag{5.89}$$

$$\left. - \frac{1}{32\lambda^4} A\left(\frac{3}{2}\right)\left(1 + \frac{1}{2\lambda}\right)\right] + \pi c_0 \left\langle \frac{1}{4\lambda} + \frac{1}{8\lambda^2} + \frac{5}{128\lambda^3} + \frac{7}{768\lambda^4} + \frac{7}{4096\lambda^5} \right.$$

$$+ \frac{1}{4\lambda^3} A\left(\frac{5}{4}\right)\left(1 + \frac{1}{2\lambda} + \frac{1}{8\lambda^2}\right) + \frac{1}{128\lambda^5} A\left(\frac{4}{3}\right)$$

$$\left. + \frac{1}{2\lambda^5}\left[\frac{3}{2} B\left(\frac{5}{4}\right) + \frac{3}{8} b_1 A\left(\frac{23}{18}\right) + \frac{1}{2} A\left(\frac{1}{2}\right) A\left(\frac{5}{4}\right)\right]\right\rangle$$

$$+ \pi c_1 \left(\frac{1}{32\lambda^2} + \frac{1}{64\lambda^3} + \frac{7}{1536\lambda^4} - \frac{b_1}{96\lambda^4}\right) + \pi c_2 \left[\frac{3}{16\lambda} + \frac{3}{64\lambda^2}\right.$$

$$\left.\left. + \frac{1}{32\lambda^3} + \frac{3}{16\lambda^3} A\left(\frac{4}{3}\right)\right] + \pi c_3 \frac{1}{32\lambda^2} + \pi c_4 \frac{5}{32\lambda} + O\left(\frac{1}{\lambda^6}\right)\right\},$$

and also the torques

$$\frac{M_x}{\theta b^3 \beta_*} = \frac{M_y}{\theta b^3 \alpha_*} = \frac{\pi}{\lambda} \exp\left(-\frac{6}{\lambda}\right) \left\{ P \left[1 + \frac{3}{2\lambda} + \frac{27}{16\lambda^2} + \frac{45}{32\lambda^3} \right.\right. \tag{5.90}$$

$$\left. + \frac{945}{1024\lambda^4} + \frac{5103}{10240\lambda^5} - \frac{9}{32\lambda^4} A\left(\frac{3}{2}\right)\left(1 + \frac{3}{2\lambda}\right)\right] + \pi c_0 \left\langle \frac{3}{4\lambda} + \frac{9}{8\lambda^2} \right.$$

$$+ \frac{135}{128\lambda^3} + \frac{189}{256\lambda^4} + \frac{8505}{20480\lambda^5} + \frac{3}{4\lambda^3} A\left(\frac{5}{4}\right)\left(1 + \frac{3}{2\lambda} + \frac{9}{8\lambda^2}\right)$$

$$\left. + \frac{27}{128\lambda^5} A\left(\frac{4}{3}\right) + \frac{3}{2\lambda^5}\left[\frac{3}{2} B\left(\frac{5}{4}\right) + \frac{3}{8} b_1 A\left(\frac{23}{18}\right) + \frac{1}{2} A\left(\frac{1}{2}\right) A\left(\frac{5}{4}\right)\right]\right\rangle$$

$$+ \pi c_1 \left(\frac{9}{32\lambda^2} + \frac{27}{64\lambda^3} + \frac{189}{512\lambda^4} - \frac{3b_1}{32\lambda^4}\right) + \pi c_2 \left[\frac{9}{16\lambda} + \frac{27}{32\lambda^2}\right.$$

$$\left.\left. + \frac{27}{32\lambda^3} + \frac{9}{16\lambda^3} A\left(\frac{4}{3}\right)\right] + \pi c_3 \frac{9}{32\lambda^2} + \pi c_4 \frac{15}{32\lambda} + O\left(\frac{1}{\lambda^6}\right)\right\}.$$

5.3.2. SOLUTION FOR A WIDE ANNULAR PUNCH

Now consider the case of small values of the parameter λ, i.e., a relatively wide annulus. Here, the principal term of asymptotics of solution of the integral equations (5.84), (5.85) for small λ must be constructed using a *boundary layer solution* describing a quick change of the contact stress close to the contours $r = a$ and $r = b$, and a *confluent* solution which is correct at a distance from the contours $r = a$ and $r = b$. The technique for constructing such solutions was described in subsection 3.2.4. This construction can have the following multiplicative form (3.138) (in dimensional notation):

$$q_0(r) = \frac{4\theta\delta}{\pi^2\sqrt{b^2 - r^2}}\left(\frac{a}{\sqrt{r^2 - a^2}} + \arccos\frac{a}{r}\right), \quad \frac{Q}{\theta b\delta} = 4, \tag{5.91}$$

$$\frac{q_1^+(r)}{\alpha_*} = \frac{q_1^-(r)}{\beta_*} = \frac{8\theta r}{\pi^2\sqrt{b^2 - r^2}}\left[\frac{a}{\sqrt{r^2 - a^2}}\left(1 - \frac{a^2}{3r^2}\right) + \arccos\frac{a}{r}\right], \tag{5.92}$$

$$\frac{M_y}{\theta b^3\alpha_*} = \frac{M_x}{\theta b^3\beta_*} = \frac{8}{3}, \tag{5.93}$$

or additive form (3.118)

$$q_0(r) = \frac{2\theta\delta}{\pi\sqrt{b^2 - r^2}}\left(1 - \frac{1}{\pi}\arccos\frac{1 + \varepsilon^2 - 2\eta^2}{1 - \varepsilon^2}\right) + \frac{2\theta\delta}{\pi^2\sqrt{r^2 - a^2}}\ln\frac{1 + \varepsilon}{1 - \varepsilon}, \tag{5.94}$$

$$\frac{Q}{\theta b\delta} = 4\left(\frac{2}{\pi}\arccos\varepsilon + \frac{1}{\pi}\sqrt{1 - \varepsilon^2}\ln\frac{1 + \varepsilon}{1 - \varepsilon}\right), \tag{5.95}$$

$$\frac{q_1^+(r)}{\alpha_*} = \frac{q_1^-(r)}{\beta_*} = \frac{4\theta r}{\pi\sqrt{b^2 - r^2}}\left(1 - \frac{1}{\pi}\arccos\frac{1 + \varepsilon^2 - 2\eta^2}{1 - \varepsilon^2}\right) \tag{5.96}$$

$$+\frac{4\theta r}{\pi^2\sqrt{r^2 - a^2}}\left[\ln\frac{1 + \varepsilon}{1 - \varepsilon} + \frac{b^2}{r^2}\left(\frac{1 - \varepsilon^2}{2}\ln\frac{1 + \varepsilon}{1 - \varepsilon} - \varepsilon\right)\right],$$

$$\frac{M_y}{\theta b^3\alpha_*} = \frac{M_x}{\theta b^3\beta_*} = \frac{8}{3}\left[\frac{2}{\pi}\arccos\varepsilon - \frac{1}{2\pi}\varepsilon\sqrt{1 - \varepsilon^2}\right. \tag{5.97}$$

$$\left.+\frac{1}{4\pi}(5 + \varepsilon^2)\sqrt{1 - \varepsilon^2}\ln\frac{1 + \varepsilon}{1 - \varepsilon}\right],$$

where we denote $\varepsilon = a/b$ and $\eta = a/r$, and the expressions for Q, M_y and M_x were obtained in accordance with formulae (5.78), (5.79).

Suppose the slope of the punch is caused by an eccentric force Q and there was no initial slope (Fig. 5.6). Then, using formulae (5.88), (5.89) and (5.91)–(5.97), one can formulate the condition of absence of separation of the punch base from the elastic solid. For example, applying formulae (5.91)–(5.93) for $\beta_* = 0$, $M_x = 0$ and $q_1^-(r) \equiv 0$, we find the inequality

$$e_* \leq b\left(\frac{\varepsilon}{\sqrt{1 - \varepsilon^2}} + \arccos\varepsilon\right)\left[\frac{\varepsilon(3 - \varepsilon^2)}{\sqrt{1 - \varepsilon^2}} + 3\arccos\varepsilon\right]^{-1},$$

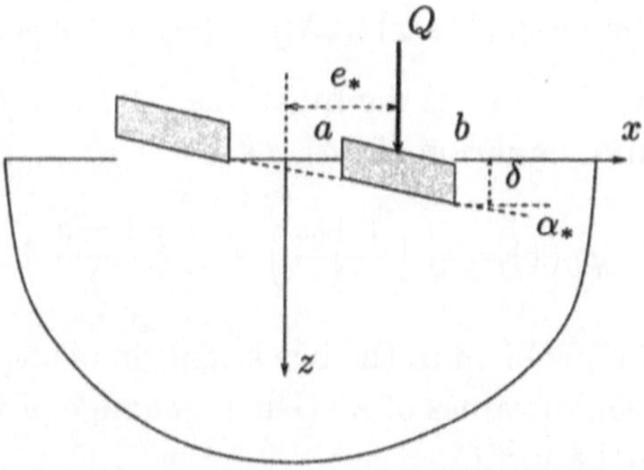

Figure 5.6: Separation of the left-hand part of the plane annular punch from a half-space (an axial section)

where e_* is the eccentricity of the force Q (distance between its point of application and the axis of symmetry of the punch; $e_* = e_x$ for this case).

5.3.3. ANOTHER SOLUTION FOR A WIDE ANNULAR PUNCH

Let us describe one more way to construct an asymptotic solution of the problem for small λ.

Let us start from the integral equations (5.84), (5.85). We present the meromorphic function $L_n(s) = sK_n(s, \pi/2)$ in the complex plane s as a sum of its principal values (see formula (*) in footnote 6 in subsection 3.3.2). Then we obtain the following representation for the kernels $M_n(t)$ ($n = 0, 1$):

$$M_n(t) = \pi \sum_{k=0}^{\infty} \frac{(2k-1+2n)!!(2k-1)!!}{(2k+2n)!!(2k)!!} \exp\left[-\left(2k+n+\frac{1}{2}\right)|t|\right]. \quad (5.98)$$

First we consider equation (5.84). We use the functions $\zeta_1[(1+x)/\lambda]$ and $\zeta_2[(1-x)/\lambda]$, which bring the solution of the following system of integral equations:

$$\int_{-1}^{\infty} \zeta_1\left(\frac{1+\xi}{\lambda}\right) M_0\left(\frac{x-\xi}{\lambda}\right) d\xi = \int_{-\infty}^{-1} \zeta_2\left(\frac{1-\xi}{\lambda}\right) M_0\left(\frac{x-\xi}{\lambda}\right) d\xi \quad (5.99)$$

$$(-1 \le x < \infty),$$

$$\int_{-\infty}^{1} \zeta_2\left(\frac{1-\xi}{\lambda}\right) M_0\left(\frac{x-\xi}{\lambda}\right) d\xi = \int_{1}^{\infty} \zeta_1\left(\frac{1+\xi}{\lambda}\right) M_0\left(\frac{x-\xi}{\lambda}\right) d\xi \quad (5.100)$$

$$+\pi\lambda \exp[(1+x)/(2\lambda)] \quad (-\infty < x \le 1),$$

One can check that the sum of the functions $\zeta_1[(1+x)/\lambda]$ and $\zeta_2[(1-x)/\lambda]$ is also the solution of equation (5.84), i.e.,

$$\varphi_0(x) = \zeta_1\left(\frac{1+x}{\lambda}\right) + \zeta_2\left(\frac{1-x}{\lambda}\right). \tag{5.101}$$

Thus, we reduce the problem to the investigation of the system of equations (5.99), (5.100) for small values of λ. Using a change of variables, we rewrite these equations in the form

$$\int_0^\infty \zeta_1(\tau) M_0(t-\tau)\, d\tau = \int_{2/\lambda}^\infty \zeta_2(\tau) M_0\left(t+\tau-\frac{2}{\lambda}\right) d\tau, \tag{5.102}$$

$$\int_0^\infty \zeta_2(\tau) M_0(t-\tau)\, d\tau = \pi \exp\left(\frac{1}{\lambda}-\frac{t}{2}\right) + \int_{2/\lambda}^\infty \zeta_1(\tau) M_0\left(t+\tau-\frac{2}{\lambda}\right) d\tau, \tag{5.103}$$

where $0 \le t < \infty$. Apparently, we have $\zeta_1(t) \sim \exp(-\kappa t)$ as $t \to \infty$, $\kappa > 0$. Therefore, the integral on the right-hand side of equation (5.103) is exponentially small for small values of λ and we can neglect it in the zero (main) approximation in comparison with the first term on the right-hand side. Then we come to a Wiener–Hopf integral equation for determination of the function $\zeta_2(t)$, where the function $L_0(u)$ for $\alpha = \pi/2$ is easily factorizable. Solving this Wiener–Hopf equation, we obtain

$$\zeta_2(t) = 2\pi^{-1} \exp(1/\lambda) \exp(-3t/2)[1-\exp(-2t)]^{-1/2}. \tag{5.104}$$

By substituting (5.104) into equation (5.102) and calculating the integral on its right-hand side with the help of expansion (5.98), we arrive again at the Wiener–Hopf integral equation for the unknown function $\zeta_1(t)$ in the form

$$\int_0^\infty \zeta_1(\tau) M_0(t-\tau)\, d\tau = \exp\left(-\frac{1}{\lambda}\right) \sum_{n=0}^\infty \left[\frac{(2n-1)!!}{(2n)!!}\right]^2 \frac{\exp(-t[2n+1/2])}{n+1}, \tag{5.105}$$

where $0 \le t < \infty$. The solution of equation (5.105) has the form

$$\zeta_1(t) = \frac{2}{\pi^2} \exp\left(-\frac{2}{\lambda}\right) \exp\left(-\frac{3t}{2}\right) [1-\exp(-2t)]^{-1/2} \sum_{n=0}^\infty \frac{(2n-1)!!}{(n+1)(2n)!!} \tag{5.106}$$

$$\times \mathcal{F}(-n, 1, 1/2; 1 - \exp(-2t)).$$

Thus, the principal term of asymptotic solution of equation (5.84) for small λ is supplied by formulae (5.101), (5.104) and (5.106).

Then we obtain the quantity Q (5.78) in the form

$$\frac{Q}{\theta b \delta} = \frac{2\pi}{\lambda} \exp\left(-\frac{2}{\lambda}\right) \int_{-1}^{1} \varphi_0(\xi) \exp \frac{1+\xi}{2\lambda}\, d\xi = 4 \left[\sqrt{1-\varepsilon^2} + \frac{1}{\pi}\varepsilon^2 \arccos \varepsilon \right. \tag{5.107}$$

$$\left. + \frac{1}{\pi}\varepsilon^3 \sqrt{1-\varepsilon^2} \sum_{n=1}^{\infty} \frac{(2n-1)!!}{(n+1)(2n)!!} \mathcal{F}(-n+1, 2, 3/2; 1-\varepsilon^2)\right].$$

When deriving (5.107) we have used integral 1.15.3.8 in in Prudnikov, Brychkov and Marichev (1990)

$$\int (1-x)^{-1/2} x^{-1/2} \mathcal{F}(-n, 1, 1/2; x)\, dx = 2x^{1/2}(1-x)^{1/2}$$

$$\times \mathcal{F}(-n+1, 1, 3/2; x).$$

The principal term of asymptotical solution of equation (5.85) for small λ can be similarly constructed, and it has the form

$$\varphi_1(x) = \zeta_1\left(\frac{1+x}{\lambda}\right) + \zeta_2\left(\frac{1-x}{\lambda}\right),$$

$$\zeta_2(t) = 4\pi^{-1} \exp(3/\lambda) \exp(-5t/2)[1-\exp(-2t)]{-1/2},$$

$$\zeta_1(t) = \frac{4}{\pi^2} \exp\left(-\frac{2}{\lambda}\right) \exp\left(-\frac{5t}{2}\right) [1-\exp(-2t)]^{-1/2} \sum_{n=0}^{\infty} \frac{(2n-1)!!}{(n+2)(2n)!!}$$

$$\times \mathcal{F}(-n, 1, 1/2; 1-\exp(-2t)),$$

$$\frac{M_y}{\theta b^3 \alpha_*} = \frac{M_x}{\theta b^3 \beta_*} = \frac{8}{3} \left[\sqrt{1-\varepsilon^2} + \frac{1}{2}\varepsilon^2\sqrt{1-\varepsilon^2} + \frac{3}{4\pi}\varepsilon^4 \arccos \varepsilon \right. \tag{5.108}$$

$$\left. + \frac{3}{2\pi}\varepsilon^5 \sqrt{1-\varepsilon^2} \sum_{n=1}^{\infty} \frac{(2n-1)!!}{(n+2)(2n)!!} \mathcal{F}(-n+1, 1, 3/2; 1-\varepsilon^2)\right].$$

Table 5.9 presents certain numerical results for the quantities

$$g_1 = \frac{Q}{\theta b \delta}, \qquad g_2 = \frac{M_y}{\theta b^3 \alpha_*} = \frac{M_x}{\theta b^3 \beta_*} \tag{5.109}$$

calculated by these formulae.

Table 5.9: Quantities (5.109)

λ	1/2	1	2	4	8
g_1 (5.89)		3.99	3.97	3.86	3.63
g_1 $(5.91)_2$	4.00	4.00	4.00		
g_1 (5.95)	4.00	4.00	3.96		
g_1 (5.107)	4.00	4.00	3.96		
g_2 (5.90)			2.64	2.64	2.53
g_2 (5.93)	2.67	2.67	2.67	2.67	
g_2 (5.97)	2.67	2.67	2.66	2.63	
g_2 (5.108)	2.67	2.67	2.66	2.64	

One can see that the convergence of the asymptotic solutions for small and large λ takes place on the interval $\lambda \in [1;2]$ for equation (5.84), and on the interval $\lambda \in [2;4]$ for equation (5.85).

Suppose the boundary point $r = b$ of the contact zone is fixed. Then the smaller the value of $r = a$, or the wider the annular punch, the smaller is the value of λ because $b/a = \exp(2/\lambda)$, due to formula $(5.80)_3$. Let the dimensional quantity b be 1 (m). Then $a = 0.779$ (m) for $\lambda = 8$, $a = 0.135$ (m) for $\lambda = 1$, and $a = 0.0183$ (m) for $\lambda = 1/2$, so that the punch is almost circular for the last value of λ. We can see from Table 5.9 that the dimensionless force g_1 and moment g_2 decrease slightly when λ increases, whereas near a conical point we had the external total load 3.5 times greater for $\lambda = 1/4$ than for $\lambda = 4$ (Table 5.8). This means that only for the regular elastic surface like a half-space does the annular punch cause a contact stress distribution which changes slowly with changing punch width, providing a stability of such constructions; near a conical point there are large gradients of the contact stress distribution.

5.4 Wedge-Shaped Punch on a Cone

The integral equation governing the contact problem of a wedge-shaped punch of angle 2β acting on an elastic infinite cone of angle 2α can be written in the form

$$\frac{\theta^{-1}}{4\pi^2 i}\int_{-\beta}^{\beta} d\psi \int_0^{\infty} q(x,\psi)dx \int_{\Gamma} \sum_{k=-\infty}^{\infty} K_k(s,\alpha)\exp(ik[\varphi-\psi])\left(\frac{x}{r}\right)^{s+1/2} ds$$
$$= f(r,\varphi), \tag{5.110}$$

where $0 \le r < \infty$, $|\varphi| \le \beta$. Here we have used formula (5.8). The function $q(\rho,\psi)$ is the unknown normal contact stress under the punch, and $f(r,\varphi)$

is a function depending on the shape of the base of the punch and on its settling. The contour Γ is a straight line parallel to the imaginary axis of the complex plane s and located in the domain $|\operatorname{Re} s| < 1/2$, and the function $K_k(s, \alpha)$ is defined by formulae (5.12)–(5.21) in which one should substitute s by $-is$. The contact domain and the punch are shown in Fig. 5.7*a*,*b* .

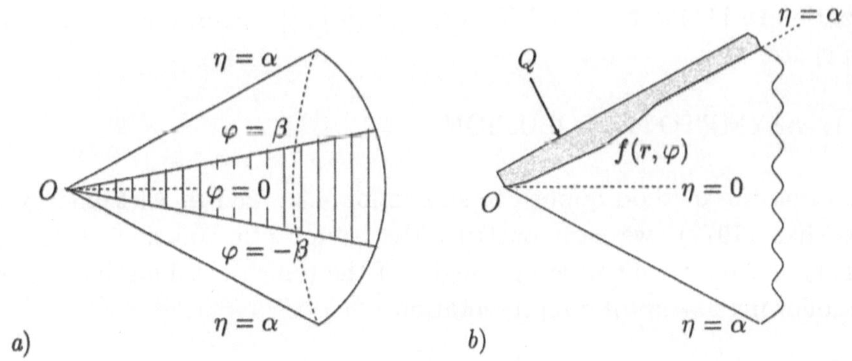

Figure 5.7: The contact between a cone and a wedge-shaped punch: *a*) the contact domain, hatched; *b*) an axial section

Using the direct and inverse Mellin transformations

$$\int_0^\infty q(\rho, \psi)\rho^{s+1/2}\, d\rho = q_s^*(\psi), \quad q(\rho, \psi) = \frac{1}{2\pi i}\int_\Gamma q_s^*(\psi)\rho^{-s-3/2}\, ds,$$

$$\int_0^\infty f(r, \varphi)r^{s-1/2}\, dr = f_s^*(\varphi), \quad f(r, \varphi) = \frac{1}{2\pi i}\int_\Gamma f_s^*(\varphi)r^{-s-1/2}\, ds,$$

and the fact that the function $K_k(s, \alpha)$ is even with respect to the exponent, we reduce the two-dimensional equation (5.110) to a one-dimensional equation. We introduce the new notation

$$q_s(\xi) = \frac{q_s^*(\psi)}{\theta}, \quad f_s(x) = \frac{f_s^*(\varphi)}{\beta}, \quad x = \frac{\psi}{\beta}, \quad \xi = \frac{\varphi}{\beta}, \quad \lambda = \frac{1}{\beta},$$

to represent the latter equation in the form

$$\int_{-1}^{1} q_s(\xi) K_s\left(\frac{\xi - x}{\lambda}\right) d\xi = \pi f_s(x), \quad |x| \le 1, \qquad (5.111)$$

$$K_s(t) = \frac{1}{2}K_0(s,\alpha) + \sum_{k=1}^{\infty} K_k(s,\alpha)\cos(kt). \tag{5.112}$$

For the case when $\alpha = \pi/2$ we have (see (5.22))

$$K_k\left(s, \frac{\pi}{2}\right) = \frac{\Gamma(s/2+k/2+1/4)\Gamma(-s/2+k/2+1/4)}{2\Gamma(s/2+k/2+3/4)\Gamma(-s/2+k/2+3/4)}. \tag{5.113}$$

Equation (5.111) with kernel (5.112) and (5.113) corresponds to an elastic half-space.

5.4.1. ASYMPTOTIC SOLUTION

Following the method applied in subsection 4.3.1 and in Aleksandrov and Babeshko (1972), we can construct the asymptotic solution of equation (5.111) as $\lambda \to \infty$ (small wedge angles of the punch). Taking into account the following asymptotic representation

$$kK_k(s,\alpha) = 1 + \sum_{n=1}^{2N} c_n k^{-n} + O(k^{-2N-1}) \quad (k \to \infty),$$

we can represent the kernel (5.112) in the form

$$K_s(t) = \ln|t| l_1(t) + |t| l_2(t) + l_3(t), \tag{5.114}$$

$$l_m(t) = \sum_{n=0}^{N} d_{mn} t^{2n} + O(t^{2N+2}), \quad d_{10} = -1.$$

If the radius of convergence for these series is equal to $\rho(\alpha)$ then, for $\lambda > 2\rho^{-1}(\alpha)$, we can solve equation (5.92) with (5.114) by the asymptotic method for *large* λ. Let $f(r,\varphi) = fr^{\mu}\exp(-\gamma r)$ ($\mu \geq \delta - 1$, $\delta > 0$, $\gamma > 0$). If $\mu = 0$ and $\gamma \to 0$, then such a punch degenerates into a flat one. Then we have $f_s^*(\varphi) = f\gamma^{-(s+1/2+\mu)}\Gamma(s+1/2+\mu)$, $\mathrm{Re}\, s > -1/2-\mu$. Let β be so small that the terms of the order λ^{-1} and the higher-order terms can be neglected. Then solution of the problem is (cf. (4.113), (4.114))

$$q(\rho,\psi) = \frac{f\gamma^{-\mu+1}}{2\pi i\sqrt{\beta^2-\psi^2}} \int_{\Gamma} \frac{(\rho\gamma)^{-s-3/2}\Gamma(s+1/2+\mu)}{g(s)}\, ds,$$

where

$$g(s) = \ln(4\lambda) - \mathrm{C} - \frac{1}{2}\psi(1/2+s) - \frac{1}{2}\psi(1/2-s) + d_*(s), \tag{5.115}$$

$$d_*(s) = \frac{1}{2}\left[K_0(s,\alpha) - K_0\left(s, \frac{\pi}{2}\right)\right] + \sum_{k=1}^{\infty}\left[K_k(s,\alpha) - K_k\left(s, \frac{\pi}{2}\right)\right],$$

C is Euler's constant and $\psi(x)$ is the psi function.

The distribution of zeros of the function $g(s)$ (5.115) in the domain $|\operatorname{Re} s| \le 3/2$ is similar to that of the corresponding function in (4.114). Specially, we have (see (5.24) with substitution of s by $-is$)

$$\lim_{s\to 1/2}\left(s - \frac{1}{2}\right) g(s) = \lim_{s\to 1/2}\left(s - \frac{1}{2}\right)\frac{1}{2} K_0(s,\alpha)$$

$$= V(\alpha) = \frac{-\sin\alpha\,\cos^2(\alpha/2)}{1 + 2\nu\cos\alpha + \cos^2\alpha}.$$

When deriving these relations, we used the fact that

$$\lim \mathcal{P}_s^1(\cos\alpha)/s = -\tan(\alpha/2) \quad (s \to 0).$$

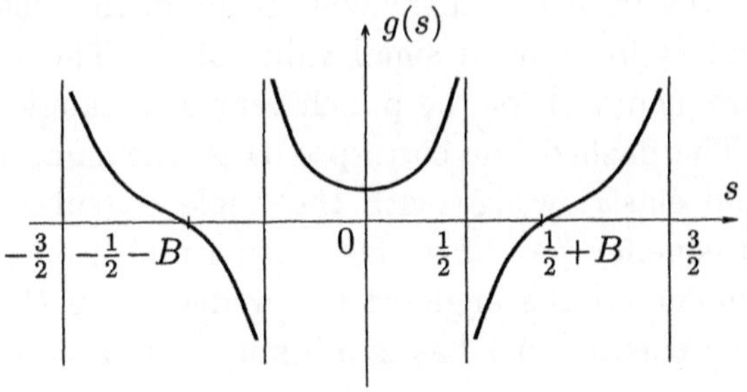

Figure 5.8: A typical graph of $g(s)$ on the real axis for large λ

It is apparent that the function $V(\alpha)$ is negative for $0 < \nu \le 1/2$ and $0 < \alpha < \pi$, hence, $g(1/2 \pm 0) = \mp\infty$. One can verify that $g(i\infty) = -\infty$. Moreover, if λ is sufficiently large, then $g(0) > 0$ and $g(3/2 - 0) > 0$ (Fig. 5.8). Thus, for a fixed angle α and $\lambda \to \infty$, the function $g(s)$ has two pairs of simple roots in the domain $|\operatorname{Re} s| \le 3/2$ one of which, $s_{1,2} = \pm iA$, $A = A(\alpha, \lambda) = O(\lambda)$ $(\lambda \to \infty)$, lies on the imaginary axis and the other, $s_{3,4} = \pm(1/2 + B)$, $B = B(\alpha,\lambda) \in (0;1)$, on the real axis. Using Rouché's Theorem 4.3 one can show that such a pattern of distribution of zeros does not change if one retains more terms in the expansions of of the desired contact pressures when using the method of *large* λ. The asymptotic behaviour of the function $q(\rho, \psi)$ as $\rho \to 0$ in this problem is similar to that described in subsection 4.3.1. Thus, irrespective of the form of the punch base, the expansion of this function as $\rho \to 0$ will contains terms of the form $\rho^{-iA-3/2}$ (oscillations) and ρ^{B-1}.

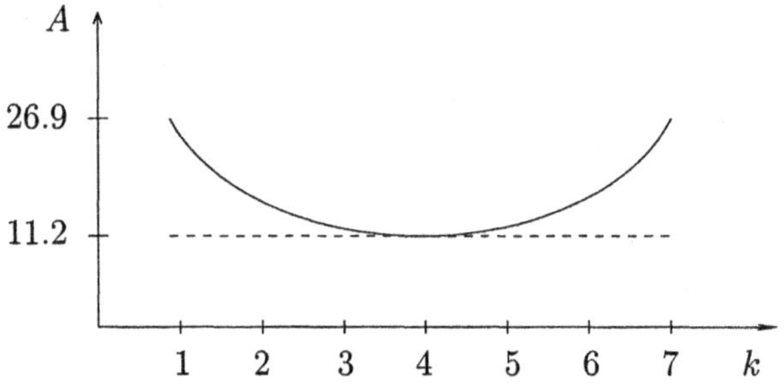

Figure 5.9: The oscillation frequency A

Let us study the influence of the cone angle of the elastic solid on the exponents A and B for a fixed small value of β. The Fig. 5.9 presents the oscillation frequency A for the punch versus the angle $\alpha = \pi k/8$ ($\beta = 0.2$, $\nu = 0.3$). The dashed line corresponds to the same problem for the three-dimensional elastic wedge with the angle α (subsection 4.3.1). If the oscillation frequency for the wedge-shaped punch on an elastic wedge is weakly dependent on the angle of the wedge, then the corresponding frequency for the elastic cone has a minimum at $\alpha = \pi/2$ (half-space). Moreover, at $\alpha = \pi k/8$ and $\alpha = \pi - \pi k/8$, $k = 1, 2, 3$, the difference of the corresponding values of A does not exceed 3%.

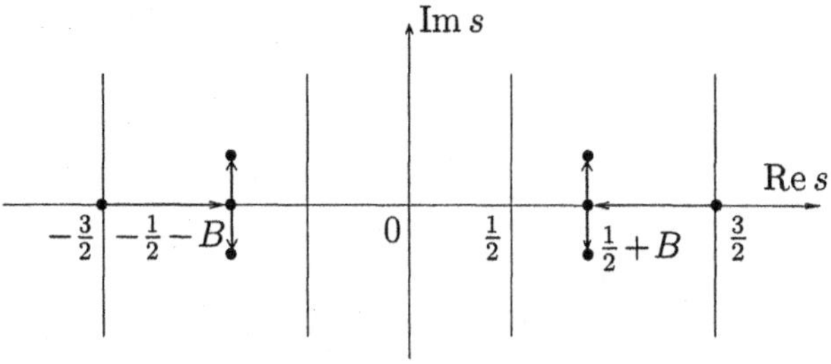

Figure 5.10: New roots for small fixed β

Calculations for a fixed small β show that as α decreases, two real roots of the equation $g(s) = 0$ enter the domain $|\,\mathrm{Re}\, s| \leq 3/2$ from outside. Then these roots merge with the roots $\pm(1/2 + B)$ and leave the real axis for the complex plane (Fig. 5.10). Thus, for small cone angles, additional oscillations appear (there are no real roots in this case). The double real root corresponds to the singularity of the form $\rho^{B-1}(c_1 + c_2 \ln \rho)$. To illustrate this process we present numerical values of the roots $s = \pm 0.5 \pm 0.01n$ for $\beta = \lambda^{-1} = 0.05$ and $\nu = 0.3$ versus the angle $\alpha = \pi t/8$:

t	6	4	2	1.75	1.712	1.71	1
n	7	12	36	66 98	82 86	—	—

It should be noted that when computing the values of the associated Legendre function $\mathcal{P}_s^k(\cos\alpha)$, one should not use the recursive formulae with respect to the superscript k, since they lead to substantial errors as $k \to \infty$, especially for acute angles α. To calculate $\mathcal{P}_s^k(\cos\alpha)$, we used formula (5.6) and the relation 8.339.4 in Gradshteyn and Ryzhik (1994)

$$\mathcal{P}_{s-1/2}^k(x) = (-1)^k \frac{\Gamma(s + 1/2 + k)(1 - x^2)^{k/2}}{\Gamma(s + 1/2 - k)\, 2^k\, k!}$$

$$\times \mathcal{F}\left(k + \frac{1}{2} - s, k + \frac{1}{2} + s, k + 1; \frac{1 - x}{2}\right),$$

$$\frac{\Gamma(s + 1/2 + k)}{\Gamma(s + 1/2 - k)} = \left[s^2 - \left(\frac{1}{2}\right)^2\right]\left[s^2 - \left(\frac{3}{2}\right)^2\right] \cdot \ldots \cdot \left[s^2 - \left(k - \frac{1}{2}\right)^2\right].$$

5.4.2. BUBNOV–GALERKIN METHOD

If the angle β is not too small, then the Bubnov–Galerkin method, used in subsection 4.3.2, is effective in the analysis of the asymptotic behaviour of the contact stress at the wedge edge. We can prove (see subsection 4.3.2) that if the determinant $D(s)$ of the infinite-dimensional matrix with entries $(m, n = 0, 1, \ldots)$

$$a_{mn}(s) = \sum_{k=0}^{\infty}{}' K_k(s, \alpha) \mathcal{J}_m(k\beta) \mathcal{J}_n(k\beta) \cos\left(\frac{m - n}{2}\pi\right) \tag{5.116}$$

vanishes for $s = s_*$, then $q(\rho, \psi) \sim \rho^{-\gamma}$, $\gamma = 3/2 + s_*$ as $\rho \to 0$. Here, the prime on the summation symbol indicates that the first term ($k = 0$) of series (5.116) has the coefficient 1/2. If $\beta > 0.1\pi$, then to calculate the zeros of the determinant of the matrix with the entries of (5.116) with the accuracy up to three significant digits it is sufficient to use the truncation of this matrix to the dimension 6×6.

Since even number of row (column) permutations in a matrix does not change its determinant and we have the zeroth entries (5.116) for the odd $m-n$, the determinant $D(s)$ is factorized as $D(s) = d^{+}(s)d^{-}(s)$, where $d^{\pm}(s)$ are the determinants of the matrices with the entries $(m, n = 0, 1, ...)$

$$a^{+}_{mn}(s) = a_{2m,2n}(s), \quad a^{-}_{mn}(s) = a_{2m+1,2n+1}(s).$$

Table 5.10: Values of maximum exponent γ

n / m	1	2	3	4	5	6	7
1	0.999	0.989	0.938	0.798	0.551	0.266	0.065
2	0.999	0.991	0.951	0.848	0.665	0.463	0.375
3	0.999	0.999	0.996	0.989	0.980	0.971	0.965
4	0.788	0.703	0.611	0.500	0.357	0.184	0.043
5	0.999	0.998	0.994	0.983	0.966	0.945	0.930
6	0.999	0.998	0.992	0.976	0.944	0.899	0.860
7	0.999	0.999	0.997	0.989	0.973	0.950	0.936

The values of the maximum exponent γ of the singularity corresponding to real zeros of $D(s)$, $s \in (-3/2; -1/2)$, are presented in Table 5.10 for different angles $\alpha = \pi m/8$ and $\beta = \pi n/8$ for $\nu = 0.3$. Throughout this table, the main (nearest to $s = -0.5$) roots are those of the equation $d^{+}(s) = 0$. For $m = 4$ the values in Table 5.10 correspond to a half-space. According to the implicit function theorem, the function $s_*(\alpha, \beta)$ describing the dependence of the roots of the equation $D(s) = 0$ on α and β is analytic. Therefore, these roots trace continuous lines as α and β vary continuously. For $s = -1/2$, the function $D(s)$ has a multiple pole, since this pole is single for each entry of the matrix (5.116). This pole can absorb and emit zeros of the function $D(s)$ as the angles α and β vary continuously. At a small deviation of the cone angle 2α from π (half-space) toward either of the sides, with β being fixed, a new zero of the function $D(s)$ appears from the pole $s = -1/2$. This new zero provides the function $q(\rho, \psi)$ with a stronger singularity $\rho^{-\gamma}$, $\gamma \approx 1$, as $\rho \to 0$, compared with the singularity corresponding to the root of the equation $D(s) = 0$ on the interval $s \in (-3/2; -1/2)$, which is unique for $2\alpha = \pi$. If α continues to vary smoothly, this zero varies continuously generating singularities with the exponent γ presented in the table. There is Galin's exact solution (Galin (1961)) with the similar singularity ρ^{-1} for the contact between a wedge-shaped punch and a half-space, loaded outside the punch, on the continuation of the punch bisectrix, as shown in Fig 5.11*a*. Now we can explain Galin's solution. The physical implication here is that this external load on the line causes a small perturbation of the regular elastic surface near the punch tip. The perturbed half-space becomes a cone with angle π+0, for which we have obtained the close singularity. We can conclude that the contact between such a cone and a wedge-shaped punch is equivalent to that between a punch and the loaded regular elastic surface. The external load increases the singularity for a half-space at the punch tip.

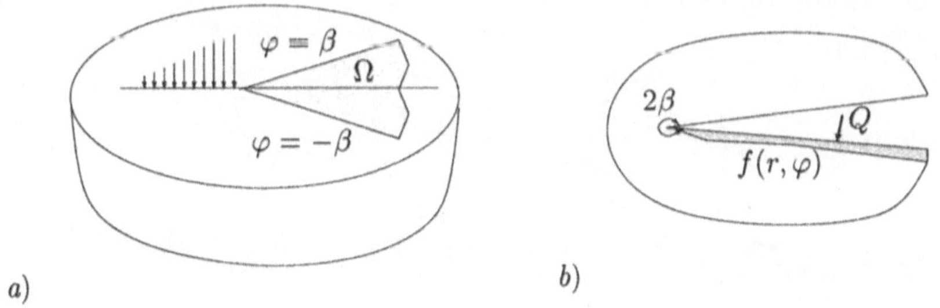

Figure 5.11: *a)* Contact domain Ω and the load outside this domain for Galin's solution; *b)* there is no separation of the apex of a wedge-shaped punch in a conical cavity (an axial section)

If the cone angle $2\beta > \pi$, we have a conical cavity, with angle $2\pi - 2\beta$, in a whole space. If 2β is close to 2π and $\alpha = \pi m/8$, $m = 2, 3, 5, 6$, then a double root of the equation $D(s) = 0$ occurs due to the simultaneous vanishing of the functions $d^+(s)$ and $d^-(s)$. For example, for $2\beta = 7.3\pi/4$, $\alpha = 3\pi/8$ and $\nu = 0.3$, we have $\gamma = 0.963$. The singularity of the form $\rho^{-\gamma}(c_1 + c_2 \ln \rho)$ occurring in this case can lead to the change of sign of the contact stress function, as $\rho \to 0$, but the shape of the conical cavity prevents separation of the punch apex, as shown in Fig 5.11*b*.

The numerical results show that, for $s \in (-1/2; 0]$ or pure imaginary s, we have $D(s) \neq 0$, i.e., this numerical method fails to determine a root generating the singularity of the form $\rho^{-iA-3/2}$, $A \in \Re$.

5.5 Contact Problem for a Cone with Unknown Contact Area

Let us study the 3-D contact problem for a frictionless indentation of a rigid punch into an elastic cone of angle 2α (Fig. 5.12). The kernel of the two-dimensional integral equation of the problem contains a logarithmic singularity $\ln R$ (outside the vertex of the cone) in addition to the familiar leading singular term of the order of $1/R$. The identification of all singularities of the kernel is problematic. This complicates the application of known analytical methods to solve this problem. In this section, we use the numerical Galanov–Newton method (see section 2.9) proposed for solving nonlinear boundary integral equations of the Hammerstein type. This method allows one to find the normal contact pressures and the unknown contact area simultaneously and is effective for the case where the contact area is far apart from the cone vertex.

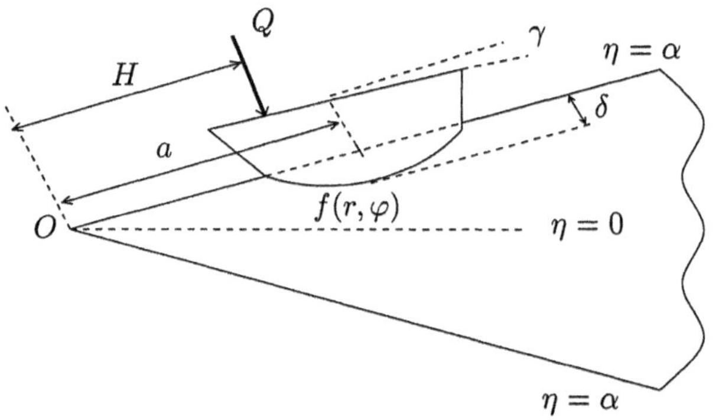

Figure 5.12: Contact between a cone and an elliptic paraboloid (an axial section)

To be specific, we have assumed that the surface of the punch is an elliptic paraboloid, decsribed by the function

$$f(r,\varphi) = (r\cos\varphi - a)^2/(2R_1) + (r\sin\varphi)^2/(2R_2), \quad R_2 \geq R_1.$$

Due to formula (5.11), we can write the integral equation of the 3-D problem in the form

$$\frac{\theta^{-1}}{2\pi}\int\limits_{\Omega} K(r,\varphi,\rho,\psi)q(\rho,\psi)\,\rho\,d\rho d\psi = \delta + \gamma(r\cos\varphi - a) - f(r,\varphi), \quad (5.117)$$

where $(r,\varphi) \in \Omega$,

$$K(r,\varphi,\rho,\psi) = \frac{2}{\pi\sqrt{r\rho}}\sum_{k=0}^{\infty}{}'\int\limits_0^{\infty} K_k(s,\alpha)\cos\left(s\ln\frac{\rho}{r}\right)ds\cos(k[\varphi-\psi]), \quad (5.118)$$

$$K_k(s,\alpha) = \frac{f_1 + f_2 p + f_3 p^2 + f_4 p^3}{g_1 + g_2 p + g_3 p^2 + g_4 p^3}, \quad p = p(s) = \frac{\mathcal{P}^k_{is-1/2}(\cos\alpha)}{\mathcal{P}^{k-1}_{is-1/2}(\cos\alpha)}, \tag{5.119}$$

the functions $f_m = f_m(s)$ and $g_m = g_m(s)$ $(m = 1,2,3,4)$ are defined by formulae (5.14)–(5.21) and the function $q(\rho,\psi)$ is the unknown contact pressure applied to the unknown contact region Ω. When acted upon by the force P applied at the distance H from the cone vertex in the plane $\varphi = 0$, the punch settles by δ, and γ characterizes the skewness of the punch with respect to the line passing through the initial contact point $r = a$, $\varphi = 0$ perpendicular to the cone surface. The prime on the sum in (5.118) means that the first term (for $k = 0$) has coefficient 1/2. The use of the ratio $p(s)$ of associated Legendre functions in the expression of the symbol of the kernel $K_k(s,\alpha)$ in (5.119) is justified by the fact that the spherical function $\mathcal{P}^k_{is-1/2}(\cos\alpha)$ is nonzero for $\alpha \in (0,\pi)$, $k \geq -1$, $k \in \mathcal{Z}$, $s \in \Re$ (Zhurina and Karmazina (1966)).

To find the relationship between P, H and δ, γ, one should complete equation (5.117) by the equilibrium conditions

$$\int_\Omega q(\rho,\psi)\,\rho\,d\rho d\psi = P, \qquad \int_\Omega q(\rho,\psi)(\rho\cos\psi - a)\,\rho\,d\rho d\psi = P(H-a) = M.$$

By using the asymptotic expansions of the Legendre functions of high degree or order (Abramowitz and Stegun (1972, formulae 8.10), Zhurina and Karmazina (1966)), we can prove that

$$K_k(s,\alpha) = \frac{c_1(k,\alpha)}{s} + \frac{c_2(k,\alpha)}{s^2} + \frac{c_3(k,\alpha)}{s^3} + O\left(\frac{1}{s^4}\right) \tag{5.120}$$

as $s \to +\infty$ for fixed k, and

$$K_k(s,\alpha) = \frac{d_1(s,\alpha)}{k} + \frac{d_2(s,\alpha)}{k^2} + \frac{d_3(s,\alpha)}{k^3} + O\left(\frac{1}{k^4}\right) \tag{5.121}$$

as $k \to +\infty$ for fixed s, where $c_1(k,\alpha) \equiv 1$, $d_1(s,\alpha) \equiv \sin\alpha$, $c_2(k,\pi/2) \equiv d_2(s,\pi/2) \equiv 0$.

Also, taking formula (5.22) into account, one can readily establish that

$$K_k\left(s,\frac{\pi}{2}\right) = \frac{e_1(s,k)}{\sqrt{s^2+k^2}} + \frac{e_2(s,k)}{(\sqrt{s^2+k^2})^2} + \frac{e_3(s,k)}{(\sqrt{s^2+k^2})^3} + O\left(\frac{1}{(\sqrt{s^2+k^2})^4}\right), \tag{5.122}$$

$$e_1(s,k) \equiv 1, \quad e_2(s,k) \equiv 0, \quad e_3(s,k) = \frac{s^2-k^2}{8(s^2+k^2)}$$

as $\sqrt{s^2+k^2} \to \infty$.

As follows from Prudnikov, Brychkov and Marichev (1992, Vol. 1, integral 2.5.6.4 and Vol. 2, series 5.9.1.4), the first term in the asymptotic expansion (5.122) completely defines the familiar singularity of the form $1/R$, $R = (r^2+\rho^2-2r\rho\cos(\varphi-\psi))^{1/2}$, in the kernel of the integral equation of the contact problem for a half-space. Since $c_2(k,\alpha) \neq 0$, $d_2(k,\alpha) \neq 0$ in formulae (5.120) and (5.121) for $\alpha \neq \pi/2$, it is clear that the variable e_2 in the expansion of type (5.122) for the function $K_k(s,\alpha)$ satisfies the condition $e_2(s,k) \neq 0$ for $\alpha \neq \pi/2$. Using the relation (Prudnikov, Brychkov and Marichev (1992, Vol. 1, integral 2.5.6.6 and series 5.4.12.6))

$$\sum_{k=1}^{\infty} \int_0^{\infty} \frac{\cos(s\ln[r/\rho])}{s^2+k^2} ds \cos(k[\varphi-\psi]) = \frac{\pi}{4}\left[\left|\ln\frac{r}{\rho}\right| - \ln\frac{R^2}{r\rho}\right],$$

we find that for $\alpha \neq \pi/2$, additional singularities of the order $\ln R$ occur in the kernel of the integral equation (5.117) outside the vertex of the cone $(r,\rho > 0)$. The exact determination of all singularities in this kernel for $\alpha \neq \pi/2$ is problematic, since in this case we have $c_n(k,\alpha) \neq d_n(s,\alpha)$ $(n = 1,2)$, and the calculation of the required integrals and series is complicated because the expressions for $e_n(s,k)$ $(n=1,2)$ are cumbersome.

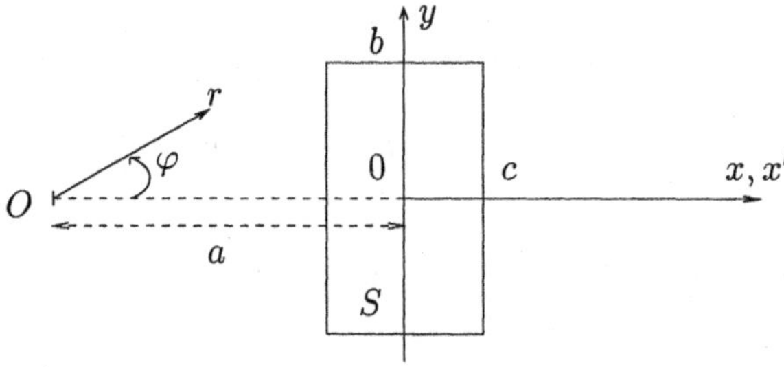

Figure 5.13: The contact plane and rectangle S

To solve the integral equation (5.117) approximately, we apply the Galanov–Newton method, which enables one to find the function $q(r,\varphi)$ and the area Ω simultaneously. Due to the smallness of the settlement δ in linear elasticity, the contact area will be small too, and we can suppose that the contact takes place over a plane surface, as in the Hertz problem. The spherical coordinates r and φ are polar coordinates in this plane of contact, with center at the cone vertex. We perform a change of variables in equation (5.117) with (5.118) according to the relations

$$r = \sqrt{x^2 + y^2}, \quad \rho = \sqrt{s^2 + t^2}, \quad \varphi = \arctan(y/x), \quad \psi = \arctan(t/s) \tag{5.123}$$

and assume that the contact region Ω entirely belongs to a rectangle S in the plane (x, y) the center of which lies at the point $r = a$, $\varphi = 0$ and the semiaxes are equal to b and c $(b \geq c)$. The contact plane and rectangle are shown in Fig. 5.13.

Then we introduce the dimensionless variables and parameters

$$x - a = x'b, \quad s - a = s'b, \quad y = y'b, \quad t = t'b, \quad \delta = \delta' b, \quad \lambda = a/b, \tag{5.124}$$

$$A = b/(2R_1), \quad B = b/(2R_2), \quad \varepsilon = c/b, \quad M = 2\pi\theta b^3 M', \tag{5.125}$$

$$q(x, y) = 2\pi\theta q'(x', y'), \quad P = 2\pi\theta b^2 P', \quad S' \to S, \quad \Omega' \to \Omega. \tag{5.126}$$

The dimensionless parameter λ characterizes the relative distance of the punch from the vertex of the cone. Extending equation (5.117), in which relations (5.123)–(5.126) are taken into account, to the area S, we can rewrite this equation as a system consisting of an equation and an inequality (in what follows, the primes are omitted)

$$\int_S K(M, N) q(N)\, d\Omega_N = g(M), \quad q(M) \geq 0, \quad M \in \Omega, \tag{5.127}$$

$$\int_S K(M, N) q(N)\, d\Omega_N > g(M), \quad q(M) = 0, \quad M \in (S \setminus \Omega), \tag{5.128}$$

where

$$M = (x, y), \quad N = (s, t), \quad g(x, y) = \delta + \gamma x - Ax^2 - By^2,$$

$$K(x, y, s, t) = 1/R + F(x, y, s, t), \quad R = \sqrt{(x - s)^2 + (y - t)^2}, \tag{5.129}$$

$$F(x, y, s, t) = \frac{2\pi^{-1}}{[(x + \lambda)^2 + y^2]^{1/4} [(s + \lambda)^2 + t^2]^{1/4}} \tag{5.130}$$

$$\times \sum_{k=0}^{\infty}{}' \cos\left[k\left(\arctan\frac{y}{x + \lambda} - \arctan\frac{t}{s + \lambda}\right)\right]$$

$$\times \int_0^{\infty} \left\{K_k(u, \alpha) - K_k\left(u, \frac{\pi}{2}\right)\right\} \cos\left[\frac{u}{2} \ln\frac{(x + \lambda)^2 + y^2}{(s + \lambda)^2 + t^2}\right] du.$$

We introduce the nonlinear operators by formulae (2.252) and consider the operator equation (2.254) with (2.255). The problem of the equivalence of the equation and system (5.127), (5.128) and the problem of the existence and uniqueness of the solution are solved as in subsection 2.9.1. To solve equation (2.254) numerically, we use the method of successive approximations based on formulae (2.260), (2.261). Since the problem is symmetric with respect to y, it is sufficient to consider only the upper half of the rectangle S $(y \geq 0)$ and cover it by a grid with m nodes spaced with step h_1 along the x-axis and with step h_2 along the y-axis (for computations, we set $m \geq 81$). When calculating the function $K(M,N)$ defined by (5.129) at these nodes, we smooth the function singularities by replacing R by $R_* = [(x-y)^2 + (s-t)^2 + \delta_*]^{1/2}$ and by adding the term $\sqrt{\delta_*}$ to the argument of the logarithmic function in formula (5.130). It can be shown that the regularizing parameter δ_* must be related to the grid steps h_1 and h_2 (in calculations, we set $\delta_* = h_1 h_2/16$). Note that one can calculate the values of the function $F(M,N)$ of the form (5.130) with high accuracy and moderate CPU time if the values of the difference $\lambda - \varepsilon$ that defines the area of the effective operation of the method are not too small (i.e., $\lambda - \varepsilon \sim 1$). The ratio $p(u)$ of the Legendre functions, defined in $(5.119)_2$, was computed by using formula (5.6) and representation 8.704 in Gradshteyn and Ryzhik (1994):

$$p(u) = \tan\frac{\alpha}{2}\,\frac{(k-1/2)^2+u^2}{k}\,\frac{\mathcal{F}(1/2-iu, 1/2+iu, 1+k; \sin^2(\alpha/2))}{\mathcal{F}(1/2-iu, 1/2+iu, k; \sin^2(\alpha/2))}.$$

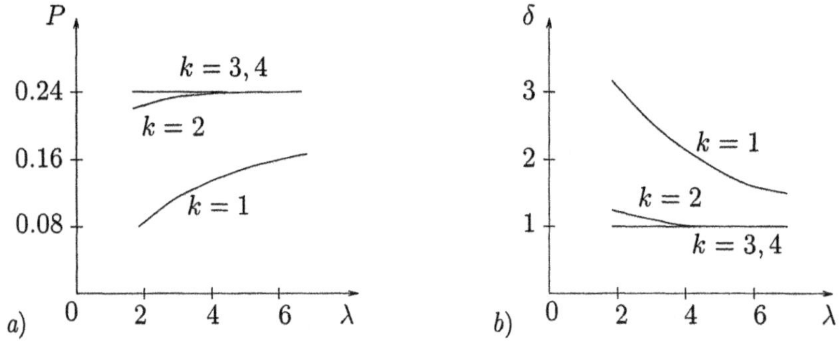

Figure 5.14: *a*) The force P versus λ; *b*) the settlement δ versus λ

Figures 5.14*a*,*b* show the plots of the force P versus λ for a constant settlement $\delta = 1$ and of the settlement δ versus λ for a constant $P = 0.254$ for different angles $\alpha = \pi k/8$ with $\varepsilon = 0.5$, $\gamma = 0$, $A = 2$, and $B = 1$.

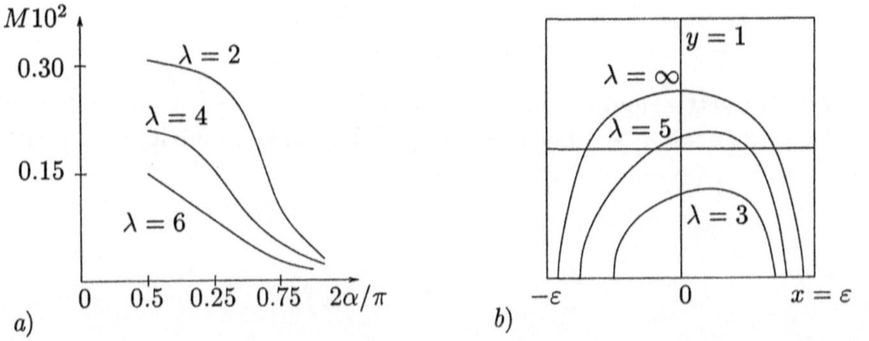

Figure 5.15: *a)* The torque M versus α/π; *b)* the upper boundary of the contact domain Ω

Figures 5.15*a*,*b* present the plots of the torque M versus α for different λ and the boundary of the upper part of the contact domain Ω versus λ for $\alpha = \pi/8$ ($\varepsilon = 0.5$, $\delta = 0.5$, $\gamma = 0$, $A = 2$, $B = 1$). It is apparent that as the punch approaches the vertex of a cone having an acute vertex angle, the value of λ and the area of the contact region decrease, while the settlement under a constant indenting force increases. The indenting force P decreases as the punch approaches the cone vertex, while the torque M, showing rotation of the punch, increases. The values of the contact stress for such a cone are smaller than those for a half-space.

5.6 Contact Problems for a Spherical Layer

5.6.1. A SPHERICAL LAYER

In spherical (r, η, φ) coordinates we consider a spherical layer $R_1 \le r \le R_2$, $0 \le \eta \le \pi$, $0 \le \varphi < 2\pi$, whose surface $r = R_2$ is fixed and whose surface $r = R_1$ is subject to an indentation of a rigid spherical punch, of radius R_0, with initial contact point $\eta = 0$, $r = R_1$ (Fig. 5.16). We assume that there is no friction between the punch and the spherical layer, the indenting force P lies along the line $\eta = 0$; and $\varepsilon = R_1 - R_0$ is a small quantity. In this case we arrive at the solution of an axisymmetric boundary-value problem for Navier's equations in spherical coordinates with the following boundary conditions:

$$u_r = \delta \cos\eta - \varepsilon(1 - \cos\eta) \quad (r = R_1, \quad 0 \le \eta \le \gamma),$$

$$\sigma_r = 0 \quad (r = R_1,\ \gamma < \eta < \pi),$$
$$\tau_{r\eta} = 0 \quad (r = R_1); \quad u_r = u_\eta = 0 \quad (r = R_2),$$

where δ is the settlement of the punch, and $0 \le \eta \le \gamma$ is the contact region (we take the values $\eta \ge 0$ because the problem is symmetric with respect to η).

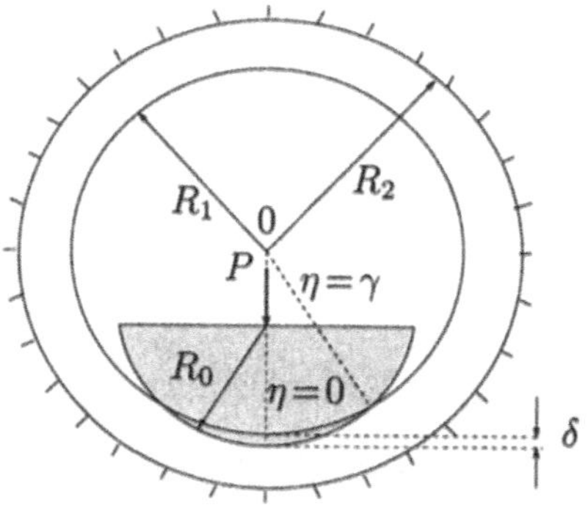

Figure 5.16: Contact between a spherical layer and a spherical punch (a radial section)

Following Lur'e (1964, section 6.2), we seek the solution of the Navier's equations in the form of expansions ($\alpha_n = n + 1/2$):

$$u_r = \sum_{n=0}^{\infty} a_n W_n(r) \mathcal{P}_{\alpha_n - 1/2}(\cos\eta), \quad u_\eta = \sum_{n=0}^{\infty} a_n V_n(r) \frac{d}{d\eta} \mathcal{P}_{\alpha_n - 1/2}(\cos\eta)$$

and satisfying the boundary conditions, we obtain the unknown functions $W_n(r)$ and $V_n(r)$, while for the unknown coefficients a_k we obtain the paired series equations (Chebakov (1992))

$$\sum_{n=0}^{\infty} a_n K(\alpha_n) \mathcal{P}_{\alpha_n - 1/2}(\cos\eta) = \frac{\theta}{R_1}[\delta\cos\eta - \varepsilon(1 - \cos\eta)] \quad (0 \le \eta \le \gamma), \tag{5.131}$$

$$\sum_{n=0}^{\infty} a_n \mathcal{P}_{\alpha_n - 1/2}(\cos\eta) = 0 \quad (\gamma < \eta < \pi), \tag{5.132}$$

where $\theta = G/(1-\nu)$,

$$K(\alpha) = t_1(\alpha)/t(\alpha), \quad \beta = R_1/R_2, \tag{5.133}$$

$$t_1(\alpha) = 2(1-2\nu)[B_1(\alpha)\cosh(2\alpha\ln\beta) - B_2(\alpha)\sinh(2\alpha\ln\beta)$$
$$+\beta^2 B_3(\alpha) + \beta^{-2}B_4(\alpha) + B_5(\alpha)],$$
$$t(\alpha) = (1-\nu)[A_1(\alpha)\cosh(2\alpha\ln\beta) + A_2(\alpha)\sinh(2\alpha\ln\beta)$$
$$+\beta^2 A_3(\alpha) + \beta^{-2}A_4(\alpha) + A_5(\alpha)],$$
$$B_1(\alpha) = 2[\alpha^2(-48\nu^2 + 64\nu - 20) + 2\nu - 1],$$
$$B_2(\alpha) = 2[\alpha^3(-32\nu^2 + 56\nu - 24) + \alpha(16\nu^2 - 26\nu + 10)],$$
$$B_3(\alpha) = \alpha^4(16\nu - 12) + \alpha^2(32\nu^2 - 52\nu + 19),$$
$$B_4(\alpha) = 4\alpha^4 + \alpha^2(64\nu^2 - 96\nu + 31), \quad B_5(\alpha) = (1-2\nu)(8\alpha^4 - 10\alpha^2 + 2),$$
$$A_1(\alpha) = 2[\alpha^4(128\nu^2 - 160\nu + 48) + \alpha^2(-384\nu^3 + 544\nu^2 - 224\nu + 24) + 18\nu - 9],$$
$$A_2(\alpha) = 2[\alpha^3(-256\nu^3 + 512\nu^2 - 320\nu + 64) + \alpha(96\nu^2 - 144\nu + 48)],$$
$$A_3(\alpha) = 16(1-2\nu)\alpha^6 - 40(1-2\nu)\alpha^4 + \alpha^2(128\nu^3 - 64\nu^2 - 146\nu + 73),$$
$$A_4(\alpha) = 16(1-2\nu)\alpha^6 + \alpha^4(-512\nu^3 + 1024\nu^2 - 560\nu + 88)$$
$$+\alpha^2(1152\nu^3 - 2304\nu^2 + 1422\nu - 279),$$
$$A_5(\alpha) = (1-2\nu)(-32\alpha^6 + 112\alpha^4 - 98\alpha^2 - 18).$$

Using these relations, one can easily obtain the quantity $A = \lim K(\alpha)$ as $\alpha \to 0$, which will be required in what follows:

$$A = -2K_1/[K_2(1-\nu)], \tag{5.134}$$

$$K_1 = -4\beta^2(1-2\nu)\ln^2\beta - 8\beta^2\ln\beta(8\nu^2 + 13\nu - 5) + \beta^4(32\nu^2 - 52\nu + 19)$$
$$+\beta^2(-96\nu^2 + 148\nu - 50) + 32(2\nu - 3)\nu + 31,$$
$$K_2 = 36\beta^2\ln^2\beta + 192\beta^2(1-\nu)\ln\beta + \beta^4(64\nu^2 - 73)$$
$$+\beta^2(-384\nu^2 + 352\nu + 50) + 576\nu^2 - 864\nu + 279.$$

The unknown contact pressure under the punch is determined via the solution of paired equations (5.131), (5.132) from the expression

$$q(\eta) = \sum_{n=0}^{\infty} a_n \mathcal{P}_{\alpha_n - 1/2}(\cos\eta) \quad (0 \le \eta \le \gamma).$$

To solve the pair of equations (5.131), (5.132), we reduce them to an infinite system of linear algebraic equations with singular coefficient matrix (the technique was described in subsection 3.6.1). If we introduce the pair of equations

$$\sum_{n=0}^{\infty} a_n^{iw} K(\alpha_n) \mathcal{P}_{\alpha_n-1/2}(\cos\eta) = \mathcal{P}_{iw-1/2}(\cos\eta) \quad (0 \le \eta \le \gamma), \tag{5.135}$$

$$\sum_{n=0}^{\infty} a_n^{iw} \mathcal{P}_{\alpha_n-1/2}(\cos\eta) = 0 \quad (\gamma < \eta < \pi), \tag{5.136}$$

then the contact stress under the punch will be given in term of their solution via the formulae

$$q(\eta) = \frac{\theta}{R_1}(\delta+\varepsilon) q_{3/2}(\eta) - \varepsilon q_{1/2}(\eta), \tag{5.137}$$

$$q_{iw}(\eta) = \sum_{n=0}^{\infty} a_n^{iw} \mathcal{P}_{\alpha_n-1/2}(\cos\eta).$$

The pair of equations (5.135), (5.136) can be reduced to the following infinite system:

$$BX^{iw} = D, \tag{5.138}$$

where $X^{iw} = \{x_n^{iw}\}$ is unknown column, $B = \{b_{mn}\}$ is matrix, and $D = \{d_m\}$ is column with elements $(n, m = 1, 2, ...)$

$$b_{mn} = \frac{\mathcal{P}^1_{i\delta_n-1/2}(c)\mathcal{P}_{i\gamma_m-1/2}(-c) + \mathcal{P}_{i\delta_n-1/2}(c)\mathcal{P}^1_{i\gamma_m-1/2}(-c)}{(\delta_n^2-\gamma_n^2)\mathcal{P}_{iw-1/2}(c)\mathcal{P}_{i\gamma_m-1/2}(-c)},$$

$$d_m = \frac{\mathcal{P}^1_{i\delta_m-1/2}(c)\mathcal{P}_{i\gamma_m-1/2}(-c) + \mathcal{P}_{iw-1/2}(c)\mathcal{P}^1_{i\gamma_m-1/2}(-c)}{K(iw)(\gamma_m^2-w^2)}, \qquad c = \cos\gamma.$$

Here, $i\delta_n$ and $i\gamma_n$ are the zeros and poles of the meromorphic function $K(\alpha)$ in a complex plane, respectively ($\mathrm{Im}\,(i\delta_n) > 0$, $\mathrm{Im}\,(i\gamma_n) > 0$, $n = 1, 2, ...$).

The function $q_{iw}(\eta)$ can be obtained in terms of the solution of infinite system (5.138) from the expression

$$q_{iw}(\eta) = K^{-1}(iw)\mathcal{P}_{iw-1/2}(\cos\eta) + \sum_{n=1}^{\infty} x_n \frac{\mathcal{P}_{i\delta_n-1/2}(\cos\eta)}{\mathcal{P}_{i\delta_n-1/2}(\cos\gamma)}.$$

5.6.2. A SPHERICAL BEARING

The problem in question offers a fairly good model of the operation of a spherical self-lubricating bearing, particularly under loads such that the size of the contact area is commensurable with the width of the bearing (Chebakov (1992)). In actual self-lubricating sliding bearings, the thickness of the antifriction coating is usually much less than the size of the contact region. In the elastic problem, therefore, it is natural to assume that the thickness of the elastic layer is relatively small ($h = R_2 - R_1 \ll R_1 \sin\gamma$). In this case function $K(\alpha)$ in the form $(5.133)_1$ can be approximated by the function

$$K^*(\alpha) = \frac{\tanh(A\alpha)}{\alpha}, \tag{5.139}$$

where A is given by (5.134). The limiting values of $K^*(\alpha)$ and $K(\alpha)$ coincide for large and small α. Numerical calculations show that, with decreasing layer thickness, the greatest approximation error using function (5.139) occurs for large α. We can assert, therefore, that with decreasing elastic layer thickness the degree of accuracy of the solution after using the approximation will increase. In what follows, when mention is made of $K(\alpha)$, as well as of its zeros and poles, we will have in mind $K^*(\alpha)$ and its zeros and poles, respectively, with no change in notation.

We take note of the fact that as $\beta \to 1$ ($R_1 \to R_2$) we have $A \to 0$, and hence as $\beta \to 1$ we have $\delta_n \to \infty$ and $\gamma_n \to \infty$ ($n = 1, 2, ...$), because in approximation (5.139) we have $\delta_n = \pi n/A$ and $\gamma_n = \pi(n - 1/2)/A$. The growth in the poles and zeros of $K(\alpha)$ with decreasing relative elastic layer thickness makes possible substantial asymptotic simplification of system (5.138) and of other expressions. Using the asymptotic form of Legendre functions for large indices, we can show that system (5.138) is asymptotically equivalent to the system

$$TX^{iw} = D^{iw}, \tag{5.140}$$

where the elements t_{mn} of the matrix T and the elements d_m^{iw} of the column D^{iw} have the form

$$t_{mn} = \frac{1}{\delta_n - \gamma_m}, \quad d_m^{1/2} = \frac{\gamma_m}{\gamma_m^2 + 1/4}, \quad d_m^{3/2} = \frac{\gamma_m \cos\gamma - \sin\gamma}{\gamma_m^2 + 9/4}. \tag{5.141}$$

The solutions of system (5.140) with (5.141) are:

$$x_n^{1/2} = s_n \operatorname{Re}\left(\frac{K_-(1/2)}{\delta_n - i/2}\right),$$

$$x_n^{3/2} = 2s_n \operatorname{Re}\left[\left(\frac{1}{2}\cos\gamma - \frac{i}{3}\sin\gamma\right)\frac{K_-(3/2)}{\delta_n - 3i/2}\right],$$

$$s_n = \frac{\pi}{A\sqrt{A}}\frac{(2n-1)!!}{(2n-2)!!}, \quad K_-(u) = \sqrt{\frac{A}{\pi}}\frac{\Gamma(1/2+iuA/\pi)}{\Gamma(1+iuA/\pi)}.$$

The formulae for the contact stress also become asymptotically simpler ($0 < \eta < \gamma$):

$$q_{1/2}(\eta) = \frac{\gamma}{K(1/2)}\left[1 + \sqrt{\frac{\sin\gamma}{\sin\eta}}\sum_{n=1}^{\infty} x_n^{1/2}\exp(-[\gamma-\eta]\delta_n)\right], \tag{5.142}$$

$$q_{3/2}(\eta) = \frac{\gamma}{K(3/2)}\left[\cos\eta + \sqrt{\frac{\sin\gamma}{\sin\eta}}\sum_{n=1}^{\infty} x_n^{3/2}\exp(-[\gamma-\eta]\delta_n)\right]. \tag{5.143}$$

Note that expressions (5.142), (5.143) are valid for $\eta > 0$. For $\eta = 0$ we have $q_{1/2}(0) = \gamma/K(1/2)$, $q_{3/2}(0) = \gamma/K(3/2)$. In asymptotic relations (5.137), (5.142) and (5.143), the size of the contact region γ and the punch settlement δ relative to the applied force P are still undetermined. To obtain them, we use the formula

$$P = 2\pi R_1^2 \int_0^{\gamma} q(\eta)\cos\eta\,\sin\eta\,d\eta \tag{5.144}$$

and the fact that the series in (5.142), (5.143) diverge for $\eta \to \gamma$, and after summation for $\eta \neq \gamma$ they yield a square root singularity if $\eta \to \gamma$. Assuming the contact stress to be bounded, we set the coefficient at this singularity equal to zero. Then we obtain the equation

$$\frac{\delta+\varepsilon}{K(3/2)} 2\,\mathrm{Re}\left[\left(\frac{1}{2}\cos\gamma - \frac{i}{3}\sin\gamma\right)K_-(3/2)\right] - \frac{\varepsilon}{K(1/2)}\mathrm{Re}\,(K_-(1/2)) = 0. \tag{5.145}$$

Expression (5.144) yields another equation:

$$\delta + \varepsilon = \frac{1}{p_{3/2}}\left(\frac{P}{\theta R_1 \gamma} + \varepsilon p_{1/2}\right), \tag{5.146}$$

where

$$p_{1/2} = \frac{2\pi}{\gamma K(1/2)}\left[\frac{\sin^2\gamma}{2} + \Sigma_{1/2}\right], \quad p_{3/2} = \frac{2\pi}{\gamma K(3/2)}\left[\frac{1-\cos^3\gamma}{3} + \Sigma_{3/2}\right], \tag{5.147}$$

$$\Sigma_{iw} = \sin\gamma\sum_{n=1}^{\infty}\frac{x_n}{\delta_n^2 + 5/2}(\delta_n\cos\gamma + \sin\gamma). \tag{5.148}$$

Thus, the system of equation (5.145), (5.146) enables us to determine the settlement δ of the punch and the size γ of the contact region in relation to the specified force P. Expression (5.145) also enables us to eliminate divergent series terms from the formulae (5.142), (5.143), thus making it possible to present the contact stress in terms of exponentially converging series. They coincide formally with (5.142), (5.143), in which x_n^{iw} should be replaced by $\hat{x}_n^{iw}$:

$$\hat{x}_n^{1/2} = \frac{s_n}{2\delta_n} \operatorname{Im} \left[\frac{K_-(1/2)}{\delta_n + i/2} \right], \tag{5.149}$$

$$\hat{x}_n^{3/2} = \frac{3s_n}{\delta_n} \operatorname{Im} \left[\left(\frac{1}{2} \cos\gamma - \frac{i}{3} \sin\gamma \right) \frac{K_-(3/2)}{\delta_n + 3i/2} \right]. \tag{5.150}$$

Series (5.147), (5.148) are poorly convergent. Improving the convergence, we replace them by finite sums:

$$\Sigma_{1/2} = \sin^2\gamma \sum_{n=1}^{N} \frac{x_n^{1/2}}{\delta_n^2 + 5/2} - \frac{1}{2} F_1 \sin(2\gamma), \tag{5.151}$$

$$\Sigma_{3/2} = \sin^2\gamma \sum_{n=1}^{N} \frac{x_n^{3/2}}{\delta_n^2 + 5/2} + \frac{1}{2} F_2 \sin(2\gamma), \tag{5.152}$$

$$F_1 = \frac{2\sqrt{A}}{\pi} \ln 2 \operatorname{Re}(K_-(1/2)) - \sum_{n=1}^{N} \frac{s_n y_n^{1/2}}{\delta_n^2(\delta_n^2 + 5/2)},$$

$$y_n^{1/2} = \operatorname{Re} \left[K_-(1/2) \left(\delta_n^2 \frac{i}{2} + \delta_n \frac{9}{4} + \frac{9i}{8} \right) \Big/ \left(\delta_n + \frac{i}{2} \right) \right],$$

$$F_2 = \frac{2\sqrt{A}}{\pi} \ln 4 \operatorname{Re} \left[\left(\frac{1}{2} \cos\gamma - \frac{i}{3} \sin\gamma \right) K_-(3/2) \right] - \sum_{n=1}^{N} \frac{s_n y_n^{3/2}}{\delta_n^2(\delta_n^2 + 5/2)},$$

$$y_n^{3/2} = \operatorname{Re} \left[\left(\frac{1}{2} \cos\gamma - \frac{i}{3} \sin\gamma \right) K_-(3/2) \right.$$
$$\left. \times \left(\delta_n^{3/2} 3i + \delta_n \frac{9}{2} + \frac{27i}{4} \right) \Big/ \left(\delta_n + \frac{3i}{2} \right) \right].$$

Here $N = E([5/(2\varepsilon_0)]^{2/5})$, where $E(...)$ denotes the integer part of the number; ε_0 is the specified computational accuracy for the series.

The initial data in the numerical analysis were as follows: P (*kg*), ν, G (*kg/mm*), R_0 (*mm*), h (*mm*) and ε (*mm*). Then the quantities δ and γ were obtained from the system of transcendental equations (5.145) and

(5.146), in which relations (5.151), (5.152) were employed. Then the contact stress under the punch was calculated using (5.137), (5.142) and (5.143). In using the last two equations we must replace the x_n^{iw} by $\hat{x}_n^{iw}$ from (5.149), (5.150). Table 5.11 gives, for some initial parameter values ($R_0 = 13$, $h = 0.5$, $\nu = 0.4$, $G = 71$, $\varepsilon = 0.005$), calculated values of the punch settlement δ in *mm*, the dimension γ of the contact region in degrees, and the contact pressure $q(\eta_n)$ in *kg/mm*, at four points $\eta_n = n\gamma/4$ ($n = 0, 1, 2, 3$). For $\eta = \gamma$ the contact pressures vanish.

Table 5.11: Contact characteristics

$P \cdot 10^{-3}$	γ	$q(\eta_0)$	$q(\eta_1)$	$q(\eta_2)$	$q(\eta_3)$	$\delta \cdot 10^2$
1	59.6	4.696	4.379	3.448	1.968	0.4947
2	67.7	8.973	8.351	6.538	3.692	0.8319
3	72.1	13.26	12.33	9.616	5.396	1.154
4	75.0	17.57	16.32	12.70	7.095	1.470
5	77.1	21.91	20.33	15.79	8.794	1.783
6	78.6	26.25	24.36	18.89	10.49	2.095

We can see from this table that, as the force P increases 6 times, the contact area, determining by γ, increases 30 %, and the contact pressure at the central point $\eta = 0$ increases 5.5 times. Analyzing the numerical dependence between P and δ, we conclude that for a spherical bearing one cannot apply the Hertz contact theory, where $P = C\delta^{3/2}$, $C = \text{const}$. We have here a more complicated monotonic dependence, which cannot be present as $P = C\delta^{\alpha}$. Due to the fixedness of the spherical surface $r = R_2$, the settlement δ for the bearing is considerably less than for the Hertz theory.

5.6.3. SECTOR OF A SPHERICAL LAYER

In spherical coordinates (r, η, φ) we now consider an elastic solid $R_1 \leq r \leq R_2$, $0 \leq \eta \leq \gamma$. Suppose that on the boundaries $r = R_1$, $\eta = \gamma$ there are no normal displacements nor tangential stresses, and the boundary $r = R_2$ in the domain $\eta \leq \beta$ interacts without friction with a rigid punch which settles vertically by an amount δ (Fig. 5.17*a*,*b*). The base surface of the punch is a sphere of radius $r = R_2$.

The formulated problem can be reduced to Navier's equations in spherical coordinates with the following boundary conditions:

$$u_r = \tau_{r\eta} = 0 \quad (r = R_1), \tag{5.153}$$

$$u_\eta = \tau_{r\eta} = 0 \quad (\eta = \gamma), \tag{5.154}$$

$$\tau_{r\eta} = 0 \quad (r = R_2), \tag{5.155}$$

$$\sigma_r = 0 \quad (r = R_2,\ \beta < \eta < \gamma), \tag{5.156}$$

$$u_r = -\delta \cos\eta \quad (r = R_2,\ 0 \le \eta \le \beta). \tag{5.157}$$

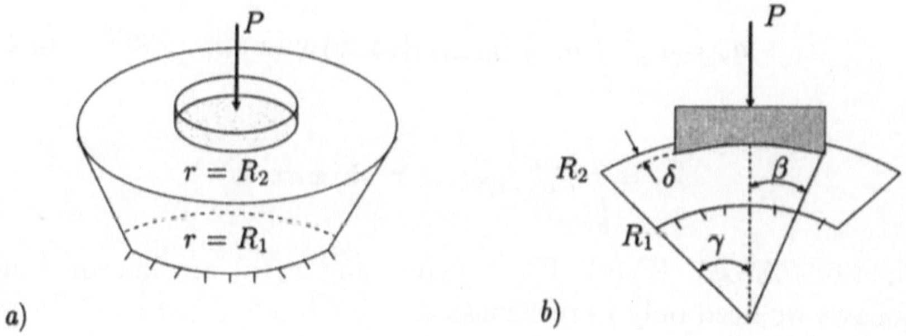

Figure 5.17: Contact between a sector of a spherical layer and a spherical punch; *a*) a 3-D view; *b*) a radial section

If we replace conditions (5.156) and (5.157) by the condition

$$\sigma_r = \begin{cases} 0 & \text{for } \beta < \eta < \gamma \\ -q(\eta) & \text{for } 0 \le \eta \le \beta \end{cases},$$

then the components of the displacement vector and of the stress tensor will be determined from the expressions (Chebakov (1987))

$$u_r = \sum_{k=0}^{\infty} \frac{a_k}{\Delta(\alpha_k)} W_k(r) \mathcal{P}_{\alpha_k - 1/2}(\cos\eta),$$

$$u_\eta = \sum_{k=0}^{\infty} \frac{a_k}{\Delta(\alpha_k)} V_k(r) \frac{d}{d\eta} \mathcal{P}_{\alpha_k - 1/2}(\cos\eta),$$

$$\tau_{r\varphi} = \sum_{k=0}^{\infty} \frac{a_k}{\Delta(\alpha_k)} \tau_k(r) \frac{d}{d\eta} \mathcal{P}_{\alpha_k - 1/2}(\cos\eta),$$

$$\sigma_r = \sum_{k=0}^{\infty} \frac{a_k}{\Delta(\alpha_k)} \sigma_k(r) \mathcal{P}_{\alpha_k - 1/2}(\cos\eta).$$

Here α_k are roots of the equation

$$\frac{d}{d\gamma} \mathcal{P}_{\alpha_k - 1/2}(\cos\gamma) = 0,$$

and $\alpha_0 = 1/2$. The coefficients a_k are used in the expansion

$$\sum_{k=0}^{\infty} a_k \mathcal{P}_{\alpha_k - 1/2}(\cos\eta) = \begin{cases} 0 & \text{for } \beta < \eta < \gamma \\ q(\eta) & \text{for } 0 \le \eta \le \beta \end{cases},$$

$$a_k = \omega_k^{-1} \int_0^{\beta} q(x) \mathcal{P}_{\alpha_k - 1/2}(\cos x) \sin x \, dx, \tag{5.158}$$

$$\omega_k = \int_0^{\gamma} \mathcal{P}_{\alpha_k - 1/2}(\cos x) \sin x \, dx.$$

The functions $\Delta(\alpha_k)$, $W_k(r)$, $V_k(r)$, $\tau_k(r)$, and $\sigma_k(r)$ are known, but in what follows we need only in relations:

$$\frac{W_0(R_2)}{\Delta(\alpha_0)} = \frac{R_2}{\theta} k_0, \quad \frac{W_k(R_2)}{\Delta(\alpha_k)} = \frac{R_2}{\theta} K(\alpha_k) \quad (k \ge 1), \tag{5.159}$$

$$K(\alpha) = \frac{B_1(\kappa^{2\alpha} + \kappa^{-2\alpha}) + B_2(\kappa^2 + \kappa^{-2}) + B_3}{(1-\nu)[A_1(\kappa^{2\alpha} - \kappa^{-2\alpha}) + A_2(\kappa^{2\alpha} + \kappa^{-2\alpha}) + 2A_3]}, \tag{5.160}$$

$$B_1 = -16(1-\nu)^2\alpha^4 + 23 - 68\nu + 48\nu^2,$$

$$B_2 = \alpha^2[4(3-4\nu)\alpha^2 - 19 + 52\nu - 32\nu^2,$$

$$B_3 = 2(1-2\nu^2)^2(4\alpha^4 - 5\alpha^2 + 1),$$

$$A_1 = -(1-\nu)[16\alpha^5 - 16\alpha^3 - 3(7-8\nu)\alpha],$$

$$A_2 = 8(1-2\nu)^2\alpha^4 + 4\nu(17-22\nu)\alpha^2 - 4.5(1-2\nu),$$

$$A_3 = [-4\kappa^2 + 4(3-4\nu)\kappa^{-2} - 8(1-2\nu)]\alpha^6$$

$$+[10\kappa^2 - 2(23 - 44\nu + 16\nu^2)\kappa^{-2} + 28(1-2\nu)]\alpha^4$$

$$+[(-18.25 + 16\nu^2)\kappa^2 + 9(4.75 - 13\nu + 8\nu^2)\kappa^{-2} - 24.5(1-2\nu)]\alpha^2$$

$$+4.5(1-2\nu), \qquad \kappa = R_2/R_1.$$

Satisfying conditions (5.156), (5.157), we reduce the problem to a dual series equation in unknown coefficients a_k in the form

$$\sum_{k=1}^{\infty} a_k K(\alpha_k) \mathcal{P}_{\alpha_k - 1/2}(\cos\eta) + a_0 k_0 = \frac{\theta\delta}{R_2} \cos\eta \quad (0 \le \eta \le \beta), \tag{5.161}$$

$$\sum_{k=0}^{\infty} a_k \mathcal{P}_{\alpha_k - 1/2}(\cos\eta) = 0 \quad (\beta < \eta < \gamma). \tag{5.162}$$

The contact stress under the punch is determined by the formula

$$q(\eta) = \sum_{k=0}^{\infty} a_k \mathcal{P}_{\alpha_k - 1/2}(\cos\eta) \quad (0 \le \eta \le \beta). \tag{5.163}$$

Consider the dual system of equations (5.161), (5.162) in general, assuming that

$$\lim_{\alpha \to 1/2} K(\alpha) \ne k_0, \quad k_0 = \frac{(1-2\nu)(\kappa^3 - 1)}{2(1-\nu)[\kappa^3(1+\nu) + 2(1-2\nu)]},$$

and transform this system to the more convenient form

$$\sum_{k=0}^{\infty} a_k K(\alpha_k) \mathcal{P}_{\alpha_k - 1/2}(\cos\eta) = f(\eta) \quad (0 \le \eta \le \beta), \tag{5.164}$$

$$\sum_{k=0}^{\infty} a_k \mathcal{P}_{\alpha_k - 1/2}(\cos\eta) = 0 \quad (\beta < \eta < \gamma), \tag{5.165}$$

$$f(\eta) = a_0 n_0 + \theta\delta R_2^{-1} \cos\eta, \qquad n_0 = K(1/2) - k_0.$$

Keeping in mind that the function $f(\eta)$ can be presented in the form

$$f(\eta) = a_0 n_0 \mathcal{P}_0(\cos\eta) + \theta\delta R_2^{-1} \mathcal{P}_1(\cos\eta),$$

we consider dual series relations (5.164), (5.165) for the case

$$f(\eta) = \mathcal{P}_{i\varepsilon - 1/2}(\cos\eta). \tag{5.166}$$

For such a dual equation, we denote the function of type (5.163) as

$$q_{i\varepsilon}(\eta) = \sum_{k=0}^{\infty} a_k^{i\varepsilon} \mathcal{P}_{\alpha_k - 1/2}(\cos\eta) \quad (0 \le \eta \le \beta),$$

taking into account that the column $a_k^{i\varepsilon}$ presents the solution of the dual system (5.164), (5.165) with the right-hand side (5.166). Then the contact stress under the punch will be determined by the formula

$$q(\eta) = a_0 n_0 q_{1/2}(\eta) + \theta\delta R_2^{-1} q_{3/2}(\eta). \tag{5.167}$$

To solve the dual series relations (5.164), (5.165) with (5.166), we use reduction to the infinite system employed in subsection 3.6.1. Note that this dual series relation possesses all the necessary properties for the application of this method.

Omitting the calculations, we finally obtain

$$q_{i\varepsilon}(\eta) = \frac{\mathcal{P}_{-1/2+i\varepsilon}(\cos\eta)}{K(i\varepsilon)} + \sum_{n=1}^{\infty} x_n^{i\varepsilon} \frac{\mathcal{P}_{-1/2+i\delta_n}(\cos\eta)}{\mathcal{P}_{-1/2+i\delta_n}(\cos\beta)}, \tag{5.168}$$

where $x_n^{i\varepsilon}$ is the solution of the following infinite system:

$$X^{i\varepsilon} = A^{-1} D^{i\varepsilon} - A^{-1} B X^{i\varepsilon}, \tag{5.169}$$

$$D^{i\varepsilon} = \{d_m^{i\varepsilon}\}, \quad B = \{b_{mn}\}, \quad X^{i\varepsilon} = \{x_m^{i\varepsilon}\}.$$

The elements of the matrix A^{-1} are determined by formulae (3.313) and (3.314);

$$d_m^{i\varepsilon} = \frac{\mathcal{P}^1_{-1/2+i\varepsilon}(a)\, \mathcal{P}_{-1/2+i\gamma_m}(-a) + \mathcal{P}_{-1/2+i\varepsilon}(a)\, \mathcal{P}^1_{-1/2+i\gamma_m}(-a)}{K(i\varepsilon)(\gamma_m^2 - \varepsilon^2)\mathrm{P}_{-1/2+i\gamma_m}(-a)}$$

$$+B(\gamma_m)\frac{\mathcal{P}^1_{-1/2+i\varepsilon}(a)\, \mathcal{P}_{-1/2+i\gamma_m}(a) - \mathcal{P}_{-1/2+i\varepsilon}(a)\, \mathcal{P}^1_{-1/2+i\gamma_m}(a)}{K(i\varepsilon)(\gamma_m^2 - \varepsilon^2)\mathcal{P}_{-1/2+i\gamma_m}(-a)},$$

$$B(\gamma_m) = \mathcal{P}^1_{-1/2+i\gamma_m}(-\cos\gamma_m)[\mathcal{P}^1_{-1/2+i\gamma_m}(\cos\gamma_m)]^{-1}, \quad a = \cos\beta,$$

$$b_{mn} = (\delta_n^2 - \gamma_m^2)^{-1}\left\{\frac{\mathcal{P}^1_{-1/2+i\delta_n}(a)}{\mathcal{P}_{-1/2+i\delta_n}(a)} - \delta_n + \frac{\mathcal{P}^1_{-1/2+i\gamma_m}(a)}{\mathcal{P}_{-1/2+i\gamma_m}(a)} - \gamma_m\right\}$$

$$+B(\gamma_m)\frac{\mathcal{P}^1_{-1/2+i\delta_n}(a)\, \mathcal{P}_{-1/2+i\gamma_m}(a) - \mathcal{P}_{-1/2+i\delta_n}(a)\, \mathcal{P}^1_{-1/2+i\gamma_m}(a)}{(\delta_n^2 - \gamma_m^2)\mathcal{P}_{-1/2+i\delta_n}(a)\, \mathcal{P}_{-1/2+i\gamma_m}(a)}.$$

Here $\pm i\delta_n$ and $\pm i\gamma_m$ are zeros and poles of the meromorphic function $K(\alpha)$, respectively.

After determination of the function $q_{i\varepsilon}(\eta)$ by formula (5.168), we can find the contact pressure by formula (5.167), if we determine the constant a_0. Using formula (5.158), we obtain

$$a_0 = (1 - \cos\gamma)^{-1} \int_0^{\beta} q(\eta) \sin\eta \, d\eta.$$

Let us denote

$$a_{i\varepsilon} = \int_0^{\beta} q_{i\varepsilon}(\eta) \sin\eta \, d\eta.$$

Then by specially integrating equality (5.167), we find

$$a_0 = \theta\delta R_2^{-1} a_{3/2}(1 - \cos\gamma - n_0\, a_{1/2})^{-1},$$

$$a_{1/2} = (1 - \cos\beta)K^{-1}(1/2) + a_{1/2}^*, \quad a_{3/2} = \sin^2\beta(2K(3/2))^{-1} + a_{3/2}^*,$$

$$a_{i\varepsilon}^* = \sin\beta \sum_{n=1}^{\infty} \frac{x_n^{i\varepsilon}}{\delta_n^2 + 1/4} \mathcal{P}_{-1/2+i\delta_n}^1(\cos\beta)\,[\mathcal{P}_{-1/2+i\delta_n}(\cos\beta)]^{-1}.$$

Let us determine the centrally applied force corresponding to the settlement δ:

$$P = 2\pi R_2^2 \int_0^{\beta} q(\eta)\cos\eta\,\sin\eta\,d\eta. \tag{5.170}$$

By integrating expression (5.167) and using (5.170), we obtain

$$P = 2\pi\theta\delta R_2 \left(\frac{n_0\, a_{3/2}}{1 - \cos\gamma - n_0\, a_{1/2}} P_{1/2} + P_{3/2} \right), \tag{5.171}$$

$$P_{1/2} = \sin^2\beta[2K(1/2)]^{-1} + P_{1/2}^*, \quad P_{3/2} = (1 - \cos^3\beta)[3K(3/2)]^{-1} + P_{3/2}^*,$$

$$P_{i\varepsilon}^* = \sin\beta \sum_{n=1}^{\infty} \frac{x_n^{i\varepsilon}}{\delta_n^2 + 9/4} \left[\cos\beta \frac{\mathcal{P}_{-1/2+i\delta_n}^1(\cos\beta)}{\mathcal{P}_{-1/2+i\delta_n}(\cos\beta)} + \sin\beta \right].$$

To construct the inverse matrix A^{-1}, it is necessary to factorize the function $K(\alpha)$, but this involves some computational difficulties. That is why we will approximate the function $K(\alpha)$ by the function $(3.316)_1$ and will use elements (3.317).

Let us obtain an asymptotic solution of the contact problem for small values of parameter $\lambda = \ln\kappa$. Taking into account that the quantities δ_n and γ_m are inversely proportional λ and

$$\mathcal{P}_{-1/2+i\alpha}(\cos\eta) \sim \frac{\Gamma(1/2 + i\alpha)}{\Gamma(1 + i\alpha)} \frac{(1 + i)\exp(\eta\alpha)}{2\sqrt{\pi\sin\eta}},$$

$$\mathcal{P}_{-1/2+i\alpha}^1(\cos\eta) \sim \frac{\Gamma(3/2 + i\alpha)}{\Gamma(1 + i\alpha)} \frac{(1 - i)\exp(\eta\alpha)}{2\sqrt{\pi\sin\eta}}$$

as $\alpha \to \infty$, one can show that an asymptotic solution of infinite system (5.169) for small λ has the form (the quantities τ_{nm} see in (3.317))

$$x_n^{i\varepsilon} \approx \sum_{m=1}^{\infty} \tau_{nm} d_m^{i\varepsilon\,*},$$

$$d_m^{3/2*} \approx \frac{-\sin\beta + \gamma_m \cos\beta - \exp(-2[\gamma-\beta]\gamma_m)(\sin\beta + \gamma_m\cos\beta)}{K(3/2)(\gamma_m^2 + 9/4)},$$

$$d_m^{1/2*} \approx \frac{\gamma_m(1-\exp(-2[\gamma-\beta]\gamma_m))}{K(1/2)(\gamma_m^2+1/4)}, \quad d_m^{i\varepsilon} \sim d_m^{i\varepsilon *} \quad (|\gamma_m| \to \infty).$$

Similarly, we obtain the following asymptotic relations for the contact stress and the indenting force for small λ $(0 \le \eta \le \beta)$:

$$q_{3/2}(\eta) \approx \frac{\cos\eta}{K(3/2)} + \sqrt{\frac{\sin\beta}{\sin\eta}} \sum_{n=1}^{\infty} x_n^{3/2} \exp(-[\beta-\eta]\delta_n), \tag{5.172}$$

$$q_{1/2}(\eta) \approx \frac{1}{K(1/2)} + \sqrt{\frac{\sin\beta}{\sin\eta}} \sum_{n=1}^{\infty} x_n^{1/2} \exp(-[\beta-\eta]\delta_n),$$

$$P_{1/2} \approx \frac{1-\cos^3\beta}{3K(3/2)} + \sin\beta \sum_{n=1}^{\infty} \frac{x_n^{3/2}[\delta_n\cos\beta + \sin\beta]}{\delta_n^2 + 9/4},$$

$$P_{3/2} \approx \frac{\sin^2\beta}{2K(1/2)} + \sin\beta \sum_{n=1}^{\infty} \frac{x_n^{1/2}[\delta_n\cos\beta + \sin\beta]}{\delta_n^2 + 9/4}, \tag{5.173}$$

$$a_{1/2} \approx \frac{1-\cos\beta}{K(1/2)} + \sin\beta \sum_{n=1}^{\infty} \frac{x_n^{1/2}\delta_n}{\delta_n^2 + 1/4},$$

$$a_{3/2} \approx \frac{\sin^2\beta}{2K(3/2)} + \sin\beta \sum_{n=1}^{\infty} \frac{x_n^{3/2}\delta_n}{\delta_n^2 + 1/4}.$$

For problem (5.153)–(5.157) we have $K(1/2) = k_0$, therefore, the general formulae (5.167) and (5.171) can be simplified into the form

$$q(\eta) = \theta\delta R_2^{-1} q_{3/2}(\eta), \quad P = 2\pi\theta\delta R_2 P_{3/2}. \tag{5.174}$$

Table 5.12 presents the dimensionless quantities

$$q^*(\eta/\beta) = R_2\beta(\theta\delta)^{-1} q(\eta), \quad P^* = (R_2\beta\theta\delta)^{-1} P,$$

calculated by formulae (5.174) with (5.172) and (5.173) for $\beta = 1$ and $\lambda = \ln 2$.

Table 5.12: The contact pressure and force

γ	$q^*(0)$	$q^*(0.4)$	$q^*(0.6)$	$q^*(0.8)$	$q^*(0.9)$	P^*
1.50	1.15	1.15	1.16	1.21	1.34	4.06
1.20	1.15	1.15	1.16	1.18	1.26	3.96
1.10	1.15	1.15	1.15	1.14	1.14	3.81

The contact pressure has a square root singularity on the boundary $\eta = \beta$ of the contact region; the series in (5.172) diverges for this value of η. We can see from Table 5.12 that, as the lateral boundary $\eta = \gamma$ of the elastic solid approaches the boundary of the contact zone, γ is near β, the contact pressure (see values of $q^*(0.8)$ and $q^*(0.9)$ for $\gamma = 1.10$) decreases near the boundary, and then begins to increase. This is an effect of the lateral boundary of the sector of the spherical layer. We have already seen such an effect several times, for instance for a finite cylinder, the end-walls of which are close to the contact zone (see the bold dashed line in Fig. 3.18 in subsection 3.7.2). In practice, to diminish the contact stress for such a part as a sector of spherical layer, one should make a correct choice of the geometrical parameters R_2/R_1, γ and β. By choosing these dimensions, one can optimize the distribution of the contact pressure for a given punch, which is useful in applications. Note that the applications may be diverse, for instance, a contact of a leg with an artificial titanium limb.

5.6.4. SUPERPOSITION OF HOMOGENEOUS SOLUTIONS

The other way of solving the contact problem for a sector of a spherical layer is the method of homogeneous solutions based on the generalized orthogonality relationships for a spherical layer (Chebakov (1987)). This method was already considered in subsection 3.6.2, that is why we will give only final formulae. The contact stress under the punch is calculated by the formula

$$q(\eta) = \delta q_0(\eta) + \sum_{n=1}^{\infty} D_n W_n^0(R_2) q_n(\eta), \tag{5.175}$$

where the functions $q_n(\eta)$ $(n = 0, 1, ...)$ are determined from the following integral equations

$$K_\eta\{q_0\} = \cos\eta, \quad K_\eta\{q_n\} = \mathcal{P}_{-1/2+i\gamma_n}(\cos\eta), \quad 0 \le \eta \le \beta, \ n = 1, 2, ..., \tag{5.176}$$

$$K_\eta\{q\} = \int_0^\beta q(x)\sin x\, dx \sum_{k=0}^{\infty} \frac{W_k(R_2)}{\Delta(\alpha_k)} \frac{2k+1}{2} \mathcal{P}_k(\cos x)\mathcal{P}_k(\cos\eta),$$

and the coefficients D_n can be found from the infinite system

$$x_k = b_k + \sum_{n=1}^{\infty} a_{kn} x_n, \quad k = 1, 2, ..., \tag{5.177}$$

$$D_k = \frac{x_k}{W_k^0(R_2)\cosh(\gamma_k \gamma)}, \quad b_k = -\delta c_k T_{0k},$$

$$a_{kn} = \frac{-c_k T_{nk}}{\cosh(\gamma_n \gamma)}, \quad c_k = \frac{\pi R_2 [W_k^0(R_2)]^2 \cosh(\gamma_k \gamma)}{2W_{kk}^0 \cosh(\gamma_k \pi)},$$

$$T_{nk} = \frac{1}{\theta} \int_0^\beta q_n(x) \mathcal{P}_{-1/2+i\gamma_k}(\cos x) \sin x \, dx, \quad T_{kn} = T_{nk}.$$

Here the function $W_k(R_2)/\Delta(\alpha_k)$ corresponds to the inhomogeneous problem (see (5.159)), $W_k^0(R_2)$ to the homogeneous problem, and $i\gamma_k$ to the poles of the function $K(\alpha)$ (5.160) in the right half-plane. The coefficients W_{kk}^0 are determined from the generalized orthogonality condition. One can show that the system (5.177) is of the Poincaré–Koch type, i.e., the elements b_k and a_{kn} exponentially decrease as their indices increase; the series in (5.175) converges no more slowly than the sum of the terms of an infinitely decreasing geometric progression. The solution of integral equations (5.176) can be obtained using a variety of effective methods including the asymptotic methods developed for a similar class of equations.

Chapter 6

CONTACT PROBLEMS FOR SPHERICAL LENS

In this chapter the method for reducing a problem in the theory of elasticity to a generalized, in the sense of I.N.Vekua, Hilbert boundary-value problem, is extended to a mixed 3-D problem for a truncated sphere with a rigidly fixed spherical surface. The normal stresses at the cut are specified. The problems for a half-space with a spherical hollow or jut are treated similarly. The systems of functional equations of these problems are transformed to systems of singular integral equations. Then contact problems of a circular and annular punch punch are considered. Integral equations of these problems are reduced, supposing a geometric symmetry of contact regions, to Fredholm integral equations of the second kind with help of the method of paired equations.

It is important that the methods used below are applicable for solving 3-D contact problems for an arbitrary elastic spherical lens, i.e., for a solid formed by intersection of two spheres with different radii. Plane problems relating to a circle were treated in Gladwell (1980).

The contact problem of a torsion of an elastic truncated sphere, with fixed spherical surface, by a circular punch lying on the cut face, was considered by Babloian (1964) and Aleksandrov and Chebakov (1973). When deriving an integral equation of this problem, a Mehler–Fock integral transform in the real axis (Sneddon (1972)) was used:

$$F(x) = \int_0^\infty \mathcal{P}_{i\tau-1/2}(x) f(\tau)\, d\tau \quad f(0) = 0 \quad (1 \le x < \infty),$$

$$f(\tau) = \tau \tanh(\pi\tau) \int_1^\infty \mathcal{P}_{i\tau-1/2}(x) F(x)\, dx,$$

where $\mathcal{P}_\mu(x)$ is the Legendre function of the first kind (8.842.1 in Gradshteyn and Ryzhik (1994))

$$\mathcal{P}_{i\tau-1/2}(\cosh\xi) = \frac{2}{\pi}\cosh(\pi\tau) \int_0^\infty \frac{\cos(\tau s)}{\sqrt{2(\cosh s + \cosh\xi)}} ds \qquad (\xi \geq 0).$$

The generalized Mehler–Fock transform and the corresponding inversion formula are:

$$F(x) = \int_0^\infty \mathcal{P}^m_{i\tau-1/2}(x) f(\tau)\, d\tau \quad f(0) = 0 \quad (1 \leq x < \infty),$$

$$f(\tau) = \frac{1}{\pi}\tau \sinh(\pi\tau)\, \Gamma\left(\frac{1}{2} - m + i\tau\right) \Gamma\left(\frac{1}{2} - m - i\tau\right) \int_1^\infty \mathcal{P}^m_{i\tau-1/2}(x) F(x)\, dx,$$

where $\mathcal{P}^m_\mu(x)$ is the associated Legendre function of the first kind 8.751.2 in Gradshteyn and Ryzhik (1994). For $m = 0$ these formulae reduce to previous ones; for $m = 1/2$, $x = \cosh\xi$ they lead to the Fourier cosine transform, and for $m = -1/2$, $x = \cosh\xi$ to the Fourier sine transform.

For solving the axisymmetric stress (or Neumann) boundary-value problem for a spherical lens, one should take a generalized Mehler–Fock integral transform in the complex μ-plane:

$$F(\xi) = \frac{1}{\pi i} \int_{c-i\infty}^{c+i\infty} \mathcal{Q}_{\mu-1/2}(\cosh\xi) \mu f(\mu)\, d\mu \qquad (\xi \geq 0),$$

$$f(\mu) = (-1)^m \int_0^\infty \mathcal{P}^{-m}_{\mu-1/2}(\cosh\xi) F(x) \sinh\xi\, d\xi,$$

where m is integer, and $\mathcal{Q}^m_\mu(x)$ is the associated Legendre function of the second kind (8.703 in Gradshteyn and Ryzhik (1994)). The $\mathcal{P}^m_\mu(x)$ and $\mathcal{Q}^m_\mu(x)$ are often called spherical functions. The complex integral transform is based on the formulae 8.731.5, 8.736.1, 8.736.7 and 8.334.2 in Gradshteyn and Ryzhik (1994):

$$\mathcal{P}^{-m}_{\mu-1/2}(x) = \mathcal{P}^{-m}_{-\mu-1/2}(x),$$

$$\mathcal{P}^{-m}_{\mu-1/2}(x) = \frac{\Gamma(\mu - m + 1/2)}{\Gamma(\mu + m + 1/2)} \mathcal{P}^{m}_{\mu-1/2}(x) \quad (x \geq 1),$$

$$\pi \tan(\pi\mu)\, \mathcal{P}^{m}_{\mu-1/2}(x) = \mathcal{Q}^{m}_{-\mu-1/2}(x) - \mathcal{Q}^{m}_{\mu-1/2}(x) \quad (x \geq 1),$$

$$\Gamma\left(\frac{1}{2} + \mu + m\right) \Gamma\left(\frac{1}{2} - \mu - m\right) = \frac{(-1)^m\, \pi}{\cos(\pi\mu)}.$$

We will use this transform below.

The final two sections are devoted to contact problems for a space with a spherical cavity, and for a sphere. These solids are the special cases of a lens, but they are treated in spherical (not toroidal) coordinates.

6.1 Equilibrium under Given Normal Loads

6.1.1. GENERAL SCHEME FOR THE 3-D CASE

We consider a 3-D problem for an elastic truncated sphere of radius R, with a fixed spherical surface. This truncated sphere is loaded by concentrated forces on its cut face. We use toroidal coordinates ξ, η, φ (Fig. 6.1), which are associated with cylindrical coordinates r, φ, z, divided by radius R_* of the section (or cut), by the relations (Korn, G.A. and Korn. T.M. (1968, p. 186))

$$r = \frac{\sinh \xi}{\cosh \xi + \cos \eta}, \qquad z = \frac{\sin \eta}{\cosh \xi + \cos \eta}, \tag{6.1}$$

$$\xi = \frac{1}{2} \ln \frac{(r+1)^2 + z^2}{(r-1)^2 + z^2}, \qquad \eta = \pi + \frac{i}{2} \ln \frac{r^2 + (z+i)^2}{r^2 + (z-i)^2}. \tag{6.2}$$

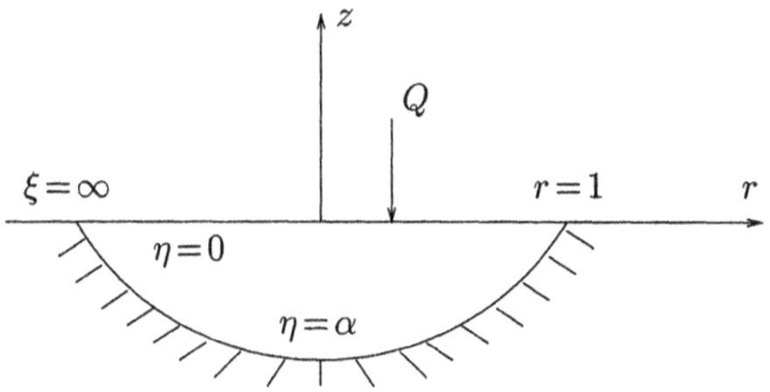

Figure 6.1: An axial section of a truncated sphere

Suppose the problem is even with respect to φ (the odd case can be treated similarly). We write the boundary conditions in the form ($R_*/R = \sin\alpha$)

$$\eta = 0: \quad \sigma_z/(2G) = -Q\delta_*(\xi - a)[\delta_*(\varphi - \varphi_0) + \delta_*(\varphi + \varphi_0)], \tag{6.3}$$

$$\tau_{rz} = \tau_{\varphi z} = 0, \tag{6.4}$$

$$\eta = \alpha: \quad u_r = u_\varphi = u_z = 0, \tag{6.5}$$

where $\delta_*(x)$ is the Dirac function.

We use the Papkovich–Neuber representation for the displacements in cylindrical coordinates via four harmonic functions, one of which can be chosen arbitrarily (Ufliand (1965)). If one of these harmonic functions is equal to zero we have the following expressions for the displacements and stresses:

$$\frac{u_r}{R_*} = \frac{\partial\Phi_1}{\partial r} - z\frac{\partial\Phi_2}{\partial r} + \sin\varphi\left[r\frac{\partial\Phi_3}{\partial r} - \varepsilon_3\Phi_3\right], \tag{6.6}$$

$$\frac{u_\varphi}{R_*} = \frac{1}{r}\frac{\partial\Phi_1}{\partial\varphi} - \frac{z}{r}\frac{\partial\Phi_2}{\partial\varphi} + \sin\varphi\frac{\partial\Phi_3}{\partial\varphi} - \varepsilon_3\cos\varphi\,\Phi_3, \tag{6.7}$$

$$\frac{u_z}{R_*} = \frac{\partial\Phi_1}{\partial z} - z\frac{\partial\Phi_2}{\partial z} + \varepsilon_3\Phi_2 + r\sin\varphi\frac{\partial\Phi_3}{\partial z}, \tag{6.8}$$

$$\frac{\sigma_z}{2G} = \frac{\partial^2\Phi_1}{\partial z^2} - z\frac{\partial^2\Phi_2}{\partial z^2} + \varepsilon_2\frac{\partial\Phi_2}{\partial z} + \sin\varphi\left[r\frac{\partial^2\Phi_3}{\partial z^2} - 2\nu\frac{\partial\Phi_3}{\partial r}\right] \tag{6.9}$$

$$-2\nu\frac{\cos\varphi}{r}\frac{\partial\Phi_3}{\partial\varphi},$$

$$\frac{\tau_{rz}}{2G} = \frac{\partial^2\Phi_1}{\partial r\partial z} - z\frac{\partial^2\Phi_2}{\partial r\partial z} + \varepsilon_1\frac{\partial\Phi_2}{\partial r} + \sin\varphi\left[r\frac{\partial^2\Phi_3}{\partial r\partial z} - \varepsilon_1\frac{\partial\Phi_3}{\partial z}\right], \tag{6.10}$$

$$\frac{\tau_{\varphi z}}{2G} = \frac{1}{r}\frac{\partial^2\Phi_1}{\partial\varphi\partial z} - \frac{z}{r}\frac{\partial^2\Phi_2}{\partial\varphi\partial z} + \frac{\varepsilon_1}{r}\frac{\partial\Phi_2}{\partial\varphi} + \sin\varphi\frac{\partial^2\Phi_3}{\partial\varphi\partial z} - \varepsilon_1\cos\varphi\frac{\partial\Phi_3}{\partial z}, \tag{6.11}$$

where $\varepsilon_1 = 1 - 2\nu$, $\varepsilon_2 = 2(1-\nu)$, $\varepsilon_3 = 3 - 4\nu$, G is the shear modulus and ν is Poisson's ratio.

We expand the harmonic functions Φ_n, $n = 1, 2, 3$, into a complex Mehler–Fock integral with respect to ξ and a Fourier series with respect to φ:

$$\Phi_k(\xi,\eta,\varphi) = \frac{\sqrt{\cosh\xi + \cos\eta}}{\pi i}\sum_{m=-\infty}^{\infty}\exp(im\varphi)\int_{c-i\infty}^{c+i\infty}[A_k^m(\mu)\cos(\mu\eta) \tag{6.12}$$

$$+B_k^m(\mu)\sin(\mu\eta)]\, \mathcal{Q}_{\mu-1/2}^m(\cosh\xi)\, d\mu \quad (k=1,2,3; c>0),$$

$$A_k^m(-\mu) = -A_k^m(\mu), \quad B_k^m(-\mu) = B_k^m(\mu).$$

Due to the evenness of the problem with respect to φ, the harmonic functions (6.12) are even with respect to φ for $k = 1, 2$, and the function $\Phi_3(\xi, \eta, \varphi)$ is odd with respect to φ. Using the formula 8.736.4 in Gradshteyn and Ryzhik (1994)

$$\mathcal{Q}_{\mu-1/2}^{-m}(x) = p(\mu, m)\, \mathcal{Q}_{\mu-1/2}^m(x), \quad p(\mu, m) = \frac{\Gamma(\mu + 1/2 - m)}{\Gamma(\mu + 1/2 + m)},$$

we obtain the relations ($e_{1,2} = 1$, $e_3 = -1$)

$$A_k^{-m}(\mu) = \frac{e_k A_k^m(\mu)}{p(\mu, m)}, \quad B_k^{-m}(\mu) = \frac{e_k B_k^m(\mu)}{p(\mu, m)}, \quad k = 1, 2, 3. \tag{6.13}$$

Note that $p(-\mu, m) = p(\mu, m)$ (Gradshteyn and Ryzhik (1994, formula 8.339.4)). Then formulae (6.13) for $k = 3$ give

$$A_3^0(\mu) = B_3^0(\mu) \equiv 0. \tag{6.14}$$

In what follows, we use the relations

$$\frac{\partial}{\partial r} = \sinh\xi\, \sin\eta \frac{\partial}{\partial \eta} + (1 + \cosh\xi\, \cos\eta)\frac{\partial}{\partial \xi}, \tag{6.15}$$

$$\frac{\partial}{\partial z} = (1 + \cosh\xi\, \cos\eta)\frac{\partial}{\partial \eta} - \sinh\xi\, \sin\eta \frac{\partial}{\partial \xi}, \tag{6.16}$$

which can be obtained from (6.1), (6.2); the relations 8.732 in Gradshteyn and Ryzhik (1994) for the spherical functions

$$2\mu \sinh\xi \frac{d}{d\xi}\, \mathcal{Q}_{\mu-1/2}^m(\cosh\xi) = \left(\mu + \frac{1}{2} - m\right)\left(\mu - \frac{1}{2}\right) \mathcal{Q}_{\mu+1/2}^m(\cosh\xi)$$
$$- \left(\mu - \frac{1}{2} + m\right)\left(\mu + \frac{1}{2}\right) \mathcal{Q}_{\mu-1/2}^m(\cosh\xi), \tag{6.17}$$

$$\frac{d}{d\xi}\, \mathcal{Q}_{\mu-1/2}(\cosh\xi) = \mathcal{Q}_{\mu-1/2}^1(\cosh\xi), \tag{6.18}$$

$$2\mu \sinh\xi\, \mathcal{Q}_{\mu-1/2}^{m+1}(\cosh\xi) = \left(\mu - \frac{1}{2} - m\right)\left(\mu + \frac{1}{2} - m\right) \mathcal{Q}_{\mu+1/2}^m(\cosh\xi) \tag{6.19}$$

$$-\left(\mu-\frac{1}{2}+m\right)\left(\mu+\frac{1}{2}+m\right)\mathcal{Q}^m_{\mu-3/2}(\cosh\xi),$$

$$2\mu\sinh\xi\,\mathcal{Q}^{m-1}_{\mu-1/2}(\cosh\xi)=\mathcal{Q}^m_{\mu+1/2}(\cosh\xi)-\mathcal{Q}^m_{\mu-3/2}(\cosh\xi), \tag{6.20}$$

$$2\mu\cosh\xi\,\mathcal{Q}^m_{\mu-1/2}(\cosh\xi)=\left(\mu+\frac{1}{2}-m\right)\mathcal{Q}^m_{\mu+1/2}(\cosh\xi) \tag{6.21}$$

$$+\left(\mu-\frac{1}{2}+m\right)\mathcal{Q}^m_{\mu-3/2}(\cosh\xi),$$

and also the expansion of two-dimensional Dirac function

$$\delta_*(\xi-a)\delta_*(\varphi-\varphi_0)=\frac{\sinh a}{2\pi^2 i}\sum_{m=-\infty}^{\infty}\exp[im(\varphi-\varphi_0)] \tag{6.22}$$

$$\times\int_{c-i\infty}^{c+i\infty}\mu(-1)^m\,\mathcal{P}^{-m}_{\mu-1/2}(\cosh a)\,\mathcal{Q}^m_{\mu-1/2}(\cosh\xi)\,d\mu.$$

We multiply the boundary condition (6.5) for u_r (6.6) termwise by $\sinh\xi(\cosh\xi+\cos\alpha)$ and represent $\sin\varphi$ in an exponential form using Euler's formula. Then we use formulae (6.15)–(6.21) and shift the integration contour in (6.12) in the complex μ-plane to a domain of regularity of functions $A^m_k(\mu)$ and $B^m_k(\mu)$ $(k=1,2,3)$; we also shift the summation in m in formula (6.12) for $\Phi_3(\xi,\eta,\varphi)$ by ±1. Equating to zero the terms in $\exp(im\varphi)\,\mathcal{Q}^m_{\mu-1/2}(\cosh\xi)$, we get a functional-difference equation in the functions $A^m_{1,2}(\mu)$, $B^m_{1,2}(\mu)$, $A^{m\pm1}_3(\mu)$, and $B^{m\pm1}_3(\mu)$, in which these functions depend also on $\mu\pm h$, $h=1,2,3$. Note that due to relations (6.17)–(6.21), the shift of the integration contour takes place symmetrically to the left and to the right, so that if an integrand had a pole, then the two corresponding residues would annihilate each other. Similarly, we multiply the boundary condition (6.5) for u_φ (6.7) by $\sinh\xi$ (the unknown functions will depend on μ, $\mu\pm1$), the boundary condition (6.5) for u_z (6.8) by $(\cosh\xi+\cos\alpha)$ (the functions will depend on $\mu\pm h$, $h=0,1,2$), the boundary condition (6.3) for σ_z (6.9) by $\sinh^2\xi(\cosh\xi+1)$ (the functions will depend on $\mu\pm h$, $h=0,1,2,3,4,5$), the boundary condition for (6.4) τ_{rz} (6.10) by $\sinh\xi(\cosh\xi+1)$ (the functions will depend on $\mu\pm h$, $h=0,1,2,3,4$), and the boundary condition (6.4) for $\tau_{\varphi z}$ (6.11) by $\sinh\xi$ (the functions will depend on $\mu\pm h$, $h=0,1,2$). When multiplying the left-hand side of (6.22) by $t(\xi)=\sinh^2\xi(\cosh\xi+1)$, we should multiply the right-hand side by $t(a)$. As an example, we give the formula

$$\frac{\partial}{\partial z}\Phi_k(\xi,\eta,\varphi) = \frac{\sqrt{\cosh\xi+\cos\eta}}{2\pi i}\sum_{m=-\infty}^{\infty}\exp(im\varphi)\int_{c-i\infty}^{c+i\infty}\left\{-\left[\left(\mu-\frac{1}{2}-m\right)\right.\right.$$
$$\times A_k^m(\mu-1)+2\mu A_k^m(\mu)+\left(\mu+\frac{1}{2}+m\right)A_k^m(\mu+1)\Big]\sin(\mu\eta) \tag{6.23}$$
$$+\left[\left(\mu-\frac{1}{2}-m\right)B_k^m(\mu-1)+2\mu B_k^m(\mu)\right.$$
$$\left.\left.+\left(\mu+\frac{1}{2}+m\right)B_k^m(\mu+1)\right]\cos(\mu\eta)\right\}\mathcal{Q}_{\mu-1/2}^m(\cosh\xi)\,d\mu.$$

As a result, we obtain a linear system of six functional-difference equations in the form

$$F_l\{A_1^m(\mu\pm h), B_1^m(\mu\pm h), A_2^m(\mu\pm h_1), \tag{6.24}$$
$$B_2^m(\mu\pm h_1), A_3^{m\pm1}(\mu\pm h), B_3^{m\pm1}(\mu\pm h)\} = f_l,$$

where $l=1,...,6$; $h=0,1,...,5$; $h_1=0,1,2,3,4$; $m\in\mathcal{Z}$.

Taking relations (6.13), (6.14) into account and putting $m=0,1,...$ in (6.24), we see that system (6.24) splits into a sequence of systems of six functional-difference equations in six unknown functions. We put $\mu=i\tau$, $\tau\in\Re$, in every such system. Taking the oddness of $A_k^m(\mu)$ and the evenness of $B_k^m(\mu)$ $(k=1,2,3)$ with respect to μ into account and using the Schwarz–Palatini formulae (Lawrentjew and Schabat (1967), see below), we reduce the resulting sequence of generalized, in the sence of I.N.Vekua, Hilbert boundary-value problems (see section 4.1) to a sequence of six singular integral equations in six new functions $\varphi_l(\tau)$, $l=1,...,6$, where (we omit index m)

$$\varphi_1(\tau)=\operatorname{Im}A_1^m(5+i\tau),\quad \varphi_2(\tau)=\operatorname{Re}B_1^m(5+i\tau),$$
$$\varphi_3(\tau)=\operatorname{Im}A_2^m(4+i\tau)+\tanh(\pi\tau)w(\tau), \tag{6.25}$$
$$\varphi_4(\tau)=\operatorname{Re}B_2^m(4+i\tau)-w(\tau), \tag{6.26}$$
$$\varphi_5(\tau)=\operatorname{Im}A_3^m(5+i\tau),\quad \varphi_6(\tau)=\operatorname{Re}B_3^m(5+i\tau),$$
$$w(\tau)=\operatorname{Re}W(4+i\tau),$$
$$W(\mu)=\frac{(-1)^{m+1}Q}{\pi}\frac{\cos(m\varphi_0)\sinh a}{(1+\cosh a)^{3/2}}\mathcal{P}_{\mu-1/2}^{-m}(\cosh a).$$

The introduction of new functions (6.25) and (6.26) provides an exponential decay of the right-hand sides of equations in the systems as $\tau \to \infty$. For $\alpha = \pi$ (a half-space) and $\nu = 1/2$ we have $\varphi_l(\tau) \equiv 0$, $l = 1, ..., 6$; for $\alpha = \pi$ and $\nu \neq 1/2$ the exact solution of the problem can be obtained by the method given in Ufliand (1965). We recall that the Schwarz–Palatini formulae allow us to recover a function, which is analytical in a strip of width $2h$, if we only know its real (imaginary) part on the boundary of the strip $|\operatorname{Re} z| = h$. If $A(-z) = -A(z)$, then

$$A(z) = \frac{2}{h} \sin\frac{\pi z}{2h} \int_0^\infty \frac{\operatorname{Im} A(h+it) \sinh[\pi t/(2h)]}{\cosh(\pi t/h) + \cos(\pi z/h)}\, dt \tag{6.27}$$

for $|\operatorname{Re} z| \leq h$. If $B(-z) = B(z)$, then

$$B(z) = \frac{2}{h} \cos\frac{\pi z}{2h} \int_0^\infty \frac{\operatorname{Re} B(h+it) \cosh[\pi t/(2h)]}{\cosh(\pi t/h) + \cos(\pi z/h)}\, dt \tag{6.28}$$

for $|\operatorname{Re} z| \leq h$. It is assumed that the integrals (6.27) and (6.28) converge. Singular integrals appear if we put $z = h + i\tau$ in (6.27), (6.28). One can regularize the systems of singular equations using the well-known method in Muskhelishvili (1953a), Gakhov (1966). The investigation of problem (6.1)–(6.3) is completed by solving the regularized systems, which can be carried out numerically. Note also that the numerical solution of the system of singular integral equations can be also obtained without regularization, using the well-known quadrature formulae for singular integrals in Erdogan (1969), Ivanov (1976) or Belotserkovsky and Lifanov (1993).

6.1.2. AN AXISYMMETRIC CASE

We consider in detail the axisymmetric case $m = 0$ (Aleksandrov and Pozharskii (1997)). For $m = 0$ we obtain $\Phi_3 \equiv 0$ from the boundary conditions (6.4) for $\tau_{\varphi z}$ and (6.5) for u_φ; the Papkovich–Neuber representation reduces to the Boussinesq representation for the axisymmetric problem (Timoshenko and Goodier (1951)). We integrate the boundary condition $\tau_{rz} = 0$ for $\eta = 0$ with respect to r and eliminate the first derivative with respect to z in the boundary condition $\sigma_z/(2G) = -\delta_*(\xi - a)$ for $\eta = 0$, using formula (6.23) and the relation (see Gradshteyn and Ryzhik (1994, formulae 8.731))

$$(\mu - 1/2 - m)\, \mathcal{P}^{-m}_{\mu-3/2}(\cosh a) + 2\mu\, \mathcal{P}^{-m}_{\mu-1/2}(\cosh a)$$

$$+(\mu+1/2+m)\,\mathcal{P}^{-m}_{\mu+1/2}(\cosh a) = 2\mu(1+\cosh a)\,\mathcal{P}^{-m}_{\mu-1/2}(\cosh a)$$

for $m=0$. For two-dimensional harmonic functions

$$\Phi_k(\xi,\eta) = \frac{\sqrt{\cosh\xi+\cos\eta}}{\pi i}\int\limits_{c-i\infty}^{c+i\infty}[A_k(\mu)\cos(\mu\eta) \tag{6.29}$$

$$+B_k(\mu)\sin(\mu\eta)]\,\mathcal{Q}_{\mu-1/2}(\cosh\xi)\,d\mu,$$

$k=1,2$, $c>0$, $A_k(-\mu)=-A_k(\mu)$, $B_k(-\mu)=B_k(\mu)$, the partial derivative with respect to r can be expressed using (6.15)–(6.21) by the relation

$$\frac{\partial}{\partial r}\Phi_k(\xi,\eta) = \frac{\sqrt{\cosh\xi+\cos\eta}}{2\pi i}\int\limits_{c-i\infty}^{c+i\infty}\{[A_k(\mu-1)+2A_k(\mu)+A_k(\mu+1)]$$

$$\times\cos(\mu\eta)+[B_k(\mu-1)+2B_k(\mu)+B_k(\mu+1)]\sin(\mu\eta)\}\,\mathcal{Q}^1_{\mu-1/2}(\cosh\xi)\,d\mu.$$

Therefore, in contrast to subsection 6.1.1 we should not multiply the boundary conditions for τ_{rz}, σ_z for $\eta=0$. Here, it is sufficient to multiply the boundary conditions for displacements in the Boussinesq representation by $(\cosh\xi+\cos\alpha)$. Then shifting the contour of integration in the integrals (6.29), we get a system of four functional equations in unknown functions $A_k(\mu)$, $B_k(\mu)$, $k=1,2$, in the form

$$\varepsilon_2 B_2(\mu) - 1/2[(\mu-1/2)A_1(\mu-1)+2\mu A_1(\mu) \tag{6.30}$$

$$+(\mu+1/2)A_1(\mu+1)] = W_0(\mu),$$

$$\varepsilon_1 A_2(\mu) + 1/2[(\mu-1/2)B_1(\mu-1)+2\mu B_1(\mu) \tag{6.31}$$

$$+(\mu+1/2)B_1(\mu+1)] = 0,$$

$$\cos(\mu\alpha)\{\cos\alpha\,[A_1(\mu-1)+2A_1(\mu)+A_1(\mu+1)]-\sin\alpha\,[A_2(\mu-1) \tag{6.32}$$

$$+2A_2(\mu)+A_2(\mu+1)]\}+\sin(\mu\alpha)\{\cos\alpha\,[B_1(\mu-1)+2B_1(\mu)$$

$$+B_1(\mu+1)]-\sin\alpha\,[B_2(\mu-1)+2B_2(\mu)+B_2(\mu+1)]\}$$

$$+\cos[(\mu-1)\alpha]\,(\mu-1)^{-1}[\varepsilon_2 B_2(\mu-1)-(A_1(\mu-1)+A_1(\mu))/2]$$

$$+\cos[(\mu+1)\alpha]\,(\mu+1)^{-1}[\varepsilon_2 B_2(\mu+1)+(A_1(\mu)+A_1(\mu+1))/2]$$

$$-\sin[(\mu-1)\alpha]\,(\mu-1)^{-1}[\varepsilon_1 A_2(\mu-1)+(B_1(\mu-1)+B_1(\mu))/2]$$

$$-\sin[(\mu+1)\alpha]\,(\mu+1)^{-1}[\varepsilon_1 A_2(\mu+1)+(B_1(\mu)+B_1(\mu+1))/2]$$

$$=W_0(\mu-1)\cos[(\mu-1)\alpha]\,(\mu-1)^{-1}+W_0(\mu+1)\cos[(\mu+1)\alpha]\,(\mu+1)^{-1},$$

$$\varepsilon_2 \{2A_2(\mu)\cos\alpha\,\cos(\mu\alpha) + (\mu-1/2)(\mu-1)^{-1}A_2(\mu-1)\cos[(\mu-1)\alpha] \quad (6.33)$$
$$+(\mu+1/2)(\mu+1)^{-1}A_2(\mu+1)\cos[(\mu+1)\alpha]\} + \varepsilon_1\{2B_2(\mu)\cos\alpha\,\sin(\mu\alpha)$$
$$+(\mu-1/2)(\mu-1)^{-1}B_2(\mu-1)\sin[(\mu-1)\alpha]$$
$$+(\mu+1/2)(\mu+1)^{-1}B_2(\mu+1)\sin[(\mu+1)\alpha]\}$$
$$+\sin\alpha\,\sin(\mu\alpha)\,[(\mu-1/2)A_2(\mu-1) + 2\mu A_2(\mu) + (\mu+1/2)A_2(\mu+1)]$$
$$-\sin\alpha\,\cos(\mu\alpha)\,[(\mu-1/2)B_2(\mu-1) + 2\mu B_2(\mu) + (\mu+1/2)B_2(\mu+1)]$$
$$= -2\cos\alpha\,\sin(\mu\alpha)\,W_0(\mu) - (\mu-1/2)(\mu-1)^{-1}W_0(\mu-1)\sin[(\mu-1)\alpha]$$
$$-\frac{\mu+1/2}{\mu+1}W_0(\mu+1)\sin[(\mu+1)\alpha], \quad W_0(\mu) = \frac{\sinh a\, Q}{(\cosh a+1)^{3/2}}\mathcal{P}_{\mu-1/2}(\cosh a).$$

We now introduce the new analytical functions $A^*(\mu)$, $B^*(\mu)$ using the formulae

$$A^*(\mu) = \mu[A_1(\mu-1/2) + A_1(\mu+1/2)], \quad (6.34)$$

$$B^*(\mu) = \mu[B_1(\mu-1/2) + B_1(\mu+1/2)]. \quad (6.35)$$

Equations (6.30) and (6.31) can then be represented in the form

$$A^*(\mu-1/2) + A^*(\mu+1/2) = 2[\varepsilon_2 B_2(\mu) - W_0(\mu)], \quad (6.36)$$

$$B^*(\mu-1/2) + B^*(\mu+1/2) = -2\varepsilon_1 A_2(\mu). \quad (6.37)$$

Due to the evenness of $A^*(\mu)$ and the oddness of $B^*(\mu)$, we can reduce equations (6.36) and (6.37) putting $\mu = i\tau$, $\tau \in \Re$, to Dirichlet problems for the functions $\operatorname{Re} A^*(\mu)$, $\operatorname{Im} B^*(\mu)$, in the strip $|\operatorname{Re}\mu| \leq 1/2$. The solutions of these problems are found using a Fourier integral transformation. The functions $\operatorname{Im} A^*(\mu)$, $\operatorname{Re} B^*(\mu)$ are then recovered using formulae (6.27) and (6.28). After equations (6.36) and (6.37) have been solved, the functions $A_1(\mu)$, $B_1(\mu)$, which can therefore be eliminated from the system, are determined as before from relations (6.34) and (6.35).

We now consider the system (6.30)–(6.33), having eliminated the functions $A_1(\mu)$, $B_1(\mu)$ from it, when $\mu = i\tau$, $\tau \in \Re$. Using formulae (6.27) and (6.28), we now reduce the system to a system of two singular integral equations in the new functions $\varphi_1(\tau)$, $\varphi_2(\tau)$ which are related to the functions $A_2(\mu)$, $B_2(\mu)$ by the equations

$$\varphi_1(\tau) = \operatorname{Im} A_2(1+i\tau) + \tanh(\pi\tau)\, w_0(\tau), \quad (6.38)$$

$$\varphi_2(\tau) = \operatorname{Re} B_2(1+i\tau) - w_0(\tau), \quad w_0(\tau) = \operatorname{Re} W_0(1+i\tau). \quad (6.39)$$

We write the system in the matrix form

$$A(\tau)\varphi(\tau) + \frac{1}{\pi i}\int_0^\infty \frac{K(\tau,t)\,\varphi(t)}{\cosh\pi t - \cosh\pi\tau}\,d\cosh\pi t \;=\; f(\tau), \qquad (6.40)$$

where $0 \leq \tau < \infty$; $A(\tau) = (a_{mn}(\tau))$ and $K(\tau,t) = i(k_{mn}(\tau,t))$ are matrices; $\varphi(\tau) = (\varphi_n(\tau))$, $f(\tau) = (f_n(\tau))$, $n, m = 1, 2$, are column vectors with entries

$$a_{11}(\tau) = (1+\varepsilon_1+\tau^2)\sin\alpha + \varepsilon_1\tau\cos\alpha\,\tanh(\tau\alpha),$$

$$a_{12}(\tau) = (1+\varepsilon_2+\tau^2)\sin\alpha\,\tanh(\tau\alpha) + \varepsilon_2\tau\cos\alpha,$$

$$a_{21}(\tau) = \varepsilon_2(1+2\tau^2)\cos\alpha - (2-\varepsilon_2+2\tau^2)\tau\sin\alpha\,\tanh(\tau\alpha),$$

$$a_{22}(\tau) = \varepsilon_1(1+2\tau^2)\cos\alpha\,\tanh(\tau\alpha) - (2-\varepsilon_1+2\tau^2)\tau\sin\alpha,$$

$$k_{11}(\tau,t) = \frac{1}{\cosh(\pi t/2)}\left\{b_{11}(\tau)\cosh\frac{\pi\tau}{2} + \left[g_1(\tau)\sinh\frac{\pi\tau}{2}\right.\right.$$

$$\left.\left.+g_0(\tau)\frac{\cosh(\pi t/2)\cosh(\pi\tau/2) - \sinh^2(\pi\tau/2)}{\cosh(\pi t/2)+\cosh(\pi\tau/2)}\right]d(\tau,t)\right\},$$

$$k_{12}(\tau,t) = \frac{1}{\sinh(\pi t/2)}\left\{b_{12}(\tau)\sinh\frac{\pi\tau}{2} + \left[h_1(\tau)\cosh\frac{\pi\tau}{2}\right.\right.$$

$$\left.\left.-h_0(\tau)\frac{\sinh(\pi\tau/2)[\cosh(\pi t/2)\cosh(\pi\tau/2) - \sinh^2(\pi t/2)]}{\cosh(\pi t/2)[\cosh(\pi t/2)+\cosh(\pi\tau/2)]}\right]d(\tau,t)\right\},$$

$$k_{21}(\tau,t) = b_{21}(\tau)\frac{\cosh(\pi\tau/2)}{\cosh(\pi t/2)} + g_2(\tau)\frac{\sinh(\pi\tau/2)}{\cosh(\pi t/2)}d(\tau,t),$$

$$k_{22}(\tau,t) = b_{22}(\tau)\frac{\sinh(\pi\tau/2)}{\sinh(\pi t/2)} + h_2(\tau)\frac{\cosh(\pi\tau/2)}{\sinh(\pi t/2)}d(\tau,t),$$

$$b_{11}(\tau) = \varepsilon_1(\cos\alpha\,\tanh(\tau\alpha) - \tau\sin\alpha),$$

$$b_{12}(\tau) = \varepsilon_2[\cos\alpha - \tau\sin\alpha\,\tanh(\tau\alpha)],$$

$$b_{21}(\tau) = \varepsilon_2\tau\cos\alpha - (\varepsilon_2 - 1 + (2\varepsilon_2-1)\tau^2)\sin\alpha\,\tanh(\tau\alpha),$$

$$b_{22}(\tau) = \varepsilon_1\tau\cos\alpha\,\tanh(\tau\alpha) - (\varepsilon_1 - 1 + (2\varepsilon_1-1)\tau^2)\sin\alpha,$$

$$d(\tau,t) = [\cosh(\pi t) - \cosh(\pi\tau)]/[\cosh(\pi t) + \cosh(\pi\tau)],$$

$$g_1(\tau) = \varepsilon_1[(1-2\tau^2)\sin\alpha + (7+4\tau^2)\tau\cos\alpha\,\tanh(\tau\alpha)]/(1+4\tau^2)$$

$$+(1+\tau^2)\sin\alpha, \quad g_0(\tau) = 3\varepsilon_1(\cos\alpha\,\tanh(\tau\alpha) - \tau\sin\alpha),$$

$$h_1(\tau) = \varepsilon_2[(1-2\tau^2)\sin\alpha\,\tanh(\tau\alpha) + (7+4\tau^2)\tau\cos\alpha]/(1+4\tau^2)$$

$$+(1+\tau^2)\sin\alpha\,\tanh(\tau\alpha), \quad h_0(\tau) = 3\varepsilon_2[\cos\alpha - \tau\sin\alpha\,\tanh(\tau\alpha)],$$

$$g_2(\tau) = 2(1+\tau^2)[\varepsilon_2\cos\alpha - \tau\sin\alpha\,\tanh(\tau\alpha)],$$

$$h_2(\tau) = 2(1+\tau^2)[\varepsilon_1\cos\alpha\,\tanh(\tau\alpha) - \tau\sin\alpha],$$

$$f_n(\tau) = f_{n1}(\tau)w_0(\tau) + f_{n2}(\tau)\sinh\frac{\pi\tau}{2}\int_0^\infty \frac{w_0(t)\cosh(\pi t/2)}{\cosh(\pi t) - \cosh(\pi\tau)}\,dt$$

$$+f_{n3}(\tau)\cosh\frac{\pi\tau}{2}\int_0^\infty \frac{w_0(t)\cosh(\pi t/2)}{\cosh(\pi t) + \cosh(\pi\tau)}\,dt + f_{n4}(\tau)\sinh\frac{\pi\tau}{2}$$

$$\times\int_0^\infty \frac{w_0(t)[\sinh^2(\pi t/2) - \cosh(\pi t/2)\cosh(\pi\tau/2)]}{[\cosh(\pi t) + \cosh(\pi\tau)][\cosh(\pi t/2) + \cosh(\pi\tau/2)]}\,dt \quad (n = 1, 2),$$

$$f_{11}(\tau) = (1 + \varepsilon_1 + \tau^2)\sin\alpha\, t_1(\tau) + \varepsilon_1\tau\cos\alpha\, t_2(\tau),$$

$$f_{12}(\tau) = 2\varepsilon_1[\cos\alpha\, t_2(\tau) - \tau\sin\alpha\, t_1(\tau)],$$

$$f_{13}(\tau) = 2\varepsilon_1[(1-2\tau^2)\sin\alpha\, t_1(\tau) + (7+4\tau^2)\tau\cos\alpha\, t_2(\tau)]$$

$$\times(1+4\tau^2)^{-1} + 2(1+\tau^2)\sin\alpha\, t_1(\tau), \quad f_{14}(\tau) = 3f_{12}(\tau),$$

$$f_{21}(\tau) = \varepsilon_2(1+2\tau^2)\cos\alpha\, t_1(\tau) - (2 - \varepsilon_2 + 2\tau^2)\tau\sin\alpha\, t_2(\tau),$$

$$f_{22}(\tau) = 2\varepsilon_2\tau\cos\alpha\, t_1(\tau) + 2(\varepsilon_2 - 1 + (2\varepsilon_2 - 1)\tau^2)\sin\alpha\, t_2(\tau),$$

$$f_{23}(\tau) = 4(1+\tau^2)[\varepsilon_2\cos\alpha\, t_1(\tau) - \tau\sin\alpha\, t_2(\tau)], \quad f_{24}(\tau) \equiv 0.$$

$$t_1(\tau) = \tanh(\pi\tau) - \tanh(\alpha\tau), \quad t_2(\tau) = \tanh(\pi\tau)\tanh(\alpha\tau) - 1.$$

System (6.40) has the exact solution $\varphi_1(\tau) \equiv \varphi_2(\tau) \equiv 0$ for $\alpha = \pi$, $\nu = 1/2$. For $\alpha = \pi$ and $\nu \neq 1/2$ the exact solution of the problem is given in Ufliand (1965). The exponential decay of the functions $t_1(\tau)$ and $t_2(\tau)$ on the right-hand side of system (6.40), as $\tau \to \infty$, enables one to assert that formulae (6.38) and (6.39) separate out the principal parts of the functions $\operatorname{Im} A_2(1+i\tau)$ and $\operatorname{Re} B_2(1+i\tau)$, as $\tau \to \infty$.

We investigate the normality of the system which is characteristic for (6.40), that is, check the condition (Muskhelishvili (1953a))

$$\det S_\pm(\tau) = \det(A(\tau) \pm B(\tau)) \neq 0, \quad (0 \leq \tau < \infty), \qquad (6.41)$$

where $B(\tau) = i(b_{nm}(\tau))$, $n, m = 1, 2$, and $S_\pm(\tau)$ are the basic matrices. We checked that condition (6.41) holds for $\alpha \in [\pi/4, 3\pi/4]$ and $\nu \neq \cos^2\alpha$; for $\alpha = \pi n/8$, $n = 0, 1, 7, 8$, and any value of ν; for $n = 4$ and $\nu > 0.092$; for $n = 2, 6$ and $\nu \neq 1/2$; for $n = 3, 5$ and $\nu = 0.3$.

Using the well-known method in Muskhelishvili (1953a), we regularize system (6.40) and reduce it to the following matrix Fredholm integral equation of the second kind:

$$[A_*(\tau)A(\tau) - B_*(\tau)B(\tau)]\varphi(\tau) \tag{6.42}$$

$$+\frac{1}{\pi}\int_0^\infty \frac{A_*(\tau)[K(\tau,t) - B(\tau)] + B_*(\tau)[A(t) - A(\tau)]}{\cosh(\pi t) - \cosh(\pi\tau)}\varphi(t)\,d\cosh(\pi t)$$

$$-\frac{B_*(\tau)}{\pi^2}\int_0^\infty\int_0^\infty \frac{K(t_0,t)\varphi(t)\,d\cosh(\pi t_0)\,d\cosh(\pi t)}{[\cosh(\pi t_0) - \cosh(\pi t)][\cosh(\pi t_0) - \cosh(\pi\tau)]}$$

$$= A_*(\tau)f(\tau) + \frac{B_*(\tau)}{\pi}\int_0^\infty \frac{f(t)\,d\cosh(\pi t)}{\cosh(\pi t) - \cosh(\pi\tau)},$$

$$A_*(\tau) = {}^1\!/\!_2\,[S_+^{-1}(\tau) + S_-^{-1}(\tau)], \quad B_*(\tau) = {}^1\!/\!_2\,[S_+^{-1}(\tau) - S_-^{-1}(\tau)].$$

One can obtain a numerical solution of equation (6.42), using a collocation method. For calculating the singular integrals on the right-hand side of (6.42), one may use formula 3.228.5 in Gradshteyn and Ryzhik (1994) (see also Pykhteev (1959))

$$\int_0^\infty \frac{f(t)\,d\cosh(\pi t)}{\cosh(\pi t) - \cosh(\pi\tau)}$$

$$= \pi\int_0^\infty \frac{f(t)\cosh(\pi t) - f(\tau)\cosh(\pi\tau)}{\cosh(\pi t) - \cosh(\pi\tau)}\tanh(\pi t)\,dt + af(\tau),$$

$$a = -\frac{1}{\pi}\left[\int_1^\infty \frac{\exp(-t)}{t}dt + \sum_{n=1}^\infty \frac{(-1)^n}{n!\,n}\right] \approx 0.184.$$

A 3-D problem for a half-space with spherical hollow or jut (Fig. 6.2) can be solved similarly if we specify the stresses for $\eta = \pi$ and the displacements for $\eta = -\alpha$ (hollow) or $\eta = \alpha$ (jut). If the hollow (jut) is stress-free, then the boundary conditions for $\eta = \alpha$ ($\eta = -\alpha$), in the case of an axial symmetry, have the form $\sigma_\eta = \tau_{\xi\eta} = 0$, where

$$\sigma_\eta = \left(\frac{\partial r}{\partial \eta}\right)^2 \sigma_r + \left(\frac{\partial z}{\partial \eta}\right)^2 \sigma_z + 2\frac{\partial r}{\partial \eta}\frac{\partial z}{\partial \eta}\tau_{rz},$$

$$\tau_{\xi\eta} = \frac{\partial r}{\partial \xi}\frac{\partial r}{\partial \eta}\sigma_r + \frac{\partial z}{\partial \xi}\frac{\partial z}{\partial \eta}\sigma_z + \left(\frac{\partial r}{\partial \xi}\frac{\partial z}{\partial \eta} + \frac{\partial r}{\partial \eta}\frac{\partial z}{\partial \xi}\right)\tau_{rz}.$$

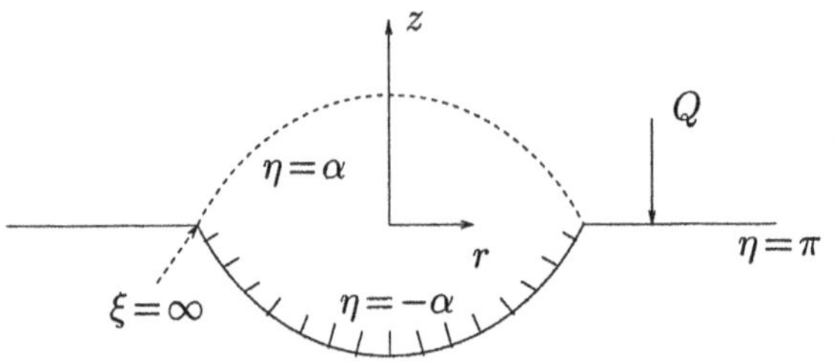

Figure 6.2: An axial section of a half-space with spherical hollow or jut (dashed line)

6.2 A Truncated Sphere

Solving the 3-D problem for an elastic truncated sphere, considered above, for the normal displacement on the cut of the sphere, we get the formula

$$u_z(\xi,0,\varphi) = \frac{4(1-\nu)Q\sqrt{\cosh\xi+1}\,\sinh a}{\pi(\cosh a+1)^{3/2}} \sum_{m=0}^{\infty} e_m \cos(m\varphi)\cos(m\varphi_0) \tag{6.43}$$

$$\times \int_0^\infty [(-1)^m(1-g(\tau))\,\mathcal{P}^{-m}_{i\tau-1/2}(\cosh a) + T^m(\tau,a)]\,\mathcal{P}^{m}_{i\tau-1/2}(\cosh\xi)\,d\tau,$$

$$g(\tau) = 1/\cosh^2(\pi\tau), \quad e_0 = 1/2, \quad e_m = 1\ (m \neq 0).$$

Here we used formula (6.25) for $\varphi_3(\tau)$, which allows us to separate out the principal part of the normal displacement, and also integral (6.27) for $z = i\tau$, $h = 4$ and $A(z) = \mathcal{P}^{-m}_{z-1/2}(\cosh a)\tan(\pi z)$. The functions $T^m(\tau,a)$, $m = 0, 1, \ldots$, in (6.43) are determined from the systems of singular integral equations mentioned above; we give below an explicit expression only for $T^0(\tau,a)$.

Using the formulae (Ufliand (1965))

$$(-1)^m\,\mathcal{P}^{-m}_{i\tau-1/2}(\cosh\xi)\,\mathcal{P}^{m}_{i\tau-1/2}(\cosh a) \tag{6.44}$$

$$= \frac{2\cosh(\pi\tau)}{\pi^2\sqrt{\sinh a\,\sinh\xi}} \int_0^\infty \mathcal{Q}_{m-1/2}(\cosh u)\cos(\tau s)\,ds,$$

$$\cosh u = (\cosh s + \cosh a\,\cosh\xi)/(\sinh a\,\sinh\xi)$$

$$\frac{2}{\pi}\sum_{m=0}^{\infty} e_m\, \mathcal{Q}_{m-1/2}(\cosh u)\cos(m\varphi) = \frac{1}{\sqrt{2(\cosh u - \cos\varphi)}},$$

we get for the principal term in (6.43) the formula

$$\sum_{m=0}^{\infty}(-1)^m e_m \cos m\varphi \int_0^{\infty} \mathcal{P}_{i\tau-1/2}^{-m}(\cosh a)\,\mathcal{P}_{i\tau-1/2}^{m}(\cosh\xi)\,d\tau \tag{6.45}$$

$$= \frac{1}{4}\frac{\sqrt{1-\tanh^2(a/2)}\sqrt{1-\tanh^2(\xi/2)}}{\sqrt{\tanh^2(a/2)+\tanh^2(\xi/2)-2\tanh^2(a/2)\tanh^2(\xi/2)\cos\varphi}}.$$

If we introduce new variables $\tanh(a/2) = \rho$ and $\tanh(\xi/2) = r$ in (6.45), we can see that the principal part in (6.43) coincide with the kernel $1/R$ of the integral equation (1.29) of the contact problem for a half-space, written in polar coordinates. Therefore, for a smooth punch in contact with a truncated sphere the Galanov–Newton method can be used to solve the contact problem.

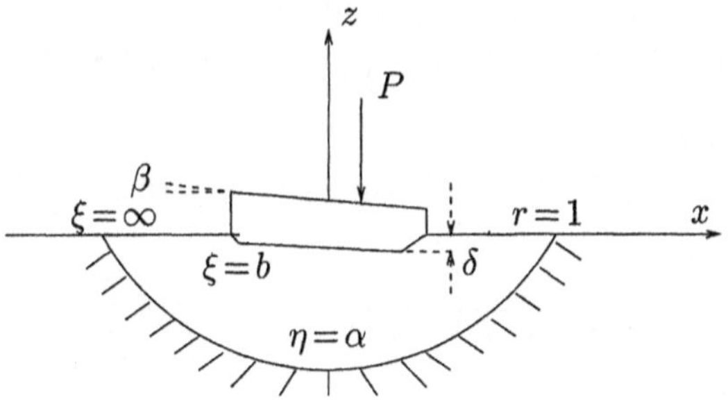

Figure 6.3: A punch on a truncated sphere (an axial section)

6.2.1. THE METHOD OF DUAL EQUATIONS

We study the contact problem for a rigid circular punch impressed by a force P into the cut face of a truncated sphere with a fixed spherical surface (Fig. 6.3). The shape of the base of this punch is described by the function $g(\xi,\varphi)$ which is even with respect to φ and related to R_*. Suppose that b is the radius of the circular contact domain with respect to the coordinate ξ with the center at the point $\xi = 0$. It is necessary to determine the distribution of the normal contact pressure $\sigma_z(\xi,0,\varphi) = -\psi(\xi,\varphi)$ $(0 \le \xi \le$

b, $0 \le \varphi < 2\pi$) and to find the values of the force P and moment M, acting onto the punch with respect to the Cartesian y-axis, where $y = r\sin\varphi$, for specified function $g(\xi, \varphi)$, values of b, of the angle β of rotation of the punch around the y-axis and of the settlement δ of the punch, related to R_*.

We expand the functions $f(\xi,\varphi) = \delta + \beta\tanh(\xi/2)\cos\varphi - g(\xi,\varphi)$, $\psi(\xi,\varphi)$ into the Fourier series with respect to φ

$$f(\xi,\varphi) = \sum_{m=0}^{\infty} f_m(\xi)\cos(m\varphi), \qquad \psi(\xi,\varphi) = \sum_{m=0}^{\infty} \psi_m(\xi)\cos(m\varphi).$$

On the basis of (6.43) the integral equations in unknown functions $\psi_m^*(\xi) = \theta^{-1}\psi_m(\xi)(\cosh\xi + 1)^{-3/2}$, $m = 0, 1, \ldots$, can be written in the form

$$\int_0^b \psi_m^*(a)\sinh a\, da \int_0^\infty [(-1)^m(1 - g(\tau))\,\mathcal{P}_{i\tau-1/2}^{-m}(\cosh a) \tag{6.46}$$

$$+T^m(\tau, a)]\,\mathcal{P}_{i\tau-1/2}^{m}(\cosh\xi)\,d\tau = f_m^*(\xi) \quad (0 \le \xi \le b),$$

where we denote $f_m^*(\xi) = f_m(\xi)/\sqrt{\cosh\xi + 1}$.

With the help of representation (6.44), one can show that

$$\int_0^\infty (-1)^m\,\mathcal{P}_{i\tau-1/2}^{-m}(\cosh a)\,\mathcal{P}_{i\tau-1/2}^{m}(\cosh\xi)\,d\tau \tag{6.47}$$

$$= \frac{\sqrt{(1-r^2)(1-\rho^2)}}{\sqrt{r\rho}}\,\mathcal{Q}_{m-1/2}\left(\frac{r^2+\rho^2}{2r\rho}\right),$$

$$r = \tanh(\xi/2), \qquad \rho = \tanh(a/2).$$

The spherical function on the right-hand side of (6.47) can be expressed via the complete elliptic integrals (Abramowitz and Stegun (1972, formulae (8.13)), giving known formulae (5.77). Thus, the function $T^m(\tau, a)$ gives a regular part of the kernel of equation (6.46).

Note that in frictionless contact problems the kernels of integral equations are always symmetric due to Betti's reciprocal theorem (see Gladwell (1980)). Replacing the variables a and ξ in the kernel of (6.47), we obtain the dual integral equations, which are equivalent to (6.47):

$$\int_0^\infty C(\tau)\{[1 - g(\tau)]\,\mathcal{P}_{i\tau-1/2}^{m}(\cosh\xi) + p(\tau, m)T^m(\tau,\xi)\}\,d\tau = f_m^*(\xi) \tag{6.48}$$

$$(0 \le \xi \le b),$$

$$\int_0^\infty C(\tau)\tau \tanh(\pi\tau)\, \mathcal{P}^m_{i\tau-1/2}(\cosh\xi)\, d\tau = 0 \quad (b < \xi < \infty), \tag{6.49}$$

where

$$C(\tau) = p(\tau, m) \int_0^b \psi^*_m(a) \sinh a\, \mathcal{P}^m_{i\tau-1/2}(\cosh a)\, da, \tag{6.50}$$

$$p(\tau, m) = (-1)^m\, \Gamma(i\tau + 1/2 + m)/\Gamma(i\tau + 1/2 - m).$$

Consider dual equations (6.48), (6.49) for $m = 0$ or for the axial symmetry. For $m = 0$ we have

$$T^0(\tau, \xi) = 2\tanh(\pi\tau)\, \sinh\frac{\pi\tau}{2} \int_0^\infty \frac{\varphi_1^*(t, \xi)\sinh(\pi t/2)}{\cosh(\pi t) + \cosh(\pi\tau)}\, dt,$$

$$\sinh\xi\, \varphi_1^*(t, \xi) = \varphi_1(t)(\cosh\xi + 1)^{3/2},$$

where the function $\varphi_1(t)$ satisfies system (6.42).

Looking for the function $C(\tau)$ in the integral form

$$C(\tau) = \int_0^b \varphi_*(x)\cos(\tau x)\, dx \tag{6.51}$$

and using a well-known technique (Babloian (1964)) and the application of Schwarz formula (6.27)

$$2\sinh\frac{\pi\tau}{2} \int_0^\infty \sinh\frac{\pi t}{2}\, \frac{\tanh(\pi t)\, \cos(tx)}{\cosh(\pi t) + \cosh(\pi\tau)}\, dt = \frac{\tanh(\pi\tau)\, \cos(\tau x)}{\cosh x},$$

we reduce the paired equations for $m = 0$ to a Fredholm integral equation of the second kind in $\varphi_*(x)$:

$$\varphi_*(x) - \frac{1}{\pi}\int_0^b \varphi_*(y)\, K_*(x, y)\, dy = F(x) \quad (0 \le x \le b), \tag{6.52}$$

where

$$K_*(x, y) = \frac{1}{2\pi}\left[\frac{x + y}{\sinh[(x + y)/2]} + \frac{x - y}{\sinh[(x - y)/2]}\right] - K_1(x, y), \tag{6.53}$$

$$K_1(x,y) = \frac{2}{\cosh y}\frac{d}{dx}\int_0^x\int_0^\infty \frac{\varphi_1^*(t,\xi)\,\sinh\xi}{\sqrt{2(\cosh x - \cosh\xi)}}\tanh(\pi t)\,\cos(ty)\,dt\,d\xi,$$

$$F(x) = \frac{2}{\pi}\frac{d}{dx}\int_0^x \frac{f_0^*(\xi)\sinh\xi\,d\xi}{\sqrt{2(\cosh x - \cosh\xi)}}. \tag{6.54}$$

Using formula 8.715 in Gradshteyn and Ryzhik (1994) and Abel inversion formula (Sneddon (1966)), we get

$$\frac{d}{dx}\int_0^x \frac{\operatorname{Re}\mathcal{P}_{i\tau+1/2}(\cosh\xi)\sinh\xi}{\sqrt{2(\cosh x - \cosh\xi)}}d\xi = \cosh x\cos(\tau x),$$

we simplify expression (6.54) to

$$K_1(x,y) = 2\frac{\cosh x}{\cosh y}\int_0^\infty \Phi_1^*(t,x)\tanh(\pi t)\cos(ty)\,dt,$$

where the function $\Phi_1^*(\tau,x) = \varphi_1(\tau)$ satisfies system (6.42) with the function $w_0(\tau)$ (6.39) replaced by $w^*(\tau) = \cos(\tau x)$.

The physical meaning of the function $\varphi_*(x)$ lies in the fact that its value at the point $x = b$ determines the coefficient for the square root singularity of the contact pressure at the same point, the so-called stress intensity factor. In fact, using formula (6.50), we obtain, for $m = 0$,

$$\psi_0^*(x) = \int_0^\infty C(\tau)\tau\tanh(\pi\tau)\,\mathcal{P}_{i\tau-1/2}(\cosh x)\,d\tau. \tag{6.55}$$

Substituting (6.51) into (6.55) and and integrating by parts, we get

$$\lim_{x\to b}\sqrt{2(\cosh b - \cosh x)}\psi_0^*(x) = \varphi_*(b).$$

The values of the quantity $\varphi_*(b)/(\sqrt{2}\delta)$ are given in Table 6.1 for $\nu = 0.3$, $f_0^*(\xi) = \delta$ (a plane punch, $F(x) = \sqrt{2}\delta/[\pi\cosh(x/2)]$), different b and angles $\alpha = \pi n/6$. The angle α characterizes a level of truncation of the sphere.

Table 6.1:, The stress intensity factors

n	1	2	3	4	5	6
b=0.1	0.300	0.302	0.312	0.324	0.324	0.325
b=1	0.240	0.146	0.194	0.347	0.350	0.359
b=2	0.0943	0.0757	0.127	0.310	0.344	0.359

The dimensionless b depends on the ratio $\rho \le 1$ of the dimensional radius of the circular contact domain and the radius R_* of the cut; we have $\rho = 0.050$ for $b = 0.1$, $\rho = 0.462$ for $b = 1$, and $\rho = 0.762$ for $b = 2$. In other words, the greater b, the closer is the punch to the circular line $r = R_*$ of a truncated sphere.

The case when $\alpha = \pi$, $n = 6$ in the table, corresponds to a half-space whose surface is fixed for $r \ge R_*$. We can see from the table that for this case of an infinite solid the stress intensity factor is greater than for a finite truncated sphere. The value of $\alpha = \pi/2$, $n = 3$ in the table defines a hemisphere. For $b = 2$ or a fairly large punch we have the minimal stress intensity factor in the table for $n = 2$ or $\alpha = \pi/3$, i.e., for a sphere which is truncated more than half.

Calculations for a plane punch show that, as b considerably increases, the value of $\varphi_*(b)$ changes its sign, i.e., the elastic medium of the truncated sphere can depart from the edge of the punch. This occurs, for example, for $b = 3$ ($\rho = 0.905$) and $\alpha = \pi/6$, and is explained by an increase of the resistivity of the elastic material near the fixed surface of the solid.

6.2.2. A NON-AXIALLY SYMMETRIC CASE

We now consider dual equations (6.48), (6.49) for $m \ne 0$. Using the formula 8.752.1 in Gradshteyn and Ryzhik (1994)

$$\mathcal{P}_\mu^m(x) = (x^2-1)^{m/2}\,\frac{d^m\,\mathcal{P}_\mu(x)}{dx^m} \qquad (x \ge 1), \tag{6.56}$$

we satisfy equation (6.49) with help of (6.51). We integrate equation (6.48) m times with respect to variable $z = \cosh\xi$ on the interval $(1, z)$. Introducing the notation

$$T_*(\tau,\xi) = p(\tau,m)\int_1^z\int_1^z\cdots\int_1^z T^m(\tau,\xi)(z^2-1)^{-m/2}\,dz^m,$$

$$F_*(\xi) = \int_1^z\int_1^z\cdots\int_1^z f_m^*(\xi)(z^2-1)^{-m/2}\,dz^m,.$$

we obtain the equation (we will omit the index m)

$$\int_0^\infty C(\tau)\{[1-g(\tau)]\,\mathcal{P}_{i\tau-1/2}(\cosh\xi) + T_*(\tau,\xi)\}\,d\tau = \chi(\xi),$$

$$\chi(\xi) = F_*(\xi) + \sum_{k=0}^{m-1} c_k z^k \quad (0 \leq \xi \leq b).$$

The function $\chi(\xi)$ contains m arbitrary constants c_0, c_1, ..., c_{m-1}. We suppose that the function $\chi(\xi)$ is known and apply the similar technique, as for $m = 0$, getting the Fredholm integral equation (6.52) with kernel (6.53), in which we must put

$$K_1(x, y) = 2\frac{d}{dx}\int_0^x\int_0^\infty T_*(\tau, \xi)\frac{\cos(\tau y)\sinh\xi\, d\tau\, d\xi}{\sqrt{2(\cosh x - \cosh\xi)}},$$

$$F(x) = \frac{2}{\pi}\frac{d}{dx}\int_0^x \frac{\chi(\xi)\sinh\xi\, d\xi}{\sqrt{2(\cosh x - \cosh\xi)}}. \tag{6.57}$$

We determine the unknown constants c_k $(k = 0, 1, ..., m - 1)$ from the conditions of integrability of the contact pressure

$$\psi_m^*(x) = \int_0^\infty C(\tau)\tau\, \tanh(\pi\tau)\, \mathcal{P}_{i\tau-1/2}^m(\cosh x)\, d\tau \tag{6.58}$$

on the interval $0 \leq x \leq b$. Substituting (6.51) into (6.58), using formula (6.56), discontinuous integrals (0.4) in Babloian (1964) and integrating m times by parts, we conclude that the function $\psi_m^*(x)$ is intergable, as $x \to b - 0$, if the conditions

$$\psi_m^{*\,(k)}(b) = 0, \qquad k = 0, 1, ..., m - 1, \tag{6.59}$$

hold, where (k) means k-derivative. Besides, one can show that the function $\psi_m^*(x)$ behaves like $(b - x)^{-1/2}$ for $x \to b - 0$, i.e., it has the well-known singularity in frictionless contact problems.

To determine the constants c_k, we represent the $\varphi_*(x)$ from (6.52) and the $F(x)$ (6.57) in the form

$$\varphi_*(x) = \varphi_1^*(x) + \varphi_2^*(x), \qquad F(x) = F_1(x) + F_2(x),$$

$$F_1(x) = \frac{2}{\pi}\frac{d}{dx}\int_0^x \frac{F_*(\xi)\sinh\xi\, d\xi}{\sqrt{2(\cosh x - \cosh\xi)}}, \qquad \varphi_2^*(x) = \sum_{k=0}^{m-1} c_k\omega_k(x),$$

$$F_2(x) = \sum_{k=0}^{m-1} c_k\frac{dI_k}{dx}, \qquad I_k(x) = \frac{2}{\pi}\int_0^x \frac{\cosh^k\xi\, \sinh\xi\, d\xi}{\sqrt{2(\cosh x - \cosh\xi)}}.$$

Then, we obtain separable Fredholm integral equations for $\varphi_1^*(x)$ and $\omega_k(x)$ with the right-hand sides $F_1(x)$ and $I_k(x)$, respectively. Doing so, we reduce conditions (6.59) to the following linear algebraic system:

$$\psi_m^{*(n)}(b) + \sum_{k=0}^{m-1} c_k \omega_k^{(n)}(b) = 0, \quad n = 0, 1, ..., m-1.$$

Note that using the change of variables $\sinh(\xi/2) = \sinh(x/2)\sin t$, we can represent the integrals $I_k(x)$ in the form of polynomials of the order $2k+1$ with respect to $\sinh(x/2)$:

$$I_k(x) = \frac{4}{\pi}\sinh\frac{x}{2}\int_0^{\pi/2} \mathcal{P}_{2k}(\sinh\frac{x}{2}\sin t)\sin t\, dt,$$

where $\mathrm{P}_{2k}(\sinh(\xi/2)) = \cosh^k(\xi)$ are the Legendre polynomials 8.910.2 in Gradshteyn and Ryzhik (1994).

6.3 A Half-Space with Spherical Hollow or Jut

Consider the axisymmetric contact problem of impression of an annular rigid punch into an elastic half-space with spherical hollow $\eta = -\alpha$ or jut $\eta = \alpha$, $0 < \alpha < \pi$ (Fig. 6.4). Suppose the interior circle of the annular contact zone coincides with the boundary of the hollow (jut) of radius R_* in the plane of the half-space boundary.

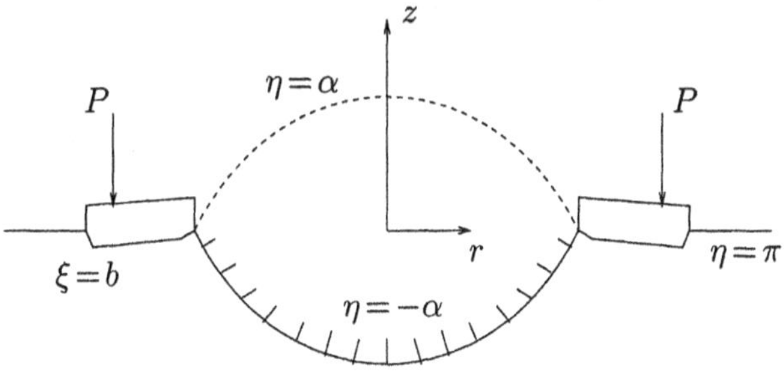

Figure 6.4: An annular punch on a half-space with spherical hollow or jut (dashed line): an axial section

As before, we will assume that the spherical surface of the elastic solid is rigidly fixed. We use the two Papkovich–Neuber functions (6.29), $k = 1, 2$, and introduce new functions

$$A_2(\mu)\cos(\pi\mu) + B_2(\mu)\sin(\pi\mu) = A_2^*(\mu),$$

$$-A_2(\mu)\sin(\pi\mu) + B_2(\mu)\cos(\pi\mu) = B_2^*(\mu).$$

Then we can write an integral equation of the contact problem in the unknown contact pressure $\varphi(\xi) = -\theta^{-1}\sigma_z(\xi, \pi)(\cosh\xi - 1)^{-3/2}$ in the form

$$\int_b^\infty \sinh a\,\varphi(a)\,da \int_0^\infty \tanh(\pi\tau)\,\mathrm{Im}\,A_2^*(i\tau)\,\mathcal{P}_{i\tau-1/2}(\cosh\xi)\,d\tau = f(\xi), \quad (6.60)$$

where $b \leq \xi < \infty$, $f(\xi) = g(\xi)(\cosh\xi - 1)^{-1/2}$, and the $g(\xi)$ defines the shape of the punch base, is related to R_* and vanishes at infinity. The function $A_2^*(\mu)$ depends also on a and is determined from the following system of four functional equations:

$$R(\mu)\sin(\pi\mu) - S(\mu)\cos(\pi\mu) - 2\varepsilon_1 A_2^*(\mu) = 0, \quad (6.61)$$

$$-R(\mu)\cos(\pi\mu) - S(\mu)\sin(\pi\mu) + 2\varepsilon_2 B_2^*(\mu) = 2W(\mu), \quad (6.62)$$

$$\begin{aligned}(\mu - 1/2)[q_1(\mu-1) + q_0(\mu)]A_2^*(\mu-1) + 2\mu[q_1(\mu-1) - q_0(\mu)]A_2^*(\mu) \quad (6.63)\\ +(\mu+1/2)[q_1(\mu+1) + q_0(\mu)]A_2^*(\mu+1)\\ +(\mu-1/2)[q_2(\mu-1) + p_0(\mu)]B_2^*(\mu-1)\\ +2\mu[q_2(\mu) - p_0(\mu)]B_2^*(\mu) + (\mu+1/2)[q_2(\mu+1) + p_0(\mu)]B_2^*(\mu+1)\\ = -W(\mu-1)(\mu-1/2)(\mu-1)^{-1}\sin[\pi\mu - (\mu-1)\alpha] + W(\mu)\cos\alpha\\ \times\sin[(\pi-\alpha)\mu] - W(\mu+1)(\mu+1/2)(\mu+1)^{-1}\sin[\pi\mu - (\mu+1)\alpha],\end{aligned}$$

$$\begin{aligned}p_-(\mu)A_1(\mu-1) + [p_-(\mu) + p_+(\mu)]A_1(\mu) + p_+(\mu)A_1(\mu+1) \quad (6.64)\\ +q_-(\mu)B_1(\mu-1) + [q_-(\mu) + q_+(\mu)]B_1(\mu) + q_+(\mu)B_1(\mu+1)\\ -[q_2(\mu-1) - p_0(\mu)]A_2^*(\mu-1) - 2p_0(\mu)A_2^*(\mu)\\ -[q_2(\mu+1) - p_0(\mu)]A_2^*(\mu+1) + [q_1(\mu-1) - q_0(\mu)]B_2^*(\mu-1)\\ +2q_0(\mu)B_2^*(\mu) + [q_1(\mu+1) - q_0(\mu)]B_2^*(\mu+1)\\ = -\frac{W(\mu-1)}{\mu-1}\cos[\pi\mu - (\mu-1)\alpha] - \frac{W(\mu+1)}{\mu+1}\cos[\pi\mu - (\mu+1)\alpha],\end{aligned}$$

where $W(\mu) = \mathcal{P}_{\mu-1/2}(\cosh a)$,

$$R(\mu) = (\mu - 1/2)A_1(\mu - 1) + 2\mu A_1(\mu) + (\mu + 1/2)A_1(\mu + 1),$$

$$S(\mu) = (\mu - 1/2)B_1(\mu - 1) + 2\mu B_1(\mu) + (\mu + 1/2)B_1(\mu + 1),$$

$$q_0(\mu) = \sin\alpha\, \sin[(\pi - \alpha)\mu], \quad p_0(\mu) = \sin\alpha \cos[(\pi - \alpha)\mu],$$

$$\mu p_1(\mu) = [\cos^2(\pi\mu) + \varepsilon_1]\cos(\mu\alpha) + \sin(\pi\mu)\cos(\pi\mu)\sin(\mu\alpha),$$

$$\mu p_2(\mu) = [\sin^2(\pi\mu) + \varepsilon_1]\sin(\mu\alpha) + \sin(\pi\mu)\cos(\pi\mu)\cos(\mu\alpha),$$

$$q_1(\mu) = p_1(\mu)\cos(\pi\mu) + p_2(\mu)\sin(\pi\mu),$$

$$q_2(\mu) = -p_1(\mu)\sin(\pi\mu) + p_2(\mu)\cos(\pi\mu),$$

$$q_\pm(\mu) = \cos\alpha\sin(\mu\alpha) \pm \sin[(\mu \pm 1)\alpha]/[2(\mu \pm 1)],$$

$$p_\pm(\mu) = \cos\alpha\cos(\mu\alpha) \pm \cos[(\mu \pm 1)\alpha]/[2(\mu \pm 1)].$$

The functions $A_1(\mu)$ and $B_1(\mu)$ can be expressed via $A_2^*(\mu)$ and $B_2^*(\mu)$ with help of equations (6.61) and (6.62). So we eliminate them and separate out the principal parts from the functions $A_2^*(\mu)$ and $B_2^*(\mu)$:

$$A_2^*(\mu) = \tan(\pi\mu)\,W(\mu) + \overline{A}_2(\mu), \quad B_2^*(\mu) = W(\mu) + \overline{B}_2(\mu). \tag{6.65}$$

The first term in formula $(6.65)_1$ determines the principal singular part in the integral equation (6.60). Apply the operator

$$T^s\{...\} = -\frac{d}{ds}\int_s^\infty \frac{...\sinh a\, da}{\sqrt{2(\cosh a - \cosh s)}}, \tag{6.66}$$

to equations (6.63) and (6.64), taking the relation

$$T^s\{W(\mu)\} = \cot(\pi\mu)\sinh(\mu s)$$

(see Babloian (1964)) into account. Rename the required functions in the resulting system of two equations by the formulae

$$iT^s\{\overline{A}_2(\mu)\} = A_2(\mu, s)\sinh(\mu s), \quad T^s\{\overline{B}_2(\mu)\} = B_2(\mu, s)\cosh(\mu s),$$

and put $\mu = i\tau$, $\tau \in \Re$ in the system. Dividing the equations of the system by $\cosh(\alpha\tau)\cosh s$, we can see that the right-hand sides of the equations have an exponential decay, as $\tau \to \infty$, and are bounded as $s \to \infty$. Using the evenness of the new functions $A_2(\mu, s)$, $B_2(\mu, s)$ with respect to μ and reintroducing the new unknown functions

$$\Psi_1(\tau, s) = \operatorname{Re} A_2(i\tau + 1, s), \qquad \Psi_2(\tau, s) = \operatorname{Re} B_2(i\tau + 1, s),$$

we obtain a system of two singular integral equations which is similar to system (6.40). With help of the inverse operator to operator (6.66) (Abel inversion formula, see Babloian (1964) and Sneddon (1966)), we write the integral equation (6.60) in the form

$$\int_b^\infty \sinh a\, \varphi(a)\, da \int_0^\infty \Big\{ \tanh^2(\pi\tau)\, \mathcal{P}_{i\tau-1/2}(\cosh a) \tag{6.67}$$

$$+\frac{4}{\pi}\tanh(\pi\tau) \int_a^\infty \frac{\sin(\tau s)\, ds}{\sqrt{\cosh s - \cosh a)}} \cosh\frac{\pi\tau}{2}$$

$$\times \int_0^\infty \cosh\frac{\pi t}{2} \frac{\Psi_1(t,s)\, dt}{\cosh(\pi t) + \cosh(\pi\tau)} \Big\} \mathcal{P}_{i\tau-1/2}(\cosh\xi)\, d\tau = f(\xi) \quad (b \le \xi < \infty).$$

Taking the symmetry of the kernel of equation (6.67) into account, we write the equivalent dual equations, in the new function

$$D(\tau) = \tanh(\pi\tau) \int_b^\infty \sinh a\, \varphi(a)\, \mathcal{P}_{i\tau-1/2}(\cosh a)\, da, \tag{6.68}$$

in the form

$$\int_0^\infty \tau D(\tau)\, \mathcal{P}_{i\tau-1/2}(\cosh\xi)\, d\tau = 0, \qquad 0 \le \xi < b, \tag{6.69}$$

$$\int_0^\infty \tanh(\pi\tau)\, D(\tau)\, \mathcal{P}_{i\tau-1/2}(\cosh\xi)\, d\tau = f(\xi) - \frac{4}{\pi}\int_0^\infty D(\tau) \cosh\frac{\pi\tau}{2}\, d\tau \tag{6.70}$$

$$\times \int_\xi^\infty \frac{\sin(\tau s)\, ds}{\sqrt{2(\cosh s - \cosh\xi)}} \int_0^\infty \cosh\frac{\pi t}{2} \frac{\Psi_1(t,s)\, dt}{\cosh(\pi t) + \cosh(\pi\tau)}, \quad b \le \xi < \infty.$$

Looking for the function $D(\tau)$ in the form

$$D(\tau) = \int_b^\infty H(y) \sin(y\tau)\, dy,$$

we identically satisfy equation (6.69). Using the Abel inversion formula and the integral

$$4\cosh\frac{\pi t}{2}\int_0^\infty \frac{\sin(y\tau)\sin(s\tau)}{\cosh(\pi t)+\cosh(\pi\tau)}\cosh\frac{\pi\tau}{2}d\tau$$
$$=\frac{\cos[(y-s)t]}{\cosh(y-s)}-\frac{\cos[(y+s)t]}{\cosh(y+s)},$$

which is a particular case of formula (6.28), we transform equation (6.70) to the following integral equation of the second kind

$$H(x)-\frac{1}{\pi}\int_b^\infty H(y)\,K(x,y)\,dy=-\frac{2}{\pi}\frac{d}{dx}\int_x^\infty \frac{f(\xi)\sinh\xi\,d\xi}{\sqrt{2(\cosh\xi-\cosh x)}}, \tag{6.71}$$

$$K(x,y)=\int_0^\infty \Psi_1(t,x)\left[\frac{\cos[(y+x)t]}{\cosh(y+x)}-\frac{\cos[(y-x)t]}{\cosh(y-x)}\right]dt, \tag{6.72}$$

on the semi-infinite interval $b\le x<\infty$.

Due to the asymptotic behaviour of the function $\Psi_1(t,x)$ for $t,x\to\infty$, the kernel $K(x,y)$ (6.72) of equation (6.71) has the same properties for $x,y\to\infty$ as the kernel (6.8) of integral equation (6.7) in Babloian (1964). To solve (6.71), we can use a collocation method.

To investigate the singularity of the contact pressure $\varphi(a)$ on the edge line, $a\to\infty$, of the elastic solid, we must use complex zeros of the determinants of the basic matrices of the characteristic system of singular integral equations in $\Psi_n(\tau,s)$, $n=1,2$ (see formula (134.11) in Muskhelishvili (1953a)). Using formula (6.68), we obtain

$$\varphi(a)=\frac{1}{\pi}\int_{c-i\infty}^{c+i\infty} \mu\cot(\pi\mu)\,D(-i\mu)\,\mathcal{Q}_{\mu-1/2}(\cosh a)\,d\mu,\quad c>0. \tag{6.73}$$

Taking the asymptotics (8.776.2) in Gradshteyn and Ryzhik (1994)

$$\mathcal{Q}_{\mu-1/2}(x)\sim x^{-\mu-1/2},\ \text{as}\ x\to\infty,$$

into account and closing the contour of integration in formula (6.73) into the right half-plane, we can show that the contact stress on the edge line tends to zero for an elastic half-space with a hollow, and to infinity for an elastic half-space with a jut. This result is similar to the well-known asymptotics of the normal stress at the vertex of a 2-D wedge (Ufliand (1965)), because for a hollow the local wedge angle is less than 2π, and for a jut it is greater than 2π.

6.4 A Space with Spherical Cavity

6.4.1. SOLUTION FOR ONE PUNCH

Using spherical coordinates r, η, φ, we consider the axisymmetric problem of impression of a rigid punch, whose surface is defined by the equation $r = R(1 - \rho(\eta))$, $\rho(0) = 0$, into an elastic space $r \geq R$ with spherical cavity (Fig. 6.5). The indenting force P is aligned along the axis of symmetry of the punch. We suppose that the cavity surface $r = R$ is free of stresses outside the contact region and there are no friction forces in the contact region, i.e., $\sigma_r(R, \eta) = -\sigma(\eta)$, $0 \leq \eta \leq \gamma$; $\sigma_r(R, \eta) = 0$, $\gamma < \eta \leq \pi$; $\tau_{r\eta}(R, \eta) = 0$, $0 \leq \eta \leq \pi$. Here, $\sigma(\eta)$ is the required contact pressure. If the punch is smooth, we also must determine the boundary of the contact region, i.e., the quantity γ or the contact angle. We write the condition of equilibrium of the punch in the form

$$P = 2\pi R^2 \int_0^\gamma \sigma(\eta) \sin \eta \, \cos \eta \, d\eta. \tag{6.74}$$

The condition of the contact between the punch and the cavity surface has the form: $u_r(R, \eta) = \delta \cos \eta - R\rho(\eta)$, $0 \leq \eta \leq \gamma$, where δ is the settlement of the punch along the direction of the force P.

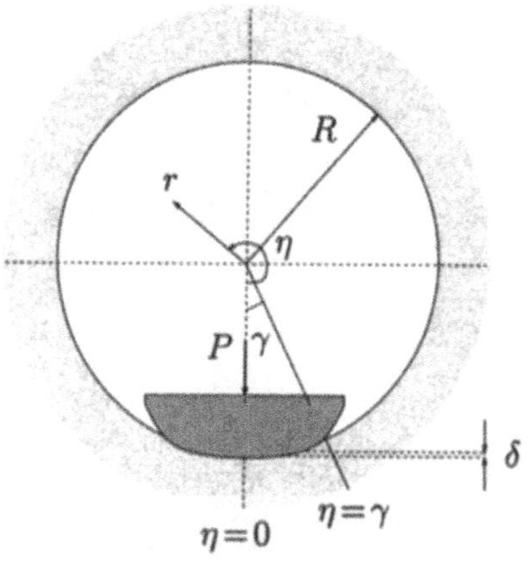

Figure 6.5: Contact between a punch and a space with spherical cavity (an axial section)

Karpenko (1975,1976b) determined the normal displacement of the cavity surface. Following his result, we write the integral equation in the contact pressure in the form

$$\int_0^\gamma \sigma(\alpha)\, H(\eta,\alpha) \sin\alpha \, d\alpha = 2\pi G[a\cos\eta - \rho(\eta)] \ \ (0 \le \eta \le \gamma), \tag{6.75}$$

where $a = \delta/R$,

$$H(\eta,\alpha) = 4(1-\nu)\frac{\mathcal{K}(k_1)}{h_1} + \operatorname{Re}\int_0^1 \left(\frac{A}{y^{1+\omega}} - y\right) U(y,\eta,\alpha)\, dy, \tag{6.76}$$

$$A = -1 + 8\nu - 8\nu^2 + i\frac{5 - 4\nu - 16\nu^2 + 16\nu^3}{\sqrt{3-4\nu^2}},$$

$$U(y,\eta,\alpha) = \mathcal{K}(k_y)h_y^{-1}, \qquad \omega = -\frac{1}{2}(1 - 2\nu - i\sqrt{3-4\nu^2}),$$

$$h_y^2 = 1 + y^2 - 2y\cos(\eta+\alpha), \qquad h_y^2 k_y^2 = 4y\sin\eta\, \sin\alpha,$$

where $\mathcal{K}(k)$ is the complete elliptic integral of the first kind.

We change the variables by the formulae

$$\varepsilon t = \tan(\alpha/2), \quad \varepsilon x = \tan(\eta/2), \quad \varepsilon = \tan(\gamma/2) \tag{6.77}$$

and rewrite equation (6.75) in the form

$$\int_0^1 q(t)\left[\frac{4t}{x+t}\mathcal{K}\left(\frac{2\sqrt{xt}}{x+t}\right) + \frac{\varepsilon}{b}S(x,t)\right] dt = \frac{\varepsilon}{b}v(x), \tag{6.78}$$

where $0 \le x \le 1$,

$$q(t) = \frac{4\sigma[2\arctan(\varepsilon t)]\varepsilon^2}{(1+\varepsilon^2 t^2)^{3/2} G}, \qquad b = \frac{1-\nu}{2\pi},$$

$$v(x) = \frac{2}{\sqrt{1+\varepsilon^2 x^2}}\left[\frac{1-\varepsilon^2 x^2}{1+\varepsilon^2 x^2}a - \rho[2\arctan(\varepsilon x)]\right], \tag{6.79}$$

$$S(x,t) = \frac{t}{\pi\sqrt{(1+\varepsilon^2 x^2)(1+\varepsilon^2 t^2)}}\operatorname{Re}\int_0^1 \left(\frac{A}{y^{1+\omega}} - y\right) \tag{6.80}$$

$$\times U(y, 2\arctan(\varepsilon x), 2\arctan(\varepsilon t))\, dy.$$

We will regularize the integral equation (6.78) of the first kind, transforming it to an integral equation of the second kind. To do this, we must invert the principal or characteristic part of the kernel of (6.78) which corresponds to the integral equation for the axisymmetric contact problem for a half-space (see formula $(5.77)_1$)

$$\int_0^1 q_*(t)\frac{4t}{x+t}\mathcal{K}\left(\frac{2\sqrt{xt}}{x+t}\right)dt = f(x), \qquad 0 \le x \le 1. \tag{6.81}$$

The general solution (1.47), (1.48) for a circular punch on a half-space for $n = 0$ can be represented for equation (6.81) in the form

$$q_*(x) = \frac{c}{\sqrt{1-x^2}} - \frac{1}{\pi^2}\mathcal{T}\left\{\frac{1}{x}f'(x) + f''(x)\right\}, \tag{6.82}$$

$$c = \frac{1}{\pi^2}\left[f(0) + \int_0^1 \frac{f'(t)}{\sqrt{1-t^2}}\,dt\right], \quad \mathcal{T}\{f(x)\} = \int_0^1 f(t)\,L(x,t)\,dt, \tag{6.83}$$

$$L(x,t) = \frac{t}{x+t}\mathcal{K}\left(\frac{2\sqrt{xt}}{x+t}\right) - \mathcal{F}\left(\frac{x}{t},t\right), \quad \mathcal{F}(k,x) = \int_0^x \frac{dt}{\sqrt{(1-t^2)(1-k^2t^2)}}. \tag{6.84}$$

Using formulae (6.82)–(6.84), we reduce (6.78) to the following integral equation of the second kind:

$$q(x) - \frac{\varepsilon}{b\pi^2}\int_0^1 q(t)Q(x,t)\,dt = \frac{c}{\sqrt{1-x^2}} \tag{6.85}$$

$$-\frac{\varepsilon}{b\pi^2}\mathcal{T}\left\{\frac{1}{x}v'(x) + v''(x)\right\}, \qquad 0 \le x \le 1,$$

where

$$Q(x,t) = \int_0^1 \left[\frac{1}{y}S'_y(y,t) + S''_{yy}(y,t)\right]L(x,y)\,dy, \tag{6.86}$$

Equation (6.85) is equivalent to (6.78) under the condition (see formula $(6.83)_1$)

$$c = \frac{\varepsilon}{b\pi^2}\left[2a - \int_0^1 q(t)S(0,t)\,dt + \int_0^1 \frac{v'(t)}{\sqrt{1-t^2}}dt\right. \tag{6.87}$$

$$- \int_0^1 dx \int_0^1 q(x) \frac{S_t'(x,t)}{\sqrt{1-t^2}}\, dt \Bigg].$$

Introduce the new function

$$p(x) = \frac{b\pi^2}{\varepsilon} \left[q(x) - \frac{c}{\sqrt{1-x^2}} \right]. \tag{6.88}$$

For the function $p(x)$ we then have, in accordance with (6.85), the integral equation

$$p(x) - \frac{\varepsilon}{b\pi^2} \int_0^1 p(t)\, Q(x,t)\, dt = w(x), \qquad 0 \leq x \leq 1, \tag{6.89}$$

$$w(x) = -\mathcal{T} \left\{ \frac{1}{x} v'(x) + v''(x) \right\} + c \int_0^1 \frac{Q(x,t)dt}{\sqrt{1-t^2}}.$$

It can be shown that the kernel $Q(x,t)$ (6.86) of the integral equation (6.85) belongs to the space $\mathcal{L}_2(0,1)$ and is continuous in x on the segment $[0,1]$. If $w(x) \in \mathcal{L}_2(0,1)$, then equation (6.89) has a unique solution in $\mathcal{L}_2(0,1)$ for fairly small ε which can be constructed using the method of successive approximations.

The contact stress depends on the two unknown constans a and c. To determine them, one can use condition (6.87) and equilibrium condition (6.74), which with the notation (6.77) and (6.88) can be rewritten in the form

$$\frac{P}{2\pi R^2 G} = \frac{\varepsilon}{b\pi^2} \int_0^1 p(t) \frac{(1-\varepsilon^2 t^2)t}{(1+\varepsilon^2 t^2)^{3/2}} dt + c \left[\frac{2}{1+\varepsilon^2} - \frac{\arctan \varepsilon}{\varepsilon} \right]. \tag{6.90}$$

We can replace condition (6.87) by the simple condition that equation (6.78) is satisfied at $x = 0$. After some reductions this gives

$$2a = c \left[\frac{b\pi^2}{\varepsilon} + \int_0^1 \frac{S(0,t)}{\sqrt{1-t^2}} dt \right] + \frac{1}{\pi} \int_0^1 \left[2 + \frac{\varepsilon}{b\pi} S(0,t) \right] p(t)\, dt. \tag{6.91}$$

For a smooth punch we have $c = 0$, and relations (6.90) and (6.91) serve for determination of the relative displacement $a = \delta/R$ of the punch and the quantity ε which characterizes the size of the contact region. If we know the size of the contact region a priori (for a non-smooth punch), relations (6.90) and (6.91) allow us to determine the quantities a and c.

We now construct a solution of the integral equation (6.89) using the asymptotic method for *large* $\lambda = \varepsilon^{-1}$. It can be shown that the function $S(x,t)$ (6.80) can be presented by a series which uniformly converges for $0 < \varepsilon < 1$, $0 \le x \le 1$, and $0 \le t \le 1$:

$$S(x,t) = t\left[\sum_{n=0}^{\infty} f_n(x,t)\varepsilon^n + \ln\varepsilon \sum_{n=0}^{\infty} g_{2n}(x,t)\varepsilon^{2n}\right], \tag{6.92}$$

where the functions $f_n(x,t)$ and $g_{2n}(x,t)$ are continuous and have integrable second derivatives. In particular, we have

$$f_0(x,t) = c_0 + \kappa_* \ln\max(x,t), \quad f_1(x,t) = \frac{c_1}{\pi}(x+t)\mathcal{E}\left(\frac{2\sqrt{xt}}{x+t}\right), \tag{6.93}$$

$$f_2(x,t) = c_2(x^2+t^2)\ln\max(x,t) + \left(c_2 + \frac{\kappa_*}{2}\right)\min(x^2,t^2) \tag{6.94}$$

$$+c_3(x^2+t^2), \quad g_0(x,t) = \kappa_*, \quad g_2(x,t) = c_2(x^2+t^2), \tag{6.95}$$

where $\mathcal{E}(k)$ is the complete elliptic integral of the second kind,

$$c_0 = -1 + 2\nu - \frac{1}{2}\mathrm{Re}\int_0^1 \frac{A(1-y^{-\omega})}{1-y}dy, \tag{6.96}$$

$$c_1 = 2(1-\nu)(1-16\nu^2), \qquad \kappa_* = (1-2\nu)^2, \tag{6.97}$$

$$c_2 = -1 - 3\nu/2 - \nu(1-\nu)(3-16\nu^2), \tag{6.98}$$

$$c_3 = -c_0/2 - \frac{1}{4}(5 + 3\nu - 34\nu^2 + 32\nu^4) \tag{6.99}$$

$$+\mathrm{Re}\int_0^1 A[1 - y^{-\omega} + \omega(1-y) + \frac{1}{2}\omega(1+\omega)(1-y)^2](1-y)^{-3}\,dy.$$

The values of coefficients c_m are given in columns 2–5 of Table 6.2 for various values of Poisson's ratio.

Table 6.2: The factors (6.96)–(6.99), (6.121) and (6.122)

ν	c_0	c_1	c_2	c_3	c_0^*	c_3^*
0	−1.567	2.000	−1.000	−1.484	−1.775	−0.378
0.1	−1.642	1.512	−1.406	−1.313	−1.855	−0.187
0.2	−1.557	0.576	−1.678	−0.660	−1.774	0.491
0.3	−1.328	−0.616	−1.778	0.040	−1.550	1.224
0.4	−0.977	−1.872	−1.706	0.634	−1.203	1.860
0.5	−0.532	−3.000	−1.500	0.963	−0.758	2.243

Taking expansion (6.92) into account, we can write the kernel $Q(x,t)$ (6.86) of the integral equation (6.89) in the form

$$Q(x,t) = \sum_{n=0}^{\infty} F_n(x,t)\varepsilon^n + \ln\varepsilon \sum_{n=1}^{\infty} G_{2n}(x,t)\varepsilon^{2n}. \tag{6.100}$$

Here,

$$F_0(x,t) = \kappa_* L(x,t), \qquad G_2(x,t) = 4c_2 t\sqrt{1-x^2},$$

$$F_1(x,t) = \frac{c_1}{\pi}\int_0^1 \frac{t}{y+t}\mathcal{K}\left(\frac{2\sqrt{yt}}{y+t}\right) L(x,y)\,dy,$$

$$F_2(x,t) = t\Big[2(2c_2+2c_3+\kappa_*)\sqrt{1-x^2} - \kappa_* t\, L(x,t)$$

$$-2\kappa_* \int_0^1 L(x,y)\,\mathrm{sign}\,(y-t)\,dy + 4c_2 \int_0^1 L(x,y)\ln\max(y,t)\,dy\Big].$$

The function $L(x,t)$ is defined by (6.84). We expand the function $v(x)$ (6.79) into the power series

$$v(x) = a\left\{2 + \sum_{n=1}^{\infty}\left[(-1)^n \frac{(2n-1)!!\,(1+4n)}{n!\,2^{n-1}} - d_{2n}\right](\varepsilon x)^{2n}\right\} \tag{6.101}$$

and write the right-hand side of (6.89), taking expansions (6.100) and (6.101) into account, in the form

$$w(x) = w_0(x) + a\,w_1(x) + c\,w_2(x), \tag{6.102}$$

where

$$w_0(x) = \sqrt{1-x^2}\sum_{n=1}^{\infty} \frac{2n+1}{2n-1}\beta_n\,\mathrm{E}_{n-1}(x^2) d_{2n}\varepsilon^{2n}, \tag{6.103}$$

$$w_1(x) = \sqrt{1-x^2}\sum_{n=1}^{\infty}(-1)^{n+1}\frac{(2n-1)!!\,(1+4n)}{n!\,2^{n-1}}\beta_n \mathrm{E}_{n-1}(x^2)\varepsilon^{2n}, \tag{6.104}$$

$$w_2(x) = \sum_{n=0}^{\infty} M_n(x)\varepsilon^n + \ln\varepsilon \sum_{n=1}^{\infty} N_{2n}(x)\varepsilon^{2n}. \tag{6.105}$$

The $\mathrm{E}_n(x) = \mathcal{F}(-n, 1, 1/2 - n; x)$ is the hypergeometric function, $\beta_n = \sqrt{\pi} n \Gamma(n+1)/\Gamma(n+3/2)$,

$$M_n(x) = \int_0^1 \frac{F_n(x,t)}{\sqrt{1-t^2}} dt, \quad N_{2n}(x) = \int_0^1 \frac{G_{2n}(x,t)}{\sqrt{1-t^2}} dt,$$

$$M_0(x) = \kappa_* J_1(x), \quad M_1(x) = \frac{1}{4}\pi c_1 \sqrt{1-x^2}, \quad N_2(x) = 4c_2\sqrt{1-x^2},$$

$$M_2(x) = 4(c_2+c_3)\sqrt{1-x^2} - \kappa_*[J_1(x) - 3J_3(x)] + 4c_2[J_2(x) - J_4(x)],$$

$$J_1(x) = \mathcal{T}\{(1-x^2)^{-1/2}\}, \quad J_2(x) = \mathcal{T}\{\ln(1+\sqrt{1-x^2})\},$$

$$J_3(x) = \mathcal{T}\{(1-x^2)^{1/2}\}, \quad J_4(x) = \frac{1}{3}\mathcal{T}\{(1-x^2)^{3/2}\}.$$

In the derivation of (6.103) and (6.104) we used the formula

$$\int_0^1 t^{2n-2} L(x,t)\, dt = \frac{2n+1}{(2n-1)(2n)^2}\beta_n \sqrt{1-x^2}\, \mathrm{E}_{n-1}(x^2).$$

Note that $J_m(x) \to 0$ as $x \to 0$.

Due to (6.102), we can seek the solution of (6.89) in the form

$$p(x) = p^0(x) + a\, p^1(x) + c\, p^2(x), \tag{6.106}$$

where $p^m(x)$ are solutions corresponding to $w_m(x)$, $m = 0, 1, 2$. We expand the functions $p^m(x)$ in the series

$$p^m(x) = e_m \left[\sum_{n=0}^{\infty} p_n^m(x)\varepsilon^n + \sum_{n=1}^{\infty}\sum_{k=0}^{\infty} p_{kn}^m(x)\varepsilon^{3n+k-1} \ln^n \varepsilon \right], \tag{6.107}$$

where $m = 0, 1, 2$, $e_0 = 4d_2$, $e_1 = 20$, $e_2 = 1$. We substitute series (6.100) into equation (6.89) for $p^m(x)$ (6.107) with the right-hand side $w_m(x)$, $m = 0, 1, 2$, (see (6.102)–(6.105)), and and equate the coefficients at the terms $\varepsilon^n \ln^m \varepsilon$. As a result, we determine the coefficients in series (6.107):

$$p_0^0(x) = p_1^0(x) = p_{0n}^1(x) = p_{1n}^1(x) = p_{2n}^1(x) = 0, \tag{6.108}$$

$$p_2^0(x) = \sqrt{1-x^2}, \quad p_3^0(x) = \frac{\kappa_*}{b\pi^2} J_3(x), \quad p_{31}^0(x) = \frac{4c_2}{3b\pi^2}\sqrt{1-x^2}, \tag{6.109}$$

$$p_4^0(x) = \left[\frac{8d_4}{9d_2}(1+2x^2) + \frac{c_1}{36b\pi}(4-x^2)\right]\sqrt{1-x^2} + \left(\frac{\kappa_*}{b\pi^2}\right)^2 \mathcal{T}\{J_3(x)\}, \tag{6.110}$$

$$p_5^0(x) = (3b\pi^2)^{-1}\{4(c_2+c_3)\sqrt{1-x^2} + 4c_2[J_2(x) - J_3(x) - J_4(x)] \quad (6.111)$$

$$-\kappa_*[(3 - 8d_4d_2^{-1})J_3(x) - (15 - 16d_4d_2^{-1})J_4(x)]\}$$

$$+\frac{\kappa_* c_1}{4b^2\pi^5}\mathcal{T}\Big\{\int_0^1 \frac{4t}{x+t}\mathcal{K}\left(\frac{2\sqrt{xt}}{x+t}\right) J_3(t)\,dt + \frac{\pi^2}{3}[J_3(x)+J_4(x)]\Big\}$$

$$+\kappa_*^3(b\pi^2)^{-3}\mathcal{T}\Big\{\int_0^1 J_3(t)\,L(x,t)\,dt\Big\}, \quad p_{01}^2(x) = 4c_2\sqrt{1-x^2},$$

$$p_0^2(x) = \kappa_* J_1(x), \quad p_1^2(x) = M_1(x) + \kappa_*^2(b\pi^2)^{-1}\mathcal{T}\{J_1(x)\}, \quad (6.112)$$

$$p_2^2(x) = M_2(x) + \frac{\kappa_* c_1}{b\pi}\left[\frac{1}{\pi^2}\mathcal{T}\Big\{\int_0^1 \frac{4t}{x+t}\mathcal{K}\left(\frac{2\sqrt{xt}}{x+t}\right) J_1(t)\,dt\Big\}\right. \quad (6.113)$$

$$\left.+J_3(x)\right] + \kappa_*^3(b\pi^2)^2\mathcal{T}\Big\{\int_0^1 J_1(t)\,L(x,t)\,dt\Big\}.$$

The functions p_n^1 and p_{kn}^1 are determined by the formulae for the corresponding functions $p_n^0(x)$ and $p_{kn}^0(x)$, in which we must put

$$d_{2n} = (-1)^{n+1}\frac{(2n-1)!!\,(1+4n)}{n!\,2^{n-1}}.$$

Finally, for the contact pressure we obtain the formula

$$\sigma[2\arctan(\varepsilon x)] = \frac{(1+\varepsilon^2x^2)^{3/2}G}{4\varepsilon^2}\Big\{\frac{c}{\sqrt{1-x^2}} + \frac{\varepsilon}{b\pi^2}[p^0(x) \quad (6.114)$$

$$+a\,p^1(x) + c\,p^2(x)]\Big\}.$$

To determine the quantities a, ε or c, as was said earlier, one should use relations (6.90) and (6.91). For a smooth punch we have $c = 0$ and, using formulae (6.92) and (6.107), we can rewrite relation (6.91) as follows

$$a = d_2\varepsilon^2\left\{1 + \frac{2\varepsilon}{3b\pi^2}\left[c_0 - \kappa_*\left(\frac{5}{6} - \ln(4\varepsilon)\right)\right]\right\} + O(\varepsilon^4). \quad (6.115)$$

For this case we determine the quantity ε from relation (6.90) which, due to (6.115), takes the form

$$P = \frac{8R^2Gd_2\varepsilon^3}{3b\pi}\Big\{1 + \frac{3\kappa_*\varepsilon}{4b\pi^2} + \varepsilon^2\Big[4 + \frac{c_1}{10b\pi} + \frac{8d_4}{5d_2} \quad (6.116)$$

$$+3\left(\frac{\kappa_*}{b\pi^2}\right)^2\int\limits_0^1 t\sqrt{1-t^2}J_3(t)\,dt\Bigg]$$

$$+\frac{\varepsilon^3}{3b\pi^2}\Bigg[10c_0+c_2(8\ln 2-3)+4c_3+\kappa_*\left(\frac{13}{12}+20\ln 2+\frac{11c_1}{24b\pi}+\frac{10d_4}{3d_2}\right)$$

$$+9\kappa_*^3(b\pi^2)^{-2}\int\limits_0^1 tJ_3^2(t)\,dt+(4c_2+10\kappa_*)\ln\varepsilon\Bigg]\Bigg\}+O(\varepsilon^7).$$

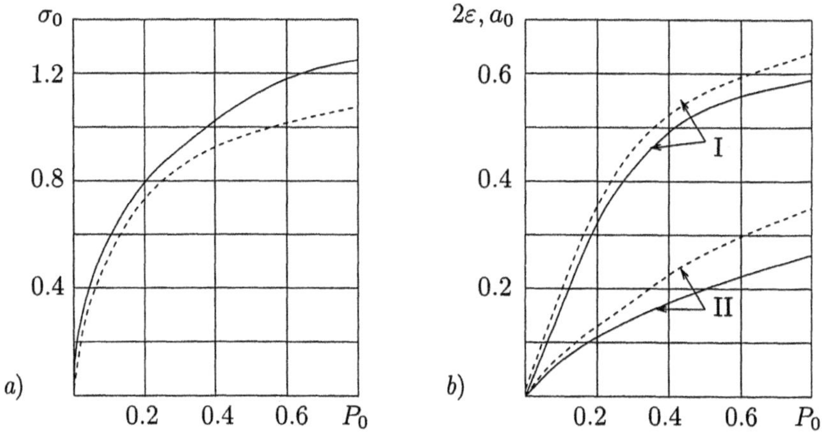

Figure 6.6: *a*) Contact stress σ_0; *b*) related size 2ε of the contact zone (lines I) and settlement a_0 (lines II). Dashed lines show the Hertz solution

Calculations were carried out for a spherical punch of radius $(1-\delta_*)R$. For this case the punch surface is defined by the equation

$$r = R(\delta_*\cos\eta+\sqrt{1-2\delta_*+\delta_*^2\cos^2\eta}).$$

Due to (6.79) and (6.101), we obtain

$$d_2=\frac{4\delta_*}{1-\delta_*},\quad d_4=-\frac{2(3-4\delta_*-\delta_*^2)\delta_*}{(1-\delta_*)^3}. \tag{6.117}$$

The plots of dependence between the contact stress $\sigma_0 = \sigma(0)(1-\delta_*)/(G\delta_*)$, related size 2ε of the contact zone, related settlement $a_0 = a(1-\delta_*)/\delta_*$ of the punch and the impressing force $P_0 = P(1-\delta_*)/(R^2G\delta_*)$ are shown in fig. 6.6*a*,*b*. Formulae (6.114)–(6.116) were used for $\nu = 0.3$ and $\delta_* \ll 1$. For comparison, the dashed lines show corresponding dependences calculated by using the well-known Hertz solution (see, e.g. appendix A3 in Johnson (1985)). The Hertz solution governs an elastic half-space. For a given force P_0 we can see from the figures that for a space with a cavity the contact pressure σ_0 is greater and the related size 2ε of the contact zone and the related settlement a_0 of the punch are smaller than for a half-space. This is to be expected because our elastic solid is larger and more massive.

6.4.2. TWO IDENTICAL PUNCHES

We now investigate the axisymmetric frictionless problem for two identical rigid punches with shapes of bases defined by the equation $r = R[1-\rho(\eta)]$, where $\rho(0) = 0$ and $\rho(\pi-\eta) = \rho(\eta)$. The punches are symmetrically impressed by force P into an elastic space $r \geq R$ with a spherical cavity (Fig. 6.7). For the unknown contact pressure $\sigma(\eta) = -\sigma_r(R,\eta)$, $0 \leq \eta \leq \gamma$, $\pi-\gamma \leq \eta \leq \pi$, we have, instead of (6.75), the following integral equation on two intervals:

$$\left(\int_0^{\gamma} + \int_{\pi-\gamma}^{\pi}\right)\sigma(\alpha)\, H(\eta,\alpha)\sin\alpha\, d\alpha = 2\pi G[a|\cos\eta| - \rho(\eta)], \qquad (6.118)$$

where $0 \leq \eta \leq \gamma$ and $\pi-\gamma \leq \eta \leq \pi$. Changing the variable in the second integral in (6.118) by formula $\alpha_1 = \pi - \alpha$ and using the symmetry relation $\sigma(\alpha) = \sigma(\pi-\alpha)$, we obtain

$$\int_0^{\gamma} \sigma(\alpha)[H(\eta,\alpha) + H(\eta,\pi-\alpha)]\sin\alpha\, d\alpha = 2\pi G[a|\cos\eta| - \rho(\eta)], \quad (6.119)$$

$$0 \leq \eta \leq \gamma, \qquad \pi-\gamma \leq \eta \leq \pi.$$

Due to relations $\rho(\pi-\eta) = \rho(\eta)$, $H(\eta,\pi-\alpha) = H(\pi-\eta,\alpha)$ and $H(\pi-\eta,\pi-\alpha) = H(\eta,\alpha)$, it is sufficient to consider equation (6.119) for $0 \leq \eta \leq \gamma$. The equation in notation (6.77) takes the form (cf. (6.78))

$$\int_0^1 q(t)\left[\frac{4t}{x+t}\mathcal{K}\left(\frac{2\sqrt{xt}}{x+t}\right) + \frac{\varepsilon}{b}S_1(x,t)\right]dt = \frac{\varepsilon}{b}v(x),$$

where $0 \leq x \leq 1$,

$$S_1(x,t) = S(x,t) + \frac{4bt}{1+\varepsilon^2 xt}\mathcal{K}\left(\frac{2\varepsilon\sqrt{xt}}{1+\varepsilon^2 xt}\right) \qquad (6.120)$$

$$+\frac{t}{\pi\sqrt{(1+\varepsilon^2x^2)(1+\varepsilon^2t^2)}}\,\mathrm{Re}\int_0^1\left[\frac{A}{y^{1+\omega}}-y\right]$$

$$\times U[y, 2\arctan(\varepsilon x), \pi - 2\arctan(\varepsilon t)]\,dy.$$

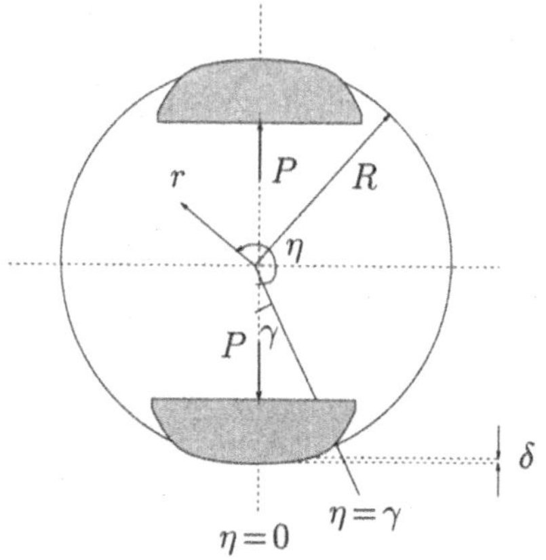

Figure 6.7: Contact between two punches and a space with spherical cavity (an axial section)

Using expansion (6.92) and an expansion of the complete elliptic integral of the first kind, we can expand the function $S_1(x,t)$ (6.120) like $S(x)$ (6.92) and also prove that the expansion converges uniformly for $0 < \varepsilon < 1$, $0 \le x \le 1$ and $0 \le t \le 1$. The few first terms of this expansion are determined by formulae (6.93)–(6.95), in which one must replace the constants c_m, $m = 0, 3$, by the constants

$$c_0^* = c_0 - 1 + \nu + \frac{1}{2}\ln 2 - \frac{1}{2}\,\mathrm{Re}\int_0^1 \frac{A\,dy}{(1+y)y^{\omega}}, \tag{6.121}$$

$$c_3^* = c_3 + \frac{21}{8} - \nu - \frac{5}{4}\ln 2 + \frac{1}{4}\,\mathrm{Re}\int_0^1 \frac{A\,dy}{(1+y)y^{\omega}} + \mathrm{Re}\int_0^1 \frac{A\,dy}{(1+y)^3 y^{\omega}}. \tag{6.122}$$

The values of the coefficients c_m^*, $m = 0, 3$, are given in the last two columns of Table 6.2.

Comparing equations (6.119) and (6.78), we conclude that the solution of (6.119) is determined by formulae (6.88) and (6.106)–(6.113), where we must replace the coefficients c_m, $m = 0, 3$, by c_m^* (6.121), (6.122). To find the relative displacement a of the punch and the quantities ε or c, we use relations (6.90) and (6.91), where we must replace $S_(0,t)$ by $S_1(0,t)$. For a smooth punch we use formulae (6.115) and (6.116) to determine a and ε, replacing c_m by c_m^*, $m = 0, 3$.

For two identical spherical punches of radius $(1 - \delta_*)R$ the surfaces of the punches are defined by the equation

$$r = R(\delta_* |\cos\eta| + \sqrt{1 - 2\delta_* + \delta_*^2 \cos^2\eta}),$$

and formulae (6.117) hold. A numerical analysis shows that, for $\gamma < \pi/6$, the displacement of the punches and the maximal contact pressure differ less than 5% from the corresponding quantities for one punch, considered in the previous subsection. When the contact angle γ decreases, the difference between the indicated quantities also diminishes. This is because the contact zones are far apart for $\gamma < \pi/6$.

6.5 Sphere, Spherical Hinge

6.5.1. A SPHERE

Using spherical coordinates r, η, φ, we consider the axisymmetric contact between an elastic sphere $r \le R$ and two identical rigid punches, the surfaces of which are defined by the equation $r = R[1 + \rho(\eta)]$, where $\rho(0) = 0$ and $\rho(\pi - \eta) = \rho(\eta)$ (Fig. 6.10). Equal forces P act onto the punches; it is necessary to determine the contact pressure $\sigma(\eta)$ $(0 \le \eta \le \gamma)$, the total settlement 2δ of the punches, and, for smooth punches only, the angle γ which characterizes the size of the contact region. Due to the symmetry, we have $\sigma(\pi - \eta) = \sigma(\eta)$.

The boundary conditions on the sphere have the form: $\sigma_r(R, \eta) = -\sigma(\eta)$, $0 \le \eta \le \gamma$, $\pi - \gamma \le \eta \le \pi$; $\sigma_r(R, \eta) = 0$, $\gamma < \eta < \pi - \gamma$; $\tau_{r\eta}(R, \eta) = 0$, $0 \le \eta \le \pi$; $u_r(R, \eta) = -\delta|\cos\eta| + R\rho(\eta)$, $0 \le \eta \le \gamma$, $\pi - \gamma \le \eta \le \pi$. Bondareva (1969) obtained the analytical solution for the problem of normal loads acting onto a shpere. Using her formula for the normal displacement of the sphere, we write the integral equation for the contact pressure in the form (Karpenko (1976a))

$$\int_0^\gamma \sigma(\alpha)[H_0(\eta, \alpha) + H_0(\eta, \pi - \alpha)] \sin\alpha \, d\alpha = 2\pi G[a \cos\eta - \rho(\eta)], \quad (6.123)$$

where $a = \delta/R$, $0 \leq \eta \leq \gamma$,

$$H_0(\eta,\alpha) = \frac{\pi(1-2\nu)}{2(1+\nu)} + 4(1-\nu)V(1,\eta,\alpha) \tag{6.124}$$

$$-\mathrm{Re}\int_0^1 (\bar{A}y^{\bar{\omega}} - y^{-2})V(y,\eta,\alpha)\,dy,$$

$$V(y,\eta,\alpha) = h_y^{-1}\,\mathcal{K}(k_y) - \frac{1}{2}\pi(1 + y\cos\eta\,\cos\alpha).$$

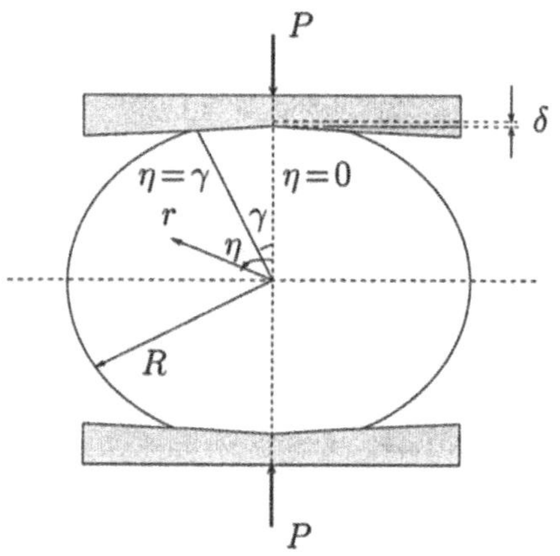

Figure 6.8: Contact between two punches and a sphere (an axial section)

The notation in (6.123), (6.124) is the same as in section 6.4; the upper bar denotes complex conjugation. For variables (6.77) equation (6.123) takes the form

$$\int_0^1 q(t)\left[\frac{4t}{x+t}\mathcal{K}\left(\frac{2\sqrt{xt}}{x+t}\right) + \frac{\varepsilon}{b}S_0(x,t)\right]dt = \frac{\varepsilon}{b}v(x), \tag{6.125}$$

where $0 \leq x \leq 1$,

$$S_0(x,t) = \frac{4bt}{1+\varepsilon^2 xt}\mathcal{K}\left(\frac{2\varepsilon\sqrt{xt}}{1+\varepsilon^2 xt}\right) \tag{6.126}$$

$$+\frac{t}{\sqrt{(1+\varepsilon^2x^2)(1+\varepsilon^2t^2)}}\left\{\frac{-3-2\nu+4\nu^2}{\pi(1+\nu)}\,\mathrm{Re}\int_0^1 (\bar{A}y^{\bar{\omega}} - y^{-2})\right.$$

$$\times[V(y, 2\arctan(\varepsilon x), 2\arctan(\varepsilon t)) + V(y, 2\arctan(\varepsilon x), \pi - 2\arctan(\varepsilon t))]\, dy\Big\}.$$

The function $S_0(x,t)$ can be expanded into the series (6.92)–(6.95), where we must replace the coefficients κ_*, c_m by $-\kappa_*$, c_m^0, respectively, where

$$c_1^0 = c_1, \quad c_2^0 = -c_2, \tag{6.127}$$

$$c_0^0 = -2 + \frac{3\nu^2}{1+\nu} + \frac{1}{2}\ln 2 - \frac{1}{2}\operatorname{Re}\int_0^1 \frac{\bar{A}y^{2+\bar{\omega}}}{1+y}\,dy \tag{6.128}$$

$$+\frac{1}{2}\operatorname{Re}\int_0^1 \bar{A}(1-y^{2+\bar{\omega}})(1-y)^{-1}\,dy,$$

$$c_3^0 = -\frac{1}{2}c_0^0 + \frac{1}{8} - \ln 2 - \nu - \nu(1-\nu)[13 - 16\nu(1+\nu)] \tag{6.129}$$

$$+\operatorname{Re}\int_0^1 \bar{A}y^{2+\bar{\omega}}\frac{3+3y+y^2}{(1+y)^3}\,dy + \operatorname{Re}\int_0^1 \bar{A}\frac{3-3y+y^2}{(1-y)^3}$$

$$\times[y^{2+\bar{\omega}} - 1 + (2+\bar{\omega})(1-y) - \frac{1}{2}(2+\bar{\omega})(1+\bar{\omega})(1-y)^2]\,dy.$$

The values of the coefficients c_m^0 $(m = 0, 3)$ are given in Table 6.3 for various values of Poisson's ratio.

Table 6.3: The factors (6.128) and (6.129)

ν	0	0.1	0.2	0.3	0.4	0.5
c_0^0	−2.931	−2.389	−1.850	−1.350	−0.922	−0.589
c_3^0	−0.255	−0.552	−0.515	−0.292	−0.016	−0.211

Comparing equation (6.125) with (6.78), we conclude that these two have solutions of the same structure, with the replacement of κ_*, c_m by $-\kappa_*$, c_m^0. Finally, the solution of (6.125) is determined by formulae (6.106)–(6.116), where we must do the indicated replacement.

As an example, we consider two spherical punches of radius $(1+\delta_*)R$. The contact surfaces of the punches are defined by the equations

$$r = R(\sqrt{1 + 2\delta_* + \delta_*^2\cos^2\eta} - \delta_*|\cos\eta|).$$

The dependences between the contact stress $\sigma^0 = \sigma(0)(1-\delta_*)/(G\delta_*)$, the related size 2ε of the contact domain, the related settlement $a^0 = a(1-\delta_*)/\delta_*$ of the punch and the indenting force $P^0 = P(1-\delta_*)/(R^2G\delta_*)$ are shown in

Fig. 6.9*a*,*b* for $\nu = 0.3$ and $\delta_* \ll 1$. The dashed lines show the corresponding dependences calculated by the Hertz solution. A comparison shows that the Hertz solution has a substantial error for large contact domains, which take place for large indenting forces. For a specified force P, the contact stress is greater and the contact area and settlement of the punch are smaller for our solution than for the Hertz one.

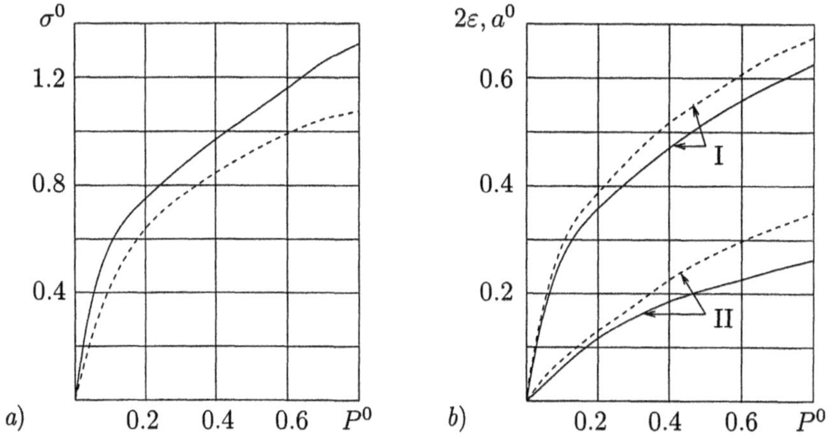

Figure 6.9: *a*) Contact stress σ^0; *b*) related size 2ε of the contact zone (lines I) and settlement a^0 (lines II). Dashed lines show the Hertz solution

6.5.2. A SPHERICAL HINGE

We now consider the axisymmetric problem of impression of an elastic sphere of radius R_0 into an elastic space with a spherical cavity of nearly equal radius $R = (1 + \delta_*)R_0$, $\delta_* \ll 1$ (Fig. 6.10). Suppose the space is fixed at infinity. Such a construction may be called a spherical hinge (Shtaerman (1949)). Let the elastic characteristics of the space be ν, G and of the sphere ν_0, G_0. We place the center of spherical coordinate system (r, η, φ) at the center of the punch, so that the line $\eta = 0$ coincides with the axis of symmetry of the given external load $Q(\eta)$, $\gamma < \eta \leq \pi$.

The relation $u_r^0 + u_r = \delta_0 \cos\eta - (1 - \cos\eta)R_0\delta_*$ holds in the contact region $0 \leq \eta \leq \gamma$, as for the plane analog of the problem (Shtaerman (1949)), where u_r^0 and u_r are the normal displacements of the sphere and the spherical cavity, respectively, and δ_0 is the reciprocal settling of the solids. Substituting the expressions for the displacements in the last relation, we get the following integral equation for the contact pressure:

$$\int_0^\gamma \sigma(\alpha)\left[\frac{RG_0}{R_0G}H(\eta,\alpha)+H_0(\eta,\alpha)\right]\sin\alpha\,d\alpha = 2\pi G_0[a\cos\eta-\rho(\eta)], \quad (6.130)$$

where $a = \delta_0/R_0$, $0 \le \eta \le \gamma$ and

$$\rho(\eta) = (1-\cos\eta)\delta_* + \frac{1}{2\pi G}\int_\gamma^\pi Q(\alpha)H_0(\eta,\alpha)\sin\alpha\,d\alpha. \quad (6.131)$$

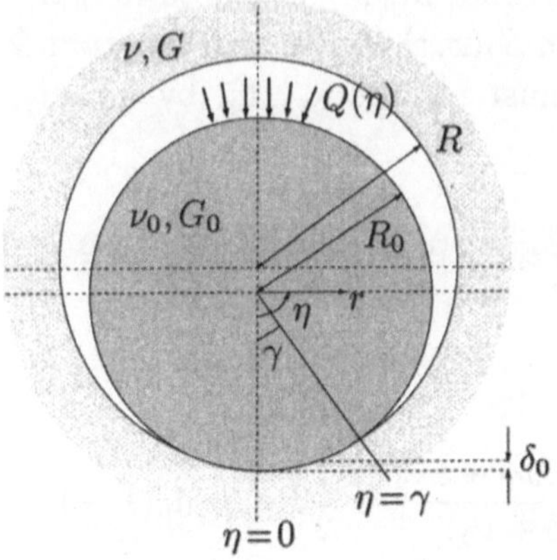

Figure 6.10: A spherical hinge (an axial section)

The functions $H(\eta,\alpha)$ and $H_0(\eta,\alpha)$ are defined by formulae (6.76) and (6.124). With the notation (6.77), equation (6.130) takes the form

$$\int_0^1 q_0(t)\left[\frac{4t}{x+t}\mathcal{K}\left(\frac{2\sqrt{xt}}{x+t}\right)+\frac{\varepsilon}{b_0}S_2(x,t)\right]dt = \frac{\varepsilon}{b_0}v(x), \quad (6.132)$$

where $0 \le x \le 1$,

$$q_0(t) = \frac{4\varepsilon^2\sigma[2\arctan(\varepsilon t)]}{(1+\varepsilon^2t^2)^{3/2}G_0}, \quad b_0 = \frac{1-\nu_0}{2\pi}+\frac{RG_0}{R_0G}\frac{1-\nu}{2\pi}, \quad (6.133)$$

$$S_2(x,t) = \frac{RG_0}{R_0G}S(x,t)+\frac{t}{\sqrt{(1+\varepsilon^2x^2t)(1+\varepsilon^2t^2)}} \quad (6.134)$$

$$\times\left[\frac{-3-2\nu_0+4\nu_0^2}{2(1+\nu_0)}-2(1-\nu_0)\frac{(1-\varepsilon^2x^2)(1-\varepsilon^2t^2)}{(1+\varepsilon^2x^2)(1+\varepsilon^2t^2)}\right.$$

$$-\frac{1}{\pi}\operatorname{Re}\int_0^1 (\bar{A}_0 y^{\bar{\omega}_0} - y^{-2})V(y, 2\arctan(\varepsilon x), 2\arctan(\varepsilon t))\,dy\Bigg],$$

$$\bar{A}_0 = -1 + 8\nu_0 - 8\nu_0^2 - i\frac{5 - 4\nu_0 - 16\nu_0^2 + 16\nu_0^3}{\sqrt{3 - 4\nu_0^2}},$$

$$\bar{\omega}_0 = -\frac{1}{2} + \nu_0 - i\frac{\sqrt{3 - 4\nu_0^2}}{2}.$$

The functions $v(x)$ and $S(x,t)$ are determined by formulae (6.79) and (6.80). The function $S_2(x,t)$ (6.134) can be expanded into the series (6.92)–(6.95), where we must replace κ_*, c_m by κ_0, c_m^1, respectively, using the formulae

$$c_m^1 = \frac{RG_0}{R_0G}c_m + \tilde{e}_m, \qquad \kappa_0 = \frac{RG_0}{R_0G}\kappa_* - (1 - 2\nu_0)^2, \tag{6.135}$$

where

$$\tilde{e}_0 = \frac{-7 - 2\nu_0 + 8\nu_0^2}{2(1 + \nu_0)} + \frac{1}{2}\operatorname{Re}\int_0^1 \bar{A}_0(1 - y^{2+\bar{\omega}_0})(1 - y)^{-1}\,dy,$$

$$\tilde{e}_1 = 2(1 - \nu_0)(1 - 16\nu_0^2), \quad \tilde{e}_2 = 1 + 3\nu_0/2 + \nu_0(1 - \nu_0)(3 - 16\nu_0^2),$$

$$\tilde{e}_3 = -\frac{1}{2}\tilde{e}_0 + 2 + (1 - \nu_0)(3 - 8\nu_0 + 16\nu_0^2) + \operatorname{Re}\int_0^1 \bar{A}_0(3 - 3y + y^2)$$

$$\times(1 - y)^{-3}[y^{2+\bar{\omega}_0} - 1 + (2 + \bar{\omega}_0)(1 - y) - \frac{1}{2}(2 + \bar{\omega}_0)(1 + \bar{\omega}_0)(1 - y)^2]\,dy.$$

We note that the solution of equation (6.132) can be constructed by formulae (6.106)–(6.116) for $c = 0$, where we must do the indicated replacement and also change b and $q(x)$ in formula (6.88), $c = 0$, to b_0 and $q_0(x)$, respectively, using formulae (6.133). The coefficients d_{2n} from formula (6.101) depend on the distribution law of the external load in accordance with formulae (6.79) and (6.131).

Consider the particular case when the external load $Q(\eta)$, $\gamma < \eta \leq \pi$, is distributed over the surface of the sphere by the same law as the required contact pressure, i.e., $Q(\eta) = \sigma(\pi - \eta)$ for $\pi - \gamma \leq \eta \leq \pi$ and $Q(\eta) = 0$ for

$\gamma < \eta < \pi - \gamma$. Note that the external load can be presented by the vector $\mathbf{P}$, $|\mathbf{P}| = P$. Then for the contact pressure we obtain the integral equation

$$\int_0^\gamma \sigma(\alpha)\left[\frac{RG_0}{R_0G}H(\eta,\alpha) + H_0(\eta,\alpha)\right. \tag{6.136}$$

$$\left. + H_0(\eta, \pi - \alpha)\right] \sin\alpha\, d\alpha = 2\pi G_0[a\cos\eta - \rho(\eta)],$$

where $\rho(\eta) = (1 - \cos\eta)\delta_*$ and $0 \le \eta \le \gamma$.

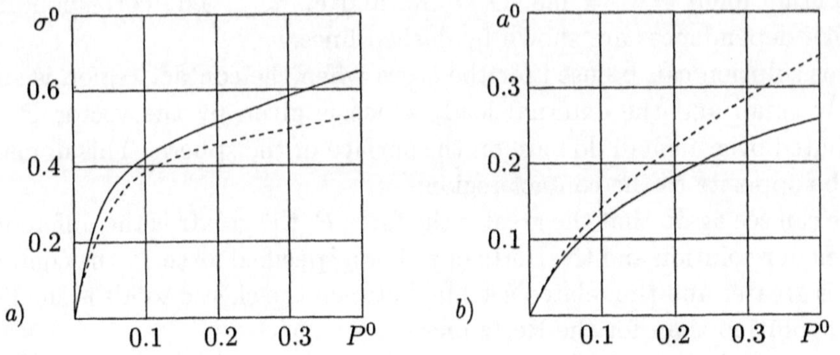

Figure 6.11: *a)* Contact stress σ^0; *b)* related settlement a^0. Dashed lines show the Hertz solution

With the notation (6.77), equation (6.136) takes the form

$$\int_0^1 q(t)\left[\frac{4t}{x+t}\mathcal{K}\left(\frac{2\sqrt{xt}}{x+t}\right) + \frac{\varepsilon}{b}S_3(x,t)\right] dt = \frac{\varepsilon}{b}v(x), \tag{6.137}$$

where

$$S_3(x,t) = \frac{RG_0}{R_0G}S(x,t) + S_0(x,t). \tag{6.138}$$

The functions $v(x)$, $S(x,t)$ and $S_0(x,t)$ are determined by formulae (6.79), (6.80) and (6.126). We expand the function $S_3(x,t)$ (6.138) into series (6.92)–(6.95), where we must replace κ_* and c_m by κ_0 $(6.135)_2$ and

$$\tilde{c}_m = RG_0(R_0G)^{-1}c_m + c_m^0,$$

respectively, where c_m and c_m^0 are determined by formulae (6.96)–(6.99) and (6.127)–(6.129). Comparing equations (6.137) with (6.78), we see that the solution of equation (6.137) can be presented by formulae (6.106)–(6.116) for $c = 0$, where we must do the indicated replacement and also change the function $q(x)$ in formula (6.88), for $c = 0$, to the function $q_0(x)$ $(6.133)_1$. The coefficients d_{2n} from formula (6.101) can be now written in explicit form

$$d_{2n} = (-1)^{n+1} \frac{4(2n-1)!!}{(2n-2)!!} \delta_*.$$

The dependences between the contact stress $\sigma^0 = \sigma(0)(1 - \delta_*)/(G\delta_*)$, the related settling $a^0 = a(1 - \delta_*)/\delta_*$, between the elastic sphere and the surface of the cavity, and the force $P^0 = P(1 - \delta_*)/(R^2 G\delta_*)$ are shown in Fig. 6.11*a*,*b* for $\nu = \nu_0 = 0.3$, $G = G_0$, and $\delta_* \ll 1$, The corresponding Hertzian dependences are shown by dashed lines.

This solution can be used for the cases when the contact region is sufficiently small and the external load, which is given by the vector $\mathbf{P}$, is distributed over a small domain on the surface of the sphere. This domain must be opposite to the contact region.

We can see again that the greater the force P, the greatr is the difference between our solution and the Hertz one. For a specified force P, the contact stress is greater and the related settling between the elastic solids is smaller for our solution than for the Hertz one.

We conclude that the investigation of the contact problems for solids with complicated geometry is improtant, because the solutions differ considerably from those for a half-space. Of course, the well-known finite element and boundary element methods are applicable to practically any elastic body, but it is often useful to have an approximate analytic solution. The methods presented in the book will enable to obtain such alternative solutions.

Appendix

GALANOV'S COMPUTER PROGRAM

Below we give Galanov's computer program (section 2.9), which is written in FORTRAN-77 and realizes Newton's method for the integral equation

$$\int_{\Omega} \frac{q(x,y)}{R}\, dxdy = \delta - (Ar^2 + Bz^2), \qquad R = \sqrt{(r-x)^2 + (z-y)^2}, \qquad (r,z) \in \Omega,$$

which corresponds to the contact between an elliptic paraboloid and an elastic half-space.

The file newton.rez contains the input parameters and the results: arrays $v(n)$ (this is the contact pressure for $v(n) > 0$ see Theorem 2.3), $d(n)$ (this is the normal displacement (2.258) outside the contact region, i.e., at the points where $v(n) < 0$), the coordinates x and y for the node n, in the contact rectangle, and also the force and moment, applied to the punch, and the normal displacement s_p at a specified point (x_p, y_p).

```
      PROGRAM Newton
      INTEGER*2 n,i,j,k,l,nx,ny,iter,n1,n2,n3,n4,i1,j1
      PARAMETER(nx=15,ny=15,n=nx*ny)
C nx, ny are numbers of nodes on x- and y-axes, respectively
      REAL*4 eps,delz,x,y,xi,eta,r1,r2,del,A,B
      REAL*4 br(n,n+1),r(n,n),e(n),xx(n),yy(n),ds(n),q1(n)
      REAL*4 vp(n),qv(n),v1(n),bb(n),p(n),v(n),d(n)
      REAL*4 Q,epsi,alfa,am,sigma,hx,hy,g
      REAL*4 rs,s,s1,s2,ss1,vv,Force,Momnt,f,xp,yp,sp
      PARAMETER(epsi=1.E-5,alfa=3.E-1,am=1.0,sigma=1.0)
      EXTERNAL Q
      COMMON/q1/del,A,B
```

```
C  Input parameters:
              eps=1.0
              del=1.0
              A=1.0
              B=1.0
              xp=1.0
              yp=0.0
C  eps≤ 1 is the ratio of dimensions of the contact rectangle
C  del, A and B determine the right-hand side of the integral equation
C  xp and yp are the coordinates of the point outside the contact zone,
C  where we want to check the normal displacement sp
              hx=2.0*eps/(nx-1)
              hy=1.0/(ny-1)
              delz=hx*hy*0.0675
              iter=0
C  Formation of the rectangular mesh with spaces hx and hy
              DO 1 i=1,ny
              DO 1 j=1,nx
              j1=j+(i-1)*nx
              xx(j1)=(j-1)*hx-eps
   1          yy(j1)=(i-1)*hy
C  Formation of the elementary areas ds(n)
              DO 4 i=1,n
   4          ds(i)=hx*hy
              ds(n)=hx*hy*0.25
              ds(n-nx+1)=hx*hy*0.25
              n1=nx*(ny-2)+1
              g=hx*hy*0.5
              DO 5 i=1,n1,nx
   5          ds(i)=g
              n2=nx*(ny-1)
              DO 6 i=nx,n2,nx
   6          ds(i)=g
              n3=n-nx+2
              n4=n-1
              DO 7 i=n3,n4
   7          ds(i)=g
C  Formation of the matrix of the operator K
              DO 14 l=1,ny
              y=hy*(l-1)
              DO 14 j=1,nx
              x=-eps+hx*(j-1)
              i1=j+(l-1)*nx
              DO 14 k=1,ny
              eta=hy*(k-1)
```

```
          DO 14 i=1,nx
          xi=-eps+hx*(i-1)
          j1=i+(k-1)*nx
          r1=SQRT((x-xi)*(x-xi)+(y-eta)*(y-eta)+delz)
          r2=SQRT((x-xi)*(x-xi)+(y+eta)*(y+eta)+delz)
          f=1.0/r1+1.0/r2
          IF(yy(j1).LE.1.E-7) GO TO 12
          rs=f
          GO TO 14
12        rs=f*0.5
14        r(i1,j1)=rs*ds(j1)
C Formation of the right-hand side of the equation
          DO 15 i=1,n
          q1(i)=Q(xx(i),yy(i))
15        vp(i)=q1(i)
35        DO 31 i=1,n
31        v1(i)=vp(i)*am
C Calculating the positive part of the function v
          DO 28 i=1,n
          IF(vp(i).GE.1.E-15) GO TO 29
          qv(i)=0.0
          GO TO 28
29        qv(i)=vp(i)
28        CONTINUE
C Calculating the residual
          DO 18 i=1,n
          s=0.0
          DO 19 j=1,n
          vv=0.0
          IF(i.EQ.j) vv=am
19        s=s+(vv-r(i,j))*qv(j)
18        bb(i)=s
          DO 20 i=1,n
20        p(i)=v1(i)-bb(i)-q1(i)
C Approximate derivative of the positive part of the function v
          DO 21 i=1,n
          IF(vp(i).GT.alfa) bb(i)=1.0
          IF(vp(i).GE.(-alfa).AND.vp(i).LE.alfa) bb(i)=(1.+vp(i)/alfa)/2.0
          IF(vp(i).LT.(-alfa)) bb(i)=0.0
21        CONTINUE
C Formation of Krasnoselskii's matrix
55        DO 23 i=1,n
          DO 23 j=1,n
23        br(i,j)=r(i,j)*bb(j)
          DO 24 i=1,n
```

```
 24         bb(i)=am-bb(i)*am
            DO 25 i=1,n
            br(i,i)=bb(i)+br(i,i)
 25         br(i,n+1)=-p(i)
C  Next iteration
            n4=n+1
            CALL GAUSS(n,n4,br,e)
            DO 36 i=1,n
 36         v(i)=vp(i)+sigma*e(i)
C  Check of the criterion to stop iterations
            s=0.0
            s1=0.0
            DO 33 i=1,n
            s=s+ABS(v(i)-vp(i))
 33         s1=s1+ABS(vp(i))
            iter=iter+1
            WRITE(*,*) iter
            ss1=s/s1
            IF(ss1.LE.epsi) GO TO 34
 50         DO 37 i=1,n
 37         vp(i)=v(i)
            GO TO 35
 34         OPEN(UNIT=2,FILE='newton.rez',STATUS='unknown',
     *      ACCESS='sequential',FORM='formatted')
            DO 43 i=1,n
            IF (v(i).GE.0.0) GO TO 43
            d(i)=-(v(i)-q1(i))
 43         CONTINUE
            WRITE(2,81) nx,ny
 81         FORMAT(2x,'nx=',i2,',',2x,'ny=',i2)
            WRITE(2,91) eps,del,A,B
 91         FORMAT(2x,'eps=',E9.3,',',2x,'del=',E9.3,',',
     *      2x,'A=',E9.3,',',2x,'B=',E9.3)
            DO 30 i=1,n
 30         WRITE(2,22) i,v(i),i,d(i),xx(i),yy(i)
 22         FORMAT(2x,'v(',i3,')=',E9.3,',',2x,'d(',i3,')=',E9.3,',',
     *      2x,'x=',E9.3,',',2x,'y=',E9.3)
C  Calculating the normal force "Force" and moment "Momnt"
            Force=0.0
            Momnt=0.0
            DO 38 i=1,n
            IF(v(i)) 38,39,39
 39         IF(yy(i).LE.1.E-7) GO TO 41
            s1=v(i)*2.0*ds(i)
            s2=xx(i)*v(i)*2.0*ds(i)
```

```
         GO TO 42
41       s1=v(i)*ds(i)
         s2=xx(i)*v(i)*ds(i)
42       Force=Force+s1
         Momnt=Momnt+s2
38       CONTINUE
         WRITE(2,69) Force,Momnt,iter
C iter is number of iterations
69       FORMAT(2x,'Force=',E9.3,',',2x,'Momnt=',E9.3,',',2x,'iter=',i3)
C Normal displacement at the point (xp,yp) outside the contact zone
         sp=0.0
         DO 64 k=1,ny
         eta=hy*(k-1)
         DO 64 i=1,nx
         xi=-eps+hx*(i-1)
         j1=i+(k-1)*nx
         r1=SQRT((xp-xi)*(xp-xi)+(yp-eta)*(yp-eta)+delz)
         r2=SQRT((xp-xi)*(xp-xi)+(yp+eta)*(yp+eta)+delz)
         f=1.0/r1+1.0/r2
         IF(yy(j1).LE.1.E-7) GO TO 61
         rs=f
         GO TO 64
61       rs=f*0.5
64       sp=sp+rs*ds(j1)*MAX(v(j1),0.0)
         WRITE(2,51) sp,xp,yp
51       FORMAT(2x,'sp=',E9.3,',',2x,'xp=',E9.3,',',2x,'yp=',E9.3)
         CLOSE(UNIT=2)
         STOP
         END

C The function Q(r,z) is the right-hand side of the integral equation
         REAL*4 FUNCTION Q(x,y)
         REAL*4 x,y,A,B,del
         COMMON/q1/del,A,B
         Q=del-A*x*x-B*y*y
         RETURN
         END
```

The subroutine GAUSS solves a system of linear equations by Gaussian method. Input parameters: l denotes the number of equations, l1 should be l+1, a1(l,l1) is a system matrix including a column of free terms. Output parameters: z(l) is a solution vector.

```
         SUBROUTINE GAUSS(l,l1,a1,z)
         INTEGER*2 l,l1,n,j,i,k,m,j1,j2,i1,i2,i3,k1
```

```
      REAL*4 a1(l,l1),z(l),temp
      n=0
      j=l+1
1     n=n+1
      DO 2 k=n,l
      IF (a1(k,n)) 3,2,3
2     CONTINUE
      WRITE (6,4)
4     FORMAT (10HNOT SYSTEM)
      STOP
3     IF (k-n) 6,7,6
6     DO 8 m=n,j
      temp=a1(n,m)
      a1(n,m)=a1(k,m)
8     a1(k,m)=temp
7     DO 9 j1=n,j
      j2=j+n-j1
9     a1(n,j2)=a1(n,j2)/a1(n,n)
      IF (n.EQ.l) GO TO 11
      i1=k+1
      i2=n+1
      DO 10 i=i1,l
      DO 10 j1=i2,j
10    a1(i,j1)=a1(i,j1)-a1(i,n)*a1(n,j1)
      GO TO 1
11    DO 12 i1=1,l
      i=l+1-i1
      z(i)=a1(i,j)
      i3=i-1
      IF (i3.EQ.0) GO TO 13
      DO 12 k1=1,i3
      k=i3-k1+1
12    a1(k,j)=a1(k,j)-a1(k,i)*z(i)
13    RETURN
      END
```

Abbreviations for Journals

AAIM. Akademiia Nauk Armianskoi S.S.R. Izvestiia. Mekhanika.

AAIF. Akademiia Nauk Armianskoi S.S.R. Izvestiia. Seriia Fiziko-Matematicheskikh Nauk.

DP. Akademiia Nauk S.S.S.R. Doklady. (till 1992).
Rossiiskaia Akademiia Nauk. Doklady. (since 1993).
[Soviet Physics Doklady]. (till 1992). [Doklady Physics]. (since 1993).

AIM. Akademiia Nauk S.S.S.R. Izvestiia. Mekhanika.

AIMM. Akademiia Nauk S.S.S.R. Izvestiia. Mekhanika i Mashinostroenie.

AIO. Akademiia Nauk S.S.S.R. Izvestiia. Otdelenie Tekhnicheskikh Nauk.

AM. Archiwum Mechaniki Stosowanej. (till 1970). Archives of Mechanics. (since 1971).

SMS. Fiziko-Khimicheskaia Mekhanika Materialov. (till 1991).
Fiziko-Khimichna Mekhanika Materialiv. (since 1992).
[Soviet Materials Science]. (till 1992). [Materials Science]. (since 1993).

IJES. International Journal of Engineering Sceince.

IJSS. International Journal of Solids and Structures.

MS. Inzhenernyi Zhurnal. Mekhanika Tverdogo Tela. (till 1968).
Akademiia Nauk S.S.S.R. Izvestiia. Mekhanika Tverdogo Tela. (from 1969 till 1992).
Rossiiskaia Akademiia Nauk. Izvestiia. Mekhanika Tverdogo Tela. (since 1993).
[Mechanics of Solids].

JAM. Journal of Applied Mechanics.

JBIO. Journal of Biomechanics.

JE. Journal of Elasticity.

MA. Mashinovedenie.

JAMM. Prikladnaia Matematika i Mekhanika.
[Journal of Applied Mathematics and Mechanics].

IAM. Prikladnaia Mekhanika.
[Soviet Applied Mechanics]. (till 1991).
[International Applied Mechanics]. (since 1992).

QJMM. Quarterly Journal of Mechanics and Applied Mathematics.

SJAM. SIAM Journal of Applied Mathemathics.

TMMS. Trudy Moskovskogo Matematicheskogo Obschestva.
[Transactions of Moscow Mathematical Society].

RER. Vestnik Mashinostroeniia.
[Russian Engineering Research].

MUMB. Vestnik Moskovskogo Universiteta. Seriia *I*.
[Moscow University Mechanics Bulletin].

References

Abramowitz, M. and Stegun, I.,(1972), *Handbook of Mathematical Functions.* 2nd ed., New York: Dover Publications.

Aleksandrov, V.M. and Vorovich, I.I.,(1960), The action of a die on an elastic layer of finite thickness. *JAMM*, **24**, No.2, 323-333, [462-476].

Aleksandrov, V.M.,(1962), Axisymmetric contact problem for an elastic infinite cylinder. (in Russian). *AIMM*, No.5, 91-94.

Aleksandrov, V.M.,(1963), Some contact problems for the elastic layer. *JAMM*, **27**, No.4, 758-764, [1164-1174].

Aleksandrov, V.M.,(1964), The solution of a type of two-dimensional integral equation. *JAMM*, **28**, No.3, 579-581, [714-717].

Aleksandrov, V.M. and Babeshko, V.A.,(1965), Contact problems for an elastic strip of slight thickness. (In Russian). *AIM*, No.2, 95-107.

Aleksandrov, V.M., Babeshko, V.A. and Kucherov, V.A.,(1966), Contact problems for an elastic layer of slight thickness. *JAMM*, **30**, No.1, 124-142, [148-172].

Aleksandrov, V.M.,(1967a), Axisymmetric problem of the effect of an annular die on an elastic half-space. *MS*, **2**, No.4, 108-116.

Aleksandrov, V.M.,(1967b), Certain contact problem for an elastic wedge. (in Russian). *AAIF*, **20**, No.1, 3-14.

Aleksandrov, V.M.,(1967c), On the approximate solution of some integral equations of the theory of elasticity and mathematical physics. *JAMM*, **31**, No.6, 1117-1131, [1122-1136].

Aleksandrov, V.M. and Belokon', A.V.,(1967), Asymptotic solution of a class of integral equations and its application to contact problems for cylindrical elastic bodies. *JAMM*, **31**, No.4, 704-710, [718-724].

Aleksandrov, V.M.,(1968), Asymptotic methods in contact problems of elasticity theory. *JAMM*, **32**, No.4, 672-683, [691-703].

Aleksandrov, V.M. and Belokon', A.V.,(1968), Asymptotic solution of a class of integral equations encountered in the investigation of mixed problems of the mathematical physics for regions with cylindrical boundaries. *JAMM*, **32**, No.3, 401-413, [402-413].

Aleksandrov, V.M. and Kucherov, V.A.,(1968), Certain problems on action of two punches onto an elastic strip. *MS*, **3**, No.4, 110-123.

Aleksandrov, V.M.,(1969), Asymptotic solution of the contact problem for a thin elastic layer. *JAMM*, **33**, No.1, 61-73, [49-63].

Aleksandrov, V.M. and Kucherov, V.A.,(1970), On the method of orthogonal polynomials in plane mixed problems of elasticity theory. *JAMM*, **34**, No.4, 643-652, [612-621].

Aleksandrov, V.M. and Babeshko, V.A.,(1972) On the pressure on an elastic half-space by a wedge-shaped stamp. *JAMM*, **36**, No.1, 88-93, [78-83].

Aleksandrov, V.M. and Chebakov, M.I. ,(1972), Mixed problems of the mechanics of continuous media associated with Hankel and Mehler–Fock integral transforms. *JAMM*, **36**, No.3, 494-504, [465-474].

Aleksandrov, V.M.,(1973), Solution of some integral and series equation pairs. *DP*, **210**, No.1, 55-58, [**18**, 351-353].

Aleksandrov, V.M. and Chebakov, M.I. ,(1973), On a method of solving dual integral equations. *JAMM*, **37**, No.6, 1087-1097, [1031-1041].

Aleksandrov, V.M.,(1975), On a method of reducing dual integral equations and dual series equations to infinite algebraic systems. *JAMM*, **39**, No.2, 324-332, [303-311].

Aleksandrov, V.M. and Chebakov, M.I. ,(1977), The method of dual series in terms of Bessel functions in mixed problems of the theory of elasticity for a circular plate. *JAMM*, **41**, No.3, 486-492, [492-499].

Aleksandrov, V.M. and Kovalenko, E.V.,(1977a), Method of orthogonal functions in mixed problems of continuum mechanics. *IAM*, **13**, No.12, 9-17, [1186-1192].

Aleksandrov, V.M. and Kovalenko, E.V.,(1977b), Periodic contact problems for an elastic strip. (in Russian). *AAIM*, **30**, No.4, 18-33.

Aleksandrov, V.M. and Karpenko, V.A. ,(1980), Torsion of a spherical layer by a spherical annular punch. *JAMM*, **44**, No.1, 143-150, [98-103].

Aleksandrov, V.M. and Sumbatyan, M.A.,(1980), Asymptotic solution of the integral equations of convolution type with the logarithmic singularity of transform of the kernel and application into problems of mechanics. *MS*, **15**, No.2, 80-88.

Aleksandrov, V.M., Kovalenko, E.V. and Mkhitarian, S.M. ,(1982), On a method of obtaining spectral relationships for integral operators of mixed problems of mechanics of continuous media. *JAMM*, **46**, No.6, 1023-1031, [825-832].

Aleksandrov, V.M., Grishin, S.A. and Kovalenko, E.V.,(1985), Contact interaction between a thick plate and a thick elastic layer. *MS*, **20**, No.5, 64-69, [62-67].

Aleksandrov, V.M.,(1986), Contact problem taking into account the wear caused by local melting. *SMS*, **22**, No.1, 116-124, [112-119].

Aleksandrov, V.M. and Pozharskii, D.A.,(1989), Action of a strip-shaped stamp on an elastic incompressible three-dimensional wedge. *IAM*, **25**, No.8, 19-26, [749-756].

Aleksandrov, V.M. and Pozharskii, D.A.,(1992), Action of a band stamp on a three-dimensional elastic wedge. *IAM*, **28**, No.1, 56-62, [49-54].

Aleksandrov, V.M.,(1993), Asymptotic methods in the mechanics of continuous media: problems with mixed boundary conditions. *JAMM*, **57**, No.2, 102-108, [321-327].

Aleksandrov, V.M. and Pozharskii, D.A.,(1994), Contact stresses at the apex of a wedge-shaped punch pressed upon the edge of an elastic three-dimensional wedge. *JAMM*, **58**, No.1, 135-141, [143-148].

Aleksandrov, V.M.,(1996), The interaction between a plane inclined ring-shaped punch and an elastic half-space. *JAMM*, **60**, No.1, 132-139, [127-134].

Aleksandrov, V.M. and Pozharskii, D.A.,(1997), An axially symmetric contact problem for a truncated sphere in the theory of elasticity. *JAMM*, **61**, No.2, 305-311, [295-300].

Aleksandrov, V.M. and Pozharskii, D.A.,(1998), On the 3D contact problem for an elastic cone with unknown contact area. *MS*, **33**, No.2, 36-41, [29-34].

Aleksandrov, V.M., Kalker, J.J. and Pozharskii, D.A.,(1999), Three-dimensional contact problem for a two-layered elastic base with an unknown contact area. *MS*, **34**, No.4, 51-55, [41-45].

Aleksandrov, V.M. and Pozharskii, D.A.,(1999a), The problem of a wedge-shaped punch on the face of an elastic wedge. *JAMM*, **63**, No.1, 152-156, [141-144].

Aleksandrov, V.M. and Pozharskii, D.A.,(1999b), An asymptotic method in contact problems. *JAMM*, **63**, No.2, 295-302, [283-290].

Aleksandrova, G.P.,(1973), The contact problems for plates bending on an elastic foundation. *MS*, **8**, No.1, 97-106.

Antipov, Yu.A.,(1989), Analytical solution of mixed problems of mathematical physics with a change of boundary conditions over a ring. *MS*, **24**, No.3, 49-56.

Arutyunyan, N.Kh.,(1994), On the contact interaction of an elastic ring with an elastic cylinder. *MS*, **29**, No.2, 204-206, [194-197].

Babeshko, V.A. and Berkovich, V.N.,(1972), On the theory of mixed problems for the three dimensional wedge. *JAMM*, **36**, No.5, 943-947, [892-896].

Babeshko, V.A., Glushkov, Ye.V. and Zinchenko, Zh.F.,(1989), *Dynamics of Nonhomogeneous Linear Elastic Medium.* (in Russian). Moscow: Nauka.

Babloian, A.A.,(1964), Solutions of certain dual integral equations. *JAMM*, **28**, No.6, 1015-1023, [1227-1236].

Barenblatt, G.I.,(1960), On finiteness conditions in the mechanics of continuous media. Static problems of the theory of elasticity. *JAMM*, **24**, No.2, 316-322, [453-461].

Belotserkovsky, S.M. and Lifanov, I.K.,(1993), *Methods of Discrete Vortices.* English translation, Cherepanov, G.P. (Ed.). Boca Raton: CRC Press.

Belyaev, N.M.,(1957), *Studies on Elasticity and Plasticity Theory.* (in Russian). Moscow: GITTL.

Bleistein, N. and Handelsman, R.A.,(1975), *Asymptotic Expansions of Integrals.* New York: Rinehart and Winston.

Bochner, S.,(1932), *Vorlesungen über Fouriersche Integrale.* Leipzig: Akademische Verlagsgesellschaft.

Bondareva V.F.,(1969), On the effect of an axisymmetric normal loading on an elastic sphere. *JAMM*, **33**, No.6, 1029-1033, [1001-1005].

Borodachev, N.M.,(1970), Impression of a punch, whose base is a narrow rectangle, into an elastic half-space. *MS*, **5**, No.4, 135-141.

Borodachev, N.M. and Galin, L.A.,(1974), Contact problem for a stamp with a narrow rectangular base. *JAMM*, **38**, No.1, 125-130, [108-113].

Boussinesq, J.,(1885), *Application des potentials à l'étude de l'équilibre et du mouvement des solides élastiques.* Paris: Gauthier-Villars.

Cerruti, V.,(1882), Ricerche in torno all' equilibrio de corpi elastici isotropi. *Atti accad. nazl. Lincei, Mem. Classe sci. fis., mat. e nat.*, **13**, 81.

Chebakov, M.I. ,(1980), Certain dynamic and static contact problems of the theory of elasticity for a circular cylinder of finite size. *JAMM*, **44**, No.5, 923-933, [651-658].

Chebakov, M.I. ,(1987), Certain contact problems of the theory of elasticity for an annular sector and a spherical layer sector. *JAMM*, **51**, No.1, 101-108, [76-82].

Chebakov, M.I.,(1992), On the theory of design of spherical self-lubricating bearings. *MS*, **27**, No.5, 58-62, [58-63].

Cochran, J.A., (1972), *Analysis of Linear Integral Equations.* New York: McGraw-Hill.

Ditkin, V.A. and Prudnikov, A.P.,,(1965), *Integral Transforms and Operational Calculus.* English translation, Sneddon, I.N. (Ed.). Oxford: Pergamon Press.

Efimov, A.B. and Efimov, D.G.,(1987), The action of a concentrated force upon an edge of an incompressible elastic wedge. *MUMB*, **42**, No.3, 98-100, [50-53].

Erdelyi, A. (Ed.),(1954), *Tables of Integral Transforms.* (2 Vols.). New York: McGraw-Hill.

Erdelyi, A. (Ed.),(1955), *Higher Transcendental Functions.* (Vol. 1). New York: McGraw-Hill.

Erdogan, F., (1969), Approximate solutions of singular integral equations. *SJAM*, **17**, 1041-1059.

Erdogan, F., Gupta, G.D. and Cook, T.S.,(1973), Numerical solution of singular integral equations. In Sih, G.C. (Ed.) *Mechanics of Fracture* I. *Methods of Analysis and Solutions of Crack Problems.* Leyden: Noordhoff, 368-425.

Filippova, L.M., Tsvetkov, A.N. and Chebakov, M.I.,(1991) Interaction between a rigid strap and a prestressed finite elastic cylinder. *MS*, **26**, No.5, 51-56, [49-54].

Gakhov, F.D.,(1966), *Boundary Value Problems.* English translation, Sneddon, I.N. (Ed.). Oxford: Pergamon Press.

Galanov, B.A.,(1985), The method of boundary equations of the Hammerstein-type for contact problems of the theory of elasticity when the regions of contact are not known. *JAMM*, **49**, No.5, 827-835, [634-640].

Galin, L.A.,(1961), *Contact Problems in the Theory of Elasticity.* English translation, Sneddon, I.N. (Ed.). Raleigh: North Carolina State College, Department of Mathematics.

Gel'fand, I.M. and Shilov, G.E.,(1964), *Generalized Functions.* (Vol.1). English translation by Saletan, E., et al. New York: Academic Press.

Gerber, C.E.,(1968), *Contact Problems for the Elastic Quarter Plane and for the Quarter Space.* Doctoral Dissertation. Stanford University.

Gladwell, G.M.L.,(1980), *Contact Problems in the Classical Theory of Elasticity.* Alphen aan den Rijn: Sijthoff and Noordhoff.

Glushkov E.V. and Glushkova N.V.,(1992), Singularities of the elastic-stress field in the neighborhood of the point of a three-dimensional wedge-shaped crack. *MS*, **27**, No.4, 77-81.

Gohberg, I.C. and Krein, M.G.,(1970), *Theory and Applications of Volterra Operators in Hilbert Space.* English translation by Feinstein, A. Providence, RI: American Mathematical Society.

Gradshteyn, I.S. and Ryzhik, I.M.,(1994), *Tables of Integrals, Series and Products.* English translation, Jeffrey, A. (Ed.). 5th ed., New York: Academic Press Inc.

Haider, M.A. and Holmes, M.H.(1997), A mathematical approximation for the solution of a static indentation test. *JBIO*, **30**, No.7, 747-751.

Hanson, M.T. and Keer, L.M.(1990), A simplified analysis for an elastic quarter-space. *QJMM*, **43**, No.4, 561-587.

Hertz, H.,(1882), Über die Berührung fester elastischer Körper. *J. reine und angewandte Mathematik*, **92**, 156-171.

Hetenyi, M.,(1970), A general solution for the elastic quarter space. *JAM*, **37**, No.1, 70-76.

Ivanov, V.V.,(1976), *The Theory of Approximate Methods and their Application to the Numerical Solution of Singular Integral Equations.* English translation, Anderssen, R.S. (Ed.). Leyden: Noordhoff.

Jahnke, E., Emde, F. and Lösch, F.,(1966), *Tables of Higher Functions.* 7th ed., Stuttgart: Teubner.

Johnson, K.L.,(1985), *Contact Mechanics.* Cambridge: Cambridge University Press.

Kalandiya, A.I.,(1973), *Mathematical Methods in Two-Dimensional Elasticity.* (in Russian). Moscow: Nauka.

Kalker, J.J.(1972), On elastic line contact. *JAM*, **39**, No.4, 1125-1132.

Kalker J.J.,(1990), *Three-Dimensional Elastic Bodies in Rolling Contact.* Dordrecht: Kluwer Academic Publishers.

Kantorovich, L.V. and Krylov, V.I.,(1958), *Approximate Methods of Higher Analysis.* English translation, Benster, A.D. (Ed.). New York: Interscience.

Karpenko, V.A.,(1975), On the closed solution of the first boundary value problem of elasticity theory for a space with a spherical cavity. *JAMM*, **39**, No.5, 951-955, [917-922].

Karpenko, V.A.,(1976a), Axisymmetric impression of two stamps into an elastic sphere. *JAMM*, **40**, No.4, 763-766, [711-714].

Karpenko, V.A.,(1976b), Contact problems for elastic space with spherical cavity. *IAM*, **12**, No.8, 23-30, [769-776].

Kats, A.M.,(1956), *Theory of Elasticity.* (in Russian). Moscow: Gostekhizdat.

Keer, L.M. and Parihar, K.S.,(1978), A note on the singularity at the corner of a wedge-shaped punch or crack. *SJAM*, **34**, 297-302.

Keer, L.M., Lee J.C. and Mura, T.,(1983), Hetenyi's elastic quarter space problem revisited. *IJSS*, **19**, No.6, 497-506.

Keer, L.M., Lee, J.C. and Mura, T.(1984), A contact problem for the elastic quarter space. *IJSS*, **20**, No.5, 513-524.

Kondrat'ev, V.A.,(1967), Boundary value problems for elliptic equations in domains with conical or angular points. *TMMS*, **16**, 209-292, [227-313].

Korn, G.A. and Korn, T.M.,(1968), *Mathematical Handbook for Scientists and Engineers.* 2nd ed., New York: McGraw-Hill.

Korotkin, V.I.,(1997), Edge effect in calculating the contact strength of Novikov transmissions. *RER*, **77**, No.6, 8-11, [**17**, 9-14].

Kovalenko, E.V., Tarasov, D.G. and Chebakov, M.I. ,(1990), Exact solution of the antiplane contact problem for finite canonical domains. *JAMM*, **54**, No.5, 837-841, [688-691].

Krasnosel'skii, M.A., Vainikko, G.M., Zabreiko, P.P., et al,(1972), *Approximate Solution of Operator Equations.* English translation by Louvish, D., Groningen: Wolters-Noordhoff.

Kucherov, V.A.,(1973), Action of two semi-infinite punches onto a thin elastic strip. *MS*, **8**, No.1, 120-126.

Lawrentjew, M.A. and Schabat, B.W.,(1967), *Methoden der Komplexen Funktionentheorie.* German translation, Pirl, U. (Ed.). Berlin: VEB Deutscher Verlag der Wissenschaften.

Lebedev, N.N.,(1965), *Special Functions and their Applications.* English translation, Silverman, R.A. (Ed.). Englewood Cliffs, N.J.: Prentice-Hall.

Love, A.E.H.,(1927), *A Treatise on the Mathematical Theory of Elasticity.* 4th ed., London: Cambridge University Press.

Lubyagin, I.A., Pozharskii, D.A. and Chebakov, M.I.,(1991), Generalization of the Boussinesq and Cerruti problems for the case of a three-dimensional elastic wedge. *DP*, **321**, No.1, 58-62, [**36**, 797-799].

Lubyagin, I.A., Pozharskii, D.A. and Chebakov, M.I.,(1992), Embedding of a punch in the form of an elliptic paraboloid into an elastic spatial wedge. *JAMM*, **56**, No.2, 286-295, [244-252].

Lur'e, A.I.,(1964), *Three-Dimensional Problems of the Theory of Elasticity.* English translation, Radok, J.R.M. (Ed.). New York: Interscience.

McLachlan, N.W.,(1951), *Theory and Applications of Mathieu Functions.* Oxford: Clarendon Press.

Mossakovskii, V.I., Kachalovskaia, N.E. and Golikova, S.S.,(1985), *Contact Problems of the Mathematical Theory of Elasticity.* (in Russian). Kiev: Naukova Dumka.

Muskhelishvili, N.I.,(1953a), *Singular Integral Equations.* English translation, Radok, J.R.M. (Ed.). Groningen: Noordhoff.

Muskhelishvili, N.I.,(1953b), *Some Basic Problems of the Mathematical Theory of Elasticity.* English translation, Radok, J.R.M. (Ed.). Groningen: Noordhoff.

Natanson, I.P.,(1975), *Theorie der Funktionen einer reellen Veränderlichen.* German translation by Bögel. 4th ed., Zürich: Verlag Harri Deutsch.

Noble, B.,(1958), *Methods Based on the Wiener–Hopf Technique for the Solution of Partial Differential Equations.* London: Pergamon Press.

Nordling, C. and Österman, J.,(1980), *Physics Handbook: Elementary Constants and Units, Tables, Formulae and Diagrams and Mathematical Formulae.* Lund: Studentlitteratur.

Nuller, B.M.,(1970), Contact problem for an elastic infinite cone . *JAMM*, **34**, No.2, 339-348, [319-329].

Panek, C. and Kalker, J.J.,(1977), A solution for the narrow rectangular punch. *JE*, **7**, No.2, 213-218.

Parihar, K.S. and Keer, L.M.,(1979), The singularity at the apex of a bonded wedge-shaped stamp. *JAM*, **46**, No.3, 577-580.

Podil'chuk, Yu.N.,(1992), General problem of static thermoelasticity for a transversely isotropic cone. *IAM*, **28**, No.12, 39-46, [808-814].

Pozharskii, D.A.,(1993), Elliptic crack in an elastic three-dimensional wedge. *MS*, **28**, No.6, 105-112, [102-109].

Pozharskii, D.A.,(1994), The problem of action of a band stamp on a three-dimensional elastic wedge one side of which is free of stresses. *IAM*, **30**, No.5, 32-42, [363-367].

Pozharskii, D.A.,(1995), The spatial contact problem for an elastic wedge with unknown contact area. *JAMM*, **59**, No.5, 812-818, [781-787].

Pozharskii, D.A.,(1997), On the spatial contact problem for an elastic cone. *MS*, **32**, No.4, 51-60, [44-52].

Pozharskii, D.A.,(1998), Singularities of contact stresses in a problem concerning a periodic system of wedgelike dies at a cone. *DP*, **361**, No.1, 54-57, [**43**, 436-439].

Pozharskii, D.A. and Chebakov, M.I.,(1998), On singularities of contact stresses in the problem of a wedge-shaped punch on an elastic cone. *MS*, **33**, No.5, 72-77, [57-61].

Pozharskii, D.A.,(2000), The three-dimensional contact promlem for an elastic wedge taking friction forces into account. *JAMM*, **64**, No.1, 151-159, [147-154].

Prudnikov, A.P., Brychkov, Yu.A. and Marichev, O.I.,(1990), *Integrals and Series: More Special Functions.* (Vol.3). English translation by Queen, N.M., New York: Gordon and Breach Science Publishers.

Prudnikov, A.P., Brychkov, Yu.A. and Marichev, O.I.,(1992), *Integrals and Series.* (Vol.1-2). English translation by Queen, N.M. 3rd ed., New York: Gordon and Breach Science Publishers.

Pykhteev, G.N.,(1959), On the evaluation of certain singular integrals with a kernel of the Cauchy type. *JAMM*, **23**, No.6, 1074-1082, [1536-1548].

Rvachev, V.L. and Protsenko, V.S.,(1977), *Contact Problems of the Theory of Elasticity for Nonclassical Regions.* (in Russian). Kiev: Naukova Dumka.

Shail, R.(1978), Lamé polynomial solution to some elliptic crack and punch problems. *IJES*, **16**, 551-563.

Sheveleva, G.I.,(1984), Calculation of elastic contact displacements on surfaces of details of finite size. (in Russian). *MA*, No.4, 92-98.

Shtaerman, A.Ia.,(1949), *Contact Problems of the Theory of Elasticity.* (in Russian). Moscow, Leningrad: Gostekhteorizdat.

Sivashinsky, G.J.,(1975), The problem of a slender die. *JE*, **5**, No.2, 161-166.

Sneddon, I.N.,(1951), *Fourier Transforms.* New York: McGraw-Hill.

Sneddon, I.N.,(1966), *Mixed Boundary Value Problems in Potential Theory.* Amsterdam: North Holland.

Sneddon, I.N.,(1972), *The Use of Intergal Transforms.* New York: McGraw-Hill.

Sokolnikoff, I.S., (1956), *Mathematical Theory of Elasticity.* New York: McGraw-Hill.

Theocaris, P.S. and Gdoutos, E.E.,(1976), Stress singularities at vertices of composite plates with smooth or roogh interfaces. *AM*, **28**, No.4, 693-704.

Timoshenko, S.P.,(1940), *Theory of Plates and Shells.* 1nd ed., New York: McGraw-Hill.

Timoshenko, S.P.,(1951), *Strength of Materials.* Vol. 2. New York: Van Nostrand.

Timoshenko, S.P. and Goodier, J.N.,(1951), *Theory of Elasticity.* New York: McGraw-Hill.

Todhunter, I and Pearson, K.,(1893), *History of the Theory of Elasticity.* Vol. 2, Part. 2, Cambridge University Press.

Tranter, C.J.,(1948), The use of Mellin transforms in finding the stress distribution in an infinite wedge. *QJMM*, **1**, 125-130.

Tsvetkov, A.N. and Chebakov, M.I.,(1989), Contact problem for a finite body of revolution with a free lateral surface. *MS*, **24**, No.2, 77-82, [73-78].

Tsvetkov, A.N. and Chebakov, M.I.,(1990), Plane contact problem for a curvilinear trapezoid. *MS*, **25**, No.6, 43-47, [43-48].

Ufliand, Ia.S.,(1965), *Survey of Articles on the Applications of Integral Transforms in the Theory of Elasticity.* English translation, Sneddon, I.N. (Ed.). Raleigh: North Carolina State University, Department of Mathematics.

Ufliand, Ia.S.,(1972), Some three-dimensional problems of the theory of elasticity for a wedge. (in Russian). In Sedov, L.I., et al (Eds.) *Continuum Mechanics and Related Problems of Analysis.* Muskhelishvili Anniversary Volume. Moscow: Nauka, 549-554.

Ulitko, A.F.,(1979), *The Method of Vector Eigenfunctions in Spatial Problems of the Theory of Elasticity.* (in Russian). Kiev: Naukova Dumka.

Vasil'yev V.B.,(1995), An integral equation for a problem of the indentation of a wedge-shaped punch. *JAMM*, **59**, No.2, 272-279, [255-262].

Voronin, T.A.,(1957), Contact stresses arising when a rigid sleeve is tightly set onto an infinite cylinder. (in Russian). *AIO*, No.8, 153-155.

Vorovich, I.I., Aleksandrov, V.M. and Babeshko, V.A.,(1974), *Nonclassical Mixed Problems of the Theory of Elasticity.* (in Russian). Moscow: Nauka.

Zhurina, M.I. and Karmazina, L.N.,(1966), *Tables and Formulas for the Spherical Functions* $\mathcal{P}^m_{-1/2+i\tau}(z)$. English translation by Albasiny, E.L. Oxford: Pergamon Press.

Index

Mechanics

SOLID MECHANICS AND ITS APPLICATIONS

Series Editor: G.M.L. Gladwell

Aims and Scope of the Series

The fundamental questions arising in mechanics are: *Why?*, *How?*, and *How much?* The aim of this series is to provide lucid accounts written by authoritative researchers giving vision and insight in answering these questions on the subject of mechanics as it relates to solids. The scope of the series covers the entire spectrum of solid mechanics. Thus it includes the foundation of mechanics; variational formulations; computational mechanics; statics, kinematics and dynamics of rigid and elastic bodies; vibrations of solids and structures; dynamical systems and chaos; the theories of elasticity, plasticity and viscoelasticity; composite materials; rods, beams, shells and membranes; structural control and stability; soils, rocks and geomechanics; fracture; tribology; experimental mechanics; biomechanics and machine design.

1. R.T. Haftka, Z. Gürdal and M.P. Kamat: *Elements of Structural Optimization.* 2nd rev.ed., 1990 ISBN 0-7923-0608-2
2. J.J. Kalker: *Three-Dimensional Elastic Bodies in Rolling Contact.* 1990 ISBN 0-7923-0712-7
3. P. Karasudhi: *Foundations of Solid Mechanics.* 1991 ISBN 0-7923-0772-0
4. *Not published*
5. *Not published.*
6. J.F. Doyle: *Static and Dynamic Analysis of Structures.* With an Emphasis on Mechanics and Computer Matrix Methods. 1991 ISBN 0-7923-1124-8; Pb 0-7923-1208-2
7. O.O. Ochoa and J.N. Reddy: *Finite Element Analysis of Composite Laminates.* ISBN 0-7923-1125-6
8. M.H. Aliabadi and D.P. Rooke: *Numerical Fracture Mechanics.* ISBN 0-7923-1175-2
9. J. Angeles and C.S. López-Cajún: *Optimization of Cam Mechanisms.* 1991 ISBN 0-7923-1355-0
10. D.E. Grierson, A. Franchi and P. Riva (eds.): *Progress in Structural Engineering.* 1991 ISBN 0-7923-1396-8
11. R.T. Haftka and Z. Gürdal: *Elements of Structural Optimization.* 3rd rev. and exp. ed. 1992 ISBN 0-7923-1504-9; Pb 0-7923-1505-7
12. J.R. Barber: *Elasticity.* 1992 ISBN 0-7923-1609-6; Pb 0-7923-1610-X
13. H.S. Tzou and G.L. Anderson (eds.): *Intelligent Structural Systems.* 1992 ISBN 0-7923-1920-6
14. E.E. Gdoutos: *Fracture Mechanics.* An Introduction. 1993 ISBN 0-7923-1932-X
15. J.P. Ward: *Solid Mechanics.* An Introduction. 1992 ISBN 0-7923-1949-4
16. M. Farshad: *Design and Analysis of Shell Structures.* 1992 ISBN 0-7923-1950-8
17. H.S. Tzou and T. Fukuda (eds.): *Precision Sensors, Actuators and Systems.* 1992 ISBN 0-7923-2015-8
18. J.R. Vinson: *The Behavior of Shells Composed of Isotropic and Composite Materials.* 1993 ISBN 0-7923-2113-8
19. H.S. Tzou: *Piezoelectric Shells.* Distributed Sensing and Control of Continua. 1993 ISBN 0-7923-2186-3
20. W. Schiehlen (ed.): *Advanced Multibody System Dynamics.* Simulation and Software Tools. 1993 ISBN 0-7923-2192-8
21. C.-W. Lee: *Vibration Analysis of Rotors.* 1993 ISBN 0-7923-2300-9
22. D.R. Smith: *An Introduction to Continuum Mechanics.* 1993 ISBN 0-7923-2454-4
23. G.M.L. Gladwell: *Inverse Problems in Scattering.* An Introduction. 1993 ISBN 0-7923-2478-1

Mechanics

SOLID MECHANICS AND ITS APPLICATIONS

Series Editor: G.M.L. Gladwell

24. G. Prathap: *The Finite Element Method in Structural Mechanics*. 1993 ISBN 0-7923-2492-7
25. J. Herskovits (ed.): *Advances in Structural Optimization*. 1995 ISBN 0-7923-2510-9
26. M.A. González-Palacios and J. Angeles: *Cam Synthesis*. 1993 ISBN 0-7923-2536-2
27. W.S. Hall: *The Boundary Element Method*. 1993 ISBN 0-7923-2580-X
28. J. Angeles, G. Hommel and P. Kovács (eds.): *Computational Kinematics*. 1993 ISBN 0-7923-2585-0
29. A. Curnier: *Computational Methods in Solid Mechanics*. 1994 ISBN 0-7923-2761-6
30. D.A. Hills and D. Nowell: *Mechanics of Fretting Fatigue*. 1994 ISBN 0-7923-2866-3
31. B. Tabarrok and F.P.J. Rimrott: *Variational Methods and Complementary Formulations in Dynamics*. 1994 ISBN 0-7923-2923-6
32. E.H. Dowell (ed.), E.F. Crawley, H.C. Curtiss Jr., D.A. Peters, R. H. Scanlan and F. Sisto: *A Modern Course in Aeroelasticity*. Third Revised and Enlarged Edition. 1995 ISBN 0-7923-2788-8; Pb: 0-7923-2789-6
33. A. Preumont: *Random Vibration and Spectral Analysis*. 1994 ISBN 0-7923-3036-6
34. J.N. Reddy (ed.): *Mechanics of Composite Materials*. Selected works of Nicholas J. Pagano. 1994 ISBN 0-7923-3041-2
35. A.P.S. Selvadurai (ed.): *Mechanics of Poroelastic Media*. 1996 ISBN 0-7923-3329-2
36. Z. Mróz, D. Weichert, S. Dorosz (eds.): *Inelastic Behaviour of Structures under Variable Loads*. 1995 ISBN 0-7923-3397-7
37. R. Pyrz (ed.): *IUTAM Symposium on Microstructure-Property Interactions in Composite Materials*. Proceedings of the IUTAM Symposium held in Aalborg, Denmark. 1995 ISBN 0-7923-3427-2
38. M.I. Friswell and J.E. Mottershead: *Finite Element Model Updating in Structural Dynamics*. 1995 ISBN 0-7923-3431-0
39. D.F. Parker and A.H. England (eds.): *IUTAM Symposium on Anisotropy, Inhomogeneity and Nonlinearity in Solid Mechanics*. Proceedings of the IUTAM Symposium held in Nottingham, U.K. 1995 ISBN 0-7923-3594-5
40. J.-P. Merlet and B. Ravani (eds.): *Computational Kinematics '95*. 1995 ISBN 0-7923-3673-9
41. L.P. Lebedev, I.I. Vorovich and G.M.L. Gladwell: *Functional Analysis*. Applications in Mechanics and Inverse Problems. 1996 ISBN 0-7923-3849-9
42. J. Menčik: *Mechanics of Components with Treated or Coated Surfaces*. 1996 ISBN 0-7923-3700-X
43. D. Bestle and W. Schiehlen (eds.): *IUTAM Symposium on Optimization of Mechanical Systems*. Proceedings of the IUTAM Symposium held in Stuttgart, Germany. 1996 ISBN 0-7923-3830-8
44. D.A. Hills, P.A. Kelly, D.N. Dai and A.M. Korsunsky: *Solution of Crack Problems*. The Distributed Dislocation Technique. 1996 ISBN 0-7923-3848-0
45. V.A. Squire, R.J. Hosking, A.D. Kerr and P.J. Langhorne: *Moving Loads on Ice Plates*. 1996 ISBN 0-7923-3953-3
46. A. Pineau and A. Zaoui (eds.): *IUTAM Symposium on Micromechanics of Plasticity and Damage of Multiphase Materials*. Proceedings of the IUTAM Symposium held in Sèvres, Paris, France. 1996 ISBN 0-7923-4188-0
47. A. Naess and S. Krenk (eds.): *IUTAM Symposium on Advances in Nonlinear Stochastic Mechanics*. Proceedings of the IUTAM Symposium held in Trondheim, Norway. 1996 ISBN 0-7923-4193-7
48. D. Ieşan and A. Scalia: *Thermoelastic Deformations*. 1996 ISBN 0-7923-4230-5

Mechanics

SOLID MECHANICS AND ITS APPLICATIONS

Series Editor: G.M.L. Gladwell

49. J.R. Willis (ed.): *IUTAM Symposium on Nonlinear Analysis of Fracture*. Proceedings of the IUTAM Symposium held in Cambridge, U.K. 1997 ISBN 0-7923-4378-6
50. A. Preumont: *Vibration Control of Active Structures*. An Introduction. 1997 ISBN 0-7923-4392-1
51. G.P. Cherepanov: *Methods of Fracture Mechanics: Solid Matter Physics*. 1997 ISBN 0-7923-4408-1
52. D.H. van Campen (ed.): *IUTAM Symposium on Interaction between Dynamics and Control in Advanced Mechanical Systems*. Proceedings of the IUTAM Symposium held in Eindhoven, The Netherlands. 1997 ISBN 0-7923-4429-4
53. N.A. Fleck and A.C.F. Cocks (eds.): *IUTAM Symposium on Mechanics of Granular and Porous Materials*. Proceedings of the IUTAM Symposium held in Cambridge, U.K. 1997 ISBN 0-7923-4553-3
54. J. Roorda and N.K. Srivastava (eds.): *Trends in Structural Mechanics*. Theory, Practice, Education. 1997 ISBN 0-7923-4603-3
55. Yu.A. Mitropolskii and N. Van Dao: *Applied Asymptotic Methods in Nonlinear Oscillations*. 1997 ISBN 0-7923-4605-X
56. C. Guedes Soares (ed.): *Probabilistic Methods for Structural Design*. 1997 ISBN 0-7923-4670-X
57. D. François, A. Pineau and A. Zaoui: *Mechanical Behaviour of Materials*. Volume I: Elasticity and Plasticity. 1998 ISBN 0-7923-4894-X
58. D. François, A. Pineau and A. Zaoui: *Mechanical Behaviour of Materials*. Volume II: Viscoplasticity, Damage, Fracture and Contact Mechanics. 1998 ISBN 0-7923-4895-8
59. L.T. Tenek and J. Argyris: *Finite Element Analysis for Composite Structures*. 1998 ISBN 0-7923-4899-0
60. Y.A. Bahei-El-Din and G.J. Dvorak (eds.): *IUTAM Symposium on Transformation Problems in Composite and Active Materials*. Proceedings of the IUTAM Symposium held in Cairo, Egypt. 1998 ISBN 0-7923-5122-3
61. I.G. Goryacheva: *Contact Mechanics in Tribology*. 1998 ISBN 0-7923-5257-2
62. O.T. Bruhns and E. Stein (eds.): *IUTAM Symposium on Micro- and Macrostructural Aspects of Thermoplasticity*. Proceedings of the IUTAM Symposium held in Bochum, Germany. 1999 ISBN 0-7923-5265-3
63. F.C. Moon: *IUTAM Symposium on New Applications of Nonlinear and Chaotic Dynamics in Mechanics*. Proceedings of the IUTAM Symposium held in Ithaca, NY, USA. 1998 ISBN 0-7923-5276-9
64. R. Wang: *IUTAM Symposium on Rheology of Bodies with Defects*. Proceedings of the IUTAM Symposium held in Beijing, China. 1999 ISBN 0-7923-5297-1
65. Yu.I. Dimitrienko: *Thermomechanics of Composites under High Temperatures*. 1999 ISBN 0-7923-4899-0
66. P. Argoul, M. Frémond and Q.S. Nguyen (eds.): *IUTAM Symposium on Variations of Domains and Free-Boundary Problems in Solid Mechanics*. Proceedings of the IUTAM Symposium held in Paris, France. 1999 ISBN 0-7923-5450-8
67. F.J. Fahy and W.G. Price (eds.): *IUTAM Symposium on Statistical Energy Analysis*. Proceedings of the IUTAM Symposium held in Southampton, U.K. 1999 ISBN 0-7923-5457-5
68. H.A. Mang and F.G. Rammerstorfer (eds.): *IUTAM Symposium on Discretization Methods in Structural Mechanics*. Proceedings of the IUTAM Symposium held in Vienna, Austria. 1999 ISBN 0-7923-5591-1

Mechanics

SOLID MECHANICS AND ITS APPLICATIONS

Series Editor: G.M.L. Gladwell

69. P. Pedersen and M.P. Bendsøe (eds.): *IUTAM Symposium on Synthesis in Bio Solid Mechanics.* Proceedings of the IUTAM Symposium held in Copenhagen, Denmark. 1999 ISBN 0-7923-5615-2
70. S.K. Agrawal and B.C. Fabien: *Optimization of Dynamic Systems.* 1999 ISBN 0-7923-5681-0
71. A. Carpinteri: *Nonlinear Crack Models for Nonmetallic Materials.* 1999 ISBN 0-7923-5750-7
72. F. Pfeifer (ed.): *IUTAM Symposium on Unilateral Multibody Contacts.* Proceedings of the IUTAM Symposium held in Munich, Germany. 1999 ISBN 0-7923-6030-3
73. E. Lavendelis and M. Zakrzhevsky (eds.): *IUTAM/IFToMM Symposium on Synthesis of Nonlinear Dynamical Systems.* Proceedings of the IUTAM/IFToMM Symposium held in Riga, Latvia. 2000 ISBN 0-7923-6106-7
74. J.-P. Merlet: *Parallel Robots.* 2000 ISBN 0-7923-6308-6
75. J.T. Pindera: *Techniques of Tomographic Isodyne Stress Analysis.* 2000 ISBN 0-7923-6388-4
76. G.A. Maugin, R. Drouot and F. Sidoroff (eds.): *Continuum Thermomechanics.* The Art and Science of Modelling Material Behaviour. 2000 ISBN 0-7923-6407-4
77. N. Van Dao and E.J. Kreuzer (eds.): *IUTAM Symposium on Recent Developments in Non-linear Oscillations of Mechanical Systems.* 2000 ISBN 0-7923-6470-8
78. S.D. Akbarov and A.N. Guz: *Mechanics of Curved Composites.* 2000 ISBN 0-7923-6477-5
79. M.B. Rubin: *Cosserat Theories: Shells, Rods and Points.* 2000 ISBN 0-7923-6489-9
80. S. Pellegrino and S.D. Guest (eds.): *IUTAM-IASS Symposium on Deployable Structures: Theory and Applications.* Proceedings of the IUTAM-IASS Symposium held in Cambridge, U.K., 6–9 September 1998. 2000 ISBN 0-7923-6516-X
81. A.D. Rosato and D.L. Blackmore (eds.): *IUTAM Symposium on Segregation in Granular Flows.* Proceedings of the IUTAM Symposium held in Cape May, NJ, U.S.A., June 5–10, 1999. 2000 ISBN 0-7923-6547-X
82. A. Lagarde (ed.): *IUTAM Symposium on Advanced Optical Methods and Applications in Solid Mechanics.* Proceedings of the IUTAM Symposium held in Futuroscope, Poitiers, France, August 31–September 4, 1998. 2000 ISBN 0-7923-6604-2
83. D. Weichert and G. Maier (eds.): *Inelastic Analysis of Structures under Variable Loads.* Theory and Engineering Applications. 2000 ISBN 0-7923-6645-X
84. T.-J. Chuang and J.W. Rudnicki (eds.): *Multiscale Deformation and Fracture in Materials and Structures.* The James R. Rice 60th Anniversary Volume. 2001 ISBN 0-7923-6718-9
85. S. Narayanan and R.N. Iyengar (eds.): *IUTAM Symposium on Nonlinearity and Stochastic Structural Dynamics.* Proceedings of the IUTAM Symposium held in Madras, Chennai, India, 4–8 January 1999 ISBN 0-7923-6733-2
86. S. Murakami and N. Ohno (eds.): *IUTAM Symposium on Creep in Structures.* Proceedings of the IUTAM Symposium held in Nagoya, Japan, 3-7 April 2000. 2001 ISBN 0-7923-6737-5
87. W. Ehlers (ed.): *IUTAM Symposium on Theoretical and Numerical Methods in Continuum Mechanics of Porous Materials.* Proceedings of the IUTAM Symposium held at the University of Stuttgart, Germany, September 5-10, 1999. 2001 ISBN 0-7923-6766-9
88. D. Durban, D. Givoli and J.G. Simmonds (eds.): *Advances in the Mechanis of Plates and Shells The Avinoam Libai Anniversary Volume.* 2001 ISBN 0-7923-6785-5
89. U. Gabbert and H.-S. Tzou (eds.): *IUTAM Symposium on Smart Structures and Structonic Systems.* Proceedings of the IUTAM Symposium held in Magdeburg, Germany, 26–29 September 2000. 2001 ISBN 0-7923-6968-8

Mechanics

SOLID MECHANICS AND ITS APPLICATIONS

Series Editor: G.M.L. Gladwell

90. Y. Ivanov, V. Cheshkov and M. Natova: ***Polymer Composite Materials – Interface Phenomena & Processes***. 2001 ISBN 0-7923-7008-2
91. R.C. McPhedran, L.C. Botten and N.A. Nicorovici (eds.): ***IUTAM Symposium on Mechanical and Electromagnetic Waves in Structured Media*** held in Sydney, NSW, Australia, 18-22 Januari 1999. 2001 ISBN 0-7923-7038-4
92. D.A. Sotiropoulos (ed.): ***IUTAM Symposium on Mechanical Waves for Composite Structures Characterization***. Proceedings of the IUTAM Symposium held in Chania, Crete, Greece, June 14-17, 2000. 2001 ISBN 0-7923-7164-X
93. V.M. Alexandrov and D.A. Pozharskii: ***Three-Dimensional Contact Problems***. 2001 ISBN 0-7923-7165-8

Kluwer Academic Publishers – Dordrecht / Boston / London

ERCOFTAC SERIES

1. A. Gyr and F.-S. Rys (eds.): *Diffusion and Transport of Pollutants in Atmospheric Mesoscale Flow Fields*. 1995 ISBN 0-7923-3260-1
2. M. Hallbäck, D.S. Henningson, A.V. Johansson and P.H. Alfredsson (eds.): *Turbulence and Transition Modelling*. Lecture Notes from the ERCOFTAC/IUTAM Summerschool held in Stockholm. 1996 ISBN 0-7923-4060-4
3. P. Wesseling (ed.): *High Performance Computing in Fluid Dynamics*. Proceedings of the Summerschool held in Delft, The Netherlands. 1996 ISBN 0-7923-4063-9
4. Th. Dracos (ed.): *Three-Dimensional Velocity and Vorticity Measuring and Image Analysis Techniques*. Lecture Notes from the Short Course held in Zürich, Switzerland. 1996 ISBN 0-7923-4256-9
5. J.-P. Chollet, P.R. Voke and L. Kleiser (eds.): *Direct and Large-Eddy Simulation II*. Proceedings of the ERCOFTAC Workshop held in Grenoble, France. 1997 ISBN 0-7923-4687-4

KLUWER ACADEMIC PUBLISHERS – DORDRECHT / BOSTON / LONDON

GPSR Compliance

The European Union's (EU) General Product Safety Regulation (GPSR) is a set of rules that requires consumer products to be safe and our obligations to ensure this.

If you have any concerns about our products, you can contact us on ProductSafety@springernature.com

In case Publisher is established outside the EU, the EU authorized representative is:

Springer Nature Customer Service Center GmbH
Europaplatz 3
69115 Heidelberg, Germany

Batch number: 10761599

Printed by Printforce, the Netherlands